SOLUTIONS MANUAL TO ACCOMPANY

ELEMENTS OF PHYSICAL CHEMISTRY

SOLUTIONS MANUAL TO ACCOMPANY

ELEMENTS OF PHYSICAL CHEMISTRY

Fifth Edition

Chares A. Trapp

Marshall P. Cady, Jr.

OXFORD
UNIVERSITY PRESS

W.H. Freeman and Company
New York

OXFORD
UNIVERSITY PRESS

Oxford University Press, Great Clarendon Street, Oxford,
OX2 6DP, United Kingdom

Oxford is a registered trade mark of the Oxford University Press
in the UK and in certain other countries.

First printing.

ISBN: 978-0-1995-5112-5

Published in the University States and Canada by
W.H. Freeman and Company
41 Madison Avenue
New York, NY 10010
www.whfreeman.com

ISBN-10: 1-4292-2400-2 ISBN-13: 978-1-4292-2400-0

British Library Cataloguing in Publication Data
Data available

Preface

This manual provides detailed solutions to all of the discussion questions and exercises in the fifth edition of *Elements of Physical Chemistry* by Peter Atkins and Julio de Paula. We hope that these complete solutions will help in deepening your understanding of physical chemistry. Solutions to exercises carried over from the fourth edition have been reworked, modified, or corrected when needed.

The solutions to the exercises in this edition rely somewhat more heavily on the mathematical and molecular modeling software that is now generally accessible to physical chemistry students, and this is particularly true for some of the new exercises, which specifically request the use of such software for their solutions. But almost all of the exercises can still be solved with a modern hand-held scientific calculator.

In general, we have adhered rigorously to the rules for significant figures in displaying the final answers. However, when intermediate answers are shown, they are often given with one more figure than would be justified by the data. These excess digits are indicated with an overline.

The solutions were carefully cross-checked for errors not only by us, but very thoroughly by Valerie Walters, who also made helpful suggestions for improving the solutions. We would be grateful to readers who bring any remaining errors to our attention.

We warmly thank our publishers for their patience in guiding this complex, detailed project to completion.

C. A. T.
M. P. C.

Contents

Introduction

Answers to discussion questions

D0.1 A gas is a form of matter that fills the container it occupies and is compressible under ordinary atmospheric conditions. It is composed of separated particles in continuous rapid, disordered motion during which particles often travel several diameters before colliding. Consequently, the particles are separated by considerable empty space, a condition that results in compressibility. The interactions between particles are negligibly weak except when they are colliding.

The particles of liquids and solids are in continuous contact with their neighbors, a condition that causes liquids and solids to be incompressible. All particles are in constant motion but they travel only a fraction of a diameter before colliding with a neighbor. The microscopic particles of the liquid phase can slip past each other, a condition that results in a non-rigid fluidity of the macroscopic phase. Microscopic particles of a solid cannot slip past each other. They can only oscillate about an average position within the solid, a condition that results in macroscopic rigidity of the solid. The interactions between particles within the liquid or solid phases are relatively strong.

D0.2 The force F acting on any object of mass m equals the mass multiplied by the acceleration a of the object. This is Newton's second law of motion: $F = ma$.

The work done on an object when moving the object against an opposing force equals the opposing force multiplied by the distance over which the object is moved:

$$\text{Work done on an object} = \text{opposing force} \times \text{distance.}$$

Energy is the capacity to do work. Kinetic energy is the energy that a mass m has because of its speed v: $E_k = \frac{1}{2}mv^2$. The potential energy, E_p, of an object is the energy it possesses due to its position. The potential energy of an object may be due to gravitational, electrical, or magnetic forces. In particular, the 'Coulombic' potential resulting from the presence of electrical charges is especially important in chemistry.

D0.3 When there are equal gas pressures on both sides of a macroscopic object such as a movable, stationary, frictionless piston confined in a cylinder, there is no net force acting upon the object. Such an object will not accelerate in either direction and is said to be in **mechanical equilibrium**. Two objects that exhibit no net flow of energy between them upon contact are in **thermal equilibrium**. These objects have the same temperature. Mechanical equilibrium is a dynamic process because the rates of molecule-wall collisions are extremely large on each side of the movable object and it is the opposing, but equal, collision rates that produce the equilibrium. Similarly, molecular, atomic, or ionic collisions at the contact surface between two macroscopic objects produce a continuous, dynamic transfer of energy between the two objects even though the objects may be at thermal equilibrium. Equal and directionally opposing rates of energy transfer balance to produce equilibrium. When objects are not at thermal equilibrium, atomic scale collisions always cause the transfer of energy from high temperature to low temperature.

D0.4 (a) Extensive; (b) intensive; (c) intensive; (d) intensive; (e) extensive.

D0.5 By specifying the **state of matter** we identify whether matter is gaseous(g), liquid(l), or solid(s). The **state symbol** within a chemical reaction equation may indicate that water, for example, is a gas, a liquid, or a solid by writing $H_2O(g)$, $H_2O(l)$, or $H_2O(s)$, respectively.

The description of the **physical state** of bulk matter includes the state of the matter along with information about physical properties such as volume, pressure, temperature, amount of substance present, and density. Two samples of a substance that have the same physical properties are in the same **state**.

Solutions to exercises

E0.1 Estimating my mass to be 82 kg and remembering that all objects experience the gravitational attraction $F = mg$ where g is the acceleration of a freely falling body at the surface of the Earth, the gravitational force I currently experience is

$$F = mg = (82\ \text{kg})\times(9.81\ \text{m s}^{-2}) = 8.0\times10^{2}\ \text{kg m s}^{-2} = \boxed{8.0\times10^{2}\ \text{N}}.\ \text{[See eqn A3.8]}$$

E0.2 Suppose that my precise mass is 82.00 kg, then

$$F_{\text{North Pole}} = mg_{\text{North Pole}} = (82.00\ \text{kg})\times(9.832\ \text{m s}^{-2}) = 806.2\ \text{N}.\ \text{[See eqn A3.8]}$$

$$F_{\text{Equator}} = mg_{\text{Equator}} = (82.00\ \text{kg})\times(9.789\ \text{m s}^{-2}) = 802.7\ \text{N}.$$

$$\text{Percentage change} = \left(\frac{806.2-802.7}{806.2}\right)100\% = \boxed{0.43\%}.$$

E0.3

$$\text{Work}_{\text{lift}} = (\text{force}_{\text{gravity}})\times(\text{vertical displacement}) = mgh\ \ [0.2]$$

$$= (65\ \text{kg})\times(9.81\ \text{m s}^{-2})\times(4.0\ \text{m}) = 2.6\times10^{3}\ \text{kg m}^{2}\ \text{s}^{-2} = \boxed{2.6\ \text{kJ}}\ \text{[See eqn A3.2]}$$

E0.4

$$E_{\text{k}} = \tfrac{1}{2}mv^{2}\ [0.1] = \tfrac{1}{2}(58\times10^{-3}\ \text{kg})\times(35\ \text{m s}^{-1})^{2} = 36\ \text{kg m}^{2}\ \text{s}^{-2} = \boxed{36\ \text{J}}\ \text{[See eqn A3.2]}$$

E0.5

$$E_{\text{k}} = \tfrac{1}{2}mv^{2}\ [0.1]$$

$$= \tfrac{1}{2}(1.5\ \text{t})\times(50\ \text{km h}^{-1})^{2}\times\left(\frac{1000\ \text{kg}}{1\ \text{t}}\right)\times\left(\frac{1000\ \text{m}}{1\ \text{km}}\right)^{2}\times\left(\frac{1\ \text{h}}{3600\ \text{s}}\right)^{2}$$

$$= 1.4\times10^{5}\ \text{J} = \boxed{1.4\times10^{2}\ \text{kJ}}.$$

E0.6

$$m_{\text{average}} = \frac{M_{\text{average}}}{N_{\text{A}}} = \frac{29\ \text{g mol}^{-1}}{6.022\times10^{23}\ \text{mol}^{-1}}\left(\frac{1\ \text{kg}}{1000\ \text{g}}\right) = 4.8\times10^{-26}\ \text{kg}.$$

$$E_{\text{k,total}} = N_{\text{A}}\left(\tfrac{1}{2}mv^{2}\right)_{\text{average}} \approx N_{\text{A}}\left(\tfrac{1}{2}m_{\text{average}}v^{2}_{\text{average}}\right)$$

$$\approx \frac{6.022\times10^{23}}{2}(4.8\times10^{-26}\ \text{kg})\times(400\ \text{m s}^{-1})^{2}$$

$$\approx 2.3\times10^{3}\ \text{J} = \boxed{2.3\ \text{kJ}}.$$

E0.7

$$m_{\text{Hg}} = \frac{M_{\text{Hg}}}{N_{\text{A}}} = \frac{200.6\ \text{g mol}^{-1}}{6.0221\times10^{23}\ \text{mol}^{-1}}\left(\frac{1\ \text{kg}}{1000\ \text{g}}\right) = 3.331\times10^{-25}\ \text{kg}.$$

$$\Delta E_{\text{p}} = m_{\text{Hg}}\times g\times\Delta h = (3.331\times10^{-25}\ \text{kg})\times(9.807\ \text{m s}^{-2})\times(760.0\ \text{mm})\times\left(\frac{10^{-3}\ \text{m}}{1\ \text{mm}}\right)$$

$$= \boxed{2.483\times10^{-24}\ \text{J}}.$$

E0.8 $$E_{\text{min}} = mgh = (25\text{ g})\times(9.81\text{ m s}^{-2})\times(50\text{ m})\times\left(\frac{1\text{ kg}}{1000\text{ g}}\right) = \boxed{12\text{ J}}.$$

E0.9 (a) $$1\text{ eV} = e\times(1\text{ V}) = (1.602\times10^{-19}\text{ C})\times(1\text{ V}) = 1.602\times10^{-19}\text{ C V} = \boxed{1.602\times10^{-19}\text{ J}}$$

(b) $$E_{\text{m}} = N_{\text{A}}\times(1.602\times10^{-19}\text{ J})\times\left(\frac{1\text{ kJ}}{1000\text{ J}}\right)$$

$$= (6.022\times10^{23}\text{ mol}^{-1})\times(1.602\times10^{-22}\text{ kJ}) = \boxed{96.47\text{ kJ mol}^{-1}}$$

E0.10 (a) $$\text{Work} = e\phi\ [\text{A3.13}]$$

$$= (1.602\times10^{-19}\text{ C})\times(1.5\text{ V}) = 2.4\times10^{-19}\text{ C V} = \boxed{2.4\times10^{-19}\text{ J}}\ \ [1\text{ C V} = 1\text{ J}]$$

(b) $$\text{Molar Work} = N_{\text{A}}e\phi$$

$$= (6.022\times10^{23}\text{ mol}^{-1})\times(2.4\times10^{-19}\text{ J}) = \boxed{1.4\times10^{2}\text{ kJ mol}^{-1}}$$

E0.11 The required fuel energy equals the difference between the gravitational potential at $r = \infty$ and the gravitational potential at $r = r_{\text{E}}$ where r_{E} is the radius of the Earth.

$$\text{Energy} = \left(-\frac{Gmm_{\text{E}}}{r}\right)_{\text{at } r=\infty} - \left(-\frac{Gmm_{\text{E}}}{r}\right)_{\text{at } r=r_{\text{E}}} = 0 + \frac{Gmm_{\text{E}}}{r_{\text{E}}} = \frac{Gmm_{\text{E}}}{r_{\text{E}}}$$

$$m_{\text{E}} = V_{\text{E}}\rho_{\text{E}} = \left(\frac{4\pi(6371\times10^{3}\text{ m})^{3}}{3}\right)\times\left(\frac{5.517\text{ g}}{\text{cm}^{3}}\right)\times\left(\frac{1\text{ cm}}{10^{-2}\text{ m}}\right)^{3}\times\left(\frac{1\text{ kg}}{10^{3}\text{ g}}\right) = 5.976\times10^{24}\text{ kg}.$$

$$\text{Energy} = \frac{(6.673\times10^{-11}\text{ N m}^{2}\text{ kg}^{-2})\times(185\text{ kg})\times(5.976\times10^{24}\text{ kg})}{6371\times10^{3}\text{ m}}$$

$$= 1.16\times10^{10}\text{ J} = \boxed{11.6\text{ GJ}}.\ [\text{See Table A1.3}]$$

E0.12 Refer to Table 0.1 for pressure conversion factors.

(a) $$p = 108\text{ kPa}\times\frac{760\text{ Torr}}{101.325\text{ kPa}} = \boxed{810\text{ Torr}}.$$

(b) $$p = 0.975\text{ bar}\times\frac{100\text{ kPa}}{1\text{ bar}}\times\frac{1\text{ atm}}{101.325\text{ kPa}} = \boxed{0.962\text{ atm}}.$$

(c) $$p = (22.5\text{ kPa})\times\left(\frac{1\text{ atm}}{101.325\text{ kPa}}\right) = \boxed{0.222\text{ atm}}.$$

(d) $$p = 770\text{ Torr}\times\left(\frac{101.325\times10^{3}\text{ Pa}}{760\text{ Torr}}\right) = \boxed{1.03\times10^{5}\text{ Pa}}.$$

E0.13 $$p = g\rho h\ [0.6] = (9.81\text{ m s}^{-2})\times\left(\frac{1.10\text{ g}}{\text{cm}^{3}}\right)\times(11.5\text{ km})\times\frac{1\text{ kg}}{10^{3}\text{g}}\times\frac{10^{6}\text{ cm}^{3}}{1\text{ m}^{3}}\times\frac{10^{3}\text{ m}}{1\text{ km}}$$

$$= 1.24\times10^{8}\text{ kg m}^{-1}\text{ s}^{-2} = 1.24\times10^{8}\text{ Pa} = \boxed{1.24\times10^{3}\text{ bar}}.$$

E0.14 In order for a planet to have a stable atmosphere, an average atmospheric molecule of mass m at the surface must on the average exert a force upward that balances the downward pull of gravity. This force is given by Newton's law of gravity:

$$F_{\text{planet}} = \frac{GM_{\text{planet}}m}{R_{\text{planet}}^2} = g_{\text{planet}}m = (gM)_{\text{planet}} / N_{\text{A}}$$

where $M = mN_{\text{A}}$ is the average molar mass of atmospheric molecules. The mole percentage composition of the Martian atmosphere is 95% CO_2 so $M_{\text{Mars}} \sim 44\ \text{g mol}^{-1}$ while Earth's atmosphere is about 80% N_2 and 20% O_2 for an average $M_{\text{Earth}} \sim 29\ \text{g mol}^{-1}$.

Envision a square of area A on the surface of a planet and extend this square upward through the atmosphere. The surface pressure on A equals the total force of the N molecules within the vertical square parallelepiped divided by A.

$$p_{\text{planet surface}} = \frac{\text{total force}}{A} = \frac{NF_{\text{planet}}}{A} = \frac{N(gM)_{\text{planet}}}{N_{\text{A}}A}$$

Reason suggests that N is proportional not only to A but also to the gaseous state at the surface. That is, N should be proportional to the surface value of mass density ρ (doubling the density should double the number of molecules in a volume). Let the constant of proportionality be a_{planet}. Then, $N = A \times (a\rho)_{\text{planet}}$ and the pressure at the planet surface becomes

$$p_{\text{planet surface}} = \frac{a_{\text{planet}}(g\rho M)_{\text{planet}}}{N_{\text{A}}}.$$

This relationship indicates that the **surface pressure depends upon atmospheric density, gravitational attraction, and atmospheric composition** (i.e., g_{planet} gives the gravitational dependence while M depends upon atmospheric composition). A quick search of the *CRC Handbook of Chemistry and Physics* and the web yields the facts that at the surface of these planets: $\rho_{\text{Mars}} = 0.013\ \text{kg m}^{-3}$ and $\rho_{\text{Earth}} = 1.22\ \text{kg m}^{-3}$. Thus, the ratio of the surface atmospheric pressure on Earth with that on Mars is

$$\frac{p_{\text{Earth}}}{p_{\text{Mars}}} = \frac{a_{\text{Earth}}(\rho gM)_{\text{Earth}}}{a_{\text{Mars}}(\rho gM)_{\text{Mars}}} = \left(\frac{1.22\times9.81\times29}{0.013\times3.7\times44}\right)\times\left(\frac{a_{\text{Earth}}}{a_{\text{Mars}}}\right) = 16\overline{4}\times\left(\frac{a_{\text{Earth}}}{a_{\text{Mars}}}\right) \approx 16\overline{4}$$

where we have assumed that $a_{\text{Earth}} \approx a_{\text{Mars}}$ because the planet-dependent properties (gravity, density, and atmospheric composition) have been dealt with explicitly. Thus, we find that, given the surface pressure on Mars (0.0060 atm),

$$p_{\text{Earth}} = 16\overline{4}p_{\text{Mars}} = 16\overline{4}\times(0.0060\ \text{atm}) = \boxed{0.98\ \text{atm}}.$$

This is in remarkable agreement with the known surface pressure on Earth (1.0 atm).

E0.15 Identifying p_{ex} in the equation $p = p_{\text{ex}} + \rho gh$ [Derivation 0.1] as the pressure at the top of the straw and p as the atmospheric pressure on the liquid, the pressure difference is $p - p_{\text{ex}} = \rho gh$.

(a) On Earth

$$p - p_{\text{ex}} = \rho gh = (1.0\times10^3\ \text{kg m}^{-3})\times(9.81\ \text{m s}^{-2})\times(0.15\ \text{m}) = \boxed{1.5\times10^3\ \text{Pa}} = 0.015\ \text{atm}$$

(b) On Mars

$$p - p_{\text{ex}} = \rho gh = (1.0\times10^3\ \text{kg m}^{-3})\times(3.7\ \text{m s}^{-2})\times(0.15\ \text{m}) = \boxed{5.6\times10^2\ \text{Pa}} = 0.0056\ \text{atm}$$

E0.16 Using the standard gravitation acceleration of exactly $9.80665\ \text{m s}^{-2}$, the pressure in pascal of 1 mmHg when the mercury density equals $13.5951\ \text{g cm}^{-3}$ (Hg density at 0°C) is [0.6]

$$p = \rho gh = (13.5951\ \text{g cm}^{-3}) \times (9.80665\ \text{m s}^{-2}) \times (0.1\ \text{cm}) \times (1\ \text{cm} / 10^{-2}\ \text{m})^2 \times (1\ \text{kg} / 10^3\ \text{g})$$

$$= 133.322\ \text{Pa}.$$

This calculated value uses the measured value of the density of mercury and it is not an exact value. It has an uncertainty of ±0.0001 g cm^{-3} and provides 6 significant figures only. In contrast 760 Torr is defined to equal 101325 Pa exactly and, consequently,

$$1 \text{ Torr} = 101325 \text{ Pa} / 760 = 133.322 \text{ Pa exactly with no uncertainty.}$$

It is apparent that 1 mmHg and 1 Torr are identical under the above conditions of density and standard gravitational acceleration. However, because of expected uncertainty in the density of the purest sample of mercury, the two are expected to $\boxed{\text{differ by as much as 1 part in } 10^6}$. We conclude that in most practical situations the measurement of pressure in mmHg gives the same result as a measurement in SI units. However, for very precise measurements the unit mmHg, a non-SI unit, should be avoided.

E0.17 (a) Using the standard gravitation acceleration of exactly 9.80665 m s^{-2}, the pressure in pascal of 1 mmH_2O when the water density equals 1000 kg m^{-3} (H_2O density at 4°C) is [0.6]

$$p = \rho g h = (1000 \text{ kg m}^{-3}) \times (9.80665 \text{ m s}^{-2}) \times (1 \times 10^{-3} \text{ m}) = 9.80665 \text{ kg m}^{-1} \text{ s}^{-2} = \boxed{9.80665 \text{ Pa}}.$$

(b) $$p = 9.80665 \text{ Pa} \times \left(\frac{1 \text{ Torr}}{133.322 \text{ Pa}} \right) = \boxed{0.0735561 \text{ Torr}}.$$

E0.18 At $T = 0$, $\theta/°C = -273.15$.

We solve the equation for Fahrenheit temperature using $\theta_F/°F = t_F$.

$$-273.15 = \tfrac{5}{9}(t_F - 32)$$

$$t_F = -273.15 \times (\tfrac{9}{5}) + 32 = -459.67$$

$$\theta_F (\text{at } 0 \text{ K}) = \boxed{-459.67°F}$$

E0.19 (a) At the freezing point of water $\theta/°C = 0$ and $\theta'/°C' = 100$ while at the boiling point of water $\theta/°C = 100$ and $\theta'/°C' = 0$. The reverse symmetry of the values suggest that the simple guessed relation is

$$\theta'/°C' = -(\theta/°C - 100) \quad \text{or} \quad \boxed{\theta/°C = 100 - \theta'/°C'}.$$

The relation is checked by first substituting the freezing point on one scale, then the boiling point on that scale, and confirming that the relation yields the point on the other scale.

(b) Substitute the relation $\theta/°C = \tfrac{5}{9}(\theta_F/°F - 32)$, where θ_F is temperature on the Fahrenheit scale, into the above relation between θ' and θ.

$$\theta'/°C' = -\{\tfrac{5}{9}(\theta_F/°F - 32) - 100\} = 117.78 - \tfrac{5}{9}\theta_F/°F \quad \text{or} \quad \boxed{\theta_F/°F = 212 - \tfrac{9}{5}\theta'/°C'}$$

E0.20 The mathematical equation for a straight line through two points $P_1(x_1, y_1)$ and $P_2(x_2, y_2)$ is

$$y = \frac{\Delta y}{\Delta x}(x - x_1) + y_1 \quad \text{where} \quad \frac{\Delta y}{\Delta x} = \frac{y_2 - y_1}{x_2 - x_1}.$$

Using $P_1(x_1, y_1) = (-209.9°C, 0°P)$, $P_2(x_2, y_2) = (-195.8°C, 100°P)$ as the two points and θ and θ_P to be temperature in degree Celsius and degree Plutonium, respectively, the function $\theta_P(\theta)$ is

$$\theta_P = \left(\frac{100°P - 0°P}{-195.8°C - (-209.9°C)} \right) \times (\theta + 209.9°C) + 0$$

$$\theta_P = (7.092°P\ °C^{-1}) \times (\theta + 209.9°C) \quad \text{or} \quad \boxed{\theta_P/°P = 7.092 \times (\theta/°C + 209.9)}.$$

(a) Substitution of the definition $\theta/°C = T/K - 273.15$ where T is kelvin temperature into the above equation gives

$$\boxed{\theta_P / °P = 7.092\times(T / K - 63.25)}.$$

(b) Substitution of the relationship $\theta/°C = {}^5/_9\,(\theta_F/°F - 32)$ where θ_F is fahrenheit temperature into the top equation gives

$$\boxed{\theta_P / °P = 3.940\times(\theta_F / °F + 345.8°F)}.$$

E0.21 On the Rankine scale 0°F = 459.67°R (see the solution to Exercise 0.18) and the degree size is identical to that of the Fahrenheit scale. Hence, the relationship between the Rankine scale and the Fahrenheit scale is

$$T_R = (1°R\ °F^{-1})\times\theta_F + 459.67°R \qquad \text{or} \qquad T_R / °R = \theta_F / °F + 459.67$$

$$T_R\,(\text{at } 212°F) = (212 + 459.67)°R = \boxed{671.67°R}.$$

E0.22 The amount of molecules (number of molecules) is

$$N = nN_A\ [0.9] = (m / M)\times N_A\ [0.10]$$

$$= (10\ \text{g}/180.15\ \text{g mol}^{-1})\times(6.022\times10^{23}\ \text{mol}^{-1}) = \boxed{3.3\times10^{22}\ \text{glucose molecules}}.$$

E0.23 The amount of molecules (number of molecules) is

$$N = nN_A\ [0.9] = (m / M)\times N_A\ [0.10] = (\rho V / M)\times N_A\ [0.07] = \rho V N_A / M$$

$$= (0.703\ \text{g cm}^{-3})\times(1.00\ \text{dm}^3)\times(6.022\times10^{23}\ \text{mol}^{-1})\times\left(\frac{10\ \text{cm}}{1\ \text{dm}}\right)^3 / (114.23\ \text{g mol}^{-1})$$

$$= \boxed{3.71\times10^{24}\ \text{octane molecules}}.$$

E0.24 The amount of molecules (number of molecules) is

$$N = nN_A\ [0.9] = (m / M)\times N_A\ [0.10]$$

$$= (1.0\ \text{g}/16.1\times10^3\ \text{g mol}^{-1})\times(6.022\times10^{23}\ \text{mol}^{-1}) = \boxed{3.7\times10^{19}\ \text{myoglobin molecules}}.$$

E0.25 Let haemoglobin = Hb and myoglobin = Mb

$$\text{mass of Hb} = 3\times10^8\ \text{molecules}\times\frac{4\ \text{mol Mb}}{1\ \text{mol Hb}}\times\frac{1\ \text{mol Hb}}{6.02\times10^{23}\ \text{molecules}}\times\frac{16.1\times10^3\ \text{g}}{1\ \text{mol Mb}}$$

$$= 3.2\overline{1}\times10^{-11}\ \text{g}$$

$$\text{Fraction Hb} = \frac{3.2\overline{1}\times10^{-11}\ \text{g}}{3.33\times10^{-11}\ \text{g}} = \boxed{0.9\overline{7}}\ \text{or } 97\%$$

E0.26 $m = n\times M$ [0.10] and the molar volume V_m is defined by $V_m = V / n$

$$\text{Therefore, } \rho = \frac{m}{V} = \frac{nM}{V} \text{ or } \boxed{\rho = M / V_m}.$$

E0.27

$$[\text{sucrose}] = n / V\ [0.11a] = m / (MV)\ [0.10]$$

$$= \frac{5.00\ \text{g}}{(342.30\ \text{g mol}^{-1})\times(0.200\ \text{dm}^3)} = 7.30\times10^{-2}\ \text{mol dm}^{-3} = \boxed{73.0\ \text{mmol dm}^{-3}}$$

E0.28 $m = nM$ [0.10] $= [\text{NaCl}]VM$ [0.11b]

$$= (1.00 \text{ mol dm}^{-3}) \times (0.300 \text{ dm}^3) \times (58.44 \text{ g mol}^{-1}) = \boxed{17.5 \text{ g NaCl}}$$

E0.29 $n_B = (m/M)_B$ [0.10] $= (2.11 \text{ g})/(234.01 \text{ g mol}^{-1}) = 9.02 \times 10^{-3} \text{ mol}$

(a) $[B] = n_B / V$ [0.11a]

(i) In water: $[B] = (9.02 \times 10^{-3} \text{ mol})/(0.100 \text{ dm}^{-3}) = \boxed{9.02 \times 10^{-2} \text{ mol dm}^{-3}}$

(ii) In benzene: $[B] = (9.02 \times 10^{-3} \text{ mol})/(0.100 \text{ dm}^{-3}) = \boxed{9.02 \times 10^{-2} \text{ mol dm}^{-3}}$

(b) The mass of the solvent plus the mass of the solute must equal the total mass (ρV). Thus,

$m_{\text{solvent}} = \rho V - m_B$ and $b_B = n_B / m_{\text{solvent}}$ [0.12] $= n_B / (\rho V - m_B)$.

(i) In water:

$$m_{\text{solvent}} = (1.01 \text{ g cm}^{-3}) \times (100 \text{ cm}^3) - 2.11 \text{ g} = 98.89 \text{ g} = 0.09889 \text{ kg}$$

$$b_B = (9.02 \times 10^{-3} \text{ mol})/(0.09889 \text{ kg}) = \boxed{9.12 \times 10^{-2} \text{ mol kg}^{-1}}$$

(ii) In benzene:

$$m_{\text{solvent}} = (0.881 \text{ g cm}^{-3}) \times (100 \text{ cm}^3) - 2.11 \text{ g} = 85.99 \text{ g} = 0.08599 \text{ kg}$$

$$b_B = (9.02 \times 10^{-3} \text{ mol})/(0.08599 \text{ kg}) = \boxed{0.105 \text{ mol kg}^{-1}}$$

E0.30 $n_{\text{benzene}} = (m/M)_B$ [0.10] $= (56 \text{ g})/(78.11 \text{ g mol}^{-1}) = 0.72 \text{ mol}$

$$n_{\text{toluene}} = (120 \text{ g})/(92.14 \text{ g mol}^{-1}) = 1.30 \text{ mol}$$

$$x_{\text{benzene}} = \frac{n_{\text{benzene}}}{n} \text{ [0.13]} = \frac{0.72}{0.72 + 1.30} = \boxed{0.36}$$

$$x_{\text{toluene}} = \frac{1.30}{0.72 + 1.30} = \boxed{0.64}$$ Alternatively, $x_{\text{toluene}} = 1 - x_{\text{benzene}}$.

E0.31 For the sake of convenience we calculate the amount of each gas in a 100.00 g dry air sample.

$$n_{N_2} = (m/M)_{N_2} \text{ [0.10]} = (75.53 \text{ g})/(28.02 \text{ g mol}^{-1}) = 2.696 \text{ mol}$$

$$n_{O_2} = (23.14 \text{ g})/(32.00 \text{ g mol}^{-1}) = 0.7231 \text{ mol}$$

$$n_{Ar} = (1.33 \text{ g})/(39.95 \text{ g mol}^{-1}) = 0.0333 \text{ mol}$$

Thus, $n = 3.452$ and the mole fractions are

$$x_{N_2} = \frac{n_{N_2}}{n} \text{ [0.13]} = \frac{2.696}{3.452} = \boxed{0.7810}$$

$$x_{O_2} = \frac{0.7231}{3.452} = \boxed{0.2095}$$

$$x_{Ar} = \frac{0.0333}{3.452} = \boxed{0.00965}.$$

Note that $x_{N_2} + x_{O_2} + x_{Ar} = 1.000$ as required by the mole fraction definition.

E0.32 From exercise 0.31 we know that $n_{O_2} = 0.7231$ mol when $m_{N_2} = 75.53 \text{ g} = 0.07553 \text{ kg}$. Thus,

$$b_{O_2} = n_{O_2} / m_{\text{solvent}} \text{ [0.12]} = (0.7231 \text{ mol})/(0.07553 \text{ kg}) = \boxed{9.574 \text{ mol kg}^{-1}}.$$

E0.33 The complete combustion of 1 mol octane produces 8 mol $CO_2(g)$. Having the values for the volume and density of the octane, reason suggests the sequence of calculations: m_{octane}, n_{octane}, n_{CO_2}, and finally m_{CO_2}. (You may wish to devise alternative setups to the calculation.)

$$m_{\text{octane}} = (\rho V)_{\text{octane}} \quad [0.7] = (0.703\ \text{g cm}^{-3}) \times (1.00\times10^3\ \text{cm}^3) = 703\ \text{g}$$

$$n_{\text{octane}} = (m/M)_{\text{octane}} \quad [0.10] = (703\ \text{g})/(114.23\ \text{g mol}^{-1}) = 6.15\ \text{mol}$$

$$n_{CO_2} = 8\times n_{\text{octane}} \quad [\text{reaction stoichiometry}] = 8\times(6.15\ \text{mol}) = 49.2\ \text{mol}$$

$$m_{CO_2} = (nM)_{CO_2} = (49.2\ \text{mol})\times(44.01\ \text{g mol}^{-1}) = \boxed{2.17\ \text{kg}}$$

An attractive alternative to the method of sequencing a series of calculations involves a single setup of successive conversion factors, each of which explicitly shows units. This procedure avoids the magnification of successive rounding errors and allows for an easy visual check of unit cancellation to yield a quantity in the desired unit. Here's the single setup for this exercise:

$$\underbrace{\left(\frac{0.703\,\text{g octane}}{\text{cm}^3\ \text{octane}}\right)(1.00\times10^3\ \text{cm}^3\ \text{octane})\left(\frac{1\ \text{mol octane}}{114.23\,\text{g octane}}\right)}_{\text{mole octane calculation}}\underbrace{\left(\frac{8\,\text{mol}\,CO_2}{1\,\text{mol octane}}\right)}_{\text{stoichiometric ratio conversion to mole }CO_2}\underbrace{\left(\frac{44.01\,\text{g}\,CO_2}{1\,\text{mol}\,CO_2}\right)\left(\frac{0.001\,\text{kg}}{\text{g}}\right)}_{\text{conversion to kg }CO_2} = 2.17\,\text{kg}\,CO_2$$

E0.34 The balanced reaction equation is: $Fe_2O_3(s) + 3\ CO(g) \rightarrow 2\ Fe(s) + 3\ CO_2(g)$. Having the mass of iron(III) oxide ($m_{Fe_2O_3} = 1.0\ \text{t} = 1.0 \times 10^6$ g; see Table A1.4) and the reaction stoichiometry, reason suggests the sequence of calculations: $n_{Fe_2O_3}$, n_{CO}, and finally m_{CO}. (You may wish to devise alternative setups to the calculation.)

$$n_{Fe_2O_3} = (m/M)_{Fe_2O_3} \quad [0.10] = (1.0\times10^6\ \text{g})/(159.69\ \text{g mol}^{-1}) = 6.3\times10^3\ \text{mol}$$

$$n_{CO} = 3\times n_{Fe_2O_3} \quad [\text{reaction stoichiometry}] = 3\times(6.3\times10^3\ \text{mol}) = 1.9\times10^4\ \text{mol}$$

$$m_{CO} = (nM)_{CO} = (1.9\times10^4\ \text{mol})\times(28.01\ \text{g mol}^{-1}) = \boxed{5.3\times10^2\ \text{kg}}\ \text{or 0.53 t (i.e., tonne)}$$

An attractive alternative to the method of sequencing a series of calculations involves a single setup of successive conversion factors, each of which explicitly shows units. This procedure avoids the magnification of successive rounding errors and allows for an easy visual check of unit cancellation to yield a quantity in the desired unit. Here's the single setup for this exercise:

$$\underbrace{(1.0\times10^6\ \text{g}\ Fe_2O_3)\left(\frac{1\ \text{mol}\ Fe_2O_3}{159.69\ \text{g}\ Fe_2O_3}\right)}_{\text{mole }Fe_2O_3\text{ calculation}}\underbrace{\left(\frac{3\ \text{mol CO}}{1\ \text{mol}\ Fe_2O_3}\right)}_{\text{stoichiometric ratio conversion to mole CO}}\underbrace{\left(\frac{28.01\ \text{g CO}}{1\ \text{mol CO}}\right)\left(\frac{0.001\ \text{kg}}{\text{g}}\right)}_{\text{conversion to kg CO}} = 5.3\times10^2\ \text{kg CO}$$

Solutions to projects

P0.35

$$E_{\text{gravitational potential}}\Big|_{r=r_E+h} = -\frac{Gmm_E}{r_E+h} = -\left(\frac{Gmm_E}{r_E}\right)\times\left(\frac{1}{1+h/r_E}\right) = -\left(\frac{Gmm_E}{r_E}\right)\times\left(1+\frac{h}{r_E}\right)^{-1}$$

Since $h/r_E \ll 1$, the last factor may be expanded in a Taylor expansion series. Second and higher order terms may be discarded because powers of very small fractions produce yet smaller fractions. Using $x = h/r_E$, the Taylor expansion is [Appendix 2.5]:

$$(1+x)^{-1} = 1 - x + x^2 - x^3 + \cdots = 1 - x.$$

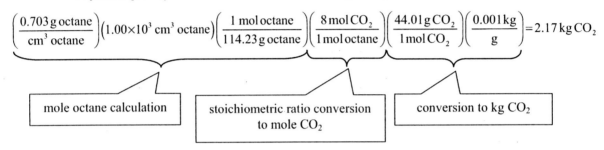

Substitution gives:

$$E_{\text{gravitational potential}}\Big|_{r=r_E+h} = -\left(\frac{Gmm_E}{r_E}\right)\times\left(1-\frac{h}{r_E}\right) = -\frac{Gmm_E}{r_E}+\frac{Gmm_E h}{r_E^{\,2}}$$

$$= -\frac{Gmm_E}{r_E} + m\times\left(\frac{Gm_E}{r_E^{\,2}}\right)\times h$$

$$= E_{\text{gravitational potential}}\Big|_{r=r_E} + mgh \qquad \text{where } g = Gm_E / r_E^{\,2}.$$

Thus, the difference between the gravitational potential at $r = r_E + h$ and the gravitation potential at $r = r_E$, when $h \ll r_E$, is mgh where $\boxed{g = Gm_E / r_E^{\,2}}$. This difference is the gravitational potential above the surface and it is normally written as $E_P = mgh$ [0.2, section A3.1]. The values of m_E and r_E (at the equator) are found in the *CRC Handbook of Chemistry and Physics* so it is possible to calculate the value of g at the equator.

$$g = \frac{(6.67259\times10^{-11}\ \text{N m}^2\ \text{kg}^{-2})(5.9763\times10^{24}\ \text{kg})}{(6378.077\times10^3\ \text{m})^2}\left(\frac{1\ \text{kg m s}^{-2}}{1\ \text{N}}\right) = 9.8027\ \text{m s}^{-2} \text{ at equator}$$

P0.36

$$E_{\text{Coulomb potential}}\Big|_{r=r_0+h} = \frac{Q_1Q_2}{4\pi\varepsilon_0(r_0+h)} \quad \text{[A3.11]}$$

$$= \left(\frac{Q_1Q_2}{4\pi\varepsilon_0 r_0}\right)\times\left(\frac{1}{1+h/r_0}\right) = \left(\frac{Q_1Q_2}{4\pi\varepsilon_0 r_0}\right)\times\left(1+\frac{h}{r_0}\right)^{-1}$$

Since $h/r_0 \ll 1$, the last factor may be expanded in a Taylor expansion series. Second and higher order terms may be discarded because powers of very small fractions produce yet smaller fractions. Using $x = h/r_0$, the Taylor expansion is (Appendix 2.5):

$$(1+x)^{-1} = 1 - x + x^2 - x^3 + \cdots = 1 - x.$$

Substitution gives:

$$E_{\text{Coulomb potential}}\Big|_{r=r_0+h} = \left(\frac{Q_1Q_2}{4\pi\varepsilon_0 r_0}\right)\times\left(1-\frac{h}{r_0}\right) = \frac{Q_1Q_2}{4\pi\varepsilon_0 r_0} - \left(\frac{Q_1Q_2}{4\pi\varepsilon_0 r_0}\right)\frac{h}{r_0}$$

$$= E_{\text{Coulomb potential}}\Big|_{r=r_0} - Q_1 g_{\text{Coulomb}} h \qquad \text{where } g_{\text{Coulomb}} = Q_2 / 4\pi\varepsilon_0 r_0^2.$$

Thus, the difference between the Coulomb potential at $r = r_0 + h$ and the Coulomb potential at $r = r_0$, when $h \ll r_0$, is $\boxed{-Q_1 g_{\text{Coulomb}} h \text{ where } g_{\text{Coulomb}} = Q_2 / 4\pi\varepsilon_0 r_0^2}$. In contrast to the gravitation potential, g_{Coulomb} may be either positive or negative because the Coulomb force may be either attractive or repulsive depending upon whether the two objects have the opposite or the same sign to their charges. This result usefully describes the interaction between point charge Q_1 that is very close to the surface of a macroscopic, charged object of radius r_0 and charge Q_2.

1 The properties of gases

Answers to discussion questions

D1.1 An equation of state is an equation that relates the variables, which define the state of a system, to each other. Boyle, Charles, and Avogadro established these relations for gases at low pressures (perfect gases) by appropriate experiments. Boyle determined how volume varies with pressure ($V \propto 1/p$), Charles how volume varies with temperature ($V \propto T$), and Avogadro how volume varies with amount of gas ($V \propto n$). Combining all of these proportionalities into one, we find

$$V \propto \frac{nT}{p}.$$

Inserting the constant of proportionality, R, yields the perfect gas equation

$$V = \frac{RnT}{P} \quad \text{or} \quad pV = nRT.$$

D1.2 The partial pressure of a gas in a mixture of gases is the pressure the gas would exert if it alone occupied the same container as the mixture at the same temperature. It is a limiting law because it holds exactly only under conditions where the gases have no effect upon each other. This can only be true in the limit of zero pressure where molecules of the gas are separated by many molecular diameters on the average. Hence, Dalton's law holds exactly only for a mixture of perfect gases; for real gases, the law is only an approximation.

D1.3 The very light molecules of hydrogen (2.02 g mol^{-1}) and helium (4.00 g mol^{-1}) are rare in the Earth's atmosphere because at atmospheric temperatures a significant fraction of these molecules travel at sufficiently high speeds to escape from the planet's gravitational attraction. A very small fraction of the heavy molecules oxygen (32.0 g mol^{-1}), carbon dioxide (44.0 g mol^{-1}), and nitrogen (28.0 g mol^{-1}) have a speed sufficiently large to escape from the Earth's gravitational pull. This reflects the basic kinetic theory relation $\bar{c} = (8RT/\pi M)^{1/2}$ [1.13 and 1.15]; the mean speed decreases by a factor of $M^{-1/2}$ as the molar mass increases. Heavy molecules move more slowly than light ones on the average.

D1.4 The r.m.s. molecular speed c is proportional to $(T/M)^{1/2}$ [1.15]. With the simple molecular kinetic theory proposition that the rate of gaseous diffusion and effusion is proportional to c it is apparent that these rates are proportional to $\sqrt{T}$ and inversely proportional to $\sqrt{M}$, which is Graham's law [1.17].

D1.5 At low pressure the attractive force between molecules of a real gas dominates and causes the compression factor to be less than that of a perfect gas ($Z = 1$) while at high pressure the repulsive force of molecules in close contact dominates and causes the compression factor to be greater than 1. An increase in temperature at a given pressure causes an increase in the mean kinetic energy relative to attractive and repulsive potential energies, which do not depend upon temperature. Thus, an increase in temperature diminishes the deviation of a real gas from perfect gas behavior and the compression factor approaches the value of a perfect gas at both low and high pressure. These qualitative characteristics are quantified in the Figure 1.1 plot of three isotherms of $Z(p)$ for gaseous ethene at 300 K, 400 K, and 500 K. (The critical temperature of ethene is 283.1 K so ethene is properly called a gas at the temperature of these isotherms.)

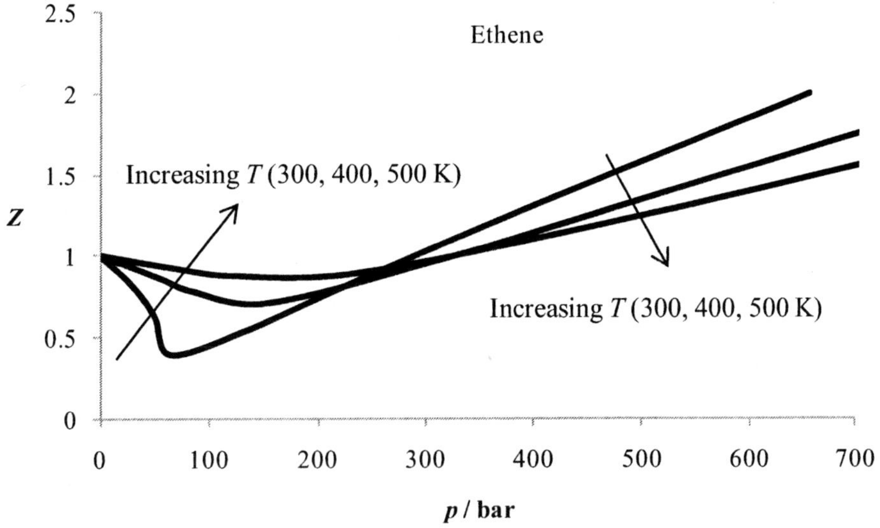

Figure 1.1

D1.6 The critical point of a substance is a single (p, T, V_m) point that is characterized by the critical constants (p_c, T_c, V_c). Table 1.4 lists the critical temperature (T_c) for select substances and the van der Waals parameters of Table 1.5 along with eqn 1.24 are used to estimate the critical pressure (p_c) and the critical volume (V_c). A liquid being heated becomes indistinguishable from the gas or vapour phase at the critical point. At temperatures above the critical temperature a gaseous substance cannot be liquefied by the application of pressure alone while at temperatures below the critical temperature the vapour can be condensed by the application of pressure.

D1.7 The van der Waals equation accounts for repulsive interactions between molecules by supposing that they cause the molecules to behave as small but impenetrable spheres. The nonzero volume of the molecules implies that instead of moving in a volume V they are restricted to a smaller volume $V - nb$, where b is approximately four times the total volume of Avogadro's number of molecules ($b = 4V_{\text{molecule}}N_A$). This argument suggests that the perfect gas law $p = nRT/V$ should be replaced by $p = nRT/(V - nb)$ when repulsions are significant.

The pressure depends on both the frequency of collisions with the walls and the force of each collision. Both the frequency of the collisions and their force are reduced by the attractive forces, which act with a strength proportional to the molar concentration, n/V, of molecules in the sample. Therefore, because the attractive forces reduce both the frequency and the force of the collisions, the pressure is reduced in proportion to the square of this concentration. If the reduction of pressure is written as $-a(n/V)^2$, where a is a positive constant characteristic of each gas, the combined effect of the repulsive and attractive forces is the van der Waals equation of state.

D1.8 The cooling of a real gas as it expands through a narrow throttle, without the loss or gain of heat to or from the environment, is called the **Joule–Thomson effect**. Attractive forces between molecules are reduced during the expansion with the requisite energy for separating the molecules being supplied by a reduction in molecular kinetic energy. The kinetic energy reduction manifests itself as a temperature reduction as $N_A\langle E_k\rangle = \tfrac{1}{2}N_A mc^2 = \tfrac{1}{2}Mc^2 = \tfrac{3}{2}RT$ [1.15].

Solutions to exercises

E1.1 Solve the perfect gas law $pV = nRT$ [1.2] for pressure. Select the Table 1.1 gas constant value that has units that both cancel with the volume unit of the data and provide the desired pressure unit. Alternatively, convert the data to units that cancel with those of a favorite gas constant value.

$$n = 3.055\ \text{g}\times\frac{1\ \text{mol N}_2}{28.02\ \text{g}} = 0.1090\ \text{mol},\quad T = (273.15+32)\ \text{K} = 305\ \text{K},\quad \text{and}\quad V = 3.00\ \text{dm}^3$$

$$p = \frac{nRT}{V} = \frac{(0.1090\ \text{mol})\times(8.3145\ \text{dm}^3\ \text{kPa K}^{-1}\ \text{mol}^{-1})\times(305\ \text{K})}{3.00\ \text{dm}^3} = \boxed{92.1\ \text{kPa}}$$

E1.2 Arranging the perfect gas law [1.2] in a form to solve for pressure and select the Table 1.1 gas constant that provides convenient unit cancellation.

$$n = \frac{0.425\ \text{g}}{20.18\ \text{g mol}^{-1}} = 2.11\times10^{-2}\ \text{mol}, \qquad T = 77\ \text{K}, \qquad V = 6.00\ \text{dm}^3$$

$$p = \frac{nRT}{V} = \frac{(2.11\times10^{-2}\ \text{mol})\times(8.3145\ \text{dm}^3\ \text{kPa K}^{-1}\ \text{mol}^{-1})\times77\ \text{K}}{6.00\ \text{dm}^3} = \boxed{2.25\ \text{kPa}}$$

E1.3 Solve the perfect gas law [1.2] for n and recognize the volume unit of the data needs to be converted to dm^3 so that units cancel conveniently in the calculation. $V = 300.0\ \text{cm}^3 = 0.3000\ \text{dm}^3$ and $T = (273.15 + 14.5)\ \text{K} = 287.65\ \text{K}$.

$$n = \frac{pV}{RT} = \frac{(34.5\ \text{kPa})\times(0.3000\ \text{dm}^3)}{(8.3145\ \text{dm}^3\ \text{kPa K}^{-1}\ \text{mol}^{-1})\times(287.65\ \text{K})} = 4.33\times10^{-3}\ \text{mol} = \boxed{4.33\ \text{mmol}}$$

E1.4 mass of CO_2 = 1.04 kg – 0.74 kg = 0.30 kg

$$n(\text{amount}) = 0.30\ \text{kg}\times\left(\frac{1\ \text{mol}}{44.01\times10^{-3}\ \text{kg}}\right) = 6.82\ \text{mol}, \qquad V = 250\ \text{cm}^3 = 0.250\ \text{dm}^3,$$

and $T = (273.15 + 20)\ \text{K} = 293\ \text{K}$. Remember that 100 kPa = 1 bar.

$$p = \frac{nRT}{V} = \frac{(6.82\ \text{mol})\times(8.3145\ \text{dm}^3\ \text{kPa K}^{-1}\ \text{mol}^{-1})\times(293\ \text{K})}{0.250\ \text{dm}^3} = 6.65\times10^4\ \text{kPa} = \boxed{665\ \text{bar}}$$

E1.5 Boyle's law states that p is inversely proportional to V at constant temperature. Therefore, $p_2 = \frac{V_1}{V_2}\times p_1$.

$V_1 = 1.00\ \text{dm}^3 = 1.00\times10^3\ \text{cm}^3$, $p_1 = 1.00\ \text{atm}$, and $V_2 = 1.00\times10^2\ \text{cm}^3$. Therefore,

$$p_2 = \left(\frac{1.00\times10^3\ \text{cm}^3}{100\ \text{cm}^3}\right)\times(1.00\ \text{atm}) = \boxed{10.0\ \text{atm}}$$

E1.6 The amount n and V are constant, hence by the perfect gas law: $nR/V = p/T$ = a constant. Thus, $\frac{p_1}{T_1} = \frac{p_2}{T_2}$ and, solving for p_2 gives $p_2 = \frac{T_2 p_1}{T_1}$.

$$p_2 = \frac{973\ \text{K}\times125\ \text{kPa}}{291\ \text{K}} = 418\ \text{kPa} = \boxed{4.18\ \text{bar}}$$

E1.7 At constant temperature equation [1.4] becomes $p_1V_1 = p_2V_2$.

$$p_2 = \frac{p_1V_1}{V_2} = \frac{101\ \text{kPa}\times7.20\ \text{dm}^3}{4.21\ \text{dm}^3} = \boxed{173\ \text{kPa}}$$

E1.8 If we assume constant pressure, then equation [1.4] becomes $\frac{T_1}{V_1} = \frac{T_2}{V_2}$

$$T_2 = \frac{T_1V_2}{V_1} = \frac{295.3\ \text{K}\times100\ \text{cm}^3}{1.00\ \text{dm}^3}\left(\frac{1\ \text{dm}^3}{1\times10^3\ \text{cm}^3}\right) = \boxed{29.5\ \text{K}}$$

E1.9 Pressure is constant, so $\frac{T_1}{V_1} = \frac{T_2}{V_2}$ [1.4].

$T_1 = 315\ \text{K}$ and the volume has increased by 25%, so $V_2 = 1.25\ V_1$.

$$T_2 = \frac{1.25\ V_1}{V_1}\times315\ \text{K} = 1.25\times315\ \text{K} = \boxed{394\ \text{K}}$$

E1.10 (a) $\dfrac{p_1V_1}{T_1}=\dfrac{p_2V_2}{T_2}$ [1.4]

$$V_2=\frac{(104\text{ kPa})\times(2.0\text{ m}^3)\times(268.2\text{ K})}{(294.3\text{ K})\times(52\text{ kPa})}=\boxed{3.6\text{ m}^3}$$

(b) $$V_2=\frac{(104\text{ kPa})\times(2.0\text{ m}^3)\times(221.2\text{ K})}{(294.3\text{ K})\times(0.880\text{ kPa})}=\boxed{178\text{ m}^3}$$

E1.11 Boyle's law states that pressure is inversely proportional to volume at constant temperature so $V_f = p_iV_i/p_f$ where the subscripts i and f reference initial and final state conditions. Additionally, there will be an increase in pressure due to the depth submerged. The total pressure at submerged depths is calculated by adding the hydrostatic pressure to the surface pressure: $p_f = p_i + \rho gh$ [0.6] where $p_i = 1.00\text{ atm} = 1.01\times10^5\text{ Pa}$.

$$p_f=p_i+\rho gh=\left(1.01\times10^5\text{ Pa}\right)+(1.025\times10^3\text{ kg m}^{-3})\times(9.81\text{ m s}^{-2})\times(50\text{ m})=6.0\times10^5\text{ Pa}$$

$$V_f=\frac{p_i}{p_f}V_i=\frac{1.01\times10^5\text{ Pa}}{6.0\times10^5\text{ Pa}}\times3.0\text{ m}^3=\boxed{0.50\text{ m}^3}$$

E1.12 Air is roughly 80% N_2(g) and 20% O_2(g). There are also some minor components, but they do not affect much the average molar mass.

$$\text{molar mass air}=\left(0.80\times28+0.20\times32\right)\text{ g mol}^{-1}\approx29\text{ g mol}^{-1}$$

$$\text{molar mass H}_2\text{(g)}=2.0\text{ g mol}^{-1}$$

Hence,

$$\frac{\text{density H}_2\text{(g)}}{\text{density air}}=\frac{2.0\text{ g mol}^{-1}}{29\text{ g mol}^{-1}}=\boxed{0.069}$$

In the same volume (the volume of the balloon), the mass of air would be $29\text{ g mol}^{-1}/2.0\text{ g mol}^{-1}$ = 14.5 times the mass of hydrogen.

$$\text{mass of air displaced}=14.5\times10\text{ kg}=145\text{ kg}$$

The payload is the difference between the mass of displaced air and the mass of the balloon (here assumed to be the mass of hydrogen).

$$\text{payload}=145\text{ kg}-10\text{ kg}=\boxed{135\text{ kg}}$$

E1.13 We use the perfect gas law, $pV = nRT$, and solve for V, the volume of SO_2 released in a day of volcanic activity. We calculate n from the mass of SO_2, which is estimated to be 250 t (= 250×10^3 kg).

$$n=\frac{\text{mass}}{\text{molar mass}}=\frac{250\times10^3\text{ kg}}{64.06\times10^{-3}\text{ kg mol}^{-1}}=3.9\times10^6\text{ mol}$$

$$V=\frac{nRT}{p}=\frac{\left(3.9\times10^6\text{ mol}\right)\times0.082057\text{ dm}^3\text{ atm K}^{-1}\text{ mol}^{-1}\times1073\text{ K}}{1.0\text{ atm}}$$

$$=\boxed{3.4\times10^8\text{ dm}^3}$$

E1.14 $pV = nRT$, with n constant, yields $\dfrac{p_f V_f}{T_f} = \dfrac{p_i V_i}{T_i}$ [1.4] where i and f reference initial and final states.

Solving for p_f and substitution of $V = \frac{4}{3}\pi r^3$ gives

$$p_f = \frac{p_i V_i T_f}{V_f T_i} = \left(\frac{r_i}{r_f}\right)^3 \frac{T_f}{T_i} \times p_i$$

$$= \left(\frac{1.5\ \text{m}}{3.5\ \text{m}}\right)^3 \times \left(\frac{248\ \text{K}}{293\ \text{K}}\right) \times 1.0\ \text{atm} = \boxed{6.7\times10^{-2}\ \text{atm}} \text{ or } 51\ \text{Torr.}$$

E1.15 (a) There is only one volume, so using the amount of nitrogen and partial pressure of nitrogen we can calculate the volume.

$$n_{N_2} = \frac{0.225\ \text{g}}{28.02\ \text{g mol}^{-1}} = 8.03\times10^{-3}\ \text{mol}, \qquad p_{N_2} = 15.2\ \text{kPa}, \qquad T = 300\ \text{K}$$

$$V = \frac{n_{N_2} RT}{p_{N_2}} \quad \text{[Each component of the mixture satisfies the perfect gas law.]}$$

$$= \frac{(8.03\times10^{-3}\ \text{mol})\times(8.3145\ \text{dm}^3\ \text{kPa K}^{-1}\ \text{mol}^{-1})\times(300\ \text{K})}{15.2\ \text{kPa}} = \boxed{1.32\ \text{dm}^3}$$

(b)

$$n_{CH_4} = \frac{0.320\ \text{g}}{16.04\ \text{g mol}^{-1}} = 2.00\times10^{-2}\ \text{mol}$$

$$n_{Ar} = \frac{0.175\ \text{g}}{39.95\ \text{g mol}^{-1}} = 4.38\times10^{-3}\ \text{mol}$$

$$n = n_{CH_4} + n_{Ar} + n_{N_2} = (2.00 + 0.438 + 0.803)\times10^{-2}\ \text{mol} = 3.24\times10^{-2}\ \text{mol}$$

Solving the perfect gas law for the total pressure p of n moles of gas, we find

$$p = \frac{nRT}{V} = \frac{(3.24\times10^{-2}\ \text{mol})\times(8.3145\ \text{dm}^3\ \text{kPa K}^{-1}\ \text{mol}^{-1})\times(300\ \text{K})}{1.32\ \text{dm}^3} = \boxed{61.2\ \text{kPa}}.$$

E1.16 The total pressure p is given by **Dalton's law**: $p = p_A + p_B + \cdots$ [1.6]. Therefore,

$$p = p_{\text{dry air}} + p_{\text{water vapour}}.$$

$$p_{\text{dry air}} = p - p_{\text{water vapour}} = (760 - 47)\ \text{Torr} = \boxed{713\ \text{Torr}}$$

E1.17 To calculate the molar mass of the compound, we need to relate density, temperature, and pressure. Using the perfect gas law:

$$pV = nRT = \left(\frac{m}{M}\right)RT \quad \text{or} \quad M = \left(\frac{m}{V}\right)\frac{RT}{p} = \frac{\rho RT}{p} \quad \text{where } \rho \text{ is density (i.e., mass/volume).}$$

$$M = \frac{(1.23\ \text{g dm}^{-3})\times(8.3145\ \text{dm}^3\ \text{kPa K}^{-1}\ \text{mol}^{-1})\times(330\ \text{K})}{(25.5\ \text{kPa})} = \boxed{132\ \text{g mol}^{-1}}$$

E1.18

$$pV = nRT = \left(\frac{m}{M}\right)RT \quad \text{or} \quad M = \frac{mRT}{pV}. \text{ Thus,}$$

$$M = \frac{(33.5\times10^{-3}\ \text{g})\times(62.364\ \text{dm}^3\ \text{Torr K}^{-1}\ \text{mol}^{-1})\times(298\ \text{K})}{(152\ \text{Torr})\times(0.250\ \text{dm}^3)} = \boxed{16.4\ \text{g mol}^{-1}}.$$

E1.19 (a) The partial pressure of the gases can be related to their mole fractions. The total amount is $n = n_{N_2} + n_{H_2} = (1.0 + 2.0)\ \text{mol} = 3.0\ \text{mol}$.

$$x_{H_2} = \frac{2.0\ \text{mol}}{3.0\ \text{mol}} = 0.67$$

$$x_{N_2} = \frac{1.0\ \text{mol}}{3.0\ \text{mol}} = 0.33$$

$$p_J = n_J \frac{RT}{V} \quad \text{where} \quad \frac{RT}{V} = \frac{(8.3145\ \text{dm}^3\ \text{kPa K}^{-1}\ \text{mol}^{-1})\times(273.15\ \text{K})}{22.4\ \text{dm}^3}$$

$$= 101\ \text{kPa mol}^{-1} = 1.01\ \text{bar mol}^{-1}$$

Thus,

$$p_{H_2} = (2.0\ \text{mol})\times(1.01\ \text{bar mol}^{-1}) = \boxed{2.0\ \text{bar}}$$

$$p_{N_2} = (1.0\ \text{mol})\times(1.01\ \text{bar mol}^{-1}) = \boxed{1.0\ \text{bar}}$$

(b) The total pressure is the sum of the partial pressures.

$$p = p_{H_2} + p_{N_2} = (2.0 + 1.0)\ \text{bar} = \boxed{3.0\ \text{bar}}$$

E1.20

$$\bar{c} = \left(\frac{8}{3\pi}\right)^{1/2} c\ [1.13] = \left(\frac{8}{3\pi}\right)^{1/2}\left(\frac{3RT}{M}\right)^{1/2}\ [1.15] = \left(\frac{8RT}{\pi M}\right)^{1/2}$$

Now we develop a convenient computational formula for various T and M.

$$\bar{c} = \left(\frac{8\times(8.3145\ \text{J K}^{-1}\ \text{mol}^{-1})\times T}{\pi M\times(10^{-3}\ \text{kg/g})}\right)^{1/2}$$

$$\bar{c}/(\text{m s}^{-1}) = 145.5\times\left(\frac{T/\text{K}}{M/\text{g mol}^{-1}}\right)^{1/2}$$

(a) $M_{He} = 4.00\ \text{g mol}^{-1};\ \bar{c}_{He}/(\text{m s}^{-1}) = 72.75\times(T/\text{K})^{1/2}$

$$\bar{c}_{He}(79\ \text{K}) = \boxed{647\ \text{m s}^{-1}}$$

$$\bar{c}_{He}(315\ \text{K}) = \boxed{1.29\ \text{km s}^{-1}}$$

$$\bar{c}_{He}(1500\ \text{K}) = \boxed{2.82\ \text{km s}^{-1}}$$

(b) $M_{CH_4} = 16.04\ \text{g mol}^{-1};\ \bar{c}_{CH_4}/(\text{m s}^{-1}) = 36.33\times(T/\text{K})^{1/2}$

$$\bar{c}_{CH_4}(79\ \text{K}) = \boxed{323\ \text{m s}^{-1}}$$

$$\bar{c}_{CH_4}(315\ \text{K}) = \boxed{645\ \text{m s}^{-1}}$$

$$\bar{c}_{CH_4}(1500\ \text{K}) = \boxed{1.41\ \text{km s}^{-1}}$$

E1.21 (a) Calculate n and solve the perfect gas law for T.

$$n = \frac{1.0\times10^{23}\ \text{molecules}}{6.02\times10^{23}\ \text{molecules mol}^{-1}} = 0.16\overline{6}\ \text{mol}$$

$$T = \frac{pV}{nR} = \frac{(100\ \text{kPa})\times(1.0\ \text{dm}^3)}{(0.16\overline{6}\ \text{mol})\times(8.3145\ \text{dm}^3\ \text{kPa K}^{-1}\ \text{mol}^{-1})} = \boxed{72\ \text{K}}$$

(b)

$$c = \left(\frac{3RT}{M}\right)^{1/2}\ [1.15] = \left\{\frac{3\times(8.3145\ \text{J K}^{-1}\ \text{mol}^{-1})\times(72\ \text{K})}{0.002016\ \text{kg mol}^{-1}}\right\}^{1/2} = \boxed{94\overline{4}\ \text{m s}^{-1}}$$

(c) The temperature would not be different if they were O_2 molecules and exerted the same pressure in the same volume, but their root mean square speed would be different.

COMMENT. This exercise could have been solved by first obtaining the root mean square speed from $pV = \frac{1}{3}nMc^2$ [1.9] and then using 1.15 to solve for the temperature. The results should be identical.

E1.22 First calculate the diameter d of the spherical vessel.

$$V = \tfrac{4}{3}\pi(d/2)^3 \qquad \text{so } d = 2\left(\frac{3V}{4\pi}\right)^{1/3} = 2\left(\frac{3(1.0\times10^{-3}\ \text{m}^3)}{4\pi}\right)^{1/3} = 0.124\ \text{m}.$$

The equation for determining the mean free path is $\lambda = RT/\sqrt{2}N_A\sigma p$ [1.19]. Solve for p with the desired condition that $\lambda = d$.

$$p = \frac{RT}{\sqrt{2}N_A\sigma\lambda} = \frac{(8.3145\ \text{Pa m}^3\ \text{K}^{-1}\ \text{mol}^{-1})\times(298.15\ \text{K})}{\sqrt{2}\times(6.022\times10^{23}\ \text{mol}^{-1})\times(0.36\times10^{-18}\ \text{m}^2)\times(0.124\ \text{m})} = \boxed{0.065\ \text{Pa}}$$

Note: 1 J = 1 Pa m^3

E1.23 First estimate the diameter d of the atoms using $\sigma = \pi d^2$ (see Section 1.8).

$$d = (\sigma/\pi)^{1/2} = (0.36\times10^{-18}\ \text{m}^2/\pi)^{1/2} = 3.39\times10^{-10}\ \text{m}$$

The formula for determining the mean free path is $\lambda = RT/\sqrt{2}N_A\sigma p$ [1.19]. Solve for p with the desired condition that $\lambda = 10d$.

$$p = \frac{RT}{\sqrt{2}N_A\sigma\lambda}$$

$$= \frac{(8.3145\ \text{Pa m}^3\ \text{K}^{-1}\ \text{mol}^{-1})\times(298.15\ \text{K})}{\sqrt{2}\times(6.022\times10^{23}\ \text{mol}^{-1})\times(0.36\times10^{-18}\ \text{m}^2)\times(33.9\times10^{-10}\ \text{m})} = \boxed{2.4\times10^6\ \text{Pa}}$$

E1.24

$$\lambda = \frac{RT}{\sqrt{2}N_A\sigma p}\ [1.19]$$

$$= \frac{(8.3145\ \text{J K}^{-1}\ \text{mol}^{-1})\times(217\ \text{K})}{\sqrt{2}\times(6.022\times10^{23}\ \text{mol}^{-1})\times(0.43\times10^{-18}\ \text{m}^2)}\times\frac{1}{(0.050\ \text{atm})\times(1.013\times10^5\ \text{Pa atm}^{-1})}$$

$$= 973\ \text{nm} = \boxed{0.97\ \mu\text{m}}$$

E1.25 We first calculate the r.m.s. speed.

$$c = \sqrt{\frac{3RT}{M}}\ [1.15] = \left(\frac{3\times(8.3145\ \text{J K}^{-1}\ \text{mol}^{-1})\times(298.15\ \text{K})}{0.03995\ \text{kg mol}^{-1}}\right)^{1/2} = 431\ \text{m s}^{-1}$$

Now we formulate an expression that conveniently allows for the calculation of the collision frequency at various pressures. (Table 1.3 provides $\sigma = 0.36\ \text{nm}^2 = 0.36\times10^{-18}\ \text{m}^2$.)

$$z = \frac{\sqrt{2}N_A\sigma c p}{RT}\ [1.19]$$

$$= \left\{\frac{\sqrt{2}(6.022\times10^{23}\ \text{mol}^{-1})\times(0.36\times10^{-18}\ \text{m}^2)\times(431\ \text{m s}^{-1})}{(8.3145\ \text{J K}^{-1}\ \text{mol}^{-1})\times(298.15\ \text{K})}\right\}\times\left\{\frac{1\times10^3\ \text{Pa}}{1\ \text{kPa}}\right\}p$$

$$z/\text{s}^{-1} = (5.3\overline{3}\times10^7)\times p/\text{kPa}$$

(a) At p = 10 bar = 1.0×10^3 kPa: $z = \boxed{5.3 \times 10^{10}\ \text{s}^{-1}}$.

(b) At p = 100 kPa: $z = \boxed{5.3 \times 10^{9}\ \text{s}^{-1}}$.

(c) At p = 1.0 Pa = 1.0×10^{-3} kPa: $z = \boxed{5.3 \times 10^{4}\ \text{s}^{-1}}$.

E1.26 Exercise 1.25 shows the calculation of the number of collisions per second for a single Ar atom. The dependence upon pressure is found to be $z/\text{s}^{-1} = (5.3\bar{3}\times10^7)\times p/\text{kPa}$. Let $Z_{\text{collisions}}$ be the total number of collisions per second in the container. Then, $Z_{\text{collisions}} = Nz/2$ where N is the number of atoms in the vessel and we have divided by 2 because each collision involves 2 atoms. The perfect gas law is used to find a formula for N.

$$N = N_{\text{A}}n = N_{\text{A}} \times \left(\frac{pV}{RT}\right) = \left(\frac{N_{\text{A}}V}{RT}\right)p$$

$$= \left\{\frac{(6.022\times10^{23}\ \text{mol}^{-1})\times(1.0\ \text{dm}^3)}{(8.3145\ \text{dm}^3\ \text{kPa K}^{-1}\ \text{mol}^{-1})\times(298.15\ \text{K})}\right\}p = (2.4\bar{3}\times10^{20})\times(p/\text{kPa})$$

Thus,

$$Z_{\text{collisions}}/\text{s}^{-1} = \tfrac{1}{2}Nz/\text{s}^{-1} = \tfrac{1}{2}(2.4\bar{3}\times10^{20})\times(5.3\bar{3}\times10^{7})\times(p/\text{kPa})^2$$

$$Z_{\text{collisions}}/\text{s}^{-1} = (6.4\bar{7}\times10^{27})\times(p/\text{kPa})^2.$$

(a) At p = 10 bar = 1.0×10^3 kPa: $Z_{\text{collisions}} = \boxed{6.5 \times 10^{33}\ \text{s}^{-1}}$.

(b) At p = 100 kPa: $Z_{\text{collisions}} = \boxed{6.5 \times 10^{31}\ \text{s}^{-1}}$.

(c) At p = 1.0 Pa = 1.0×10^{-3} kPa: $Z_{\text{collisions}} = \boxed{6.5 \times 10^{21}\ \text{s}^{-1}}$.

E1.27 $z = c/\lambda$ [1.18]. We have already calculated λ (= 0.97 μm) in problem 1.24. The r.m.s. speed is:

$$c = \sqrt{\frac{3RT}{M}}\ [1.15] = \left(\frac{3\times(8.3145\ \text{J K}^{-1}\ \text{mol}^{-1})\times(217\ \text{K})}{(28.02\times10^{-3}\ \text{kg mol}^{-1})}\right)^{1/2} = 439\ \text{m s}^{-1}$$

$$z = \frac{439\ \text{m s}^{-1}}{0.97\times10^{-6}\ \text{m}} = \boxed{4.5\times10^{8}\ \text{s}^{-1}}$$

E1.28 $\lambda = \dfrac{RT}{\sqrt{2}N_{\text{A}}\sigma p}$ [1.19]; σ is given as 0.43 nm^2

Because we want to compute z for several different p's, first develop a convenient computational formula.

$$\lambda = \frac{(8.3145\ \text{J K}^{-1}\ \text{mol}^{-1})\times(298.15\ \text{K})}{\sqrt{2}\times(6.022\times10^{23}\ \text{mol}^{-1})\times(0.43\times10^{-18}\ \text{m}^2)\times p}$$

$$= (6.8\times10^{-3}\ \text{m})\times(p/\text{Pa})^{-1}$$

(a) When p = 10 bar = 1.0×10^6 Pa, $\lambda = (6.8\times10^{-3}\ \text{m})\times(1.0\times10^{6})^{-1} = 6.8\times10^{-9}\,\text{m} = \boxed{6.8\ \text{nm}}$.

(b) When p = 103 kPa, $\lambda = (6.8\times10^{-3}\ \text{m})\times(103\times10^{3})^{-1} = 6.6\times10^{-8}\ \text{m} = \boxed{68\ \text{nm}}$.

(c) When p = 1.0 Pa, $\lambda = 6.8 \times 10^{-3}$ m = $\boxed{7\ \text{mm}}$.

E1.29 $\lambda = \dfrac{RT}{\sqrt{2}N_{\text{A}}\sigma p}$ [1.19]. Substitution of $p = \dfrac{nRT}{V}$ yields $\lambda = \dfrac{RT}{\sqrt{2}N_{\text{A}}\sigma}\times\dfrac{V}{nRT} = \dfrac{V}{\sqrt{2}N_{\text{A}}\sigma n}$. Hence, λ is $\boxed{\text{independent of temperature}}$ when V is constant.

E1.30 (a) For a perfect gas $p = nRT/V$.

(i) $n = 1.0$ mol, $T = 273.15$ K, $V = 22.414$ dm^3.

$$p = \frac{(1.0\text{ mol})\times(8.3145\text{ dm}^3\text{ kPa K}^{-1}\text{ mol}^{-1})\times(273.15\text{ K})}{22.414\text{ dm}^3} = \boxed{10\bar{1}\text{ kPa}}$$

(ii) $n = 1.0$ mol, $T = 1000$ K, $V = 100$ cm^3 = 0.100 dm^3.

$$p = \frac{(1.0\text{ mol})\times(8.3145\text{ dm}^3\text{ kPa K}^{-1}\text{ mol}^{-1})\times(1000\text{ K})}{0.100\text{ dm}^3} = 8.3\bar{1}\times10^4\text{ kPa} = \boxed{83\bar{1}\text{ bar}}$$

(b) The van der Waals equation of state is $p = \dfrac{nRT}{V-nb} - a\left(\dfrac{n}{V}\right)^2$ [1.23a].

Table 1.5 provides $a = 5.507$ dm^6 atm mol^{-2} and $b = 0.0651$ dm^3 mol^{-1} for ethane.

(i) $n = 1.0$ mol, $T = 273.15$ K, $V = 22.414$ dm^3.

$$p = \frac{(1.0\text{ mol})\times(8.2058\times10^{-2}\text{ dm}^3\text{ atm K}^{-1}\text{ mol}^{-1})\times(273.15\text{ K})}{22.414\text{ dm}^3 - (1.0\text{ mol})\times(0.0651\text{ dm}^3\text{ mol}^{-1})}$$

$$-5.507\text{ dm}^6\text{ atm mol}^{-2}\left(\frac{1.0\text{ mol}}{22.414\text{ dm}^3}\right)^2$$

$$p = \boxed{0.99\text{ atm}}$$

(ii) $n = 1.0$ mol, $T = 1000$ K, $V = 100$ cm^3 = 0.100 dm^3.

$$p = \frac{(1.0\text{ mol})\times(8.2058\times10^{-2}\text{ dm}^3\text{ atm K}^{-1}\text{ mol}^{-1})\times(1000\text{ K})}{0.100\text{ dm}^3 - (1.0\text{ mol})\times(0.0651\text{ dm}^3\text{ mol}^{-1})}$$

$$-5.507\text{ dm}^6\text{ atm mol}^{-2}\left(\frac{1.0\text{ mol}}{0.100\text{ dm}^3}\right)^2$$

$$p = \boxed{1.8\times10^3\text{ atm}}$$

E1.31

$$n = 10.00\text{ g CO}_2\times\frac{1\text{ mol CO}_2}{44.01\text{ g}} = 0.2272\text{ mol.}$$

For a perfect gas:

$$p = \frac{nRT}{V} = \frac{(0.2272\text{ mol})\times(0.08206\text{ dm}^3\text{ atm mol}^{-1}\text{ K}^{-1})\times(298.1\text{ K})}{0.100\text{ dm}^3} = \boxed{55.6\text{ atm}}$$

For a van der Waals gas:

$$p = \frac{nRT}{V-nb} - a\left(\frac{n}{V}\right)^2\ [1.23a]\text{ where } a = 3.610\text{ dm}^6\text{ atm mol}^{-2}\text{ and } b = 0.0429\text{ dm}^3\text{ mol}^{-1}$$

$$p = \frac{(0.2272\text{ mol})\times(0.08206\text{ dm}^3\text{ atm mol}^{-1}\text{ K}^{-1})(298.1\text{ K})}{(0.100\text{ dm}^3) - (0.2272\text{ mol}\times0.0429\text{ dm}^3\text{ mol}^{-1})}$$

$$-3.610\text{ dm}^6\text{ atm mol}^{-2}\times\left(\frac{0.2272\text{ mol}}{0.100\text{ dm}^3}\right)^2$$

$$= \boxed{43.0\text{ atm}}$$

The fact that $p_{\text{van der Waals}} < p_{\text{perfect}}$ indicates that attractive molecular forces are significant at these conditions. The perfect gas law overestimates the gas pressure by (55.6 – 43.0) × 100% / 43.0 = 29.3%, a deviation that is unacceptably large for most applications.

E1.32 Solve the van der Waals equation for p, giving $p=\dfrac{RT}{V_m-b}-\dfrac{a}{V_m^2}$ [1.23a] with $V_m=V/n$. This is easily factored into $p=\dfrac{RT(1-b/V_m)^{-1}}{V_m}-\dfrac{a}{V_m^2}$. Since $b/V_m \ll 1$, it is possible to expand the factor $(1-b/V_m)^{-1}$ in a Taylor series for which higher order terms become negligibly small:

$$(1-b/V_m)^{-1}=1+\frac{b}{V_m}+\left(\frac{b}{V_m}\right)^2+\cdots.$$

Thus,

$$p=\frac{RT}{V_m}\left(1+\frac{b}{V_m}+\frac{b^2}{V_m^2}+\cdots\right)-\frac{a}{V_m^2}=\frac{RT}{V_m}\left[1+\left(b-\frac{a}{RT}\right)\frac{1}{V_m}+\frac{b^2}{V_m^2}+\cdots\right].$$

Comparing this expression with the virial equation of state, $p=\dfrac{RT}{V_m}\left(1+\dfrac{B}{V_m}+\dfrac{C}{V_m^2}+\cdots\right)$ [1.22], we see that $\boxed{B=b-\dfrac{a}{RT} \text{ and } C=b^2}$.

E1.33 $$\left(p+\frac{a}{V_m^2}\right)(V_m-b)=RT \text{ [1.23b]} \qquad \text{where } V_m=V/n$$

State conditions:

$V_m=5.00\times10^{-4}\ \text{m}^3\ \text{mol}^{-1}=0.500\ \text{dm}^3\ \text{mol}^{-1}$, $T=273\ \text{K}, p=3.0\ \text{MPa}=3.0\times10^3\ \text{kPa}$,
$a=0.50\ \text{m}^6\ \text{Pa mol}^{-2}=0.50\times10^3\ \text{dm}^6\ \text{kPa mol}^{-2}$.

Solve the van der Waals equation for b, giving

$$\begin{aligned}b&=V_m-RT\left(p+\frac{a}{V_m^2}\right)^{-1}\\&=(0.500\ \text{dm}^3\ \text{mol}^{-1})\\&\quad-(8.3145\ \text{dm}^3\ \text{kPa K}^{-1}\ \text{mol}^{-1})\times(273\ \text{K})\times\{(3.0\times10^3\ \text{kPa})\\&\quad+(0.50\times10^3\ \text{dm}^6\ \text{kPa mol}^{-2})\times(0.500\ \text{dm}^3\ \text{mol}^{-1})^{-2}\}^{-1}\\&=\boxed{4.60\times10^{-2}\ \text{dm}^3\ \text{mol}^{-1}}.\end{aligned}$$

$$Z=\frac{pV_m}{RT}\text{ [1.20b]}=\frac{(3.0\times10^3\ \text{kPa})\times(0.500\ \text{dm}^3\ \text{mol}^{-1})}{(8.3145\ \text{dm}^3\ \text{kPa K}^{-1}\ \text{mol}^{-1})\times(273\ \text{K})}=\boxed{0.66}$$

We conclude that, since $Z<1$, $\boxed{\text{molecular attractions dominate}}$ at these conditions.

E1.34 In exercise 1.32 it is found that the virial and van der Waals coefficients are related by

$B=b-\dfrac{a}{RT}$ and $C=b^2$. Consequently, because $C=1200\ \text{cm}^6\ \text{mol}^{-2}$,

$b=C^{1/2}=34.6\ \text{cm}^3\ \text{mol}^{-1}=\boxed{3.46\times10^{-2}\ \text{dm}^3\ \text{mol}^{-1}}$ and

$$\begin{aligned}a&=RT(b-B)\\&=(8.206\times10^{-2}\ \text{dm}^3\ \text{atm mol}^{-1}\ \text{K}^{-1})\times(273\ \text{K})\times(3.46+2.17)\times10^{-2}\ \text{dm}^3\ \text{mol}^{-1}\\&=\boxed{1.26\ \text{dm}^6\ \text{atm mol}^{-2}}.\end{aligned}$$

E1.35 From Exercise 1.32, $B = b - \dfrac{a}{RT}$. Thus, $B = 0$ when $\dfrac{a}{RT} = b$ or $T = \dfrac{a}{bR}$. For CO_2:

$$T_{B=0} = \frac{3.610\ \text{dm}^6\ \text{atm mol}^{-2}}{0.0429\ \text{dm}^3\ \text{mol}^{-1} \times 0.08206\ \text{dm}^3\ \text{atm K}^{-1}\ \text{mol}^{-1}} = \boxed{1.03\times10^3\ \text{K}}.$$

E1.36 The critical volume of a van der Waals gas is $V_c = 3b$ [1.24] so

$$b = \tfrac{1}{3}V_c = \tfrac{1}{3}(148\,\text{cm}^3\,\text{mol}^{-1}) = 49.3\,\text{cm}^3\,\text{mol}^{-1} = \boxed{0.0493\ \text{dm}^3\ \text{mol}^{-1}}.$$

As shown in Derivation 1.1, $b \approx 4V_{\text{molecule}}N_A$ with $V_{\text{molecule}} = \tfrac{4}{3}\pi r^3$. Substituting and solving for r gives

$$r = \tfrac{1}{2}\left(\frac{3b}{2\pi N_A}\right)^{1/3} = \tfrac{1}{2}\left(\frac{3(49.3\ \text{cm}^3\ \text{mol}^{-1})}{2\pi(6.022\times10^{23}\ \text{mol}^{-1})}\right)^{1/3} = 1.70\times10^{-8}\ \text{cm} = \boxed{170\ \text{pm}}.$$

The critical pressure is $p_c = \dfrac{a}{27b^2}$ [1.24] so

$$a = 27p_cb^2 = 27(48.20\,\text{atm})\times(0.0493\ \text{dm}^3\ \text{mol}^{-1})^2 = \boxed{3.16\ \text{dm}^6\ \text{atm}\,\text{mol}^{-2}}.$$

But this problem is overdetermined. We have another piece of information: $T_c = \dfrac{8a}{27Rb}$ [1.24]. According to the constants we have already determined, T_c should be

$$T_c = \frac{8(3.16\,\text{dm}^6\,\text{atm}\,\text{mol}^{-2})}{27(0.08206\,\text{dm}^3\,\text{atm}\,\text{K}^{-1}\text{mol}^{-1})\times(0.0493\,\text{dm}^3\,\text{mol}^{-1})} = 231\ \text{K}$$

However, the reported T_c is 305.4 K, suggesting our computed a/b is about 25 percent lower than it should be.

Solutions to projects

P1.37 (a) $$\langle v\rangle = \int_0^\infty sF(s)\,\text{d}s = \int_0^\infty s\left\{4\pi\left(\frac{M}{2\pi RT}\right)^{3/2} s^2\text{e}^{-Ms^2/2RT}\right\}\text{d}s\ [1.16]$$

$$= 4\pi\left(\frac{M}{2\pi RT}\right)^{3/2}\int_0^\infty s^3\text{e}^{-Ms^2/2RT}\,\text{d}s$$

$$= 4\pi\left(\frac{M}{2\pi RT}\right)^{3/2}\left(\frac{1}{2}\right)\left(\frac{2RT}{M}\right)^2 = \left(\frac{8RT}{\pi M}\right)^{1/2}$$

$$\bar{c} = \langle v\rangle = \left(\frac{8RT}{\pi M}\right)^{1/2}$$

(b) $$\langle v^2\rangle = \int_0^\infty s^2F(s)\,\text{d}s = \int_0^\infty s^2\left\{4\pi\left(\frac{M}{2\pi RT}\right)^{3/2} s^2\text{e}^{-Ms^2/2RT}\right\}\text{d}s\ [1.16]$$

$$= 4\pi\left(\frac{M}{2\pi RT}\right)^{3/2}\int_0^\infty s^4\text{e}^{-Ms^2/2RT}\,\text{d}s$$

$$= 4\pi\left(\frac{M}{2\pi RT}\right)^{3/2}\left(\frac{3}{8}\right)\left(\frac{2RT}{M}\right)^2\pi^{1/2}\left(\frac{2RT}{M}\right)^{1/2} = \frac{3}{2}\left(\frac{2RT}{M}\right) = \frac{3RT}{M}$$

$$c = \langle v^2\rangle^{1/2} = \left(\frac{3RT}{M}\right)^{1/2}$$

(c) $$F(s)=4\pi\left(\frac{M}{2\pi RT}\right)^{3/2}s^2\mathrm{e}^{-Ms^2/2RT}\quad[1.16]$$

$$\frac{\mathrm{d}F(s)}{\mathrm{d}s}=4\pi\left(\frac{M}{2\pi RT}\right)^{3/2}\left\{2s\,\mathrm{e}^{-Ms^2/2RT}+s^2\left(\frac{-2Ms}{2RT}\right)\mathrm{e}^{-Ms^2/2RT}\right\}$$

$$=4\pi s\left(\frac{M}{2\pi RT}\right)^{3/2}\mathrm{e}^{-Ms^2/2RT}\left(2-\frac{Ms^2}{RT}\right)$$

The most probable speed c^* is located at the peak of $F(s)$ where $\mathrm{d}F(s)/\mathrm{d}s=0$. Inspection of the above equation reveals that the last factor must equal zero at the peak.

$$2-\frac{M(c^*)^2}{RT}=0\quad\text{and}\quad\boxed{c^*=\left(\frac{2RT}{M}\right)^{1/2}}$$

(d) The Maxwell distribution of speeds is $f=4\pi\left(\frac{M}{2\pi RT}\right)^{3/2}s^2\mathrm{e}^{-Ms^2/2RT}\,\Delta s$ [1.16].

At the center of the range, $s=295\ \mathrm{m\ s^{-1}}$:

$$f=4\times\pi\times\left(\frac{28.02\times10^{-3}\ \mathrm{kg\ mol^{-1}}}{2\times\pi\times(8.3145\ \mathrm{kg\ m^2\ s^{-2}\ K^{-1}\ mol^{-1}})\times500\ \mathrm{K}}\right)^{3/2}$$

$$\times(295\ \mathrm{m\ s^{-1}})^2\,\mathrm{e}^{-(28.02\times10^{-3})(295)^2/(2\times8.3145\times500)}\times10\ \mathrm{m\ s^{-1}}$$

$$=\boxed{9.06\times10^{-3}}.$$

P1.38 (a) Using the substitution $V_\mathrm{m}=V/n$, equation [1.23a] becomes $p=\frac{RT}{V_\mathrm{m}-b}-\frac{a}{V_\mathrm{m}^2}$. Thus,

$$\left(\frac{\partial p}{\partial V_\mathrm{m}}\right)_T=-\frac{RT}{(V_\mathrm{m}-b)^2}+\frac{2a}{V_\mathrm{m}^3}$$

$$\left(\frac{\partial^2 p}{\partial V_\mathrm{m}^2}\right)_T=\frac{2RT}{(V_\mathrm{m}-b)^3}-\frac{6a}{V_\mathrm{m}^4}$$

Evaluating these three equations at the critical point (p_c, V_c, T_c), where the first and second derivatives equal zero on the isotherm, yields three independent equations.

(1) $$p_\mathrm{c}=\frac{RT_\mathrm{c}}{V_\mathrm{c}-b}-\frac{a}{V_\mathrm{c}^2}$$

(2) $$-\frac{RT_\mathrm{c}}{(V_\mathrm{c}-b)^2}+\frac{2a}{V_\mathrm{c}^3}=0\quad\text{or}\quad V_\mathrm{c}^3=\frac{2a(V_\mathrm{c}-b)^2}{RT_\mathrm{c}}$$

(3) $$\frac{RT_\mathrm{c}}{(V_\mathrm{c}-b)^3}-\frac{3a}{V_\mathrm{c}^4}=0\quad\text{or}\quad V_\mathrm{c}^4=\frac{3a(V_\mathrm{c}-b)^3}{RT_\mathrm{c}}$$

Division of (3) by (2) and solving for V_c yields

(4) $$\boxed{V_\mathrm{c}=3b}.$$

Substitution of (4) into (2) and solving for T_c yields

(5) $$\boxed{T_\mathrm{c}=\frac{8a}{27bR}}.$$

Substitution of (4) and (5) into (1) and simplifying the expression yields

(6) $\boxed{p_c = \dfrac{a}{27b^2}}$.

(b) Evaluation of equation [1.20b] at the critical point and substitution of equations (4)–(6) yields and expression for Z_c.

$$Z_c = \frac{p_c V_c}{RT_c} = \frac{\left(\dfrac{a}{27b^2}\right)(3b)}{R\left(\dfrac{8a}{27bR}\right)} = \frac{3}{8}$$

$$\boxed{Z_c = \frac{3}{8}}$$

P1.39 (a) As for any perfect gas, the pressure in the interior of the Sun is related to the mass density, $\rho = m/V$, by

$$p = \frac{nRT}{V} = \frac{mRT}{MV} = \frac{\rho RT}{M}.$$

The problem is to know the molar mass to use. Atoms are stripped of their electrons in the interior of stars, so if we suppose that the interior consists of ionized hydrogen atoms and free electrons, the mean molar mass is one-half the molar mass of hydrogen, or 0.5 g mol^{-1} (the mean of the molar mass of H^+ and e^-, the latter being almost 0). Halfway to the centre of the Sun, the pressure is

$$p = \frac{(1.20\times10^3\ \text{kg m}^{-3})\times(8.3145\ \text{J K}^{-1}\ \text{mol}^{-1})\times(3.6\times10^6\ \text{K})}{0.50\times10^{-3}\ \text{kg mol}^{-1}} = 7.2\times10^{13}\ \text{Pa} = \boxed{720\ \text{Mbar}}.$$

(b) Substitution of $E_{\text{k total}} = \tfrac{1}{2}Nmc^2$ [1.11] into $p = \tfrac{1}{3}nMc^2/V$ [1.14] $= \tfrac{1}{3}Nmc^2/V$ gives $\boxed{p = \tfrac{2}{3}\rho_k}$ where $\rho_k = E_{\text{k total}}/V$.

(c) Using the equation of part (b), $\rho_k = \tfrac{3}{2}p = \tfrac{3}{2}(7.2\times10^{13}\ \text{Pa}) = \boxed{1.1\times10^{14}\ \text{J m}^{-3}}$. The ratio of the kinetic energy density halfway to the centre of the Sun to the kinetic energy density of our atmosphere is

$$\frac{1.1\times10^{14}\ \text{J m}^{-3}}{1.5\times10^5\ \text{J m}^{-3}} = 7.3\times10^8.$$ The Sun's value is a $\boxed{\text{billion times larger}}$.

(d) We use $p = \dfrac{nRT}{V} = \dfrac{\rho RT}{M}$ where M is the average molar mass of the particles. For each C^{6+} ion, there are 6 electrons. The average molar mass is therefore

$$M = \frac{(12 + 6\times0)\ \text{g mol}^{-1}}{7} = 1.7\ \text{g mol}^{-1}.$$

$$p = \frac{(1.20\times10^3\ \text{kg m}^{-3})\times(8.3145\ \text{J K}^{-1}\ \text{mol}^{-1})\times(3.5\times10^3\ \text{K})}{1.7\times10^{-3}\ \text{kg mol}^{-1}}$$

$$= 2.1\times10^{10}\ \text{Pa} = \boxed{0.21\ \text{Mbar}}$$

(e) The average molar mass is now 12 g mol^{-1}. Then if the mass density were the same, the pressure would be $p = \boxed{0.029\ \text{Mbar}}$.

2 Thermodynamics: the first law

Answers to discussion questions

D2.1 The **system** is the part of the world in which we have a special interest. It may be a reaction vessel, an engine, an electrochemical cell, a biological cell, and so on. The **surroundings** comprise the region outside the system and are where we make our measurements. The system is **open** if matter can be transferred between the system and surroundings. Otherwise, it is **closed**. If no heat can be transferred between the system and surroundings, the boundary is **adiabatic**; otherwise, it is **diathermic**. An **isolated system** can exchange neither matter nor energy with its surroundings.

The choice between what is the system of interest and what is the surroundings depends upon the phenomena and substance of interest but the division is often dictated by physical boundaries. The container wall at which a confined gas has contact logically provides the boundary between the system (the gas) and the surroundings (the contact surface and everything beyond it). Yet, if liquid properties alone are the focus of interest, the vapour of the liquid may be considered part of the surroundings and the liquid-gas interface separates the system of interest from the surroundings. In the case for which liquid-gas equilibrium is the centre of focus, both liquid and vapour become part of the system of interest and surroundings begin with the container walls. It is not always necessary to have a boundary that separates the system from surroundings. When discussing stratospheric ozone, it may be convenient to consider the nitrogen and oxygen of the stratosphere to be the surrounding environment of the ozone; energy and mass transfers between the system and surroundings may then be the focus of analysis. Always attempt to clearly define the system and its surroundings so that work, heat, energy, and mass transfers can be correctly accounted for.

D2.2 (a) **Temperature**, T, is an intensive property of a system. It is often said to measure the hotness or coldness of a system in the sense that heat always flows from high temperature to lower temperature. If two objects are at the same temperature, they are in **thermal equilibrium** and there is no net flow of heat between them when they are in contact. Temperature is the single parameter that tells us the relative molecular (and/or atomic) populations over the available energy levels of a system.

(b) **Heat**, q, is energy in transit as a result of a temperature difference. It is characterized by energy transfer that the causes or utilizes chaotic, disorderly motion in the surroundings and it depends upon molecular collisions. In contrast, **work** w is the transfer of energy that causes or utilizes uniform, orderly motion of atoms in the surroundings.

(c) **Energy** is the capacity to do work where work is by definition the process of achieving motion against an opposing force. **Internal energy**, U, is the sum of all the kinetic and potential contributions to the energy of all the atoms, ions, and molecules in the system. It is the total energy of the system. According to the First Law of Thermodynamics, internal energy is a state property and a change in internal energy, ΔU, occurs because of the surroundings doing work on the system and transferring heat to the system: $\Delta U = w + q$ [2.8]. The internal energy of an isolated system is a constant; it is conserved.

D2.3 At the molecular level, work is a transfer of energy that results in orderly motion of the atoms and molecules in a system; heat is a transfer of energy that results in disorderly motion. See Figures 2.6 and 2.7 of text.

D2.4 The **law of conservation of energy** states that energy can be neither created nor destroyed but merely converted from one form into another or moved from place to place. The law of conservation of energy, belonging to the field of classical mechanics (see text appendix section A3.1), considers energy to be the sum of kinetic and potential energies and either one of these can convert to the other. Work is the transfer mode for energy and mechanics does not deal with the concept of heat. The **first law of thermodynamics** adds the transfer of energy via heat to that of work and the sum relates to the internal energy U, which becomes the conserved state property. Furthermore, with the recognition that transfers of internal energy occur between the system and surroundings, the first law emphasizes that the transfers change the system's internal energy according to the relation: $\Delta U = q + w$ [2.8]. The heat and work transfers may be very complex but $\Delta U = 0$ for an isolated system; U is conserved.

D2.5 The most general expression for **expansion work** is $dw = -p_{ex}dV$. Both the left and right sides of this expression are infinitesimals (differentials), which are neither numbers nor functions, and the expression must be integrated with the concepts and rules of the integral calculus for the particulars of an application. When the expansion work is against a constant external pressure, the integration gives a relation between functions that is computationally practical:

$$w = -p_{ex}\Delta V = -p_{ex}\times(V_f - V_i) \quad \text{[2.2, expansion work against constant external pressure].}$$

We need only specify the values of p_{ex}, V_i, and V_f before calculating the value of w in this specific process. For example, suppose that $p_{ex} = 1.00$ bar, $V_i = 5.20\ \text{dm}^3$, and $V_f = 6.30\ \text{dm}^3$, we then find that $w = -(1.00\ \text{bar})\times(6.30-5.20)\ \text{dm}^3 = -1.10\ \text{bar dm}^3 = -110\ \text{J}$. Remarkably, the computation does not depend upon whether the system is gaseous, liquid, or solid; it does not even depend upon the composition of the system. Independence of composition and phase is a powerful generality of thermodynamics. This particular example has only the single restriction that the expansion be against a constant external pressure. Other computations have further, or different, restrictions.

When the expansion work is reversible, $p = p_{ex}$ but further information is needed before the general expression can be integrated. We can only substitute $w = \int dw$ for the integration of the left side of the general expression and substitute $-\int_{V_i}^{V_f} p dV$ for the integration of the right side, thereby, giving

$$w = -\int_{V_i}^{V_f} p dV \quad \text{[reversible expansion].}$$

This expression cannot be directly used to make a computation because the behavior of p with changing V has not been specified; the integral cannot be performed as yet to give a practical computation equation. (This expression is interpreted as saying that expansion work is the negative of the area under a curve of p against V between the limits V_i and V_f.) If the specifics of the expansion are known, the work integral and computation is possible. For example, suppose that during a particular reversible expansion the pressure and volume are maintained in the linear correspondence: $p = a + bV$ between $V_i = 5.20\ \text{dm}^3$ and $V_f = 6.30\ \text{dm}^3$ where a and b are the constants $a = 1.00$ bar and $b = -0.30\ \text{bar dm}^{-3}$. The value of the expansion work is now restricted by the reversible condition and the particulars of the $p(V)$ relation during the expansion. The work is determined by first performing the requisite integration (identify the integration rules used in the following; see text appendix section A2.6):

$$\begin{aligned} w &= -\int_{V_i}^{V_f} pdV = -\int_{V_i}^{V_f}\{a+bV\}dV \\ &= -\int_{V_i}^{V_f} adV - \int_{V_i}^{V_f} bVdV = -a\int_{V_i}^{V_f} dV - b\int_{V_i}^{V_f} VdV = -aV\Big]_{V_i}^{V_f} - (b/2)V^2\Big]_{V_i}^{V_f} \\ &= -a(V_f - V_i) - (b/2)\times(V_f^2 - V_i^2) \quad \text{[reversible expansion work with } p = a + bV\text{].} \end{aligned}$$

Now we substitute values and perform the calculation:

$$\begin{aligned} w &= -(1.00\ \text{bar})\times(6.30-5.20)\ \text{dm}^3 - \tfrac{1}{2}(-0.30\ \text{bar dm}^{-3})\times(6.30^2 - 5.20^2)\ \text{dm}^3 \\ &= -1.10\ \text{bar dm}^3 + 1.90\ \text{bar dm}^3 = +0.80\ \text{bar dm}^3 \\ &= +80\ \text{J}. \end{aligned}$$

Another work example is analyzed in text Derivation 2.2. It is the reversible, isothermal expansion work of a perfect gas, which is computed with the relation:

$$w = -nRT \ln \frac{V_f}{V_i} \quad [2.3; \text{ reversible, isothermal expansion of perfect gas}].$$

These examples demonstrate that work is not a property. Work depends upon the particulars of the process (the so called 'path' of the process); it is not a state function. Consequently, great care must always be taken to select, or derive, the work relation that is applicable to a particular process. This involves integration of the work integral along the process path between the initial and final states. Similarly, heat is also not generally a state function.

D2.6 The difference results from the definition $H = U + pV$ [2.12]; hence $\Delta H = \Delta U + \Delta(pV)$. As $\Delta(pV)$ is not usually zero, except for isothermal processes in a perfect gas, the difference between ΔH and ΔU is a non-zero quantity. As shown in section 2.8 of the text, ΔH can be interpreted as the heat associated with a process at constant pressure, and ΔU as the heat at constant volume.

D2.7

$q = nRT \ln(V_f / V_i)$	limitations: reversible, isothermal expansion of a perfect gas
$\Delta H = \Delta U + p\Delta V$	limitation: constant pressure process
$C_{p,m} - C_{V,m} = R$	limitation: perfect gas

Solutions to exercises

E2.1 (a) $w_{\text{expansion}} = -p_{ex}\Delta V = -(1.00 \times 10^5 \text{ Pa}) \times (1.0 \times 10^{-6} \text{ m}^3) = \boxed{-0.10 \text{ J}}$

(b) $w_{\text{expansion}} = -p_{ex}\Delta V = -(1.00 \times 10^5 \text{ Pa}) \times (1.0 \times 10^{-3} \text{ m}^3) = \boxed{-100. \text{ J}}$

In order to calculate the amount of work required to return the system to its original state an exact knowledge of the details of the process is required because w is not a state function. In the case for which the compression is at same constant external pressure as the above expansion:

(a) $w_{\text{compression}} = -p_{ex}\Delta V = -(1.00 \times 10^5 \text{ Pa}) \times (-1.0 \times 10^{-6} \text{ m}^3) = \boxed{+0.10 \text{ J}}$

(b) $w_{\text{compression}} = -p_{ex}\Delta V = -(1.00 \times 10^5 \text{ Pa}) \times (-1.0 \times 10^{-3} \text{ m}^3) = \boxed{+100. \text{ J}}$.

E2.2

$$w = -nRT \ln \frac{V_f}{V_i} \quad [2.3; \text{ reversible, isothermal expansion of perfect gas}].$$

$$= -(2.0 \text{ mol}) \times (8.3145 \text{ J K}^{-1} \text{ mol}^{-1}) \times (300 \text{ K}) \times \ln\left(\frac{3.0 \text{ dm}^3}{1.0 \text{ dm}^3}\right)$$

$$= \boxed{-5.5 \text{ kJ}}$$

E2.3 (a) $w = -p_{ex}\Delta V \; [2.2] = -(30.0 \times 10^3 \text{ Pa}) \times (3.3 \text{ dm}^3) \times \left(\frac{1 \times 10^{-3} \text{ m}^3}{1 \text{ dm}^3}\right) = \boxed{-99 \text{ J}}$

(b) $n = \frac{4.50 \text{ g}}{16.04 \text{ g mol}^{-1}} = 0.2805 \text{ mol}, \; V_i = 12.7 \text{ dm}^3, \; V_f = (12.7 + 3.3) \text{ dm}^3 = 16.0 \text{ dm}^3$

$$w = -nRT \ln \frac{V_f}{V_i} \quad [2.3; \text{ reversible, isothermal expansion of perfect gas}]$$

$$= -(0.2805 \text{ mol}) \times (8.3145 \text{ J K}^{-1} \text{ mol}^{-1}) \times (310 \text{ K}) \times \ln\left(\frac{16.0 \text{ dm}^3}{12.7 \text{ dm}^3}\right) = \boxed{-167 \text{ J}}$$

E2.4 $w = -nRT \ln \dfrac{V_f}{V_i}$ [2.3; reversible, isothermal expansion of perfect gas].

$$= -(0.0520\ \text{mol}) \times (8.3145\ \text{J K}^{-1}\ \text{mol}^{-1}) \times (260\ \text{K}) \times \ln\left(\frac{100\ \text{cm}^3}{300\ \text{cm}^3}\right)$$

$$= \boxed{+123\ \text{J}}$$

E2.5 $p_{ex} = 95.2\ \text{bar} \times \dfrac{10^5\ \text{Pa}}{1\ \text{bar}} = 9.52 \times 10^6\ \text{Pa}$

$$\Delta V = -(0.57) \times (0.550\ \text{dm}^3) = -0.314\ \text{dm}^3 \times \frac{10^{-3}\ \text{m}^3}{1\ \text{dm}^3} = -3.14 \times 10^{-4}\ \text{m}^3$$

$w = -p_{ex}\Delta V$ [2.2, expansion work against constant external pressure]

$$= (-9.52 \times 10^6\ \text{Pa}) \times (-3.14 \times 10^{-4}\ \text{m}^3) = 2.99 \times 10^3\ \text{Pa m}^3 = \boxed{+2.99\ \text{kJ}}$$

E2.6 $Mg(s) + 2\ HCl(aq) \rightarrow H_2(g) + MgCl_2(aq)$

$$n_{H_2\ \text{produced}} = n_{Mg\ \text{consumed}} = \frac{m_{Mg}}{M_{Mg}} = \frac{12.5\ \text{g}}{24.31\ \text{g mol}^{-1}} = 0.514\ \text{mol}$$

The volume changes due to the reaction of the solid and composition changes in the aqueous solution are negligibly small compared to the volume change from the gas production so

$$\Delta V = \Delta V_{H_2} = V_f(H_2) - V_i(H_2) = \frac{n_{H_2}RT}{p_{ex}} - 0 = \frac{n_{H_2}RT}{p_{ex}}$$

$w = -p_{ex}\Delta V$ [2.2, expansion work against constant external pressure]

$$= -p_{ex} \times \left(\frac{n_{H_2}RT}{p_{ex}}\right) = -n_{H_2}RT$$

$$= -(0.514\ \text{mol}) \times (8.3145\ \text{J K}^{-1}\ \text{mol}^{-1}) \times (293.4\ \text{K})$$

$$= \boxed{-1.25\ \text{kJ}}$$

E2.7 amount sucrose, $n_{\text{sucrose}} = 10.0\ \text{g}\ C_{12}H_{22}O_{11} \times \dfrac{1\ \text{mol}}{342.30\ \text{g}} = 2.92 \times 10^{-2}\ \text{mol sucrose}$

The balanced reaction equation for the complete combustion of sucrose is

$$C_{12}H_{22}O_{11}(s) + 12\ O_2(g) \rightarrow 12\ CO_2(g) + 11\ H_2O$$

In Part (a) the net expansion work done on the chemical system, w, is computed for the case in which water is produced as a liquid. In Part (b) water is considered to form as a gas. The difference is important because gases have a large molar volume compared to the negligibly small molar volumes of solids and liquids. When a balanced reaction indicates a large net change in the number of moles of gas per reaction, $\Delta\nu_{gas} = \nu_{\text{product gases}} - \nu_{\text{reactant gases}}$, the magnitude of w is significantly large. Should there be no change in $\Delta\nu_{gas}$ (= 0), the extremely small changes in the volumes of reactant liquids and solids to product liquids and solid can be expressed as $\Delta V \simeq 0$ and, consequently, $w = -p_{ex}\Delta V\,[2.2] \simeq 0$.

(a) When the reaction water is written as liquid, gaseous carbon dioxide is produced and gaseous oxygen is consumed. The combustion of 1 mol sucrose causes the gaseous change:

$$\Delta\nu_{gas} = \nu_{\text{product gases}} - \nu_{\text{reactant gases}} = 12 - 12 = 0$$

We conclude that $w = -p_{ex}\Delta V\,[2.2] \simeq \boxed{0}$.

(b) When the reaction water is written as gas, both gaseous carbon dioxide and gaseous water are produced while gaseous oxygen is consumed. The combustion of 1 mol sucrose causes the gaseous change: $\Delta\nu_{gas} = \nu_{product\ gases} - \nu_{reactant\ gases} = 12 + 11 - 12 = 11$. Expansion work is significant and it is caused by the appearance of 11 new moles of gas per reaction.

$$\Delta V \text{ per reaction} = \frac{\Delta\nu_{gas} \times RT}{p_{ex}}$$

$$w \text{ per reaction} = -p_{ex}\Delta V\ [2.2] = -p_{ex}\left(\frac{\Delta\nu_{gas} \times RT}{p_{ex}}\right) = -\Delta\nu_{gas} \times RT$$

$$w \text{ per mole sucrose} = \frac{-\Delta\nu_{gas} \times RT}{1 \text{ mol sucrose}}$$

$$w \text{ per amount sucrose} = \left(\frac{-\Delta\nu_{gas} \times RT}{1 \text{ mol sucrose}}\right) \times n_{sucrose}$$

$$= -\left(\frac{(11 \text{ mol})(8.3145 \text{ J K}^{-1} \text{ mol}^{-1})(293 \text{ K})}{1 \text{ mol sucrose}}\right)$$

$$\times(2.92 \times 10^{-2} \text{ mol sucrose})$$

$$= \boxed{-782 \text{ J}}$$

The reaction does 782 J of expansion work as the production of gas pushes on the surrounding atmospheric gases.

E2.8 $p_{ex} = 1.00 \times 10^5$ Pa, $\Delta V = 100 \text{ cm}^2 \times 10 \text{ cm} = 1.0 \times 10^3 \text{ cm}^3 = 1.0 \times 10^{-3} \text{ m}^3$

$$w = -p_{ex}\,\Delta V\ [2.2]$$

$$= -1.00 \times 10^5 \text{ Pa} \times 1.0 \times 10^{-3} \text{ m}^3 = -1.0 \times 10^2 \text{ Pa m}^3\ [1 \text{ Pa m}^3 = 1 \text{ J}]$$

$w = \boxed{-1.0\times10^2 \text{ J}}$ The expanding gas does 100 J of work while moving the piston.

E2.9

$$C = \frac{q}{\Delta T}\ [2.4a] = \frac{124 \text{ J}}{5.23 \text{ K}} = \boxed{23.7 \text{ J K}^{-1}}$$

E2.10 (a) The expansion work that occurs when the solid and the liquid change temperature is negligibly small compared to heat exchange. Consequently, the energy exchange is to a very good approximation equal to the heat exchange alone: the heat lost by the iron (59 g sample) equals the heat gain of the water.

$$\Delta U_{\text{isolated system}} = q_{Fe(s)} + q_{H_2O(l)} = 0$$

$$-q_{Fe(s)} = q_{H_2O(l)}$$

$$-(mC_s\Delta T)_{Fe(s)} = (mC_s\Delta T)_{H_2O(l)}$$

$$C_{s,Fe(s)} = \frac{-(mC_s\Delta T)_{H_2O(l)}}{(m\Delta T)_{Fe(s)}}$$

$$= \frac{-(100 \text{ g})\times(4.184 \text{ J K}^{-1} \text{ g}^{-1})\times(23-20)°\text{C}}{(59 \text{ g})\times(23-70)°\text{C}}$$

$$= \boxed{0.45 \text{ J K}^{-1} \text{ g}^{-1}}$$

(b) $C_m = C_s M = (0.45 \text{ J K}^{-1} \text{ g}^{-1})\times(55.85 \text{ g mol}^{-1}) = \boxed{25 \text{ J K}^{-1} \text{ mol}^{-1}}$

E2.11 $q = n\,C_{p,m}\,\Delta T$ where $n = 250 \text{ g} \times (1 \text{ mol}/18.0 \text{ g}) = 13.9$ mol

$$q = (13.9 \text{ mol}) \times (75.3 \text{ J mol}^{-1} \text{ K}^{-1}) \times (40 \text{ K}) = 4.2 \times 10^4 \text{ J} = \boxed{42 \text{ kJ}}$$

E2.12

$$q = I\mathcal{V}t\ [2.5] = (1.55\ \text{A})\times(110\ \text{V})\times(8.5\ \text{min})\times\left(\frac{60\ \text{s}}{1\ \text{min}}\right) = \boxed{8.7\times10^4\ \text{J}}$$

E2.13

$$C_V = \frac{q_V}{\Delta T}\ [2.10,\ 2.11] = \frac{229\ \text{J}}{2.55\ \text{K}} = 89.8\ \text{J K}^{-1}$$

The molar heat capacity at constant volume is therefore

$$C_{V,\text{m}} = \frac{89.8\ \text{J K}^{-1}}{3.0\ \text{mol}} = \boxed{30\ \text{J K}^{-1}\ \text{mol}^{-1}}.$$

For a perfect gas $C_{p,\text{m}} - C_{V,\text{m}} = R$ [2.19] so

$$C_{p,\text{m}} = C_{V,\text{m}} + R = (30 + 8.3)\ \text{J K}^{-1}\ \text{mol}^{-1} = \boxed{38\ \text{J K}^{-1}\ \text{mol}^{-1}}.$$

E2.14 We take the room temperature and pressure to be $T = 298$ K and $p = 1.00$ atm. The volume is

$$V = (5.5\ \text{m})\times(6.5\ \text{m})\times(3.0\ \text{m}) = 1.0\overline{7}\times10^5\ \text{dm}^3.$$

The amount of gas in the room is computed with the perfect gas law.

$$n = \frac{pV}{RT} = \frac{(1.00\ \text{atm})\times(1.0\overline{7}\times10^5\ \text{dm}^3)}{(8.206\times10^{-2}\ \text{dm}^3\ \text{atm K}^{-1}\ \text{mol}^{-1})\times(298\ \text{K})} = 4.4\times10^3\ \text{mol}$$

Thus, the energy required to raise the air temperature by 10°C is

$$\begin{aligned} q &= n\,C_{p,\text{m}}\Delta T \\ &= (4.4\times10^3\ \text{mol})\times(21\ \text{J K}^{-1}\ \text{mol}^{-1})\times(10\ \text{K}) = 9.2\times10^5\ \text{J} = \boxed{9.2\times10^2\ \text{kJ}}. \end{aligned}$$

Because $q = P\times t$ where P is the power of the heater and t is the time for which it operates,

$$t = \frac{q}{P} = \frac{9.2\times10^5\ \text{J}}{1.5\times10^3\ \text{J s}^{-1}} = \boxed{6.1\times10^2\ \text{s}}.$$

In practice, the walls and furniture of a room are also heated.

E2.15

$$q = nRT\ \ln\left(\frac{V_\text{f}}{V_\text{i}}\right)\ [2.7] = (1.00\ \text{mol})\times(8.3145\ \text{J K}^{-1}\ \text{mol}^{-1})\times(300\ \text{K})\times\ \ln\left(\frac{30.0}{22.0}\right) = \boxed{773\ \text{J}}$$

E2.16 We take the room temperature to be 25°C giving $\Delta T = (25 - 65)\ \text{K} = -40\ \text{K}$. The amount of iron is

$$n_{\text{Fe(s)}} = m/M = (1.4\times10^3\ \text{g})/(55.84\ \text{g mol}^{-1}) = 25.\overline{1}\ \text{mol}.$$

Being open to the room means that the cooling of the solid is definitely a reversible, constant pressure process for which $\Delta H = q_p$ [2.15b]. However, there is very little, even negligible, volume shrinkage of the solid during the cooling. So for all practical purposes, we can consider this to be a reversible, constant volume process for which $\Delta U = q_V$ [2.10b]. Both of these relations are valid when

$$q = q_p \simeq q_V \qquad \text{and} \qquad \Delta H \simeq \Delta U \qquad \text{and} \qquad C_p \simeq C_V\ \text{[by eqns 2.11 and 2.16]}.$$

These considerations allow us to write a practical computation relation for the internal energy of the iron.

$$\begin{aligned} \Delta U &\simeq q_p = nC_{p,\text{m}}\Delta T \\ &\simeq (25.\overline{1}\ \text{mol})\times(25.1\ \text{J K}^{-1}\ \text{mol}^{-1})\times(-40\ \text{K}) \\ &\simeq \boxed{-25\ \text{kJ}} \end{aligned}$$

E2.17 We begin by using the data in a computation of the calorimeter heat capacity C.

$$q = IVt\,[2.5] = (1.27\text{ A}) \times (12.5\text{ V}) \times (157\text{ s}) = 2.49\text{ kJ} \qquad (1\text{ A s} = 1\text{ C}, 1\text{ C V} = 1\text{ J})$$

$$C = \frac{q}{\Delta T}\,[2.4\text{a}] = \frac{2.49\text{ kJ}}{3.88\text{ K}} = 0.642\text{ kJ K}^{-1}$$

With the use of an oxygen bomb calorimeter the combustion is at constant volume giving

$$\Delta U_{\text{calorimeter}} = q_V = C\Delta T = (0.642\text{ kJ K}^{-1}) \times (2.89\text{ K}) = \boxed{+1.86\text{ kJ}}$$

which is the energy released by the combustion reaction.

E2.18

$$w_{\text{lift}} = mgh\,[2.1] = (0.250\text{ kg}) \times (9.81\text{ m s}^{-2}) \times (1.85\text{ m}) = 4.54\text{ J} = -w_{\text{animal}}$$

$$\Delta U_{\text{animal}} = w_{\text{animal}} + q_{\text{animal}}\,[2.8] = (-4.54\text{ J}) + (-10.0\text{ J}) = \boxed{-14.54\text{ J}}$$

E2.19 Assuming that the calorimeter has constant volume,

$$\Delta U = w + q_{\text{heater}} = q_{\text{heater}} = IVt\,[2.5] = (0.02222\text{ A}) \times (11.8\text{ V}) \times (162\text{ s}) = \boxed{+42.5\text{ J}}.$$

E2.20 Upon receiving the heat the parcel of gas expands against the constant external pressure of the surrounding atmosphere so this is a constant pressure process and $\Delta H_{\text{parcel}} = q_p\,[2.15\text{b}] = \boxed{20\text{ kJ}}$.

E2.21 For a perfect gas: $\Delta H_{\text{m}} = \Delta U_{\text{m}} + R\Delta T\,[2.13\text{b}]$.

Since both ΔU_{m} and ΔT equal zero for an isothermal expansion of a perfect gas, $\Delta H_{\text{m}} = 0$ for an isothermal expansion of a perfect gas.

E2.22 $H_{\text{m}} = U_{\text{m}} + pV_{\text{m}}\,[2.13\text{a}]$, hence the difference between molar enthalpy and molar energy is pV_{m}.

For a perfect gas $pV_{\text{m}} = RT = (8.3145\text{ J K}^{-1}\text{ mol}^{-1}) \times (298.15\text{ K}) = 2.479\text{ kJ mol}^{-1}$.

The van der Waals equation of state, $(p + a/V_{\text{m}}^2)(V_{\text{m}} - b) = RT\,[1.23\text{b}]$, is a cubic equation in V_{m} so we cannot conveniently use algebra to solve for V_{m}. Rather, we resort to a numerical procedure. We first rearrange the equation into the form:

$$V_{\text{m}} = \frac{RT}{(p + a/V_{\text{m}}^2)} + b$$

where $a = 3.610$ atm dm^6 mol^{-2}, $b = 0.0429$ dm^3 mol^{-1}, and $RT = 24.4654$ atm dm^3 mol^{-1} (units chosen to be compatible with those of a and b). Substitution of these values, along with $p = 1.00$ atm, gives

$$x = \frac{24.4654}{(1.00 + 3.610/x^2)} + 0.0429 \qquad \text{where } x = V_{\text{m}}/(\text{dm}^3\text{ mol}^{-1}).$$

Since $x \sim 24$ dm^3 mol^{-1}, we see that $3.610/x^2 \ll 1.00$ and we are certainly tempted to discard the term $3.610/x^2$ as negligibly small. Rather than do this, let us substitute on the right side of the equation only the perfect gas value for x at this temperature and pressure (24.4654). This gives

$$x = \frac{24.4654}{(1.00 + 3.610/24.4654^2)} + 0.0429 = 24.3616.$$

This is an improvement over the perfect gas value so let us repeat the process by substituting the new value of x into the right side of the computation equation. This gives

$$x = \frac{24.4654}{(1.00 + 3.610/24.3616^2)} + 0.0429 = 24.3604.$$

Once again we have improved on our estimate of x so let's do it again. This gives

$$x = \frac{24.4654}{(1.00 + 3.610/24.3604^2)} + 0.0429 = 24.3604.$$

The value of x used on the right side is seen to provide a calculation that yields the same value of x on the left side and we conclude that this value must be the solution to the van der Waals equation. This procedure for finding the solution is a **numerical iteration** and in this example successive iterations have quickly **converged** to the answer. The van der Waals equation gives the molar volume 24.3604 $dm^3\ mol^{-1}$ and the difference between molar enthalpy and molar energy is

$$pV_m = (1.01325 \times 10^5\ \text{Pa}) \times (0.0243604\ \text{m}^3\ \text{mol}^{-1}) = \boxed{2.468\ \text{kJ mol}^{-1}}.$$

E2.23 $q_p = \boxed{-1.2\ \text{kJ}}$ (heat leaves the sample)

At constant pressure $\Delta H = q_p$ [2.15b], hence $\Delta H = \boxed{-1.2\ \text{kJ}}$.

$$C_p = \frac{\Delta H}{\Delta T}\ [2.16] = \frac{-1.2\ \text{kJ}}{-15\ \text{K}} = \boxed{80\ \text{J K}^{-1}}$$

E2.24 $\Delta H = C_p\Delta T\ [2.16] = nC_{p,m}\Delta T = (3.0\ \text{mol})\times(29.4\ \text{J K}^{-1}\ \text{mol}^{-1})\times(25\ \text{K}) = \boxed{+2.2\ \text{kJ}}$

$q_p = \Delta H\ [2.15a] = \boxed{+2.2\ \text{kJ}}$

$$\begin{aligned}\Delta U &= \Delta H - \Delta(pV)\ [2.12] = \Delta H - \Delta(nRT)\ \text{(perfect gas)}\\ &= \Delta H - nR\Delta T\\ &= 2.2\ \text{kJ} - (3.0\ \text{mol})\times(8.3145\ \text{J K}^{-1}\ \text{mol}^{-1})\times(25\ \text{K})\\ &= 2.2\ \text{kJ} - 0.62\ \text{kJ} = \boxed{+1.6\ \text{kJ}}\end{aligned}$$

E2.25 $C_{V,m} = C_{p,m} - R\ [2.19] = 29.14\ \text{J K}^{-1}\ \text{mol}^{-1} - 8.31\ \text{J K}^{-1}\ \text{mol}^{-1} = \boxed{20.83\ \text{J K}^{-1}\ \text{mol}^{-1}}$

E2.26 (a) $\Delta H_m = C_{p,m}\Delta T\ \ [2.16]$

$$= (29.14\ \text{J K}^{-1}\ \text{mol}^{-1})\times(37-15)\ \text{K} = \boxed{641\ \text{J mol}^{-1}}$$

(b) $\Delta U_m = \Delta H_m - \Delta(pV)\ [2.14a] = \Delta H - \Delta(RT) = \Delta H - R\Delta T$

$$= 641\ \text{J mol}^{-1} - (8.31\ \text{J K}^{-1}\ \text{mol}^{-1}) \times (22\ \text{K}) = \boxed{458\ \text{J mol}^{-1}}$$

Solutions to projects

P2.27 In this exercise we find an integrated expression for the reversible, isothermal expansion work for gases that obey one-or-another equation of state. The general differential equation for a reversible expansion is $dw = -pdV$ so the general integral expression is $w = -\int_{V_i}^{V_f} p\,dV$. The desired equation of state is used to express the integrand, p, as a function of both T and V. Then the integral is analytically performed with the constraint that T is a constant. The result is the desired expression for w for a reversible, isothermal expansion.

(a) The equation of state is $p = nRT/(V - nb)$ so

$$w = -\int_{V_i}^{V_f} \frac{nRT}{V-nb}dV = -nRT\int_{V_i}^{V_f}\frac{1}{V-nb}dV \text{ [the factor } nRT \text{ is a constant for an isothermal process].}$$

Referring to a handbook compilation of indefinite integrals we find the standard integral

$$\int \frac{dx}{Ax+B} = \frac{1}{A}\ln(Ax+B)$$

With the transformations $A \to 1$, $B \to -nb$, and $x \to V$ we see that our integral is

$$\int \frac{dV}{V-nb} = \ln(V-nb).$$

Thus,

$$w = -nRT\ln(V-nb)\Big]_{V=V_i}^{V=V_f} = -nRT\left(\ln(V_f-nb)-\ln(V_i-nb)\right) = \boxed{-nRT\ln\left(\frac{V_f-nb}{V_i-nb}\right)}.$$

In the case for which $V_f > V_i$, the value of w is more negative than the value for a perfect gas.

(b) The equation of state is $p = nRT/V - n^2a/V^2$ so

$$w = -\int_{V_i}^{V_f}\left\{\frac{nRT}{V}-\frac{n^2a}{V^2}\right\}dV = -\int_{V_i}^{V_f}\frac{nRT}{V}dV + \int_{V_i}^{V_f}\frac{n^2a}{V^2}dV$$

$$= -nRT\int_{V_i}^{V_f}\frac{1}{V}dV + n^2a\int_{V_i}^{V_f}\frac{1}{V^2}dV \quad [T \text{ is constant}]$$

$$= -nRT\ln V\Big]_{V=V_i}^{V=V_f} - n^2a\left(\frac{1}{V}\right)\Big]_{V=V_i}^{V=V_f} \quad \text{[standard integrals]}$$

$$= \boxed{-nRT\ln\left(\frac{V_f}{V_i}\right) - n^2a\left(\frac{1}{V_f}-\frac{1}{V_i}\right)}$$

In the case for which $V_f > V_i$, the value of w is more positive than the value for a perfect gas because the first term to the right of the equality is that of a perfect gas while the second term is positive for $V_f > V_i$.

P2.28 Since the process is non-isothermal, we must substitute the process dependent relation $T = T_i - c(V - V_i)$ into the integrand along with the equation of state $p(V, T)$ before the integral is analytically evaluated.

(a) The equation of state is $p = nRT/V$ so

$$w = -\int_{V_i}^{V_f}\frac{nRT}{V}dV = -nR\int_{V_i}^{V_f}\frac{T}{V}dV = -nR\int_{V_i}^{V_f}\frac{T_i - c(V-V_i)}{V}dV$$

$$= -nR(T_i + cV_i)\int_{V_i}^{V_f}\frac{1}{V}dV + nRc\int_{V_i}^{V_f}\frac{V}{V}dV \quad \text{[These standard integrals are found in handbooks.]}$$

$$= -nR(T_i + cV_i)\ln(V)\Big]_{V=V_i}^{V=V_f} + nRc(V_f - V_i)$$

$$= \boxed{-nR(T_i + cV_i)\ln\left(\frac{V_f}{V_i}\right) + nRc(V_f - V_i)}$$

(b) In the case for which $V_f > V_i$ and c is positive, the value of w is made more negative than the isothermal expansion at T_i by the first term while the last term makes it more positive than the value for the isothermal expansion.

P2.29 In this exercise we need the derivative definition for $C_{V,m}$ that is presented in Section 2.7 of the text because either a sum (integration) or differences (derivatives) over a continuous range of temperatures must be examined:

$$C_{V,m} = \frac{dU_m}{dT} \text{ at constant } V_m \qquad \text{or} \qquad C_{V,m} = \left(\frac{\partial U_m}{\partial T}\right)_{V_m}.$$

We will simply remember that our results apply at constant V_m and not make further reference to that fact.

(a) We are given the molar heat capacity as a function of temperature and asked to find an expression for the internal energy as a function of temperature. Since differentials like dU_m and dT

can be manipulated as algebraic symbols, we algebraically solve the above definition for dU_m. Then we integrate the expression and recognize that $\Delta U_m = U_{m,f} - U_{m,i} = \int_{U_{m,i}}^{U_{m,f}} dU_m$.

$$dU_m = C_{V,m} dT = aT^3 dT$$

$$\int_{U_{m,i}}^{U_{m,f}} dU_m = \int_{T_i}^{T_f} aT^3 dT$$

$$\Delta U_m = a\int_{T_i}^{T_f} T^3 dT = \tfrac{1}{4} aT^4 \Big]_{T=T_i}^{T=T_f} = \tfrac{1}{4} a(T_f^4 - T_i^4)$$

Let us choose T_i to be the absolute zero of temperature (0 K) and let us write the T_f simply as T, the temperature of interest (ΔU_m becomes the internal energy change that occurs as T is raised above absolute zero). Then, $\boxed{\Delta U_m = \tfrac{1}{4} aT^4}$. The internal energy varies as the fourth power of temperature.

(b)
$$C_{V,m} = \frac{dU_m}{dT}$$

$$= \frac{d}{dT}(a + bT + cT^2) = \frac{da}{dT} + \frac{d(bT)}{dT} + \frac{d(cT^2)}{dT} = 0 + b\frac{dT}{dT} + c\frac{dT^2}{dT}$$

$$= \boxed{b + 2cT}$$

P2.30 In this exercise we need the derivative definition for $C_{p,m}$ because either a sum (integration) or differences (derivatives) over a continuous range of temperatures must be examined. In analogy to the derivative definition for $C_{V,m}$ (see Exercise 2.29) we write

$$C_{p,m} = \frac{dH_m}{dT} \text{ at constant } p \qquad \text{or} \qquad C_{p,m} = \left(\frac{\partial H_m}{\partial T}\right)_p .$$

We will simply remember that our results apply at constant p and not make further reference to that fact.

(a) We are given the molar heat capacity as a function of temperature and we are to find an expression for the enthalpy change that occurs when heating from T_i to T_f. Since differentials like dH_m and dT can be manipulated as algebraic symbols, we algebraically solve the above definition for dH_m. Then we integrate the expression and recognize that $\Delta H_m = H_{m,f} - H_{m,i} = \int_{H_{m,i}}^{H_{m,f}} dH_m$.

$$dH_m = C_{p,m} dT = \{a + bT + c/T^2\} dT$$

$$\int_{H_{m,i}}^{H_{m,f}} dH_m = \int_{T_i}^{T_f} \{a + bT + c/T^2\} dT = a\int_{T_i}^{T_f} dT + b\int_{T_i}^{T_f} T dT + c\int_{T_i}^{T_f} \frac{dT}{T^2}$$

$$\Delta H_m = \boxed{a(T_f - T_i) + \tfrac{1}{2} b(T_f^2 - T_i^2) - c\left(\frac{1}{T_f} - \frac{1}{T_i}\right)}$$

When $T_i = 288.15$ K and $T_f = 310.15$ K, ΔH_m is given by

$$\Delta H_m = \left[aT + \frac{bT^2}{2} - \frac{c}{T}\right]_{T_i = 288.15\text{ K}}^{T_f = 310.15\text{ K}}$$

$$= \left\{(44.22 \text{ J K}^{-1} \text{ mol}^{-1})(310.15 \text{ K}) + \frac{(8.79\times10^{-3} \text{JK}^{-2}\text{mol}^{-1})(310.15\text{K})^2}{2} - \frac{(-8.62\times10^5 \text{JKmol}^{-1})}{310.15 \text{ K}}\right\}$$

$$-\left\{(44.22 \text{ J K}^{-1}\text{mol}^{-1})(288.15 \text{ K}) + \frac{(8.79\times10^{-3} \text{JK}^{-2}\text{mol}^{-1})(288.15\text{K})^2}{2} - \frac{(-8.62\times10^5 \text{JKmol}^{-1})}{288.15 \text{ K}}\right\}$$

$\boxed{\Delta H_m = 818 \text{ J mol}^{-1}}$.

(b) Let $T_i = 288.15$ K and let $T = T_f$ be any temperature in the range $288.15 \text{ K} < T < 310.15$ K. The computation equation of part (a) becomes

$$\Delta H_m = aT + \tfrac{1}{2} bT^2 - \frac{c}{T} - 16.10 \text{ kJ mol}^{-1}.$$

ΔH_m is then the molar enthalpy change that occurs upon heating the substance from T_i to T at constant p. Figure 2.1 shows a plot of ΔH_m against T. Inspection of the figure shows it to be linear so we conclude that the last two terms of the computation equation (the non-linear terms) do not contribute significantly over this small temperature range. This is often the case; we may usually consider $C_{p,m}$ to be a constant over a small temperature range and, consequently, ΔH_m is linear in temperature. Figure 2.2 shows a plot of ΔH against T over the large temperature range 288.15 K < T < 500 K. Inspection of the plot shows that the non-linear terms now provide a significant contribution.

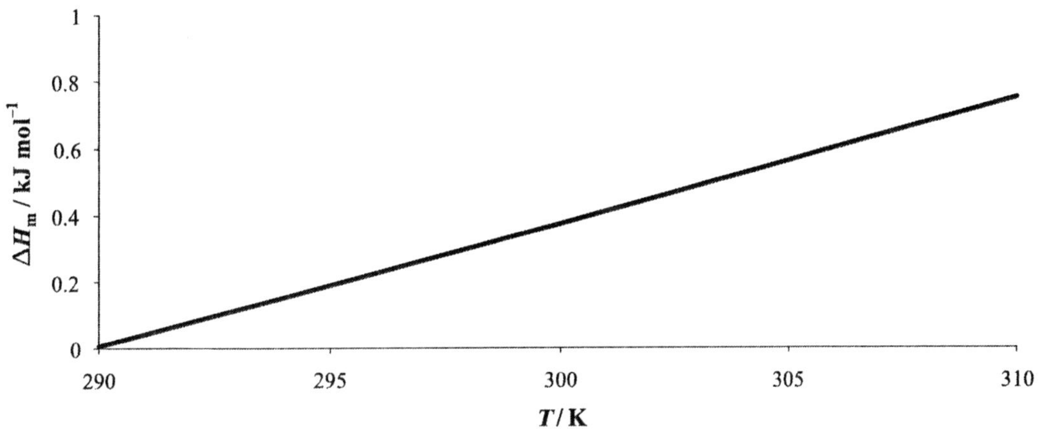

Figure 2.1

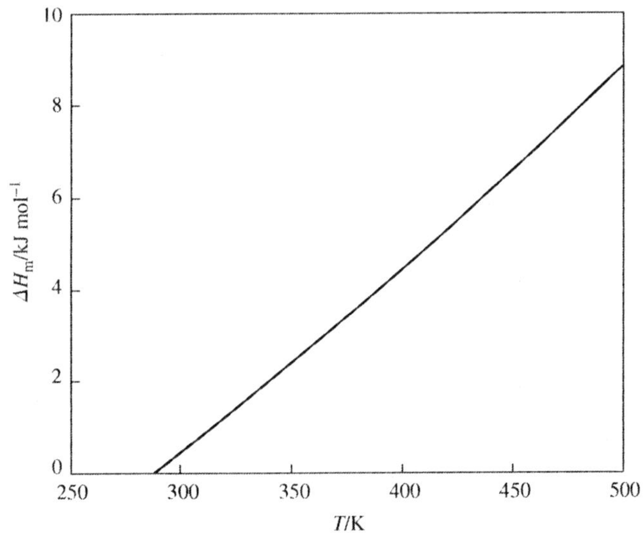

Figure 2.2

P2.31 For a perfect gas $V = nRT/p$.

$$\alpha = \frac{1}{V}\left(\frac{\partial V}{\partial T}\right)_p = \frac{p}{nRT}\left(\frac{\partial\left(\frac{nRT}{p}\right)}{\partial T}\right)_p = \left(\frac{p}{nRT}\right)\left(\frac{nR}{p}\right)\left(\frac{\partial T}{\partial T}\right)_p$$

$$\boxed{\alpha = \frac{1}{T}}$$

$$\kappa = -\frac{1}{V}\left(\frac{\partial V}{\partial p}\right)_T = -\frac{p}{nRT}\left(\frac{\partial\left(\frac{nRT}{p}\right)}{\partial p}\right)_T = -\left(\frac{p}{nRT}\right)(nRT)\left(\frac{\partial\left(\frac{1}{p}\right)}{\partial p}\right)_T = -p\left(-\frac{1}{p^2}\right)$$

$$\boxed{\kappa = \frac{1}{p}}$$

Thus, $C_p - C_V = \dfrac{\alpha^2 TV}{\kappa} = \dfrac{\left(\frac{1}{T}\right)^2 TV}{\frac{1}{p}} = \dfrac{pV}{T} = nR$

$$\boxed{C_{p,\mathrm{m}} - C_{V,\mathrm{m}} = \frac{C_p - C_V}{n} = R} \quad [2.19]$$

P2.32 (a) A difference in the thermogram baseline at two different temperatures indicates that the sample has different heat capacities at the two temperatures. The sample may have experienced a phase transition from one crystalline, or non-crystalline, structure to another. It may have melted, evaporated, or decomposed. It may have reacted with the atmosphere or, if the sample is a mixture, components may have reacted. In all of these examples the sample has experienced either a physical or chemical change while being heated from one temperature to the other and, consequently, exhibits different heat capacities at those temperatures.

Solid polymers often exhibit a DSC baseline change for a "glass transition" at a well defined temperature without the occurrence of reaction. A typical thermogram for a glass transition is shown in Figure 2.3. Upon reaching the temperature of the glass transition the polymer appears to change from a brittle glass-like state to a more rubber-like state which allows for increased rotational motion of the polymer chain and great randomness in the positions of molecules. It is a transition from a crystalline lattice to a glassy state. The glass transition of pure polystyrene, the plastic of cheap rulers and coffee spoons, occurs at about 90°C.

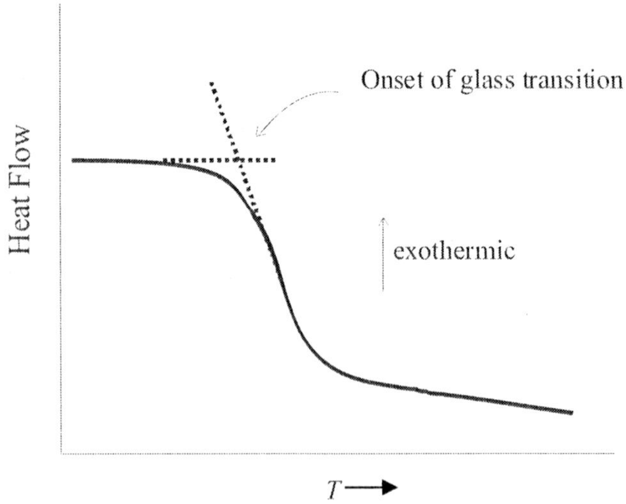

Figure 2.3

(b) Impurities lower the onset temperature observed both in the thermograms of the DSC melting of organic substances and in the observed glass transition temperatures of polymers. Samples with smaller mole fractions of the major substance give lower transition temperatures. Thus, a series of samples that are prepared with a range of mole fractions of the major constituent can be examined with DSC and a standard curve of mole fraction against onset temperatures, which are provided by the thermograms, can be prepared. The DSC of a chemical reactor sample of unknown purity gives the sample onset temperature that is checked against the standard curve to determine the purity mole fraction. Peak temperatures of DSC melting curves are used similarly.

3 Thermodynamics: applications of the first law

Answers to discussion questions

D3.1 The **standard state** of a substance is the pure substance at exactly 1 bar. A **standard reaction enthalpy** is the enthalpy change with each reactant and product at 1 bar and temperature T. We take T to be the conventional temperature 298.15 K (25°C) unless otherwise specified. The standard state is denoted with the superscript $^{\ominus}$ on the symbol for the property and the general symbol for a reaction enthalpy is $\Delta_r H_m^{\ominus}$. The symbols for standard enthalpies of phase transitions are $\Delta_{fus} H^{\ominus}$ for fusion (solid $\rightarrow$ liquid), $\Delta_{sub} H^{\ominus}$ for sublimation (solid $\rightarrow$ gas), and $\Delta_{vap} H^{\ominus}$ for vaporization (liquid $\rightarrow$ gas).

(a) Vaporization example: $CH_4(l) \rightarrow CH_4(g)$ $\quad \Delta_{vap} H_m^{\ominus}(T_b = 111.7\ \text{K}) = +8.18\ \text{kJ}$

(b) Fusion example: $CH_4(s) \rightarrow CH_4(l)$ $\quad \Delta_{fus} H^{\ominus}(T_f = 90.68\ \text{K}) = +0.941\ \text{kJ}$

(c) Sublimation example: $I_2(s) \rightarrow I_2(g)$ $\quad \Delta_{sub} H^{\ominus}(298\ \text{K}) = +62.44\ \text{kJ}$

(d) The **standard enthalpy of ionization** is the standard molar enthalpy accompanying the removal of an electron from a gas-phase atom (or ion).

Example: $H(g) \rightarrow H^+(g) + e^-(g)$ $\quad \Delta_{ion} H^{\ominus}(298\ \text{K}) = +1318\ \text{kJ}$

(e) The **standard electron-gain enthalpy** is the standard molar enthalpy accompanying the capture of an electron by a gas-phase atom (or ion).

Example: $Cl(g) + e^-(g) \rightarrow Cl^-(g)$ $\quad \Delta_{eg} H^{\ominus}(298\ \text{K}) = -354.81\ \text{kJ}$

(f) The **mean bond enthalpy** is the average bond enthalpy over a related series of compounds.

Example: $CH_4(g) \rightarrow C(g) + 4\ H(g)$ $\quad \Delta_r H \simeq 4 \times \Delta_B H(\text{C—H}) = 4 \times (412\ \text{kJ})$

$$\simeq 1648\ \text{kJ}$$

D3.2 (a) A **standard reaction enthalpy** is the enthalpy change with each reactant and product at 1 bar and temperature T. We take T to be the conventional temperature 298.15 K (25°C) unless otherwise specified. The standard state is denoted with the superscript $^{\ominus}$ on the symbol for the property and the general symbol for a reaction enthalpy is $\Delta_r H^{\ominus}$.

Example: $Fe_2O_3(s) + 3\ CO(g) \rightarrow 2\ Fe(s) + 3\ CO_2(g)$ $\quad \Delta_r H^{\ominus} = -24.74\ \text{kJ}$

(b) The **standard enthalpy of combustion** is the change in standard enthalpy per mole of combustible substance upon reaction with oxygen to produce $CO_2(g)$, $H_2O(l)$, and $N_2(g)$ from its carbon, hydrogen, and nitrogen component in addition to oxides of the remaining elements.

Example: $C_3H_8(g) + 5\ O_2(g) \rightarrow 3\ CO_2(g) + 4\ H_2O(l)$ $\quad \Delta_c H^{\ominus} = -2220\ \text{kJ}$

(c) The **standard enthalpy of formation** is the change in standard enthalpy per mole of substance formed from its elements in their standard reference state. The **reference state** of an element is its most stable form under the prevailing conditions.

Example: $3\ C(\text{s, graphite}) + 4\ H_2(g) \rightarrow C_3H_8(g)$ $\quad \Delta_f H^{\ominus} = -103.85\ \text{kJ}$

D3.3 The vaporization of the water is an endothermic transformation that cools the linen and its immediate environment: $H_2O(l) \rightarrow H_2O(g) \quad \Delta_{vap}H^{\ominus} = +44.01$ kJ.

D3.4 Standard reaction enthalpies can be calculated from a knowledge of the standard enthalpies of formation of all the substances (reactants and products) participating in the reaction. This is an exact method which involves no approximations. The only disadvantage is that standard enthalpies of formation are not known for all substances.

Approximate values can be obtained from mean bond enthalpies. See Example 3.3 for an illustration of the method of calculation. This method is often quite inaccurate, though, because the average values of the bond enthalpies used may not be close to the actual values in the compounds of interest and the method ignores attractive forces between molecules, or is applicable to perfect gases only.

Computer aided molecular modeling is now the method of choice for estimating standard reaction enthalpies, especially for large molecules with complex three-dimensional structures, but accurate numerical values are still difficult to obtain.

D3.5 The standard state of a substance at a specified temperature is its pure form at 1 bar. The reference state of a substance is its most stable state at the specified temperature and 1 bar. The distinction (the reference state must be the most stable state) is important because formation enthalpies are for formation of a substance from the elements in their reference states under prevailing conditions but other types of reactions may list a substance in a non-reference state.

D3.6 (a) $\Delta_r H = \Delta_r U + \Delta \nu_{gas} RT$

Limitations: perfect gas and negligible volume contribution from condensed phases.

(b) $\Delta_r H^{\ominus}(T') = \Delta_r H^{\ominus}(T) + \Delta_r C_p^{\ominus} \times (T' - T)$

Limitation: negligible dependence of heat capacities upon temperature.

D3.7 Expressions such as 'heat of combustion' and 'latent heat of vaporization' poorly convey information because it is not clear that the reaction heat is either for a constant pressure process ($q_p = \Delta H$) or a constant volume process ($q_V = \Delta U$). The distinction is important because generally q_p does not equal q_V.

Solutions to exercises

E3.1

$$C(\text{s, graphite}) + O_2(g) \rightarrow CO_2(g)$$

$$\Delta_f H(CO_2,g,p,T) \equiv \Delta_f H(p,T) = H_m(CO_2,g,p,T) - H_m(O_2,g,p,T) - H_m(\text{graphite},p,T)$$

We begin by pointing out that both the internal energy and enthalpy of a perfect gas depend upon T alone; they have no pressure dependence. To see this, recall that perfect gases exhibit neither attractive nor repulsive forces between molecules and, consequently, $\Delta U_m = 0$ [2.9] for the isothermal expansion of a perfect gas. Additionally, equation 2.13b indicates that $\Delta H_m = \Delta U_m + R\Delta T$ for a perfect gas; there is no pressure dependence on the right side of this equation so ΔH_m is independent of pressure. Thus, an isothermal change in the pressure of an ideal gas leaves both the internal energy and enthalpy unchanged. Because of this we can immediately write that for the above formation reaction:

$$\begin{aligned}\Delta_f H(p^{\ominus},T) - \Delta_f H(p,T) &= \{H_m(CO_2,g,p^{\ominus},T) - H_m(O_2,g,p^{\ominus},T) - H_m(\text{graphite},p^{\ominus},T)\} \\ &\quad - \{H_m(CO_2,g,p,T) - H_m(O_2,g,p,T) - H_m(\text{graphite},p,T)\} \\ &= H_m(\text{graphite},p,T) - H_m(\text{graphite},p^{\ominus},T).\end{aligned}$$

To find $\Delta_f H(p^{\ominus},T) - \Delta_f H(p,T)$, we need only find $H_m(\text{graphite},p,T) - H_m(\text{graphite},p^{\ominus},T)$.

Appling the definition $H_m = U_m + pV_m$ gives

$$\Delta_f H(p^\ominus,T) - \Delta_f H(p,T) = \{U_m(\text{graphite},p,T) + pV_m(\text{graphite},p,T)\}$$
$$-\{U_m(\text{graphite},p^\ominus,T) + p^\ominus V_m(\text{graphite},p^\ominus,T)\}.$$

We will evaluate this difference with p = 1.000 atm = 1.013 bar and $p^\ominus = 1\text{ bar} = 1\times10^5$ Pa. This is a very small difference in pressure so we will estimate that

$$U_m(\text{graphite},p,T) - U_m(\text{graphite},p^\ominus,T) \simeq 0.$$

Then, because graphite is an incompressible solid, $V_m(\text{graphite},p,T) \simeq V_m(\text{graphite},p^\ominus,T)$ and

$$\Delta_f H(p^\ominus,T) - \Delta_f H(p,T) \simeq pV_m(\text{graphite},p,T) - p^\ominus V_m(\text{graphite},p^\ominus,T)$$
$$\simeq (p - p^\ominus)V_m(\text{graphite},p^\ominus,T)$$

where $V_m(\text{graphite}) = (2.260\text{ g cm}^{-3})^{-1} \times (12.01\text{ g mol}^{-1}) = 5.314\times10^{-6}\text{ m}^3\text{ mol}^{-1}$.

$$\Delta_f H(p^\ominus,T) - \Delta_f H(1\text{ atm},T) \simeq (1.013-1)\times10^5\text{ Pa}\times(5.134\times10^{-6}\text{ m}^3\text{ mol}^{-1})$$
$$\simeq \boxed{6.91\text{ mJ mol}^{-1}}$$

This is a negligibly small difference for all practical purposes.

E3.2 $\Delta_{fus}H^\ominus(\text{Na}) = +2.60\text{ kJ mol}^{-1}$ [CRC *Handbook of Chemistry and Physics*] ?

$$n = \frac{250\times10^3\text{g}}{22.99\text{ g mol}^{-1}} = 1.09\times10^4\text{mol}$$

$$q = n\Delta_{fus}H^\ominus = 1.09\times10^4\text{ mol}\times2.60\text{ kJ mol}^{-1}$$
$$= \boxed{+2.83\times10^4\text{ kJ}}$$

E3.3 (a) $$q = n\Delta_{vap}H(298.15\text{ K}) = \frac{m}{M}\Delta_{vap}H(298.15\text{ K})$$

$$= \frac{m}{M}\{\Delta_f H(\text{g, }298.15\text{ K}) - \Delta_f H(\text{l, }298.15\text{ K})\}$$

$$= \left(\frac{1.00\times10^3\text{ g}}{18.02\text{ g mol}^{-1}}\right)\times(-241.82-(-285.83))\text{kJ mol}^{-1} = \boxed{+2.44\times10^3\text{ kJ}}\quad\text{[data tables]}$$

(b) $$q = n\Delta_{vap}H(373.15\text{ K}) = \frac{m}{M}\Delta_{vap}H(373.15\text{ K})$$

$$= \left(\frac{1.00\times10^3\text{ g}}{18.02\text{ g mol}^{-1}}\right)\times(40.7\text{ kJ mol}^{-1})\text{ [Table 3.1]} = \boxed{+2.26\times10^3\text{ kJ}}$$

E3.4 The heat supplied to the sample is

$$q = I\mathcal{V}t\ [2.5]$$
$$= (0.812\text{ A})\times(11.5\text{ V})\times(303\text{ s}) = 2.83\times10^3\text{ J}$$

$$q = \Delta H = n\Delta_{vap}H^\ominus\ \text{(pressure is constant)}$$

$$n = \frac{4.27\text{ g}}{60.04\text{ g mol}^{-1}} = 0.0711\text{ mol}$$

$$\Delta_{vap}H^\ominus = \frac{q}{n} = \frac{2.83\times10^3\text{ J}}{0.0711\text{ mol}} = 3.98\times10^4\text{ J mol}^{-1} = \boxed{+39.8\text{ kJ mol}^{-1}}$$

E3.5

$$\Delta H = q_p = n\Delta_{\text{vap}}H^{\ominus} = (2.50\ \text{mol})\times(32.0\ \text{kJ mol}^{-1}) = \boxed{+80.0\ \text{kJ}}$$

$$w = -p_{\text{ex}}\Delta V \approx -p_{\text{ex}}V(\text{g})\ [\text{because } V(\text{g}) >> V(\text{l})]$$

$$= -p_{\text{ex}}\left(\frac{nRT}{p_{\text{ex}}}\right) = -nRT\ [\text{perfect gas}]$$

$$= -(2.50\ \text{mol})\times(8.3145\ \text{J K}^{-1}\ \text{mol}^{-1})\times(250\ \text{K}) = \boxed{-5.20\ \text{kJ}}$$

$$\Delta U = q_p + w = +80.0\ \text{kJ} - 5.20\ \text{kJ} = \boxed{+74.8\ \text{kJ}}$$

E3.6

$$n = \frac{100\ \text{g ice}}{18.0\ \text{g mol}^{-1}} = 5.55\ \text{mol}$$

The heat needed to melt 100 g of ice is

$$q_1 = n\times\Delta_{\text{fus}}H^{\ominus}$$
$$= 5.55\ \text{mol}\times 6.01\ \text{kJ mol}^{-1} = +33.4\ \text{kJ}.$$

The heat needed to raise the temperature of the water from 0°C to 100°C is

$$q_2 = (100\ \text{g})\times(4.18\ \text{J K}^{-1}\ \text{g}^{-1})\times(100\ \text{K}) = 4.18\times10^4\ \text{J} = +41.8\ \text{kJ}.$$

The heat needed to vaporize the water is

$$q_3 = (5.55\ \text{mol})\times(40.7\ \text{kJ mol}^{-1}) = +226\ \text{kJ}.$$

The total heat is $q = q_1 + q_2 + q_3 = 33.4\ \text{kJ} + 41.8\ \text{kJ} + 226\ \text{kJ} = \boxed{+301\ \text{kJ}}$.

The graph of temperature against time is sketched in Figure 3.1. Note that the length of the liquid + gas, two-phase line is longer than the solid + liquid line in proportion to their $\Delta_{\text{trs}}H$ values.

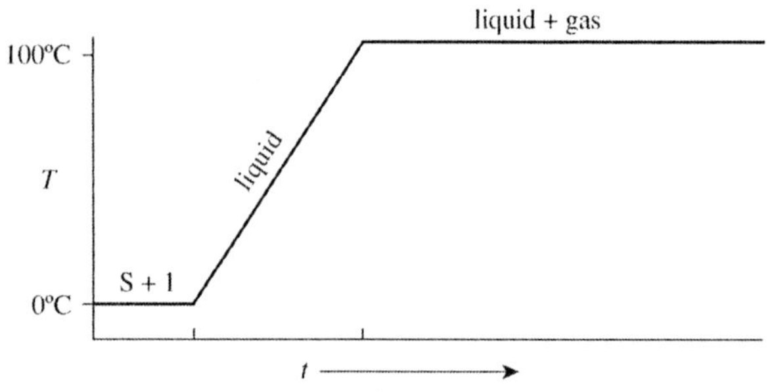

Figure 3.1

E3.7

We follow text Example 3.2, but with calcium in place of magnesium.

The overall process is $Ca(s) \rightarrow Ca^{2+}(g) + 2\ e^-(g)$.

The process is broken down as follows:

				ΔH / (kJ mol^{-1})
Sublimation:	Ca(s)	→	Ca(g)	+178.2
First ionization:	Ca(g)	→	$Ca^+(g) + e^-(g)$	+590
Second ionization:	$Ca^+(g)$	→	$Ca^{2+}(g) + e^-(g)$	+1150
Overall:	Ca(s)	→	$Ca^{2+}(g) + 2\ e^-(g)$	+1918

For the 5.0 g sample,

$$q = n\Delta H$$

$$= \left(\frac{5.0\ \text{g}}{40.08\ \text{g mol}^{-1}}\right)\times 1918\ \text{kJ mol}^{-1}$$

$$= \boxed{+239\ \text{kJ}}.$$

E3.8 $Ca(g) \rightarrow Ca^{2+}(g) + 2\,e^-(g) \qquad \Delta\nu_{gas} = 2$

$$\Delta_{ion}H = \Delta_{ion}U + \Delta\nu_{gas}RT \qquad \text{[generalization of eqn 3.3]}$$

$$\Delta_{ion}H - \Delta_{ion}U = \Delta\nu_{gas}RT = 2\times(8.3145\ \text{J K}^{-1}\ \text{mol}^{-1})\times(298.15\ \text{K}) = \boxed{+4.96\ \text{kJ mol}^{-1}}$$

E3.9 $Br(g) + e^-(g) \rightarrow Br^-(g) \qquad \Delta\nu_{gas} = -1$

$$\Delta_{eg}H = \Delta_{eg}U + \Delta\nu_{gas}RT \qquad \text{[generalization of eqn 3.3]}$$

$$\Delta_{eg}H - \Delta_{eg}U = \Delta\nu_{gas}RT = (-1)\times(8.3145\ \text{J K}^{-1}\ \text{mol}^{-1})\times(298.15\ \text{K}) = \boxed{-2.48\ \text{kJ mol}^{-1}}$$

E3.10 $Cl_2(g) \rightarrow Cl^+(g) + Cl^-(g)$

$$\Delta_m H = \Delta_{ion}H(\text{Cl}) + \Delta_{eg}H(\text{Cl}) + \Delta_B H(\text{Cl}_2) = \Delta_{ion}H(\text{Cl}) + \Delta_{eg}H(\text{Cl}) + 2\Delta_f H(\text{Cl})$$

$$= (1260 - 349 + 2\times121.68)\ \text{kJ mol}^{-1} = \boxed{+1154\ \text{kJ mol}^{-1}}$$

(Hess's law [3.4a]; data tables 3.2, 3.3 and D1.2)

$$\Delta H = n\Delta_m H = 10.0\ \text{g}\left(\frac{1\ \text{mol}}{70.90\ \text{g}}\right)(1154\ \text{kJ mol}^{-1}) = \boxed{+163\ \text{kJ}}$$

E3.11 (a) $Cl^-(g) \rightarrow Cl(g) + e^-(g) \qquad \Delta_{ion}H(Cl^-) = -\Delta_{eg}H(Cl) = \boxed{+354.8\ \text{kJ mol}^{-1}}$

(b) $$\Delta_{ion}H(\text{Cl}^-) = \Delta_{ion}U(\text{Cl}^-) + \Delta\nu_{gas}RT \qquad \text{[generalization of eqn 3.3]}$$

$$\Delta_{ion}U(\text{Cl}^-) = \Delta_{ion}H(\text{Cl}^-) - \Delta\nu_{gas}RT$$

$$= 354.8\ \text{kJ mol}^{-1} - 1\times(8.3145\ \text{J K}^{-1}\ \text{mol}^{-1})\times(298.15\ \text{K}) = \boxed{+352.3\ \text{kJ mol}^{-1}}$$

E3.12 (a) The mean bond enthalpy is the average of the three enthalpy changes given.

$$\Delta H_B(\text{N}-\text{H}) = \left(\frac{460+390+314}{3}\right)\times\text{kJ mol}^{-1}$$

$$= \boxed{388\ \text{kJ mol}^{-1}}$$

The bond dissociation energies and enthalpies refer to gas phase dissociation to atoms. Because of the dissociation, the number of moles of gas phase particles increases ($\Delta\nu_g > 0$). Therefore, since $\Delta H = \Delta U + \Delta\nu_g RT$, ΔU_B(N—H) is expected to be smaller than ΔH_B(N—H).

E3.13 (a) $C_6H_{12}O_6(aq) \rightarrow 2\,CH_3CH(OH)COOH(aq)$

The molecular formulas of these compounds are shown:

Bonds	Structure
7 C–H single bonds	HC=O
5 O–H single bonds	HC–OH
5 C–O single bonds	HO–CH
1 C=O double bond	HC–OH
5 C–C single bonds	HC–OH
	CH_2OH
	glucose

Bonds	Structure
4 C–H single bonds×2	O=C–OH
2 O–H single bonds×2	HO–C–H
2 C–O single bonds×2	CH_3
1 C=O double bond×2	lactic acid
2 C–C single bonds×2	

$$\Delta_r H = \Sigma\Delta H\text{ (bonds broken)} + \Sigma\Delta H\text{ (bonds formed)}$$

Note that the ΔH for the formation of a bond is the negative of the bond enthalpy.

$$\Sigma\Delta H\text{ (bonds broken)} = 7\ \Delta H(\text{C—H}) + 5\ \Delta H(\text{O—H}) + 5\ \Delta H(\text{C—O})$$
$$+ 1\ \Delta H(\text{C=O}) + 5\ \Delta H(\text{C—C})$$
$$= (7 \times 412) + (5 \times 463) + (5 \times 360) + 743 + (5 \times 348)$$
$$= 9482\text{ kJ mol}^{-1}$$

$$\Sigma\Delta H\text{ (bonds formed)} = -8\Delta H(\text{C—H}) - 4\ \Delta H(\text{O—H}) - 4\ \Delta H(\text{C—O})$$
$$-2\ \Delta H(\text{C=O}) - 4\ \Delta H(\text{C—C})$$
$$= -(8 \times 412) - (4 \times 463) - (4 \times 360) - (2 \times 743) - (4 \times 348)$$
$$= -9466\text{ kJ mol}^{-1}$$

$$\Delta_r H = (9482 - 9466)\text{ kJ mol}^{-1} = \boxed{16\text{ kJ mol}^{-1}}$$

The approximate nature of this kind of calculation can be seen by comparing this value of $\Delta_r H$ to the value of $\Delta_r H^{\ominus}$ calculated from standard enthalpies of formation, which is $\Delta_r H^{\ominus} = -114\text{ kJ mol}^{-1}$. Strictly speaking, bond enthalpies apply only to gas phase processes.

(b) $C_6H_{12}O_6(aq) + 6\ O_2(g) \rightarrow 6\ CO_2(g) + 6\ H_2O(l)$

$\Delta_r H = 9482\text{ kJ mol}^{-1}$ [from part (a)] $+ 6 \times \Delta H\ (\text{O=O}) - 12 \times \Delta H(\text{O=CO}) - 12 \times \Delta H(\text{H—OH})$

$$= 9482\text{ kJ mol}^{-1} + (6 \times 497) - (12 \times 799) - (12 \times 492)\text{ kJ mol}^{-1}$$
$$= \boxed{-3028\text{ kJ mol}^{-1}}$$

E3.14 (a) 1.00 t N_2 consumed is $(1.00 \times 10^6\text{ g})/(28.02\text{ g mol}^{-1}) = 3.57 \times 10^4$ mol N_2.

$$\Delta H^{\ominus} = (3.57\times10^4\text{ mol})\times(-92.22\text{ kJ mol}^{-1}) = \boxed{-3.29\text{ GJ}}$$

(b) 1.00 t NH_3 formed is $(1.00 \times 10^6\text{ g})/(17.03\text{ g mol}^{-1}) = 5.87 \times 10^4$ mol NH_3.

$$\Delta H^{\ominus} = \left(\frac{1\text{ mol N}_2}{2\text{ mol NH}_3}\right)\times(5.87\times10^4\text{ mol NH}_3)\times(-92.22\text{ kJ mol}^{-1}) = \boxed{-2.71\text{ GJ}}$$

E3.15 (a) The standard enthalpy of combustion applies to the combustion of one mole ethane, therefore

$$\Delta_c H^{\ominus} = \left(\frac{1\text{ mol}}{2\text{ mol}}\right)\times(-3120\text{ kJ mol}^{-1}) = \boxed{-1560\text{ kJ mol}^{-1}}.$$

(b) Specific enthalpy of combustion of ethane: $-\left(\dfrac{-1560\text{ kJ mol}^{-1}}{30.07\text{ g mol}^{-1}}\right) = \boxed{51.88\text{ kJ g}^{-1}}$

(c) Using data of Table 3.6 we find that the specific enthalpy of combustion of methane is $-\left(\dfrac{-890\text{ kJ mol}^{-1}}{16.04\text{ g mol}^{-1}}\right) = 55.5\text{ kJ g}^{-1}$. Because the specific enthalpy of methane is more exothermic than that of ethane, $\boxed{\text{ethane is a less efficient fuel}}$.

E3.16

$$C_6H_5C_2H_5(l) + {}^{21}\!/_{2}O_2(g) \rightarrow 8\ CO_2(g) + 5\ H_2O(l)$$
$$\Delta_c H^{\ominus} = 8\ \Delta_f H^{\ominus}(CO_2,g) + 5\ \Delta_f H^{\ominus}(H_2O,l) - \Delta_f H^{\ominus}(C_6H_5C_2H_5,l)$$
$$= 8\times(-393.51\text{ kJ mol}^{-1}) + 5\times(-285.83\text{ kJ mol}^{-1}) - (-12.5\text{ kJ mol}^{-1})$$
$$= \boxed{-4564.7\text{ kJ mol}^{-1}}$$

E3.17 The reaction wanted is: $C_6H_{10}(l) + H_2(g) \rightarrow C_6H_{12}(l)$

Use the following reactions:

$$\text{(1) } C_6H_{10}(l) + {}^{17}\!/_2 O_2 \rightarrow 6\ CO_2(g) + 5\ H_2O(l) \qquad \Delta_c H^\ominus = -3752\ \text{kJ mol}^{-1}$$
$$\text{(2) } C_6H_{12}(l) + 9\ O_2(g) \rightarrow 6\ CO_2(g) + 6\ H_2O(l) \qquad \Delta_c H^\ominus = -3953\ \text{kJ mol}^{-1}$$
$$\text{(3) } H_2(g) + {}^1\!/_2 O_2(g) \rightarrow H_2O(l) \qquad \Delta_f H^\ominus = -286\ \text{kJ mol}^{-1}$$

Now sum eqn (1), eqn (3), and the reverse of eqn (2) to get the desired equation, then

$$\Delta_r H^\ominus = -3752\ \text{kJ mol}^{-1} + 3953\ \text{kJ mol}^{-1} - 286\ \text{kJ mol}^{-1} = \boxed{-85\ \text{kJ mol}^{-1}}$$

E3.18

$$3\ C(s) + 3\ H_2(g) + O_2(g) \rightarrow CH_3COOCH_3(l) \qquad \Delta_f H^\ominus = -442\ \text{kJ mol}^{-1}$$
$$\Delta_f U^\ominus = \Delta_f H^\ominus - \Delta(pV) \quad [2.13a]$$
$$= \Delta_f H^\ominus - \Delta\nu_{\text{gas}} RT \text{ [generalization of eqn 3.3]} \qquad \text{where } \Delta\nu_{\text{gas}} = -4\ \text{mol}$$
$$= -442\ \text{kJ mol}^{-1} - (-4)\times(8.3145\ \text{J K}^{-1}\ \text{mol}^{-1})\times(298.15\ \text{K})\times(1\times10^{-3}\ \text{kJ J}^{-1})$$
$$= \boxed{-432\ \text{kJ mol}^{-1}}$$

E3.19

$$C_{14}H_{10}(s) + {}^{33}\!/_2\ O_2(g) \rightarrow 14\ CO_2(g) + 5\ H_2O(l) \qquad \Delta_c H^\ominus = -7163\ \text{kJ mol}^{-1}$$

The reverse reaction is

$$14\ CO_2(g) + 5\ H_2O(l) \rightarrow C_{14}H_{10}(s) + {}^{33}\!/_2\ O_2(g) \qquad \Delta H^\ominus = +7163\ \text{kJ mol}^{-1}$$

The CO_2 and H_2O can be replaced by adding the following two reactions and using data from Table 3.8 for $\Delta_f H^\ominus(CO_2)$ and $\Delta_f H^\ominus(H_2O)$.

$$14\ C(s) + 14\ O_2(g) \rightarrow 14\ CO_2(g) \qquad \Delta H^\ominus = 14\times(-393.5\ \text{kJ mol}^{-1}) = -5509\ \text{kJ mol}^{-1}$$
$$5\ H_2(g) + {}^5\!/_2\ O_2(g) \rightarrow 5\ H_2O(l) \qquad \Delta H^\ominus = 5\times(-285.8\ \text{kJ mol}^{-1}) = -1429\ \text{kJ mol}^{-1}$$

This gives the anthracene formation reaction:

$$14\ C(s) + 5\ H_2(g) \rightarrow C_{14}H_{10}(s).$$

$$\Delta_f H^\ominus(\text{anthracene}) = (+7163 - 5509 - 1429)\ \text{kJ mol}^{-1} = \boxed{+225\ \text{kJ mol}^{-1}}$$

E3.20

$$q = n\Delta_c H^\ominus \text{ with } \Delta_c H^\ominus(\text{naphthalene}) = -5157\ \text{kJ mol}^{-1} \text{ [Table D1.1]}$$

Calculating q,

$$|q| = \frac{320\times10^{-3}\ \text{g}}{128.18\ \text{g mol}^{-1}} \times 5157\ \text{J mol}^{-1} = 12.87\ \text{kJ}$$

$$C = \frac{q}{\Delta T} = \frac{12.87\ \text{kJ}}{3.05\ \text{K}} = \boxed{4.22\ \text{kJ K}^{-1}}$$

When phenol is used, $\Delta_c H^\ominus = -3054\ \text{kJ mol}^{-1}$ [Table D1.1]

$$|q| = \frac{100\times10^{-3}\ \text{g}}{94.12\ \text{g mol}^{-1}} \times 3054\ \text{J mol}^{-1} = 3.245\ \text{kJ}$$

$$\Delta T = \frac{q}{C} = \frac{3.245\ \text{kJ}}{4.22\ \text{kJ K}^{-1}} = \boxed{0.769\ \text{K}}$$

E3.21 (a) $q = C\Delta T$

$$\Delta_c H^\ominus = \frac{q}{n} = \frac{C\Delta T}{n} = \frac{MC\Delta T}{m}$$

where M is the molar mass of glucose and m is the mass of the sample.

$$M = 180.16\ \text{g mol}^{-1}$$

$$\left|\Delta_c H^{\ominus}\right| = \frac{180.16\ \text{g mol}^{-1} \times 641\ \text{J K}^{-1} \times 7.793\ \text{K}}{0.3212\ \text{g}} = 2802\ \text{kJ mol}^{-1}$$

Because the combustion is exothermic, $\Delta_c H^{\ominus} = \boxed{-2.80\ \text{MJ mol}^{-1}}$

(b) The combustion reaction is

$$C_6H_{12}O_6(s) + 6\ O_2(g) \rightarrow 6\ CO_2(g) + 6\ H_2O(l) \qquad \Delta\nu_{gas} = 0$$

Therefore $\Delta_f U^{\ominus} = \Delta_c H^{\ominus} = \boxed{-2.80\ \text{MJ mol}^{-1}}$

(c) For the enthalpy of formation we combine the following equations:

$$6\ CO_2(g) + 6\ H_2O(l) \rightarrow C_6H_{12}O_6(s) + 6\ O_2(g) \qquad \Delta_r H^{\ominus} = +2.80\ \text{MJ mol}^{-1}$$

$$6\ C(s) + 6\ O_2(g) \rightarrow 6\ CO_2(g) \qquad \Delta_r H^{\ominus} = 6 \times \Delta_f H^{\ominus}(CO_2, g) = -2.36\ \text{MJ mol}^{-1}$$

$$6\ H_2(g) + 3\ O_2(g) \rightarrow 6\ H_2O(l) \qquad \Delta_r H^{\ominus} = 6 \times \Delta_f H^{\ominus}(H_2O, l) = -1.71\ \text{MJ mol}^{-1}$$

The sum of the three equations is

$$6\ C(s) + 6\ H_2(g) + 3\ O_2(g) \rightarrow C_6H_{12}O_6(s)$$

$$\text{Therefore } \Delta_f H^{\ominus}(\text{glucose}) = (2.80 - 2.36 - 1.71)\ \text{MJ mol}^{-1}$$

$$= \boxed{-1.27\ \text{MJ mol}^{-1}}$$

E3.22 (a) On the assumption that the calorimeter is a constant volume bomb calorimeter such as the one described in Figure 2.17, the heat released directly equals the internal energy of combustion, $\Delta_c U^{\ominus}$. Therefore, $\Delta_c U^{\ominus} = \boxed{-1333\ \text{kJ mol}^{-1}}$.

(b) $\quad HOOCCH{=}CHCOOH(s) + 3\ O_2(g) \rightarrow 4\ CO_2(g) + 2\ H_2O(l)$

$$\Delta\nu_{gas} = +1$$

$$\Delta_c H^{\ominus} = \Delta_c U^{\ominus} + \Delta\nu_g\ RT$$

$$= -1333\ \text{kJ mol}^{-1} + (1 \times 2.5\ \text{kJ mol}^{-1}) = \boxed{-1331\ \text{kJ mol}^{-1}}$$

(c) $\quad \Delta_c H^{\ominus} = 4\Delta_f H^{\ominus}(CO_2, g) + 2\ \Delta_f H^{\ominus}(H_2O, l) - \Delta_f H^{\ominus}(\text{fumaric acid})$

$$= -1331\ \text{kJ mol}^{-1}$$

$$\Delta_f H^{\ominus}(\text{fumaric acid}) = 4 \times (-393.51\ \text{kJ mol}^{-1}) + 2 \times (-285.83\ \text{kJ mol}^{-1}) + 1331\ \text{kJ mol}^{-1}$$

$$= \boxed{-815\ \text{kJ mol}^{-1}}$$

E3.23 The specific enthalpy equals $-M\Delta_c H^{\ominus}$. When comparing a carbon-based series of compounds, all having similar molar masses, those that are least oxidized will have the most exothermic $\Delta_c H^{\ominus}$ and the higher specific enthalpies. Thus, glucose ($C_6H_{12}O_6$), being more highly oxidized than decanoic acid ($C_{10}H_{20}O_2$) has a lower specific enthalpy.

E3.24

$$AgI(s) \rightarrow Ag^+(aq) + I^-(aq)$$

$$\Delta_{soln} H^{\ominus} = \Delta_f H^{\ominus}(Ag^+, aq) + \Delta_f H^{\ominus}(I^-, aq) - \Delta_f H^{\ominus}(AgI, s)$$

$$= [(105.58) + (-55.19) - (-61.88)]\ \text{kJ mol}^{-1} = \boxed{+112.27\ \text{kJ mol}^{-1}}$$

E3.25

$$NH_3SO_2(s) \rightarrow NH_3(g) + SO_2(g) \qquad \Delta H^{\ominus} = +40\ \text{kJ mol}^{-1}$$

Therefore, $NH_3(g) + SO_2(g) \rightarrow NH_3SO_2(s) \qquad \Delta H^{\ominus} = -40\ \text{kJ mol}^{-1}$.

For the latter reaction

$$\Delta_r H^\ominus = \Delta_f H^\ominus(NH_3SO_2) - \Delta_f H^\ominus(NH_3) - \Delta_f H^\ominus(SO_2) = -40 \text{ kJ mol}^{-1}.$$

Therefore, after solving for $\Delta_f H^\ominus(NH_3SO_2)$

$$\Delta_f H^\ominus(NH_3SO_2, s) = \Delta_f H^\ominus(NH_3, g) + \Delta_f H^\ominus(SO_2, g) - 40 \text{ kJ mol}^{-1}$$

$$= (-46.11 - 296.83 - 40) \text{ kJ mol}^{-1} = \boxed{-383 \text{ kJ mol}^{-1}}.$$

E3.26 (1) $C(gr) + O_2(g) \rightarrow CO_2(g)$ $\Delta_c H^\ominus = -393.5 \text{ kJ mol}^{-1}$

(2) $C(diam) + O_2(g) \rightarrow CO_2(g)$ $\Delta_c H^\ominus = -395.41 \text{ kJ mol}^{-1}$

Subtracting (2) from (1) yields

$$C(gr) \rightarrow C(diam)$$

$$\Delta_{trs} H^\ominus = [-393.5 - (-395.41)] \text{ kJ mol}^{-1} = \boxed{+1.9 \text{ kJ mol}^{-1}}.$$

E3.27

$$\Delta_{trs} U = w + q = -p_{ex}\Delta V + q$$

$$q_p = \Delta_{trs} H = +1.9 \text{ kJ mol}^{-1}$$

$$p = 150 \text{ kbar} = 1.50 \times 10^5 \text{ bar} = 1.50 \times 10^{10} \text{ Pa}$$

For 1 mol graphite

$$V_{gr} = \frac{12.01 \text{ g/mol}}{2.250 \text{ g/cm}^3} \times \frac{1 \text{ m}^3}{10^6 \text{ cm}^3} = 5.338 \times 10^{-6} \text{ m}^3 \text{ mol}^{-1}.$$

For 1 mol diamond

$$V_{diam} = \frac{12.01 \text{ g/mol}}{3.510 \text{ g/cm}^3} \times \frac{1 \text{ m}^3}{10^6 \text{ cm}^3} = 3.422 \times 10^{-6} \text{ m}^3 \text{ mol}^{-1}.$$

$$\Delta V = V_{diam} - V_{gr} = 3.422 \times 10^{-6} \text{ m}^3 - 5.338 \times 10^{-6} \text{ m}^3 = -1.916 \times 10^{-6} \text{ m}^3 \text{ mol}^{-1}$$

$$-p_{ex}\Delta V = -1.50 \times 10^{10} \text{ Pa} \times (-1.916 \times 10^{-6} \text{ m}^3 \text{ mol}^{-1})$$

$$= 2.874 \times 10^4 \text{ J mol}^{-1} = 28.74 \text{ kJ mol}^{-1}$$

$$\Delta_{trs} U = 28.74 \text{ kJ mol}^{-1} + 1.9 \text{ kJ mol}^{-1} = \boxed{+30.6 \text{ kJ mol}^{-1}}$$

E3.28 (a) $q = mC_s\Delta T$

$$\Delta T = \frac{q}{mC_s} = \frac{10 \times 10^3 \text{ kJ}}{(65 \text{ kg}) \times (4.18 \text{ J g}^{-1} \text{ K}^{-1})} = 37 \text{ K} = \boxed{37°C}$$

(b) Let m_{vap} be the mass of water evaporated by q.

$$q = (m_{vap}/M)\Delta_{vap} H^\ominus$$

$$m_{vap} = \frac{Mq}{\Delta_{vap} H^\ominus} = \frac{(0.018016 \text{ kg mol}^{-1}) \times (10 \times 10^3 \text{ kJ})}{44 \text{ kJ mol}^{-1}} = \boxed{4.1 \text{ kg}}$$

This estimate ignores both the conduction of heat that occurs from high temperature to low temperature without evaporation and the small amount of heat used to bring ingested water to body temperature.

E3.29

$C_3H_8(l) \rightarrow C_3H_8(g)$ $\Delta_{vap} H^\ominus = +15 \text{ kJ}$

$C_3H_8(g) + 5\, O_2(g) \rightarrow 3\, CO_2(g) + 4\, H_2O(l)$ $\Delta_c H^\ominus(g) = -2220 \text{ kJ}$

(a) $\Delta_c H^\ominus(\mathrm{l}) = \Delta_{\mathrm{vap}} H^\ominus + \Delta_c H^\ominus(\mathrm{g})$

$$= 15 \text{ kJ mol} - 2220 \text{ kJ mol}^{-1} = \boxed{-2205 \text{ kJ mol}^{-1}}$$

(b) $\Delta\nu_{\mathrm{gas}} = -2$ [5 O_2(g) replaced with 3 CO_2(g)]

$\Delta_c U^\ominus(\mathrm{l}) = \Delta_c H^\ominus(\mathrm{l}) - (-2)RT$

$$= -2205 \text{ kJ mol}^{-1} + (2\times 2.5 \text{ kJ mol}^{-1}) = \boxed{-2200 \text{ kJ mol}^{-1}}$$

E3.30 (a) exothermic, $\Delta_r H^\ominus$ = negative

(b) endothermic, $\Delta H^\ominus$ = positive

(c) endothermic, $\Delta_{\mathrm{vap}} H^\ominus$ = positive

(d) endothermic, $\Delta_{\mathrm{fus}} H^\ominus$ = positive

(e) endothermic, $\Delta_{\mathrm{sub}} H^\ominus$ = positive

E3.31 (a) $\Delta_r H^\ominus = \Delta_f H^\ominus(\mathrm{N_2O_4, g}) - 2\,\Delta_f H^\ominus(\mathrm{NO_2, g})$

$$= [9.16 - 2\times 33.18] \text{ kJ mol}^{-1} = \boxed{-57.20 \text{ kJ mol}^{-1}}$$

(b) $\Delta_r H^\ominus = \tfrac{1}{2}\,\Delta_f H^\ominus(\mathrm{N_2O_4, g}) - \Delta_f H^\ominus(\mathrm{NO_2, g})$

$$= \tfrac{1}{2}(9.16) - 33.18 \text{ kJ mol}^{-1} = \boxed{-28.6 \text{ kJ mol}^{-1}}$$

(c) $\Delta_r H^\ominus = 2\times\Delta_f H^\ominus(\mathrm{HNO_3, aq}) + \Delta_f H^\ominus(\mathrm{NO, g}) - 3\times\Delta_f H^\ominus(\mathrm{NO_2, g}) - \Delta_f H^\ominus(\mathrm{H_2O, l})$

$$= [2\times(-207.36) + 90.25 - 3\times(33.18) - (-285.83)] \text{ kJ mol}^{-1}$$

$$= \boxed{-138.2 \text{ kJ mol}^{-1}}$$

(d) $\Delta_r H^\ominus = \Delta_f H^\ominus(\text{propene, g}) - \Delta_f H^\ominus(\text{cyclopropane, g})$

$$= [20.42 - 53.30] \text{ kJ mol}^{-1} = \boxed{-32.88 \text{ kJ mol}^{-1}}$$

(e) In order to calculate $\Delta_r H^\ominus$ first write the net ionic equation:

$$\mathrm{H^+(aq) + Cl^-(aq) + Na^+(aq) + OH^-(aq) \rightarrow Na^+(aq) + Cl^-(aq) + H_2O(l)}$$

Simplifying we obtain

$$\mathrm{H^+(aq) + OH^-(aq) \rightarrow H_2O(l)}$$

$\Delta_r H^\ominus = \Delta_f H^\ominus(\mathrm{H_2O, l}) - \Delta_f H^\ominus(\mathrm{H^+, aq}) - \Delta_f H^\ominus(\mathrm{OH^-, aq})$

$$= [-285.83 - 0 - (-229.99)] \text{ kJ mol}^{-1} = \boxed{-55.84 \text{ kJ mol}^{-1}}$$

E3.32 The formation of N_2O_5 is the sum of the three reactions.

	$\Delta_r H^\ominus/(\mathrm{kJ\ mol^{-1}})$
$\mathrm{2\,NO(g) + O_2(g) \rightarrow 2\,NO_2(g)}$	−114.1
$\mathrm{\tfrac{1}{2}O_2(g) + 2\,NO_2(g) \rightarrow N_2O_5(g)}$	$\tfrac{1}{2}(-110.2)$
$\mathrm{N_2(g) + O_2(g) \rightarrow 2\,NO(g)}$	180.5
$\mathrm{N_2(g) + \tfrac{5}{2}O_2(g) \rightarrow N_2O_5(g)}$	+11.3

Therefore, $\Delta_f H^\ominus(\mathrm{N_2O_5, g}) = \boxed{+11.3 \text{ kJ mol}^{-1}}$.

E3.33 We use equations 3.6 and 3.7.

$$\Delta_r H^{\ominus}(T_2) = \Delta_r H^{\ominus}(T_1) + \Delta_r C_p^{\ominus}\Delta T \quad [3.6]$$

$$\Delta_r C_p^{\ominus} = \Sigma \nu C_{p,m}^{\ominus}\,(\text{products}) - \Sigma \nu C_{p,m}^{\ominus}\,(\text{reactants}) \quad [3.7]$$

$$= (77.28 - 2 \times 37.20)\ \text{J K}^{-1}\ \text{mol}^{-1} = +2.88\ \text{J K}^{-1}\ \text{mol}^{-1}$$

$$\Delta_r H^{\ominus}(373\ \text{K}) = \Delta_r H^{\ominus}(298\ \text{K}) + \Delta_r C_p^{\ominus}\Delta T$$

$$= -57.20\ \text{kJ mol}^{-1} + (2.88\ \text{J K}^{-1} \times 75\ \text{K}) \quad [\text{See E3.31(a)}]$$

$$= (-57.20 + 0.22)\ \text{kJ mol}^{-1} = \boxed{-56.98\ \text{kJ mol}^{-1}}$$

E3.34

$$\Delta_{\text{vap}} H^{\ominus}(T') = \Delta_{\text{vap}} H^{\ominus}(T) + \Delta_r C_p^{\ominus} \times (T' - T) \quad [3.6]$$

$$\Delta_{\text{vap}} H^{\ominus}(373\ \text{K}) = 44.01\ \text{kJ mol}^{-1} + (-41.71\ \text{J K}^{-1}\ \text{mol}^{-1}) \times (373\ \text{K} - 298\ \text{K})$$

$$= \boxed{40.88\ \text{kJ mol}^{-1}}$$

E3.35 The sign of $\Delta_r C_p$ in equation 3.6 determines whether or not $\Delta_r H^{\ominus}$ will increase or decrease with increasing T. A negative value of $\Delta_r C_p$ implies a decrease, a positive value an increase with increasing T.

(a) $\Delta_r C_p = (2 \times 4R) - (3 \times \tfrac{7}{2}R) = -\tfrac{5}{2}R$, therefore $\Delta H^{\ominus}$ will $\boxed{\text{decrease}}$ with increasing T.

(b) $\Delta_r C_p = 8R - \tfrac{7}{2}R - (3 \times \tfrac{7}{2}R) = -6R$, $\boxed{\text{decrease}}$

(c) $\Delta_r C_p = 8R + \tfrac{7}{2}R - 4R - (2 \times \tfrac{7}{2}R) = +\tfrac{1}{2}R$, $\boxed{\text{increase}}$

E3.36 (a) $\Delta_r C_p = 2 \times 9R - 3 \times \tfrac{7}{2}R = +\tfrac{15}{2}R$, $\boxed{\text{increase}}$

(b) $\Delta_r C_p = \tfrac{7}{2}R + (2 \times 9R) - 4R - (2 \times \tfrac{7}{2}R) = +\tfrac{21}{2}R$, $\boxed{\text{increase}}$

Solutions to projects

P3.37 (a) $\nu_A A + \nu_B B + \cdots \rightarrow \nu_P P + \nu_Q Q + \cdots \qquad \Delta_r H(T)$

Let $dH_m(i)$ be the infinitesimal molar enthalpy change of the ith chemical species due to the infinitesimal temperature change dT. The infinitesimal reaction enthalp change due to the temperature change is $d\Delta_r H(T)$, which equals the sum of $\nu_{\text{product}} dH_m$ (product) minus the sum of $\nu_{\text{reactant}} dH_m$ (reactant).

$$d\Delta_r H(T) = \nu_P\, dH_m(P) + \nu_Q\, dH_m(Q) + \cdots - \{\nu_A\, dH_m(A) + \nu_B\, dH_m(B) + \cdots\}$$

$$= \Sigma\, \nu\, dH_m\,(\text{products}) - \Sigma\, \nu\, dH_m\,(\text{reactants})$$

Substitution of $dH_m(i) = C_{p,m}(i)\, dT$ [2.16 as infinitesimal expression] gives

$$d\Delta_r H(T) = \Sigma\, \nu\, C_{p,m}(\text{products})\, dT - \Sigma\, \nu\, C_{p,m}(\text{reactants})\, dT$$

$$= \{\Sigma\, \nu\, C_{p,m}(\text{products}) - \Sigma\, \nu\, C_{p,m}(\text{reactants})\}\, dT$$

or

$$d\Delta_r H(T) = \Delta_r C_p\, dT \qquad \text{where } \Delta_r C_p = \Sigma\, \nu\, C_{p,m}(\text{products}) - \Sigma\, C_{p,m}(\text{reactants}).$$

Integration between T and T' gives

$$\int_T^{T'} d\Delta_r H(T) = \int_T^{T'} \Delta_r C_p\, dT$$

$$\Delta_r H(T') - \Delta_r H(T) = \int_T^{T'} \Delta_r C_p\, dT \qquad \text{or} \qquad \boxed{\Delta_r H(T') = \Delta_r H(T) + \int_T^{T'} \Delta_r C_p\, dT}.$$

If $\Delta_r C_p$ is either temperature independent or negligibly dependent upon temperature over the temperature range, the integral on the right simplifies to Kirchhoff's law.

$$\Delta_r H(T') = \Delta_r H(T) + \int_T^{T'} \Delta_r C_p \mathrm{d}T = \Delta_r H(T) + \Delta_r C_p \int_T^{T'} \mathrm{d}T = \Delta_r H(T) + \Delta_r C_p \times (T' - T)$$

(b) If $\Delta_r C_p$ is temperature dependent, care must be taken with the integral on the right of the expression derived in part (a). For example, if $\Delta_r C_p = a + bT + c/T^2$ the integral is

$$\int_T^{T'} \Delta_r C_p \mathrm{d}T = \int_T^{T'} \left(a + bT + \frac{c}{T^2} \right) \mathrm{d}T = a\int_T^{T'} \mathrm{d}T + b\int_T^{T'} T\mathrm{d}T + c\int_T^{T'} \frac{1}{T^2}\mathrm{d}T = aT\Big|_T^{T'} + \frac{bT^2}{2}\Bigg|_T^{T'} - \frac{c}{T}\Bigg|_T^{T'}$$

$$= a(T' - T) + \frac{b}{2}(T'^2 - T^2) - c\left(\frac{1}{T'} - \frac{1}{T}\right)$$

and the reaction enthalpy is

$$\Delta_r H(T') = \Delta_r H(T) + \int_T^{T'} \Delta_r C_p \mathrm{d}T$$

$$= \boxed{\Delta_r H(T) + a(T' - T) + \frac{b}{2}(T'^2 - T^2) - c\left(\frac{1}{T'} - \frac{1}{T}\right)}$$

P3.38 (a) Complex carbohydrates have 17 kJ g^{-1} specific enthalpies (Box 3.1) so 40 g of carbs can provide 40 × 17 kJ = 680 kJ. Given a total daily energy requirement of 2200 Cal (9.21 × 10^3 kJ), the percentage provided by the 40 g of carbs is 680 / (9.21 × 10^3 kJ) × 100% = $\boxed{7.4\%}$.

(b) Glucose has a 16 kJ g^{-1} specific enthalpy so the combustion enthalpy of a 2.5 g tablet is 2.5 × 16 kJ = $\boxed{40\text{ kJ}}$.

(c) The effective energy available for work is 0.25 × 40 kJ = 10 kJ.

Since the work required for a climb to height h is given by $w = mgh$, the glucose tablet provides the energy for a 70 kg person to climb to

$$h = \frac{w}{mg} = \frac{10\times10^3 \text{ J}}{(70\text{ kg})\times(9.81\text{ m s}^{-2})} = \boxed{15\text{ m}}.$$

(d) $C_6H_{12}O_6(s) + 6\,O_2(g) \rightarrow 6\,CO_2(g) + 6\,H_2O(l)$

$$\Delta_r C_p^{\ominus} = \{6\times(37.11) + 6\times(75.291) - 1\times(115) - 6\times(29.355)\}\text{ J K}^{-1}\text{ mol}^{-1} = 383\text{ J K}^{-1}\text{ mol}^{-1}$$

According to Kirchhoff's law [3.6], the combustion enthalpy of glucose is $\boxed{\text{larger}}$ at blood temperature than at 25°C because $\Delta_r C_p^{\ominus} > 0$ and the blood temperature is greater than 25°C.

(e) Sucrose has a (5645 / 342.30) kJ g^{-1} = 16.5 kJ g^{-1} specific enthalpy so the combustion enthalpy of a 1.5 g tablet is 1.5 × 16.5 kJ = $\boxed{25\text{ kJ}}$.

(f) The effective energy available for work is 0.25 × 25 kJ = 6.25 kJ.

Since the work required for a climb to height h is given by $w = mgh$, the sucrose tablet provides the energy for a 70 kg person to climb to

$$h = \frac{w}{mg} = \frac{6.25\times10^3 \text{ J}}{(70\text{ kg})\times(9.81\text{ m s}^{-2})} = \boxed{9.1\text{ m}}.$$

4 Thermodynamics: the second law

Answers to discussion questions

D4.1 (a) Ceaseless disorderly motion of gas phase molecules ensures that they spread throughout the entire, available volume. There is a negligibly small probability that all molecules spontaneously move into a volume that is smaller than the total available space because to do so would be a move to greater order. The natural direction of change corresponds to the dispersal of matter. By occupying the total available space, and expanding when the container volume increases, the gas molecules maximize molecular disorder and entropy, the measure of the current state of disorder, within the constraints of volume and total energy. Eqn 4.2 quantifies the entropy change when a perfect gas expands isothermally form V_i to V_f.

$$\Delta S = nR\ln\frac{V_f}{V_i} \quad [4.2]$$

Ludwig Boltzmann's fundamental equation $S = k\ \ln W$, where W is the number of ways that molecules of the system can be arranged yet correspond to the same total energy, gives insight into the spontaneous spread of molecules into an expanding volume. As the volume expands, the quantum energy levels occupied by the molecules get closer together, and there are more ways of arranging the molecules for a given total energy. That is, as the container expands, W increases, and therefore S increases too.

(b) Entropy of a sample increases as the temperature is raised from T_i to T_f, because the thermal disorder of the system is greater at the higher temperature due to the more vigorous molecular motion. Eqn 4.3 quantifies the entropy change when the heat capacity is constant over the range of temperatures:

$$\Delta S = C\ln\frac{T_f}{T_i} \quad [4.3]$$

where C is the heat capacity of the system; if the pressure is constant during heating, we use C_p, and if the volume is constant, we use C_V. As temperature increases, a greater number of molecular quantum states become available to molecules. This increases W and makes S larger.

D4.2 Since $C_V = \dfrac{\Delta U}{\Delta T}$ at constant volume [2.11] and $C_p = \dfrac{\Delta H}{\Delta T}$ at constant pressure [2.16], we see that a high heat capacity implies that a lot of heat is required to produce a given change in temperature. We can see that this causes high entropies for substances that have high heat capacity by writing eqn 4.1 in the infinitesimal form

$$dS = \frac{dq_{rev}}{T}\ [4.1] = \frac{CdT}{T}.$$

Integration over the range of the temperature change gives

$$\Delta S = \int_{T_i}^{T_f} dS = \int_{T_i}^{T_f} \frac{dq_{rev}}{T} = \int_{T_i}^{T_f} \frac{CdT}{T}$$

which has a numerator that clearly implies that a high heat requirement (high heat capacity) for a given temperature change causes a high entropy change. In molecular terms, the reception of a large quantity of heat means that a greater number of molecular quantum states have been made available to the molecules than would be available for a substance of low heat capacity; W and S are larger by the Boltzmann equation $S = k \ln W$.

D4.3 A liquid and its vapor are in equilibrium at the boiling point. Since the process is reversible under this condition, $\Delta S = q_{rev}/T$ [4.1] $= \Delta H/T$ [2.15b, constant p] and the entropy of vaporization is given by $\Delta_{vap}S_m = \Delta_{vap}H_m/T_b$.

Trouton's rule states that the ratio $\Delta_{vap}H/T_b$, and thus the vaporization entropy, is a constant. Explore the origin of the constancy by considering that the vaporization entropy has two components, only one of which depends upon liquid phase properties. They are the molar entropy of the liquid and the molar entropy of the gas: $\Delta_{vap}S_m = S_m(g) - S_m(l)$. Under ordinary conditions the value of $S_m(g)$ is expected to be identical for all gases and relatively large with respect to $S_m(l)$ because gases behave as perfect gases for which molecular volume and intermolecular forces are negligibly small. This allows completely random, very high entropy molecular motion, which is independent of molecular properties. In addition to the large $S_m(g)$ value, should the value of $S_m(l)$ be either negligibly small or a constant value for a series of compounds, $\Delta_{vap}S_m$, and the ratio $\Delta_{vap}H/T_b$, will be a constant. Trouton's rule is followed. Small, non-polar molecules provide examples that meet these conditions. The relative absence of molecular order in their liquid states gives relatively small and constant $S_m(l)$ values. Thus, bromine, carbon tetrachloride, and cyclohexane have approximate identical vaporization entropies (~85 J K^{-1} mol^{-1}, Table 4.1).

Exceptions to Trouton's rule include liquids in which the interactions between molecules result in the liquid being less disordered than the random jumble of molecules in something like carbon tetrachloride. This includes liquids in which hydrogen bonding creates local order as in water and small alcohols. It also includes liquid metals in which the metallic bond creates atomic organization as in mercury.

It is also interesting to explore the origin of Trouton's rule with a careful analysis of the vaporization enthalpy (heat). Energy in the form of heat supplied to a liquid manifests itself as an increase in thermal motion. This is an increase in the kinetic energy of molecules. When the kinetic energy of the molecules is sufficient to overcome the attractive energy that holds them together the liquid vaporizes. The enthalpy of vaporization is the heat required to accomplish this at constant pressure. It seems reasonable that the greater the enthalpy of vaporization, the greater the kinetic energy required, and the greater the temperature needed to achieve this kinetic energy. Hence, we expect that $\Delta_{vap}H$ is proportional to T_b, which implies that their ratio is a constant.

D4.4 Ludwig Boltzmann's fundamental equation $S = k \ln W$, where W is the number of ways that molecules of the system can be arranged yet correspond to the same total energy, provides the statistical definition of entropy that associates molecular motion and molecular quantum states with the thermodynamic entropy. We justify its identification with the thermodynamic entropy $\Delta S = q_{rev}/T$ [4.1] by the finding that the entropy computed with the Boltzmann equation matches the entropy value found by thermodynamically based experiments. This includes the entropy values at the absolute zero of temperature, residual entropies, the variation of entropy with temperature and volume, and the entropy changes of phase transitions.

D4.5 The Gibbs energy, G, is the system property that is used to identify the direction of spontaneity under the conditions of constant temperature and pressure. Under these conditions: $\Delta G = -T\Delta S_{total}$ [4.17] and, since the **Second Law of Thermodynamics** summarizes evidence that spontaneity occurs when $\Delta S_{total} > 0$, we conclude that the direction for which $\Delta G < 0$ is the spontaneous direction.

D4.6 We must remember that the second law of thermodynamics states only that the total entropy of both the system (here, the molecules organizing themselves into cells) and the surroundings (here, the medium) must increase in a naturally occurring process. It does not state that entropy must increase in a portion of the universe that interacts with its surroundings. In this case, the cells grow by using chemical energy from their surroundings (the medium) and in the process the increase in the entropy of the medium outweighs the decrease in entropy of the system. Hence, the second law is not violated.

Solutions to exercises

E4.1

$$\Delta S_{\rm sur} = \frac{q_{\rm sur}}{T}\ [4.8] = \frac{120\ \rm J}{293\ \rm K} = \boxed{0.410\ \rm J\ K^{-1}}$$

E4.2 (a) We assume that the ice melts reversibly under the conditions described, therefore

$$\Delta S_{\rm ice} = \frac{q_{\rm rev}}{T}\ [4.1] = \frac{33\ \rm kJ}{273\ \rm K} = +0.12\ \rm kJ\ K^{-1}.$$

(b) $$\Delta S_{\rm sur} = \frac{q_{\rm sur}}{T}\ [4.8] = \frac{-33\ \rm kJ}{273\ \rm K} = \boxed{-0.12\ \rm kJ\ K^{-1}}$$

Note: Because this process is reversible, the total entropy change is zero.

E4.3

$$q = nC_{p,\rm m}\Delta T$$

$$q = \frac{1.00\times10^3\ \rm g}{26.98\ \rm g\ mol^{-1}}\times 24.35\ \rm J\ K^{-1}\ mol^{-1}\times(-50\ K)$$

$$= \boxed{-45.1\ \rm kJ}$$

$$\Delta S = C_p \ln\frac{T_{\rm f}}{T_{\rm i}}\ [4.3] = nC_{p,\rm m}\ln\frac{T_{\rm f}}{T_{\rm i}}$$

$$= \frac{1.00\times10^3\rm g}{26.98\ \rm g\ mol^{-1}}\times 24.35\ \rm J\ K^{-1}\ mol^{-1}\times\ln\frac{250\ K}{300\ K} = \boxed{-165\ \rm J\ K^{-1}}$$

E4.4 For the first step, melting 100 g ice:

$$\Delta_{\rm fus}S = \frac{\Delta_{\rm fus}H}{T_{\rm fus}} = \frac{6.01\ \rm kJ\ mol^{-1}}{273\ \rm K}\times 100\ \rm g\times\frac{1\ mol}{18.0\ g}$$

$$= \boxed{122\ \rm J\ K^{-1}}$$

For the second step, heating the water:

$$\Delta S = C_p \ln\frac{T_{\rm f}}{T_{\rm i}} = 4.18\ \rm J\ K^{-1}\ g^{-1}\times 100\ g\times\ln\frac{373}{273} = \boxed{130\ \rm J\ K^{-1}}$$

For the third step, vaporization:

$$\Delta_{\rm vap}S = \frac{\Delta_{\rm vap}H}{T_{\rm b}} = \frac{40.7\ \rm kJ\ mol^{-1}}{373\ \rm K}\times 100\ \rm g\times\frac{1\ mol}{18.0\ g} = \boxed{606\ \rm J\ K^{-1}}$$

$$\Delta S_{\rm total} = (122 + 130 + 606)\ \rm J\ K^{-1} = \boxed{858\ \rm J\ K^{-1}}$$

(a) A graph of temperature vs. time (Figure 4.1) shows a constant 273 K temperature until all the ice is melted. Temperature would increase until the boiling point, 373 K, is reached. Temperature again remains constant until all the liquid is vaporized.

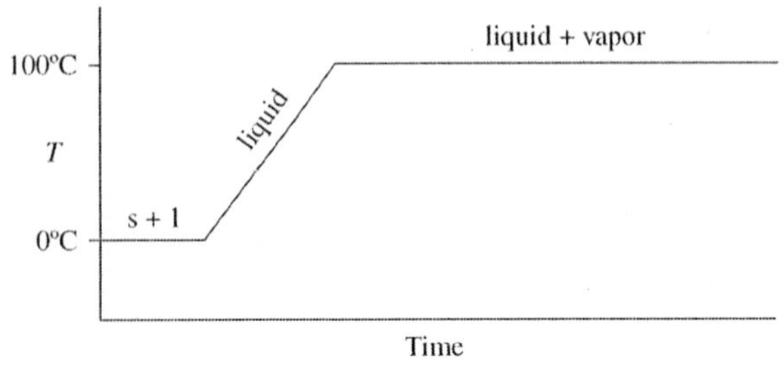

Figure 4.1

(b) A sketch of enthalpy as a function of time is shown in Figure 4.2. Note that absolute values of enthalpy are indeterminate.

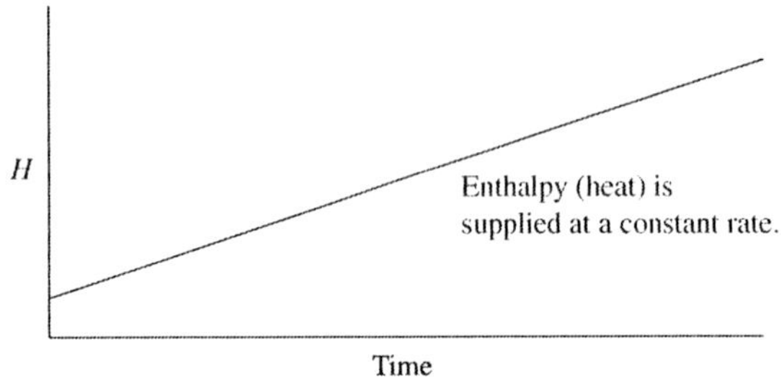

Figure 4.2

(c) A sketch of entropy as a function of time is shown in Figure 4.3.

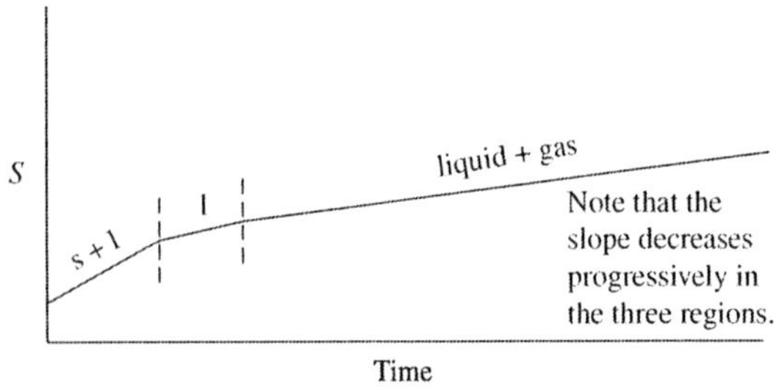

Figure 4.3

The graph of entropy against time does not look much different from that of enthalpy against time. The reason is that $\Delta t \propto \Delta H$. Therefore

$$\frac{\Delta S}{\Delta t} \propto \frac{\Delta S}{\Delta H}.$$

In the three regions, we see that this ratio is roughly a constant.

$$\frac{\Delta S}{\Delta H} = \frac{122\ \text{J K}^{-1}}{33.3\ \text{kJ}} = 3.66\ \text{J K}^{-1}\ \text{kJ}^{-1}\ (\text{s}+\text{l region})$$

$$\frac{\Delta S}{\Delta H} = \frac{130\ \text{J K}^{-1}}{41.8\ \text{kJ}} = 3.17\ \text{J K}^{-1}\ \text{kJ}^{-1}\ (\text{liquid region})$$

$$\frac{\Delta S}{\Delta H} = \frac{606\ \text{J K}^{-1}}{226\ \text{kJ}} = 2.68\ \text{J K}^{-1}\ \text{kJ}^{-1}\ (\text{liquid}+\text{gas region})$$

E4.5

$$\Delta S_{\text{m}} = R \ln \frac{V_{\text{f}}}{V_{\text{i}}}\ [4.2]\ (\text{Assume gas is perfect.})$$

$$= 8.3145\ \text{J K}^{-1}\ \text{mol}^{-1} \times \ln\left(\frac{5.5\ \text{dm}^3}{1.0\ \text{dm}^3}\right) = \boxed{+14\ \text{J K}^{-1}\ \text{mol}^{-1}}$$

E4.6

$$\Delta S = nR \ln\frac{V_f}{V_i} \text{ [4.2]} = -10.0 \text{ J K}^{-1}$$

Use the perfect gas law to calculate nR.

$$nR = \frac{p_i V_i}{T_i} = \frac{1.00 \text{ atm} \times 15.0 \text{ dm}^3}{250 \text{ K}}, \text{ converting units}$$

$$= \frac{1.013\times10^5 \text{ Pa}\times15.0\times10^{-3} \text{ m}^3}{250 \text{ K}} = 6.08 \text{ J K}^{-1}$$

$$\ln\frac{V_f}{V_i} = \frac{\Delta S}{nR} = \frac{-10.0 \text{ J K}^{-1}}{6.08 \text{ J K}^{-1}} = -1.64$$

$$\frac{V_f}{V_i} = e^{\Delta S/nR}$$

$$V_f = V_i e^{\Delta S/nR} = (15.0 \text{ dm}^3)\times e^{-1.64} = \boxed{2.91 \text{ dm}^3}$$

E4.7

$$\Delta S = nR\ln\frac{V_f}{V_i} \text{ [4.2]}$$

Because entropy is a state function, it does not matter whether the change in state occurs reversibly or irreversibly. Therefore, for (a) and (b)

Substitute $V = \frac{nRT}{p}$ into the expression for ΔS.

$$\Delta S = nR\ln\frac{p_i}{p_f}$$

$$= \frac{15 \text{ g}}{16.04 \text{ g mol}^{-1}}\times8.3145 \text{ J K}^{-1} \text{ mol}^{-1}\times \ln\frac{105}{1.50} = \boxed{33 \text{ J K}^{-1}}$$

E4.8

$$\Delta S = C_p \ln\frac{T_f}{T_i} = nC_{p,m} \ln\frac{T_f}{T_i} \text{ [4.3]}$$

$$= \frac{100 \text{ g}}{18.02 \text{ g mol}^{-1}}\times75.5 \text{ J K}^{-1} \text{ mol}^{-1}\times\ln\left(\frac{310 \text{ K}}{293 \text{ K}}\right) = \boxed{23.6 \text{ J K}^{-1}}$$

E4.9

$$\Delta S = C_p \ln\frac{T_f}{T_i} \text{ [4.3]} = nC_{p,m} \ln\frac{T_f}{T_i}$$

$$= \frac{1.00 \text{ kg}}{0.2072 \text{ kg mol}^{-1}}\times26.44 \text{ J K}^{-1} \text{ mol}^{-1}\times\ln\frac{373 \text{ K}}{773 \text{ K}} = \boxed{-93.0 \text{ J K}^{-1}}$$

E4.10 We begin by using eqn 4.4 to account for the variation of the heat capacity with temperature. From data tables we find that

$C_{p,m}/(\text{J K}^{-1} \text{ mol}^{-1}) = a + bT + c/T^2$ where $a = 22.13$, $b = 11.72\times10^{-3} \text{ K}^{-1}$, $c = 0.96\times10^5 \text{ K}^2$.

Substitute this into eqn 4.4 and perform the integrations.

$$\Delta S_m = \int_{T_i}^{T_f} \frac{C_m dT}{T} \text{ [4.4]} = \int_{T_i}^{T_f} \frac{C_{p,m} dT}{T} = \int_{T_i}^{T_f} \frac{(a + bT + c/T^2)dT}{T}$$

$$= a\int_{T_i}^{T_f} \frac{dT}{T} + b\int_{T_i}^{T_f} dT + c\int_{T_i}^{T_f} \frac{dT}{T^3}$$

$$= a\ln T\Big]_{T=T_i}^{T=T_f} + bT\Big]_{T=T_i}^{T=T_f} - \frac{c}{2}\frac{1}{T^2}\Bigg]_{T=T_i}^{T=T_f} \quad \text{(standard integrals)}$$

$$= a\ln\left(\frac{T_f}{T_i}\right) + b(T_f - T_i) - \frac{c}{2}\left(\frac{1}{T_f^2} - \frac{1}{T_i^2}\right)$$

$$\Delta S_{\mathrm{m}}/(\mathrm{J\,K^{-1}\,mol^{-1}}) = (22.13)\times\ln\left(\frac{373}{773}\right) + (11.72\times10^{-3})\times(-400) - \left(\frac{0.96\times10^{5}}{2}\right)\times\left(\frac{1}{373^{2}} - \frac{1}{773^{2}}\right)$$
$$= -21.1$$

$$\Delta S = n\Delta S_{\mathrm{m}} = \frac{1.00\ \mathrm{kg}}{0.2072\ \mathrm{kg\,mol^{-1}}}\times(-21.1\ \mathrm{J\,K^{-1}\,mol^{-1}}) = \boxed{-101.\bar{8}\ \mathrm{J\,K^{-1}}}$$

The percentage error caused by ignoring the temperature variation of heat capacity (exercise 4.9) is:

$$\left(\frac{101.\bar{8}\ \mathrm{J\,K^{-1}} - 93.0\ \mathrm{J\,K^{-1}}}{101.\bar{8}\ \mathrm{J\,K^{-1}}}\right)100\% = \boxed{8.64\%\ \text{high}}.$$

E4.11 Because entropy changes depend only on the initial and final states, it does not matter if the change is accomplished in one or more than one step. Therefore, calculate the change in two steps.

ΔS_{m} for compression:

$$\Delta S_{\mathrm{m}} = R\ln\frac{V_{\mathrm{f}}}{V_{\mathrm{i}}}\ [4.2]$$
$$= 8.3145\ \mathrm{J\,K^{-1}\,mol^{-1}}\times\ln\left(\frac{0.500\ \mathrm{dm^{3}}}{2.0\ \mathrm{dm^{3}}}\right) = -11.5\ \mathrm{J\,K^{-1}\,mol^{-1}}.$$

ΔS_{m} for heating:

$$\Delta S_{\mathrm{m}} = C_{V,\mathrm{m}}\ln\frac{T_{\mathrm{f}}}{T_{\mathrm{i}}}\ [4.3],\quad C_{V,\mathrm{m}} = \frac{3}{2}R\ \ln\frac{T_{\mathrm{f}}}{T_{\mathrm{i}}}$$
$$= \frac{3}{2}\times8.3145\ \mathrm{J\,K^{-1}\,mol^{-1}}\times\ln\left(\frac{400\ \mathrm{K}}{300\ \mathrm{K}}\right) = +3.59\ \mathrm{J\,K^{-1}\,mol^{-1}}.$$

$$\Delta S_{\mathrm{total}} = (-11.5 + 3.59)\ \mathrm{J\,K^{-1}\,mol^{-1}} = \boxed{-7.9\ \mathrm{J\,K^{-1}\,mol^{-1}}}$$

E4.12 $V_{\mathrm{f}} = 2\,V_{\mathrm{i}}$

$$\Delta S = nR\ln\frac{V_{\mathrm{f}}}{V_{\mathrm{i}}} = nR\ln 2,\ \text{for the first step (isothermal expansion)}$$

$$\Delta S = nC_{V,\mathrm{m}}\ln\frac{T_{\mathrm{f}}}{T_{\mathrm{i}}}\ \text{for the second step (cooling)}$$

ΔS for the second step is the negative of ΔS for the first step because entropy is a state property. Therefore

$$nR\ln 2 = -nC_{V,\mathrm{m}}\ln\left(\frac{T_{\mathrm{f}}}{T_{\mathrm{i}}}\right) = -n\times(\tfrac{3}{2}R)\ln\left(\frac{T_{\mathrm{f}}}{T_{\mathrm{i}}}\right)$$

$$\ln\left(\frac{T_{\mathrm{f}}}{T_{\mathrm{i}}}\right) = -\tfrac{2}{3}\ln 2 = -0.4621 \quad\text{and}\quad T_{\mathrm{f}} = T_{\mathrm{i}}\mathrm{e}^{-0.4621} = \boxed{0.6300\,T_{\mathrm{i}}}.$$

E4.13 Entropy changes occur in steps 1 and 3 and are the negatives of each other. Temperature changes in steps 2 and 4 are the negatives of each other. See Figure 4.4.

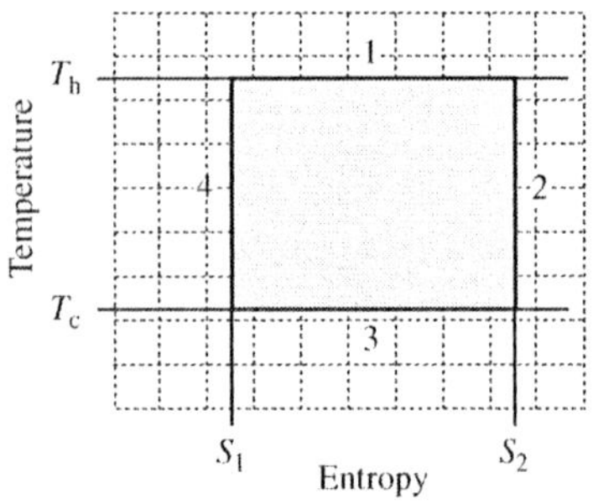

Figure 4.4

Step 1: $\Delta S_1 = nR\ln\dfrac{V_2}{V_1} = +,\ \Delta T = 0$

Step 3: $\Delta S_3 = nR\ln\dfrac{V_1}{V_2} = -,\ \Delta T = 0$

Step 2: $\Delta S = 0,\ \Delta T_2 = -$

Step 4: $\Delta S = 0,\ \Delta T_4 = +$

$$\Delta S_1 = -\Delta S_3,\ \Delta T_2 = -\Delta T_4$$

E4.14 We use the formula of Derivation 4.3.

$$S_m(T) - S_m(0) = \tfrac{1}{3}C_{V,m}(T)\ [4.11b] = \tfrac{1}{3}\times 1.2\times10^{-3}\ \mathrm{J\ K^{-1}\ mol^{-1}} = 4.0\times10^{-4}\ \mathrm{J\ K^{-1}\ mol^{-1}}$$

As $S_m(0)$ for the pure crystalline substance KCl is expected to be zero,

$$S_m(T) = \boxed{4.0\times10^{-4}\ \mathrm{J\ K^{-1}\ mol^{-1}}}.$$

E4.15 First find the common final temperature, T_f, by noting that the heat lost by the hot sample is gained by the cold sample.

$$q_{\text{cold sample 1}} = -q_{\text{warm sample 2}}$$

$$n_1C_{p,m}(T_f - T_{i1}) = -n_2C_{p,m}(T_f - T_{i2})$$

Solving for T_f

$$T_f = \frac{n_1T_{i1} + n_2T_{i2}}{n_1 + n_2}.$$

Because $n_1 = n_2 = \dfrac{100\ \mathrm{g}}{18.02\ \mathrm{g\ mol^{-1}}} = 5.55\ \mathrm{mol}$,

$$T_f = \tfrac{1}{2}(353\ \mathrm{K} + 283\ \mathrm{K}) = 318\ \mathrm{K}.$$

The total entropy change is therefore

$$\Delta S_{\text{total}} = \Delta S_1 + \Delta S_2 = n_1C_{p,m}\ln\frac{T_f}{T_{i1}} + n_2C_{p,m}\ln\frac{T_f}{T_{i2}} = n_1C_{p,m}\ln\left\{\left(\frac{T_f}{T_{i1}}\right)\times\left(\frac{T_f}{T_{i2}}\right)\right\}$$

$$= (5.55\ \mathrm{mol})\times(75.5\ \mathrm{J\ K^{-1}\ mol^{-1}})\times\ln\left\{\frac{318}{353}\times\frac{318}{283}\right\} = \boxed{5.11\ \mathrm{J\ K^{-1}}}.$$

E4.16 In a manner similar to equations 4.6 and 4.7, we may write in general,

$$\Delta_{trs}S = \frac{\Delta_{trs}H}{T_{trs}}; \qquad \text{therefore } \Delta_{trs}S = \frac{+1.9\ \mathrm{kJ\ mol^{-1}}}{2000\ \mathrm{K}} = \boxed{0.95\ \mathrm{J\ K^{-1}\ mol^{-1}}}.$$

E4.17 (a) $\Delta_{vap}S = \dfrac{\Delta_{vap}H}{T_b}$ [4.7]

$$= \frac{29.4\times10^3\ \mathrm{J\ mol^{-1}}}{334.88\ \mathrm{K}} = \boxed{+87.8\ \mathrm{J\ K^{-1}\ mol^{-1}}}$$

(b) Because the vaporization process can be accomplished reversibly, $\Delta S_{\text{total}} = 0$; hence $\Delta S_{\text{sur}} = \boxed{-87.8\ \mathrm{J\ K^{-1}\ mol^{-1}}}$.

E4.18 $\Delta_{\text{fus}}C_p^{\ominus} = C_{p,\text{m}}^{\ominus}(\text{l}) - C_{p,\text{m}}^{\ominus}(\text{s})$ [3.7]

$$= (33-17)\ \text{J K}^{-1}\ \text{mol}^{-1} = 16\ \text{J K}^{-1}\ \text{mol}^{-1}$$

$$\Delta_{\text{fus}}S^{\ominus}(T') = \Delta_{\text{fus}}S^{\ominus}(T) + \int_T^{T'} \frac{\Delta_{\text{fus}}C_p^{\ominus}}{T}\,\text{d}T \quad \text{[phase transition analog of eqn 4.4]}$$

$$= \Delta_{\text{fus}}S^{\ominus}(T) + \Delta_{\text{fus}}C_p^{\ominus}\int_T^{T'} \frac{1}{T}\,\text{d}T = \Delta_{\text{fus}}S^{\ominus}(T) + \Delta_{\text{fus}}C_p^{\ominus}\ln(T)\Big|_T^{T'}$$

$$= \Delta_{\text{fus}}S^{\ominus}(T) + \Delta_{\text{fus}}C_p^{\ominus}\ln\left(\frac{T'}{T}\right)$$

At $T = T_\text{f}$ the transition is reversible and $\Delta_{\text{fus}}S^{\ominus}(T_\text{f}) = \dfrac{\Delta_{\text{fus}}H^{\ominus}(T_\text{f})}{T_\text{f}}$ [4.6].

$$\Delta_{\text{fus}}S^{\ominus}(T') = \frac{\Delta_{\text{fus}}H^{\ominus}(T_\text{f})}{T_\text{f}} + \Delta_{\text{fus}}C_p^{\ominus}\ln\left(\frac{T'}{T_\text{f}}\right)$$

$$\Delta_{\text{fus}}S^{\ominus}(298\ \text{K}) = \frac{36\ \text{kJ mol}^{-1}}{424\ \text{K}} + (16\ \text{J K}^{-1}\ \text{mol}^{-1}) \times \ln\left(\frac{298\ \text{K}}{424\ \text{K}}\right)$$

$$= \boxed{79\ \text{J K}^{-1}\ \text{mol}^{-1}}$$

E4.19 (a) According to Trouton's rule, which applies well to hydrocarbons such as octane,

$\Delta_{\text{vap}}S = \boxed{+85\ \text{J K}^{-1}\ \text{mol}^{-1}}$.

(b) $\Delta_{\text{vap}}S = \dfrac{\Delta_{\text{vap}}H}{T_\text{b}} = +85\ \text{J K}^{-1}\ \text{mol}^{-1}$

$\Delta_{\text{vap}}H = \Delta_{\text{vap}}S \times T_\text{b} = +85\ \text{J K}^{-1}\text{mol}^{-1} \times 399\ \text{K} = \boxed{+34\ \text{kJ K}^{-1}\ \text{mol}^{-1}}$

E4.20 $S = k\ln W$ [4.12] where $W = cV^N$ (c is a constant for an isothermal system)

$$\Delta S = S(V_\text{f}) - S(V_\text{i}) = k\ln W_\text{f} - k\ln W_\text{i} = k\ln\left(\frac{W_\text{f}}{W_\text{i}}\right) = k\ln\left(\frac{cV_\text{f}^N}{cV_\text{i}^N}\right) = k\ln\left(\frac{V_\text{f}}{V_\text{i}}\right)^N = \boxed{kN\ln\left(\frac{V_\text{f}}{V_\text{i}}\right)}$$

When $N = N_\text{A}$, we replace kN with R and the equation becomes $\Delta S_\text{m} = R\ln(V_\text{f}/V_\text{i})$ in agreement with eqn 4.2.

E4.21 $S_{\text{m,residual}} = k\ln W_{\text{residual}}$ $[4.12] = k\ln 4^{N_\text{A}} = kN_\text{A}\ln 4 = R\ln 4$

$$= (8.3145\ \text{J K}^{-1}\ \text{mol}^{-1})\ln 4 = \boxed{11.5\ \text{J K}^{-1}\ \text{mol}^{-1}}$$

E4.22 (a) $\boxed{\text{positive}}$, due to greater disorder in the product, though the difference may not be large

(b) $\boxed{\text{negative}}$, less disorder (smaller number of moles of gas) in the product

(c) $\boxed{\text{positive}}$, two new substances are formed, resulting in greater disorder on the product side

E4.23 (a) $\Delta_\text{r}S^{\ominus} = 2\,S_\text{m}^{\ominus}(CH_3COOH,\ \text{l}) - 2\,S_\text{m}(CH_3CHO,\ \text{g}) - S_\text{m}^{\ominus}(O_2,\ \text{g})$

$$= [(2\times159.8) - (2\times250.3) - 205.14]\ \text{J K}^{-1}\ \text{mol}^{-1}$$

$$= \boxed{-386.1\ \text{J K}^{-1}\ \text{mol}^{-1}}$$

(b) $\Delta_r S^\ominus = 2\,S_m^\ominus(\text{AgBr, s}) + S_m^\ominus(\text{Cl}_2\text{, g}) - 2\,S_m^\ominus(\text{AgCl, s}) - S_m^\ominus(\text{Br}_2\text{, l})$

$= [(2\times107.1) + 223.07 - (2\times96.2) - 152.23]\ \text{J K}^{-1}\ \text{mol}^{-1}$

$= \boxed{+92.6\ \text{J K}^{-1}\ \text{mol}^{-1}}$

(c) $\Delta_r S^\ominus = S_m^\ominus(\text{HgCl}_2\text{, s}) - S_m^\ominus(\text{Hg, l}) - S_m^\ominus(\text{Cl}_2\text{, g})$

$= (146.0 - 76.02 - 223.07)\ \text{J K}^{-1}\ \text{mol}^{-1}$

$= \boxed{-153.1\ \text{J K}^{-1}\ \text{mol}^{-1}}$

(d) $\Delta_r S^\ominus = S_m^\ominus(\text{Zn}^{2+}\text{, aq}) + S_m^\ominus(\text{Cu, s}) - S_m^\ominus(\text{Zn, s}) - S_m^\ominus(\text{Cu}^{2+}\text{, aq})$

$= (-112.1 + 33.15 - 41.63 + 99.6)\ \text{J K}^{-1}\ \text{mol}^{-1}$

$= \boxed{-21.0\ \text{J K}^{-1}\ \text{mol}^{-1}}$

(e) $\Delta_r S^\ominus = 12\,S_m^\ominus(\text{CO}_2\text{, g}) + 11\,S_m^\ominus(\text{H}_2\text{O, l}) - S_m^\ominus(\text{C}_{12}\text{H}_{22}\text{O}_{11}\text{, s}) - 12\,S_m^\ominus(\text{O}_2\text{, g})$

$= [(12\times213.74) + (11\times69.91) - 360.2 - (12\times205.14)]\ \text{J K}^{-1}\ \text{mol}^{-1}$

$= \boxed{+512.0\ \text{J K}^{-1}\ \text{mol}^{-1}}$

E4.24 $\Delta_c H^\ominus = -2808\ \text{kJ mol}^{-1} = -q_{\text{body}}$

$$\Delta S_{\text{body}}^\ominus = \frac{q_{\text{body}}}{T} = \frac{2808\ \text{kJ mol}^{-1}\times100\ \text{g}\times\dfrac{1\ \text{mol}}{180\ \text{g}}}{273\ \text{K} + 37\ \text{K}} = \boxed{5.03\ \text{kJ K}^{-1}}$$

Note: The above calculation uses the value of $\Delta_c H^\ominus$ at 25°C. The value at 37°C should not be much different and can be calculated from knowledge of the heat capacities of all of the substances involved in the reaction.

E4.25 $N_2(g) + 3\,H_2(g) \rightarrow 2\,NH_3(g)$

$\Delta_r S^\ominus = 2\,S_m^\ominus(\text{NH}_3\text{, g}) - S_m^\ominus(\text{N}_2\text{, g}) - 3\,S_m^\ominus(\text{H}_2\text{, g})$

$= (2\times192.45 - 191.61 - 3\times130.684)\ \text{J K}^{-1}$

$= \boxed{-198.72\ \text{J K}^{-1}}$

$\Delta_r H^\ominus = 2\,\Delta_f H^\ominus(\text{NH}_3\text{, g}) = 2\times(-46.11)\ \text{kJ} = -92.22\ \text{kJ}$

$$\Delta S_{\text{sur}} = \frac{q_{\text{sur}}}{T} = \frac{-\Delta_r H^\ominus}{T} = \frac{92.22\times10^3\ \text{J}}{298\ \text{K}} = \boxed{309\ \text{J K}^{-1}}$$

E4.26 $\Delta S = nC_{p,\text{m}} \ln(T_f / T_i)$ for each substance in each reaction, therefore

$\Delta_r S = \Delta_r C_p \ln(T_f / T_i)$ for each reaction.

(a) $\Delta_r C_p = (2\ \text{mol}\times4R) - (3\ \text{mol}\times\tfrac{7}{2}R) = -\tfrac{5}{2}R\ \text{mol}$

$$\Delta_r S = -\tfrac{5}{2}\,R\ \text{mol}\times\ln\left(\frac{283\ \text{K}}{273\ \text{K}}\right) = \boxed{-0.75\ \text{J K}^{-1}}$$

(b) $\Delta_r C_p = (2\text{ mol}\times 4R)+(1\text{ mol}\times \tfrac{7}{2}R)-(1\text{ mol}\times 4R)-(2\text{ mol}\times \tfrac{7}{2}R) = +\tfrac{1}{2}R\text{ mol}$

$$\Delta_r S = +\tfrac{1}{2}R\text{ mol}\times\ln\left(\frac{283\text{ K}}{273\text{ K}}\right) = \boxed{+0.15\text{ J K}^{-1}}$$

E4.27

$$\Delta_r G^{\ominus} = \Delta_r H^{\ominus} - T\Delta_r S^{\ominus}\ [4.16]$$
$$= -92.22\text{ kJ} - (298\text{ K})\times(-0.19876\text{ kJ})$$
$$= \boxed{-32.99\text{ kJ}}$$

E4.28

(a) $\Delta G = \Delta H - T\Delta S\ [4.16]$

$$= -135\text{ kJ mol}^{-1} - 310\text{ K}\times(-136\text{ J K}^{-1}\text{ mol}^{-1}) = \boxed{-93\text{ kJ mol}^{-1}}$$

(b) Yes, ΔG is negative.

(c) $\Delta G = -T\Delta S_{\text{total}}\ [4.17]$

$$\Delta S_{\text{total}} = -\frac{\Delta G}{T} = -\left(\frac{-93\text{ kJ mol}^{-1}}{310\text{ K}}\right) = \boxed{+0.30\text{ kJ K}^{-1}\text{ mol}^{-1}}$$

E4.29 $\Delta G = w'_{\max} = -2828\text{ kJ mol}^{-1}$, so the maximum work that can be done is 2828 kJ mol^{-1}. We will assume that we will be able to extract the maximum work from the reaction.

$$w = mgh = 65\text{ kg}\times 9.81\text{ m s}^{-2}\times 10\text{ m}$$
$$= 6.4\times 10^{3}\text{ J} = 6.4\text{ kJ}$$
$$\text{amount}(n) = \frac{6.4\text{ kJ}}{2828\text{ kJ mol}^{-1}} = 2.3\times 10^{-3}\text{ mol}$$
$$\text{mass of glucose} = 2.3\times 10^{-3}\text{ mol}\times 180\text{ g mol}^{-1} = \boxed{0.41\text{ g}}$$

E4.30

$$C_{12}H_{22}O_{11}(s) + 12\,O_2(g) \rightarrow 12\,CO_2(g) + 11\,H_2O(l)$$

$$\Delta_c G^{\ominus} = 12\,\Delta_f G^{\ominus}(CO_2, g) + 11\,\Delta_f G^{\ominus}(H_2O, l) - \Delta_f G^{\ominus}(C_{12}H_{22}O_{11}, s) - 12\,\Delta_f G^{\ominus}(O_2, g)$$
$$= [12\times(-394.36) + 11\times(-237.13) - (-1543) - 12\times(0)]\text{ kJ mol}^{-1}$$
$$= -5798\text{ kJ mol}^{-1}$$

$\Delta_r G$ gives the maximum non-expansion work, w', that can be extracted from the reaction at constant temperature and pressure.

$$w'_{\max,m} = -\Delta_c G^{\ominus}\ [4.18] = 5798\text{ kJ mol}^{-1}$$

$$w'_{\max} = nw'_{\max,m} = \left(\frac{1.0\times 10^{-3}\text{ g}}{342.30\text{ g mol}^{-1}}\right)\times(5798\text{ kJ mol}^{-1}) = \boxed{17\text{ J}}$$

E4.31 (a) Yes, coupling the two reactions can give a net ΔG that is negative, hence the overall process is spontaneous. For example, for one mole of glutamate and one mole of ATP,

$$\Delta G = (14.2 - 31)\text{ kJ mol}^{-1} = -17\text{ kJ mol}^{-1}.$$

(b) The minimum amount of ATP required is $\dfrac{1\text{ mol}\times(-14.2\text{ kJ mol}^{-1})}{-31\text{ kJ mol}^{-1}} = \boxed{0.46\text{ mol ATP}}$

E4.32 For the synthesis, $\Delta G = +42\text{ kJ mol}^{-1}$; hence at least -42 kJ would need to be provided by the ATP in order to make ΔG overall negative.

$$\text{amount}(n)\text{ of ATP} = \frac{-42\text{ kJ}}{-31\text{ kJ mol}^{-1}} = 1.35\text{ mol ATP}$$

$$1.35\text{ mol}\times 6.02\times 10^{23}\text{ mol}^{-1} = \boxed{8.1\times 10^{23}\text{ molecules of ATP}}$$

E4.33

$$n(\text{ATP}) = \frac{10^6}{6.02\times10^{23}\ \text{mol}^{-1}} = 1.7\times10^{-18}\ \text{mol}$$

$$\Delta G = 1.7\times10^{-18}\ \text{mol s}^{-1}\times(-31\ \text{kJ mol}^{-1}) = -5.3\times10^{-17}\ \text{kJ s}^{-1} = -5.3\times10^{-14}\ \text{J s}^{-1}$$

$$\text{Power density of cell} = \frac{\Delta G\ \text{of cell per second}}{\text{volume of cell}}$$

$$V_{\text{cell}} = \tfrac{4}{3}\pi r^3 = \tfrac{4}{3}\pi(10\times10^{-6}\ \text{m})^3 = 4.2\times10^{-15}\ \text{m}^3$$

$$\text{Power density of cell} = \frac{5.3\times10^{-14}\ \text{J s}^{-1}}{4.2\times10^{-15}\ \text{m}^3} = \boxed{13\ \text{W m}^{-3}}$$

$$\text{Power density of battery} = \frac{15\ \text{W}}{100\ \text{cm}^3\times10^{-6}\ \text{m}^3/\text{cm}^3} = \boxed{150\ \text{kW m}^{-3}}$$

The battery has the greater power density.

Solutions to projects

P4.34

$$\Delta S = \int_{T_i}^{T_f}\frac{C}{T}\,dT\ [4.4] = \int_{T_i}^{T_f}\left(\frac{a+bT+\dfrac{c}{T^2}}{T}\right)dT$$

$$= a\int_{T_i}^{T_f}\frac{1}{T}\,dT + b\int_{T_i}^{T_f}dT + c\int_{T_i}^{T_f}\frac{1}{T^3}\,dT = a\ln(T)\Big|_{T_i}^{T_f} + bT\Big|_{T_i}^{T_f} - \frac{c}{2T^2}\Big|_{T_i}^{T_f}$$

$$\Delta S = a\ln\left(\frac{T_f}{T_i}\right) + b(T_f - T_i) - \frac{c}{2}\left(\frac{1}{T_f^2} - \frac{1}{T_i^2}\right)$$

P4.35 (a) Let $|q|$ be the heat extracted from the refrigerator at T_{cold} and $|q'|$ be the heat delivered to environment at T_{hot}. The work needed to accomplish this is $w = |q'| - |q|$. The heat transfer occurs most efficiently when

$$\Delta S_{\text{total}} = \frac{|q'|}{T_{\text{hot}}} - \frac{|q|}{T_{\text{cold}}} = 0 \quad\text{or}\quad \frac{|q|}{|q'|} = \left(\frac{T_{\text{cold}}}{T_{\text{hot}}}\right).$$

The best coefficient of cooling performance is

$$c_{\text{cool}} = \frac{|q|}{w} = \frac{|q|}{|q'|-|q|} = \frac{(|q|/|q'|)}{1-(|q|/|q'|)} = \frac{(T_{\text{cold}}/T_{\text{hot}})}{1-(T_{\text{cold}}/T_{\text{hot}})} = \boxed{\frac{T_{\text{cold}}}{T_{\text{hot}}-T_{\text{cold}}}}.$$

Example calculation:

Since $|q| = c_{\text{cool}}w$, the rate of extracting heat equals c_{cool} multiplied by the refrigerator power rating.

$$\text{Rate of heat extraction} = \left(\frac{T_{\text{cold}}}{T_{\text{hot}}-T_{\text{cold}}}\right)\times\text{power rating} = \left(\frac{278\ \text{K}}{295\ \text{K}-278\ \text{K}}\right)\times 200\ \text{W} = \boxed{3.27\ \text{kW}}$$

(b) Let $|q|$ be the heat extracted by the heat pump at T_{cold} and $|q'|$ be the heat delivered at T_{hot}. The work needed to accomplish this is $w = |q'| - |q|$. The heat transfer occurs most efficiently when

$$\Delta S_{\text{total}} = \frac{|q'|}{T_{\text{hot}}} - \frac{|q|}{T_{\text{cold}}} = 0 \quad\text{or}\quad \frac{|q|}{|q'|} = \left(\frac{T_{\text{cold}}}{T_{\text{hot}}}\right).$$

The best coefficient of heating performance is

$$c_{\text{warm}} = \frac{|q'|}{w} = \frac{|q'|}{|q'|-|q|} = \frac{1}{1-(|q|/|q'|)} = \frac{1}{1-(T_{\text{cold}}/T_{\text{hot}})} = \boxed{\frac{T_{\text{hot}}}{T_{\text{hot}}-T_{\text{cold}}}}.$$

Example calculation:

Since $|q'| = c_{\text{warm}}w$, the rate of heat delivery equals c_{warm} multiplied by the heat pump power rating.

$$\text{Rate of heat delivery} = \left(\frac{T_{\text{hot}}}{T_{\text{hot}}-T_{\text{cold}}}\right)\times\text{power rating} = \left(\frac{295\text{ K}}{295\text{ K}-291\text{ K}}\right)\times 2.5\text{ kW} = \boxed{184\text{ kW}}$$

Practical heat pumps are not reversible so the heat gain is less.

5 Physical equilibria: pure substances

Answers to discussion questions

D5.1 For a one-component system the **chemical potential**, μ (mu), is equivalent to the molar Gibbs energy, G_m: $\mu = G_m = H_m - TS_m$. The chemical potential is the difference between the total stored energy and the energy stored randomly. This means that it is available energy for doing non-expansion work, or, in an equivalent view, it is the energy stored in the orderly motion and arrangement of the molecules of the system. It is a very important physical property because the phase of a one-component system that has the lowest chemical potential is the stable phase at a given temperature and pressure and the chemical potential decreases in a spontaneous change at constant temperature and pressure (Section 4.13). Furthermore, phase 1 and phase 2 of a pure substance are in equilibrium when $\mu_1 = \mu_2$.

(a) Eqn 5.2 tells us that $d\mu = V_m dp - S_m dT$ so that at constant p (i.e., $dp = 0$) we see that $d\mu = -S_m dT$. Our interpretation is that as temperature increases, the chemical potential varies as $-S_m$. The chemical potential varies with temperature because of the entropy of the system. Since the entropy is always positive, the chemical potential decreases with increasing temperature.

(b) At constant temperature (i.e., $dT = 0$) the expression $d\mu = V_m dp - S_m dT$ reduces to $d\mu = V_m dp$. Our interpretation is that as pressure increases, the chemical potential varies as V_m. The chemical potential varies with pressure because of the molar volume of the system. Since the molar volume is always positive, the chemical potential increases with increasing pressure.

D5.2 Consider two phases of a system, labeled α and β. The phase with the lower molar Gibbs energy under the given set of conditions is the more stable phase. First, consider the variation of the molar Gibbs energy of each phase with temperature at a fixed pressure by comparing the equation [5.4] expressions:

$$\frac{\Delta G_\alpha}{\Delta T} = -S_\alpha \quad \text{and} \quad \frac{\Delta G_\beta}{\Delta T} = -S_\beta \text{ at constant } p.$$

They clearly show that, if S_β is larger in magnitude than S_α, then ΔG_β decreases to a greater extent than ΔG_α as temperature increases. β phase becomes the more stable phase at higher temperature.

Second, consider the variation of the molar Gibbs energy of each phase with pressure at a fixed temperature by comparing the equation [5.1] expressions:

$$\frac{\Delta G_\alpha}{\Delta p} = V_\alpha \quad \text{and} \quad \frac{\Delta G_\beta}{\Delta p} = V_\beta \text{ at constant } T$$

These equations clearly show that, if V_β is larger in magnitude than V_α, then ΔG_β increases to a greater extent than ΔG_α as pressure increases. β phase becomes the unstable phase at higher pressure; α phase becomes the stable phase.

D5.3 (a) Attractive interactions tend to decrease the pressure of a gas relative to its perfect value for the same volume. We may qualitatively use equation 5.1 to decide that the molar Gibbs energy will be lowered relative to its "perfect" value.

(b) Repulsive interactions have the opposite effect on the pressure of a gas, so we may qualitatively decide that they will raise the molar Gibbs energy relative to its "perfect" value.

D5.4 The Clapeyron equation is exact and applies rigorously to all first-order phase transitions. It shows how pressure and temperature vary with respect to each other (temperature or pressure) along the phase boundary line, and in that sense, it defines the phase boundary line.

The Clausius–Clapeyron equation serves the same purpose, but it is not exact; its derivation involves approximations, in particular the assumptions that the perfect gas law holds and that the volume of condensed phases can be neglected in comparison to the volume of the gaseous phase. It applies only to phase transitions between the gaseous state and condensed phases.

D5.5 $C = 1$ for the sulfur phase diagram and the phase rule is $F = C - P + 2 = 3 - P$.

See Figure 5.1. In the areas (off any line segment) labeled Solid 1, Solid 2, Liquid, and Gas there is one phase and $F = 2$. Both T and p may be independently varied within these regions.

Curve segments A and B are sublimation curves between crystal form 1 of sulfur and the gas phase, and crystal form 2 of sulfur and the gas phase, respectively. On these equilibrium curves $P = 2$ and $F = 1$. There is only one independent variable, which means that once either T or P is set the other variable has a unique value determined by the equilibrium criteria. For identical reasons curve segments C, D, E, and F also have only one independent variable. Curves C and D are fusion curves in which the liquid is in equilibrium with either crystal form 2 or crystal form 1, respectively. Curve E is the vapor pressure curve. Curve F represents the points at which the two crystal forms are in equilibrium.

Point I, II, and III are triple points at which $P = 3$ and $F = 0$. There is no independent variable at these points because triple points are fixed by equilibrium criteria.

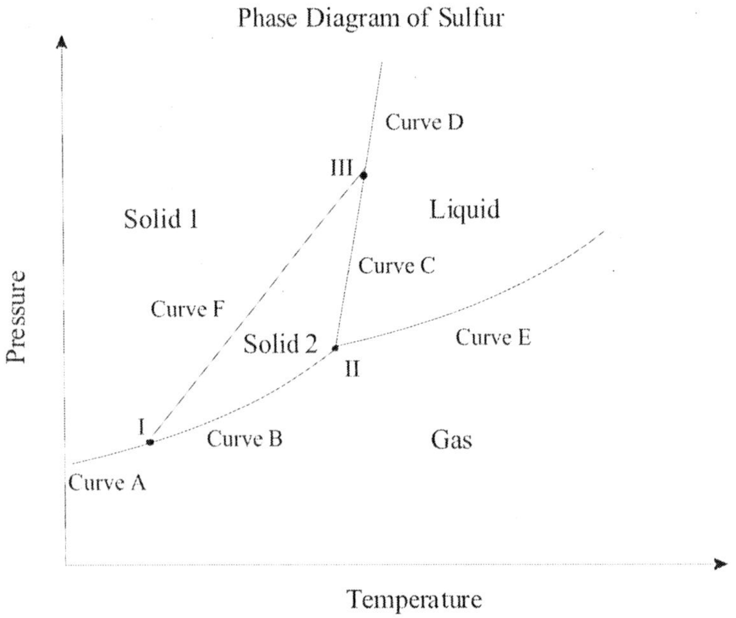

Figure 5.1

D5.6 A discussion about the '**structure of a liquid**' centers upon a molecule (or atom, or ion) in the liquid bulk and contrasts the positional arrangements of neighboring molecules at short and long intermolecular distances. Molecules are in continual, random motion but on the average there is considerable **short-range order** in which molecules are positioned with some degree of crystalline structure even though the molecules can slip past one another. The order is due largely to intermolecular forces that exert their influence over short distances. As in crystals, specific intermolecular distances are preferred. However, in contrast to crystals there is no **long-range order**. Intermolecular forces in liquids are not strong enough at large distances to overcome the randomizing effect of vigorous molecular motions and one intermolecular distance is as probable as another.

Solutions to exercises

E5.1 The substance with the lower molar Gibbs energy is the more stable; therefore, $\boxed{\text{rhombic sulfur}}$ is the more stable.

E5.2 No, the application of pressure tends to favor the substance with the smaller molar volume (higher density; see Discussion Question 5.1). Therefore, rhombic sulfur becomes even more stable relative to monoclinic sulfur as the pressure increases. This is easily seen in eqn 5.1: $\Delta G_m = V_m\,\Delta p$.

ΔG is less positive with an increase in pressure for rhombic sulfur than for monoclinic sulfur, so relative to monoclinic sulfur, the Gibbs energy of rhombic sulfur becomes more negative.

E5.3 (a) We assume that the molar volume of water is approximately constant with respect to variation in pressure. Then

$$V_m = \frac{18.02\ \text{g mol}^{-1}}{1.03\ \text{g cm}^{-3}} = 17.5\ \text{cm}^3\ \text{mol}^{-1}$$
$$= 1.75\times10^{-5}\,\text{m}^3\ \text{mol}^{-1}$$

$$\Delta p = g\rho h\ [0.6] \qquad [\Delta p = p_{\text{trench}} - p_{\text{surface}}]$$
$$= 9.81\ \text{m s}^{-2}\times1.03\ \text{g cm}^{-3}\times\frac{1\ \text{kg}}{10^3\,\text{g}}\times\frac{10^6\ \text{cm}^3}{\text{m}^3}\times11.5\times10^3\ \text{m}$$
$$= 1.16\times10^8\ \text{Pa} = 116\ \text{MPa}$$

$$\Delta G_m = V_m\Delta p = 1.75\times10^{-5}\ \text{m}^3\ \text{mol}^{-1}\times1.16\times10^8\ \text{Pa}$$
$$= 2.03\times10^3\ \text{J mol}^{-1} = \boxed{+2.03\ \text{kJ mol}^{-1}}$$

(b) The pressure at the bottom of the mercury column is 1.000 atm = 1.013×10^5 Pa.

$$\Delta p = 1.013\times10^5\ \text{Pa} - 0.160\ \text{Pa} \approx 1.013\times10^5\ \text{Pa}$$

$$V_m = \frac{200.6\ \text{g mol}^{-1}}{13.6\ \text{g cm}^3} = 14.8\ \text{cm}^3\ \text{mol} = 1.48\times10^{-5}\ \text{m}^3\ \text{mol}^{-1}$$

$$\Delta G_m = V_m\Delta p = 1.48\times10^{-5}\ \text{m}^3\ \text{mol}^{-1}\times1.013\times10^5\ \text{Pa}$$
$$= \boxed{+1.50\ \text{J mol}^{-1}}$$

E5.4
$$\Delta G_m = V_m\Delta p$$

$$V_m = \frac{891.51\ \text{g mol}^{-1}}{0.95\ \text{g cm}^{-3}} = 938\ \text{cm}^3\ \text{mol}^{-1} = 9.4\times10^{-4}\ \text{m}^3\ \text{mol}^{-1}$$

$$\Delta p = g\rho h = 9.81\ \text{m s}^{-2}\times1.03\times10^3\ \text{kg m}^{-3}\times1.5\times10^3\ \text{m} = 1.5\times10^7\ \text{Pa}$$

$$\Delta G_m = 9.4\times10^{-4}\ \text{m}^3\ \text{mol}^{-1}\times1.5\times10^7\ \text{Pa}$$
$$= +1.4\times10^4\ \text{J mol}^{-1} = \boxed{+14\ \text{kJ mol}^{-1}}$$

E5.5
$$\Delta G_m = RT\ \text{In}\ \frac{p_f}{p_i}\ [5.3b]$$

(a)
$$\Delta G_m = 8.3145\ \text{J K}^{-1}\ \text{mol}^{-1}\times293\ \text{K}\times\ln\left(\frac{3.0\ \text{bar}}{1.0\ \text{bar}}\right)$$
$$= 2.7\times10^3\ \text{J mol}^{-1} = \boxed{+2.7\ \text{kJ mol}^{-1}}$$

(b) $\Delta G_m = 8.3145\ \text{J K}^{-1}\ \text{mol}^{-1} \times 293\ \text{K} \times \ln\left(\dfrac{0.00027\ \text{bar}}{1.0\ \text{bar}}\right)$

$$= -2.0\times10^4\ \text{J mol}^{-1} = \boxed{-2.0\ \text{kJ mol}^{-1}}$$

E5.6 At these pressures, water vapor may be considered a perfect gas; therefore $p_iV_i = p_fV_f$ and $\dfrac{p_f}{p_i} = \dfrac{V_i}{V_f}$.

Therefore

$$\Delta G_m = RT\ \ln\frac{p_f}{p_i} = RT\ \ln\frac{V_i}{V_f}$$

$$= 8.3145\ \text{J K}^{-1}\ \text{mol}^{-1}\times 473\ \text{K}\times\ln\left(\frac{350\ \text{cm}^3}{120\ \text{cm}^3}\right)$$

$$= +4.2\times10^3\ \text{J mol}^{-1} = \boxed{+4.2\ \text{kJ mol}^{-1}}.$$

E5.7 For the transition S (rhombic) $\rightarrow$ S (monoclinic):

$$\Delta G_m^{\ominus}(298\ \text{K}) = +0.33\ \text{kJ mol}^{-1}\ \text{(Exercise 5.1)},$$

$$\Delta S_m^{\ominus}(298\ \text{K}) = (32.6-31.8)\ \text{J K}^{-1}\ \text{mol}^{-1} = 0.8\ \text{J K}^{-1}\ \text{mol}^{-1}\text{, and}$$

$$\Delta H_m^{\ominus}(298\ \text{K}) = \Delta G_m^{\ominus}(298\ \text{K}) + T\Delta S_m^{\ominus}(298\ \text{K})$$

$$= +0.33\ \text{kJ mol}^{-1} + (298\ \text{K})\times(0.8\times10^{-3}\ \text{kJ K}^{-1}\ \text{mol}^{-1})$$

$$= 0.57\ \text{kJ mol}^{-1}.$$

(a) If we assume that ΔH_m and ΔS_m are roughly independent of temperature, the equation $\Delta G_m = \Delta H_m - T\Delta S_m$ indicates that, since $\Delta S_m > 0$, there should be a range of temperatures above 25°C for which $\Delta G_m < 0$ and the monoclinic form of sulfur is most stable.

(b) Let T_{eq} be the temperature at which the two crystalline forms are in equilibrium at standard pressure. Rhombic sulfur is stable below T_{eq}; monoclinic sulfur is stable above T_{eq}. At equilibrium

$$\Delta G_m = \Delta H_m - T_{eq}\Delta S_m = 0 \text{ and, solving for } T_{eq},$$

$$T_{eq} \simeq \Delta H_m^{\ominus}/\Delta S_m^{\ominus} = (0.57\ \text{kJ mol}^{-1})/(0.8\ \text{J K}^{-1}\ \text{mol}^{-1})$$

$$\simeq \boxed{710\ \text{K}}.$$

E5.8 $\Delta G_m = -S_m\Delta T$ at constant p and small temperature changes [5.4]

$$\Delta G_m = -S_m(T_f - T_i) = -173.3\ \text{J K}^{-1}\ \text{mol}^{-1}\times 20\ \text{K}$$

$$= -3.5\times10^3\ \text{J mol}^{-1} = \boxed{-3.5\ \text{kJ mol}^{-1}}$$

E5.9 The slope of a graph of G_m against T is $-S_m$, that is $\dfrac{\Delta G_m}{\Delta T} = -S_m$ [5.4].

The slopes in all phases are negative, because S_m is always positive, but

$$\left|\frac{\Delta G_m}{\Delta T}\ (\text{g})\right| > \left|\frac{\Delta G_m}{\Delta T}\ (\text{l})\right| > \left|\frac{\Delta G_m}{\Delta T}\ (\text{s})\right|$$

because $S_m(\text{g}) > S_m(\text{l}) > S_m(\text{s})$.

Therefore, a graph of G_m against T appears as in Figure 5.2. Absolute values of G_m are not known, but ΔG_m in each phase could be calculated as illustrated in Exercise 5.8.

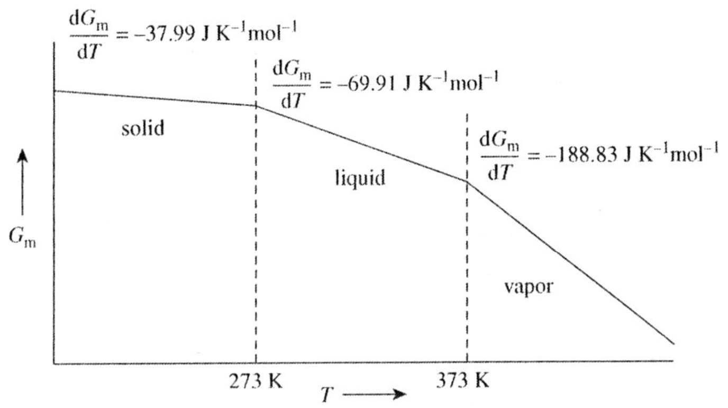

Figure 5.2

Note the discontinuous change in slopes at the transition temperatures.

E5.10 Use the perfect gas law to calculate the amount and the mass.

$$V = 5.0\ \text{m}\times 4.3\ \text{m}\times 2.2\ \text{m} = 47.\bar{3}\ \text{m}^3 = 47.\bar{3}\times 10^3\ \text{dm}^3$$

$$m = nM = \frac{pVM}{RT}$$

(a) $$m = \frac{3.2\ \text{kPa}\times(47.\bar{3}\times 10^3\ \text{dm}^3)\times(18.02\ \text{g mol}^{-1})}{(8.3145\ \text{dm}^3\ \text{kPa K}^{-1}\ \text{mol}^{-1})\times(298.15\ \text{K})} = \boxed{1.1\ \text{kg}}$$

(b) $$m = \frac{10\ \text{kPa}\times(47.\bar{3}\times 10^3\ \text{dm}^3)\times(78.11\ \text{g mol}^{-1})}{(8.3145\ \text{dm}^3\ \text{kPa K}^{-1}\ \text{mol}^{-1})\times(298.15\ \text{K})} = \boxed{15\ \text{kg}}$$

(c) $$m = \frac{(0.30\times 10^{-3}\ \text{kPa})\times(47.\bar{3}\times 10^3\ \text{dm}^3)\times(200.59\ \text{g mol}^{-1})}{(8.3145\ \text{dm}^3\ \text{kPa K}^{-1}\ \text{mol}^{-1})\times(298.15\ \text{K})} = \boxed{1.1\ \text{g}}$$

E5.11 (a) The Clapeyron equation for the solid-liquid phase boundary is

$$\frac{dp}{dT} = \frac{\Delta_{\text{fus}}H}{T_{\text{fus}}\Delta_{\text{fus}}V}\ [5.5\text{b}].$$

$$\Delta_{\text{fus}}V = V_{\text{m}}(\text{l}) - V_{\text{m}}(\text{s}) = M\left(\frac{1}{\rho_{\text{l}}} - \frac{1}{\rho_{\text{s}}}\right)$$

$$= 18.02\ \text{g mol}^{-1}\left(\frac{1}{0.99984\ \text{g cm}^{-3}} - \frac{1}{0.91671\ \text{g cm}^{-3}}\right)$$

$$= -1.634\ \text{cm}^3\ \text{mol}^{-1} = -1.634\times 10^{-6}\ \text{m}^3\ \text{mol}^{-1}$$

$$\frac{dp}{dT} = \frac{6.008\times 10^3\ \text{J mol}^{-1}}{(273.15\ \text{K})\times(-1.634\times 10^{-6}\ \text{m}^3\ \text{mol}^{-1})}$$

$$= -1.346\times 10^7\ \text{Pa K}^{-1} = \boxed{-134.6\ \text{bar K}^{-1}}$$

The slope is very steep with an unusual negative slope that is caused by the decrease in volume that occurs when ice melts. The melting process destroys some of the hydrogen bond scaffolding that holds ice in a larger molar volume.

(b) $$\frac{\Delta p}{\Delta T} = -134.6\ \text{bar K}^{-1}$$

For $\Delta T = -1$ K, $\Delta p = 134.6$ bar. Consequently, $p = p_{\text{i}} + \Delta p = 1.0\ \text{bar} + 134.6\ \text{bar} = \boxed{135.6\ \text{bar}}$.

E5.12 (a) $\log(p/\text{kPa}) = A - B/T$ [5.8]

For hexane $A = 6.849$, $B = 1655$ K from Table 5.1.

$$\Delta_{\text{vap}}H = BR\ln(10) \quad \text{(See } \textit{A brief illustration} \text{ after eqn 5.8)}$$

$$\Delta_{\text{vap}}H = 2.303 \times 8.3145 \text{ J K}^{-1} \text{ mol}^{-1} \times 1655 \text{ K}$$

$$= \boxed{31.69 \text{ kJ mol}^{-1}}$$

This value is about 10% different from the value for hexane at its boiling point, which is 28.85 kJ mol^{-1}.

(b) $$T_{\text{b}} = \frac{\Delta_{\text{vap}}H(T_{\text{b}})}{\Delta_{\text{vap}}S} \, [4.7] \approx \frac{31.69 \text{ kJ mol}^{-1}}{85 \text{ J K}^{-1} \text{ mol}^{-1}} \, (\text{Trouton's rule}) = \boxed{37\overline{3} \text{ K}}$$

Once again, this estimate is about 10% high. The normal boiling point of hexane is 342 K.

E5.13 $\log(p/\text{kPa}) = A - B/T$ where $A = 7.455$, $B = 2047$ K for methylbenzene [Table 5.1].

Since 760 Torr = 101.325 kPa,

$$\log\left(\left\{\frac{p}{\text{kPa}}\right\}\left\{\frac{101.325 \text{ kPa}}{760 \text{ Torr}}\right\}\right) = A - \frac{B}{T}$$

$$\log\left(\left\{\frac{p}{\text{Torr}}\right\}\left\{\frac{101.325}{760}\right\}\right) = A - \frac{B}{T}$$

$$\log\left(\frac{p}{\text{Torr}}\right) + \log\left(\frac{101.325}{760}\right) = A - \frac{B}{T}$$

$$\log\left(\frac{p}{\text{Torr}}\right) = A - \log\left(\frac{101.325}{760}\right) - \frac{B}{T}$$

$$\log\left(\frac{p}{\text{Torr}}\right) = A' - \frac{B}{T} \quad \text{where } A' = A - \log\left(\frac{101.325}{760}\right) = 7.455 - \log\left(\frac{101.325}{760}\right) = \boxed{8.330}.$$

B is unchanged.

E5.14 We use

$$\ln\frac{p'}{p} = \frac{\Delta_{\text{vap}}H}{R}\left(\frac{1}{T} - \frac{1}{T'}\right) \quad [5.7]$$

with $T = 293$ K, $p = 160$ mPa, and $T' = 313$ K; then solve for p'.

$$\ln\frac{p'}{p} = \frac{59.30 \times 10^3 \text{ J mol}^{-1}}{8.3145 \text{ J K}^{-1} \text{ mol}^{-1}}\left(\frac{1}{293 \text{ K}} - \frac{1}{313 \text{ K}}\right) = 1.56$$

$$\frac{p'}{p} = \text{e}^{1.56} = 4.74$$

$$p' = (4.74) \times (160 \text{ mPa}) = 758 \text{ mPa} = \boxed{0.758 \text{ Pa}}$$

E5.15 We use

$$\ln\frac{p'}{p} = \frac{\Delta_{\text{vap}}H}{R}\left(\frac{1}{T} - \frac{1}{T'}\right) \quad [5.7]$$

with $p' = 1$ atm = 101.3 kPa, $T' = 388.4$ K, $p = 50.0$ kPa, $T = 365.7$ K; then solve for $\Delta_{\text{vap}}H$.

$$\ln\frac{101.3}{50.0} = \frac{\Delta_{\text{vap}}H}{8.3145 \text{ J K}^{-1} \text{ mol}^{-1}}\left(\frac{1}{365.7 \text{ K}} - \frac{1}{388.4 \text{ K}}\right)$$

$$0.706 = 1.922 \times 10^{-5}\ \text{J}^{-1}\ \text{mol} \times \Delta_{\text{vap}}H$$

$$\Delta_{\text{vap}}H = \boxed{36.7\ \text{kJ mol}^{-1}}$$

E5.16 We first use the Clausius–Clapeyron equation, $\ln\dfrac{p'}{p} = \dfrac{\Delta_{\text{vap}}H}{R}\left(\dfrac{1}{T} - \dfrac{1}{T'}\right)$ [5.7], and the data provided to find $\Delta_{\text{vap}}H$.

$$\ln\frac{20}{50.0} = \frac{\Delta_{\text{vap}}H}{8.3145\ \text{J K}^{-1}\ \text{mol}^{-1}}\left(\frac{1}{331.95\ \text{K}} - \frac{1}{308.15\ \text{K}}\right)$$

$$-0.916 = \left(-2.80\times10^{-5}\right)\times\left(\Delta_{\text{vap}}H / \text{J K}^{-1}\ \text{mol}^{-1}\right)$$

$$\Delta_{\text{vap}}H / \text{J K}^{-1}\ \text{mol}^{-1} = 3.27\times10^{4}$$

Now we use the Clausius–Clapeyron equation and either one of the two data pairs provided to establish an equation for the vapour pressure p at any temperature T (in the range for which our equation is valid).

$$\ln\frac{p}{50.0\ \text{kPa}} = \frac{3.27\times10^{4}\ \text{J K}^{-1}\ \text{mol}^{-1}}{8.3145\ \text{J K}^{-1}\ \text{mol}^{-1}}\left(\frac{1}{331.95\ \text{K}} - \frac{1}{T}\right)$$

$$\ln\left(p/\text{kPa}\right) = 15.7\bar{8} - \frac{3.94\times10^{3}}{T/\text{K}} \quad \text{or} \quad T/\text{K} = \frac{3.94\times10^{3}}{15.7\bar{8} - \ln\left(p/\text{kPa}\right)}$$

The normal boiling point, T_{b}, occurs when p = 1 atm = 101.325 kPa.

$$T_{\text{b}} = \frac{3.94\times10^{3}\ \text{K}}{15.7\bar{8} - \ln(101.325)} = \boxed{353\ \text{K}}$$

E5.17 (a) The two components are Na_2SO_4 and H_2O (proton transfer equilibria to give HSO_4^- etc. do not change the number of independent components) so $\boxed{C = 2}$. There are three phases present (solid salt, liquid solution, vapour), so $\boxed{P = 3}$.

(b) The variance (the number of degrees of freedom) is $F = C - P + 2 = 2 - 3 + 2 = \boxed{1}$.

Either pressure or temperature may be considered the independent variable, but not both as long as the equilibrium is maintained. If the pressure is changed, the temperature must be changed to maintain the equilibrium.

E5.18 (a) As in Exercise 5.17 the number of components is still $\boxed{C = 2}$ (Na_2SO_4, H_2O), but now there is no solid phase present, so $\boxed{P = 2}$ (liquid solution, vapour).

(b) The variance is $F = 2 - 2 + 2 = \boxed{2}$. We are free to change any two of the three variables, amount of dissolved salt, pressure, or temperature, but not the third. If we change the amount of dissolved salt and the pressure, the temperature is fixed by the equilibrium condition between the two phases.

E5.19 The vapor pressure of ice at –5°C is 3.9×10^{-3} atm, or 3.0 Torr (CRC *Handbook of Chemistry and Physics*). Since the partial pressure of water is lower (2 Torr), the frost will sublime. A partial pressure of 3.0 Torr or more will ensure that the frost remains.

E5.20 (a) The volume decreases as the vapor is cooled from 400 K, at constant pressure, in a manner described by the perfect gas equation $V = nRT/p$. That is, V is a linear function of T. This continues until 373 K is reached where the vapor condenses to a liquid and there is a large decrease in volume. As the temperature is lowered further to 273 K, liquid water freezes to ice. Only a small decrease in volume occurs in the liquid as temperature is decreased, and a small (~9%) increase in volume occurs when the liquid freezes. Water remains as a solid at 260 K.

(b) The cooling curve appears roughly as sketched in Figure 5.3. The vapor and solid phases show a steeper rate of decline than for the liquid phase due to their smaller heat capacities. The temperature halt in the liquid plus vapor region is longer than for the liquid plus solid region due to its larger heat of transition.

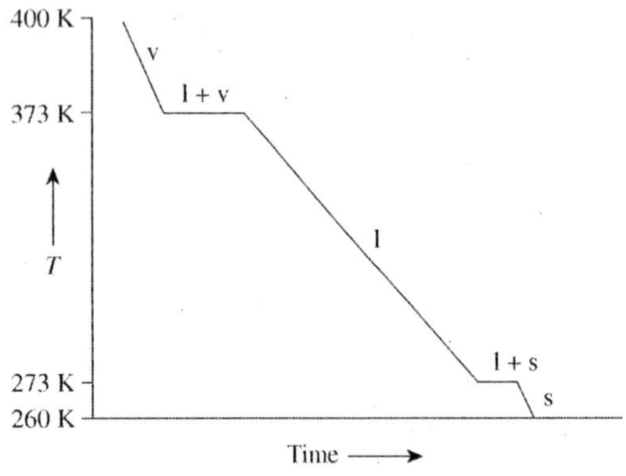

Figure 5.3

E5.21 Cooling from, say, 400 K at 0.006 bar (text Figs 5.15 and 5.16) will cause a decrease in volume of gaseous water until 273.16 K is reached, at which temperature liquid and solid water will appear. All three phases will remain in equilibrium until the constant cooling completely transforms the gas and liquid phases to ice. Then, cooling drops the temperature below 273.16 K.

E5.22 (a) The gaseous sample expands. (b) The sample contracts but remains gaseous because 320 K is greater than the critical temperature. (c) The gas contracts and forms a liquid-like substance without the appearance of a discernible surface. As the temperature lowers further to the solid phase boundary line, solid carbon dioxide forms in equilibrium with the liquid. At 210 K the sample has become all solid. (d) The solid expands slightly as the pressure is reduced and sublimes when the pressure reaches about 5 atm. (e) The gas expands as it is heated at constant pressure.

E5.23 The slope of the He-II/He-I phase boundary line appears to be negative everywhere. That is

$$\frac{\mathrm{d}p}{\mathrm{d}T} = \frac{\Delta_{\mathrm{trs}}H_{\mathrm{m}}}{T_{\mathrm{trs}}\Delta_{\mathrm{trs}}V_{\mathrm{m}}} = -.$$

If we assume that $\Delta_{\mathrm{trs}}H_{\mathrm{m}}$ for He-II $\rightarrow$ He-I is positive, it implies that $\Delta_{\mathrm{trs}}V_{\mathrm{m}}$ is negative, so He-I is expected to be more dense than He-II. This argument, which would work well for normal fluids, fails in the case of He. The transition to the superfluid shows no measurable $\Delta_{\mathrm{trs}}H_{\mathrm{m}}$ or $\Delta_{\mathrm{trs}}V_{\mathrm{m}}$. So at the transition, there is no difference in density between the two forms of He.

Solutions to projects

P5.24 The van der Waals equation of state [1.23b] for one mole of gas is

$$\left(p+\frac{a}{V_{\mathrm{m}}^2}\right)(V_{\mathrm{m}}-b) = RT \quad \text{where } V_{\mathrm{m}} = V/n \text{ is the molar volume.}$$

After neglecting attractive effects, it becomes $p(V_{\mathrm{m}}-b) = RT$.

(a) Solving for V_m we get

$$V_{\mathrm{m}} = \frac{RT}{p} + b$$ where the first term is recognized as the molar volume of a perfect gas.

Following Derivation 5.2 with temperature constant gives

$$\Delta G_{\mathrm{m}} = \int_{p_{\mathrm{i}}}^{p_{\mathrm{f}}} V_{\mathrm{m}}\mathrm{d}p = \int_{p_{\mathrm{i}}}^{p_{\mathrm{f}}} \frac{RT}{p}\,\mathrm{d}p + \int_{p_{\mathrm{i}}}^{p_{\mathrm{f}}} b\mathrm{d}p$$

$$= \boxed{RT\ln\frac{p_{\mathrm{f}}}{p_{\mathrm{i}}} + b(p_{\mathrm{f}} - p_{\mathrm{i}})} = \text{perfect value} + b(p_{\mathrm{f}} - p_{\mathrm{i}})$$

(b) For $p_{\mathrm{f}} > p_{\mathrm{i}}$, the change is $\boxed{\text{greater}}$ for this gas by the amount $b(p_{\mathrm{f}} - p_{\mathrm{i}})$.

(c) perfect gas value $= RT\ln\frac{p_{\mathrm{f}}}{p_{\mathrm{i}}}$, assume $T = 298.15$ K

$$RT\ln\frac{p_{\mathrm{f}}}{p_{\mathrm{i}}} = (8.3145\ \mathrm{J\ K^{-1}\ mol^{-1}})\times(298.15\ \mathrm{K})\times\ln\left(\frac{10.0\ \mathrm{atm}}{1.0\ \mathrm{atm}}\right)$$

$$= 5.71\times10^{3}\ \mathrm{J} = 5.71\ \mathrm{kJ}$$

$$b(p_{\mathrm{f}} - p_{\mathrm{i}}) = 0.0429\ \mathrm{dm^3\ mol^{-1}}\times\frac{1\ \mathrm{m^3}}{10^3\ \mathrm{dm^3}}\times(10-1)\ \mathrm{atm}\times\frac{1.013\times10^5\ \mathrm{Pa}}{1.0\ \mathrm{atm}}$$

$$= 39\ \mathrm{J}$$

The real gas value is then 5.71×10^3 J + 39 J = 5.75×10^3 J = 5.75 kJ.

The percentage difference is approximately

$$\frac{39\ \mathrm{J}}{5.7\times10^3\ \mathrm{J}}\times100\% = \boxed{0.68\%}.$$

P5.25 Starting with the differential form of the Clausius-Clapeyron equation [Derivation 5.5, perfect gas]

$$\frac{\mathrm{d}\ln p}{\mathrm{d}T} = \frac{\Delta_{\mathrm{vap}}H}{RT^2} \quad \text{or} \quad \mathrm{d}\ln p = \frac{\Delta_{\mathrm{vap}}H}{RT^2}\mathrm{d}T$$

integrate between the points (p,T) and (p',T').

$$\int_{\ln p}^{\ln p'}\mathrm{d}\ln p = \int_{T}^{T'}\frac{\Delta_{\mathrm{vap}}H}{RT^2}\mathrm{d}T = \int_{T}^{T'}\frac{a+bT}{RT^2}\mathrm{d}T = \frac{1}{R}\left\{a\int_{T}^{T'}\frac{1}{T^2}\mathrm{d}T + b\int_{T}^{T'}\frac{1}{T}\mathrm{d}T\right\}$$

$$\ln\left(\frac{p'}{p}\right) = \frac{1}{R}\left\{a\left(\frac{-1}{T}\right)\Bigg|_{T}^{T'} + b\ln\left(\frac{T'}{T}\right)\right\}$$

$$\ln\left(\frac{p'}{p}\right) = \frac{1}{R}\left\{a\left(\frac{1}{T} - \frac{1}{T'}\right) + b\ln\left(\frac{T'}{T}\right)\right\}$$

$$\boxed{\ln p' = \ln p + \frac{1}{R}\left\{a\left(\frac{1}{T} - \frac{1}{T'}\right) + b\ln\left(\frac{T'}{T}\right)\right\}}$$

P5.26 (a) The critical point corresponds to a point of zero slope which is simultaneously a point of inflection in a plot of pressure versus molar volume. A critical point exists if there are values of p, V, and T that result in a point which satisfies these conditions. Let $V_{\mathrm{m}} = V/n$, the equation of state becomes:

$$p = \frac{RT}{V_{\mathrm{m}}} - \frac{a}{V_{\mathrm{m}}^2} + \frac{b}{V_{\mathrm{m}}^3}$$

$$\left.\begin{aligned}\left(\frac{\partial p}{\partial V_{\mathrm{m}}}\right)_T &= -\frac{RT}{V_{\mathrm{m}}^2} + \frac{2a}{V_{\mathrm{m}}^3} - \frac{3b}{V_{\mathrm{m}}^4} = 0\\ \left(\frac{\partial^2 p}{\partial V_{\mathrm{m}}^2}\right)_T &= \frac{2RT}{V_{\mathrm{m}}^3} - \frac{6a}{V_{\mathrm{m}}^4} + \frac{12b}{V_{\mathrm{m}}^5} = 0\end{aligned}\right\}\text{at the critical point}$$

That is,

$$\left.\begin{aligned}-RT_cV_c^2+2aV_c-3b&=0\\ RT_cV_c^2-3aV_c+6b&=0\end{aligned}\right\}$$

These two independent equations describe a critical point if they have a real solution. Solving these simultaneous equations for V_c and T_c gives

$$\boxed{V_c=\frac{3b}{a}} \quad \text{and} \quad \boxed{T_c=\frac{a^2}{3Rb}}$$

Now use the equation of state to find p_c.

$$p_c=\frac{RT_c}{V_c}-\frac{a}{V_c^2}+\frac{b}{V_c^3}=\left(\frac{Ra^2}{3Rb}\right)\times\left(\frac{a}{3b}\right)-a\left(\frac{a}{3b}\right)^2+b\left(\frac{a}{3b}\right)^3=\boxed{\frac{a^3}{27b^2}}$$

It follows that

$$Z_c=\frac{p_cV_c}{RT_c}=\left(\frac{a^3}{27b^2}\right)\times\left(\frac{3b}{a}\right)\times\left(\frac{1}{R}\right)\times\left(\frac{3Rb}{a^2}\right)=\boxed{\frac{1}{3}}$$

Having found a real inflection point, we conclude that this equation of state does describe a critical point and that the critical constants are related to the parameters a and b by the above equations.

(b) The supercritical fluid extractor consists of a pump to pressurize the solvent (e.g., CO_2), an oven with extraction vessel, and a trapping vessel. Extractions are performed dynamically or statically. Supercritical fluid flows continuously through the sample within the extraction vessel when operating in dynamic mode. Analytes extracted into the fluid are released through a pressure maintaining restrictor into a trapping vessel. In static mode the supercritical fluid circulates repetitively through the extraction vessel until being released into the trapping vessel after a period of time. Supercritical carbon dioxide volatilizes when decompression occurs upon release into the trapping vessel.

Advantages	Disadvantages	Current uses
Dissolving power of SCF can be adjusted with selection of T and p	Elevated pressures are required and the necessary apparatus expensive	Extraction of caffeine, fatty acids, spices, aromas, flavors, and biological materials from natural sources
Select SCF's are inexpensive and non-toxic. They reduce pollution	Cost may prohibit large scale applications	Extraction of toxic salts (with a suitable chelation agent) and organics from contaminated water
Thermally unstable analytes may be extracted at low temperature	Modifiers like methanol (1–10%) may be required to increase solvent polarity	Extraction of herbicides from soil
The volatility of $scCO_2$ makes it easy to isolate analyte	$scCO_2$ is toxic to whole cells in biological applications (CO_2 is not toxic to the environment.)	scH_2O oxidation of toxic, intractable organic waste during water treatment
SCF's have high diffusion rates, low viscosity, and low surface tension		Synthetic chemistry, polymer synthesis and crystallization, textile processing
O_2 and H_2 are completely miscible with $scCO_2$. This reduces multi-phase reaction problems		Heterogeneous catalysis for green chemistry processes

6 The properties of mixtures

Answers to discussion questions

D6.1 A **partial molar property** is the contribution (per mole) that a substance makes to an overall property of a mixture. Two important partial molar properties of substance J are the **partial molar volume, V_J**, and the **partial molar Gibbs energy, G_J**. The total property is given by the sum of the partial molar property for all substances in the mixture weighted by the amount of each. For a mixture consisting of substance A and B:

$$V = n_A V_A + n_B V_B \ [6.1] \quad \text{and} \quad G = n_A G_A + n_B G_B \ [6.2a].$$

The partial molar property 'X_A' (e.g., V_A and G_A) is the slope of the plot of X against n_A at constant p, T, and n_B. This slope has the mathematical form of a partial derivative:

$$X_A = \left(\frac{\partial X}{\partial n_A}\right)_{p,T,n_B} \quad \text{and similarly } X_B = \left(\frac{\partial X}{\partial n_B}\right)_{p,T,n_A}.$$

A substance with a high partial molar property contributes greatly to the property and values depend upon interactions between atoms, ions, and molecules. As interactions between substances are generally different than those between particles of a pure substance, partial molar properties depend upon mixture composition.

D6.2 The partial molar Gibbs energy of substance A, G_A, is given a special name and symbol. It is the **chemical potential, μ_A** (mu) and for a mixture of A and B eqn 6.2a is written as

$$G = n_A \mu_A + n_B \mu_B \ [6.2b]$$

$$\text{where } \mu_A = G_A = \left(\frac{\partial G}{\partial n_A}\right)_{p,T,n_B} \quad \text{and} \quad \mu_B = G_B = \left(\frac{\partial G}{\partial n_B}\right)_{p,T,n_A}.$$

A substance with a high chemical potential contributes greatly to the total Gibbs energy G of a mixture and, consequently, has a high ability to drive a reaction or some other process forward. The chemical potential of a substance within a mixture depends on interactions with neighboring atoms, ions, and molecules and, consequently, it depends upon mixture composition. Chemical potentials are especially important when discussing the direction of spontaneity of constant T and p processes because spontaneity occurs in the direction for which G decreases. Should the difference between the Gibbs energy of two possible states equal zero, the two states are present in equilibrium with neither state preferred by spontaneous changes in physical or chemical properties. Applications include help in understanding the thermodynamic basis and relations of spontaneous mixing, partial vapour pressures of liquid mixtures, Raoult's law, Henry's constant, boiling point elevation and the ebullioscopic constant, freezing point depression and the cryoscopic constant, osmosis, and phase diagrams of mixtures.

D6.3 At equilibrium, the chemical potentials of any component in both the liquid and vapor phases must be equal. This is justified with the equilibrium criteria that under constant temperature and pressure conditions, with no additional work, $\Delta G = 0$. Consider the relationships $dG_{m,J(\alpha)} = V_{m,J(\alpha)}dp - S_{m,J(\alpha)}dT + \mu_J(\alpha)dn_J$ and $dG_{m,J(\beta)} = V_{m,J(\beta)}dp - S_{m,J(\beta)}dT + \mu_J(\beta)dn_J$ for a chemical species "J" that is present in the two phases α and β (see Derivations 5.1 and 6.2). The terms with dp and dT on the right side of these expressions equal zero for constant p and constant T processes, and the near equilibrium transformation $J(\beta) \rightleftharpoons J(\alpha)$ is such a process. Subtraction and application of the equilibrium criteria gives: $dG_J = dG_{m,J(\alpha)} - dG_{m,J(\beta)} = \{\mu_J(\alpha) - \mu_J(\beta)\}dn_J = 0$ or $\mu_J(\alpha) - \mu_J(\beta) = 0$ at equilibrium. The chemical potential of each chemical species must be equal in the liquid (α) and vapor (β) phases or in any two phases that are at equilibrium: $\boxed{\mu_J(\alpha) = \mu_J(\beta)}$.

D6.4 **Raoult's law**, $p_J = x_J p_J^*$, describes the partial pressure of volatile component J above a solution for which the mole fraction of J is x_J. p_J^* is the vapour pressure of pure J. The law is most reliable when the solution components have similar molecular shapes and are held together in the liquid by similar types and strengths of intermolecular forces. An **ideal** solution is a hypothetical solution of a solute B in a solvent A that obeys Raoult's law throughout the composition range from pure A to pure B. No mixture is perfectly ideal and all real mixtures show deviations from Raoult's law. Deviations are small for the solvent when the solution is very dilute and the law becomes exact in the limit of zero concentration of solute. It is a **limiting law**.

Henry's law, $p_B = x_B K_H$, describes the partial pressure of volatile solute B above a solution for which the mole fraction of B is x_B. Henry's law constant, K_H, is characteristic of the solute and chosen to equal the slope (dp_B/dx_B) at $x_B = 0$ of the empirical p_B against x_B data plot. This law is also a limiting law in that it is usually obeyed only at low concentrations of solute in ideal-dilute solutions. It is an especially useful relation between the solubility and partial pressure of low-solubility gases. In ideal-dilute solutions solvent molecules are in an environment very much like the one they have in the pure liquid. However, the solute molecules are surrounded by solvent molecules, which is entirely different from their environment when pure. Thus, the solvent behaves like a slightly modified pure liquid, but the solute behaves entirely differently from its pure state unless the solvent and solute molecules happen to be very similar. In the latter case, the solute also obeys Raoult's law.

D6.5 All the colligative properties (properties that depend only on the number of solute particles present, not their chemical identity) are a result of the lowering of the chemical potential of the solvent due to the presence of the solute. This reduction takes the form $\mu_A = \mu_A^* + RT\ln x_A$ or $\mu_A = \mu_A^* + RT\ln a_A$, depending on whether or not the solution can be considered ideal. The lowering of the chemical potential results in a freezing point depression and a boiling point elevation as illustrated in Figures 6.16 and 6.17 of the text. Both of these effects can be explained by the lowering of the vapour pressure of the solvent in solution due to the presence of the solute. The solute molecules get in the way of the solvent molecules, reducing their escaping tendency.

D6.6 The activity of a solute is that property which determines how the chemical potential of the solute varies from its value in a specified standard state. This is seen from the general definition

$$\mu_J = \mu_J^{\ominus} + RT\ln a_J \quad [6.15]$$

where $\mu_J^{\ominus}$ is the value of the chemical potential of J in the standard state for which $a_J = 1$. The relation is true at all concentrations and for both the solvent and the solute. It is well worth remembering several useful activity forms.

Ideal solutions:	$a_J = x_J$
Ideal-dilute solutions:	$a_B = [B]/c^{\ominus}$ where $c^{\ominus} = 1$ mol dm^{-3}
Solvent A of a non-ideal solution:	$a_A = \gamma_A x_A$
Solute B of a non-ideal solution:	$a_B = \gamma_B[B]/c^{\ominus}$ where $c^{\ominus} = 1$ mol dm^{-3}

The dimensionless activity coefficients, γ_J, of non-ideal solutions must be deduced from experimental data. Also, the activity of a pure solid or a pure liquid at 1 bar always equals 1 as these are standard states.

D6.7 Osmosis, the passage of a pure solvent into a solution separated from it by a semipermeable membrane through which solute cannot pass, is spontaneously driven by a diminishing total Gibbs energy until the solvent chemical potentials on each side of the membrane are equal:

$$\mu_A \text{ (pure solvent at pressure } p) = \mu_A \text{ (solvent in solution at pressure } p + \Pi)$$

where Π is the osmotic pressure. Entropy increases as the solution becomes dilute.

Vigorous molecular motion causes molecules of the pure solvent to flow through microscopic holes of the membrane into the solution. Solvent molecules also flow from the solution side to the pure solvent side. However, the rate of the latter flow is reduced by the blocking action of solute with their bulky solvent cages. This results in a net flow from the pure solvent side to the solution side. The osmotic pressure is the pressure that must be applied to the solution side of the membrane in order to stop the net flow of solvent. The blocking action of solute ions or molecules is proportional to the number of solute particles present and, consequently, osmosis is a colligative property.

D6.8 The osmotic pressure, Π, method (see Example 6.4) for determination of polymer molar mass involves measurement of Π for a series of successively more dilute mass concentrations c_{polymer}. The extrapolated intercept at $c_{\text{polymer}} = 0$ of a Π/c_{polymer} against c_{polymer} plot equals RT/M_{polymer}. Consequently, $M_{\text{polymer}} = RT/\text{intercept}$.

Solutions to exercises

E6.1 Knowing the partial molar volumes, we calculate the solution volume with $V = n_P V_P + n_T V_T$ [6.1] where the subscripts P and T denote propanone ($M_P = 58.08$ g mol^{-1}) and trichloromethane ($M_T = 119.37$ g mol^{-1}). However, we must first deduce the component moles with our knowledge of total mass m and mole fractions. The definition of mole fraction tells us that $n_P = x_P n$ and $n_T = x_T n$. Therefore,

$$(1)\quad n_P = \frac{x_P}{x_T} \times n_T.$$

The mass contributions of the mixture components are $m_P = n_P M_P$ and $m_T = n_T M_T$. By adding these to get the total mass, substituting eqn (1), solving for n_T and recognizing that the sum of all mole fractions equals 1 (so $x_P = 1 - x_T$), we arrive at an equation that is suitable for computation.

$$m = m_P + m_T = n_P M_P + n_T M_T = (x_P / x_T) n_T M_P + n_T M_T = \{(x_P / x_T) M_P + M_T\} n_T$$

$$n_T = \frac{m}{(x_P / x_T) M_P + M_T} = \frac{x_T m}{x_P M_P + x_T M_T} = \frac{x_T m}{(1 - x_T) M_P + x_T M_T}$$

$$= \frac{0.4693 \times 1000 \text{ g}}{(0.5307 \times 58.08) + (0.4693 \times 119.37) \text{ g mol}^{-1}} = 5.404 \text{ mol}$$

Substitution of this result into eqn (1) gives n_P.

$$n_P = \frac{0.5307}{0.4693} \times 5.404 = 6.111 \text{ mol}$$

Finally,

$$V = n_P V_P + n_T V_T \quad [6.1]$$
$$= (6.111 \text{ mol} \times 74.166 \text{ cm}^3 \text{ mol}^{-1}) + (5.404 \text{ mol} \times 80.235 \text{ cm}^3 \text{ mol}^{-1})$$
$$= \boxed{886.8 \text{ cm}^3}.$$

E6.2 Let the subscripts E and W denote ethanol and water

$$n_E = (V \times \rho / M)_E = (50.0 \text{ cm}^3) \times (0.789 \text{ g cm}^{-3}) \times \left(\frac{1 \text{ mol}}{46.07 \text{ g}}\right) = 0.856 \text{ mol}$$

$$n_W = (V \times \rho / M)_W = (50.0 \text{ cm}^3) \times (1.000 \text{ g cm}^{-3}) \times \left(\frac{1 \text{ mol}}{18.02 \text{ g}}\right) = 2.775 \text{ mol}$$

$$x_E = \frac{n_E}{n_E + n_W} = \frac{0.856}{0.856 + 2.775} = 0.236$$

From text Figure 6.1 we roughly estimate the partial molar volumes as $V_E = 55.8\ \text{cm}^3\ \text{mol}^{-1}$ and $V_W = 17.7\ \text{cm}^3\ \text{mol}^{-1}$ when $x_E = 0.236$. Then

$$\begin{aligned} V &= n_E V_E + n_W V_W \\ &= (0.856\ \text{mol} \times 55.8\ \text{cm}^3\ \text{mol}^{-1}) + (2.775\ \text{mol} \times 17.7\ \text{cm}^3\ \text{mol}^{-1}) \\ &= \boxed{96.9\ \text{cm}^3} \end{aligned}$$

E6.3

$$\begin{aligned} \mu_{CO_2} - \mu_{CO_2}^{\ominus} &= RT \ln(p_{CO_2}/p^{\ominus}) \quad [6.4b] \\ &= (8.3145\ \text{J K}^{-1}\ \text{mol}^{-1}) \times (310\ \text{K}) \ln(2.0\ \text{bar}/1.0\ \text{bar}) \\ &= \boxed{1.8\ \text{kJ mol}^{-1}} \end{aligned}$$

E6.4

1 atm = 1.01325 bar

$$\mu - \mu^{\ominus} = RT \ln(p/p^{\ominus}) \quad [6.4b]$$

$$\begin{aligned} \mu(1\ \text{atm}) - \mu^{\ominus} &= (8.3145\ \text{J K}^{-1}\ \text{mol}^{-1}) \times (298.15\ \text{K}) \ln(1.01325\ \text{bar}/1.0\ \text{bar}) \\ &= \boxed{32.631\ \text{J mol}^{-1}} \end{aligned}$$

E6.5 (a) $\Delta G_m = RT\,(x_A \ln x_A + x_B \ln x_b)$ [6.5] where A = N_2(g) and B = O_2(g)

$$\begin{aligned} \Delta G_m &= 2.479\ \text{kJ mol}^{-1}\{0.78 \ln(0.78) + 0.22 \ln(0.22)\} \\ &= \boxed{-1.31\ \text{kJ mol}^{-1}} \end{aligned}$$

Because ΔG_m is negative, the mixing is $\boxed{\text{spontaneous}}$.

(b)

$$\begin{aligned} \Delta S_m &= -R(x_A \ln x_A + x_B \ln x_B) \quad [6.6b] \\ &= -(8.3145\ \text{J K}^{-1}\ \text{mol}^{-1}) \times \{0.78 \ln(0.78) + 0.22 \ln(0.22)\} \\ &= \boxed{+4.38\ \text{J K}^{-1}\ \text{mol}^{-1}} \end{aligned}$$

E6.6

$\Delta G_m = RT\,(x_A \ln x_A + x_B \ln x_B + x_C \ln x_C)$ [6.5] where A = N_2(g), B = O_2(g), and C = Ar(g)

$$\begin{aligned} \Delta G_m &= 2.4790\ \text{kJ mol}^{-1}\{0.780 \ln(0.780) + 0.210 \ln(0.210) + 0.0096 \ln(0.0096)\} \\ &= \boxed{-1.40\ \text{kJ mol}^{-1}} \end{aligned}$$

Because the change in ΔG_m is negative upon the addition of argon, the mixing is $\boxed{\text{spontaneous}}$.

$$\begin{aligned} \Delta S_m &= -R(x_A \ln x_A + x_B \ln x_B + x_C \ln x_C) \\ &= \boxed{+4.71\ \text{J K}^{-1}\ \text{mol}^{-1}} \end{aligned}$$

By adding to the mixture of exercise 6.5 a third gas as about 1% of the whole, the Gibbs energy is lowered by about 10% and the entropy of mixing is increased by about 10%.

E6.7

$$n_{C_{60}} = (m/M)_{C_{60}} = (2.33\ \text{g})/(720.6\ \text{g mol}^{-1}) = 3.23 \times 10^{-3}\ \text{mol}$$

$$n_{\text{toluene}} = (m/M)_{\text{toluene}} = (100\ \text{g})/(92.14\ \text{g mol}^{-1}) = 1.085\ \text{mol}$$

$$x_{\text{toluene}} = \frac{n_{\text{toluene}}}{n_{C_{60}} + n_{\text{toluene}}} = \frac{1.085\ \text{mol}}{(3.23 \times 10^{-3}\ \text{mol}) + 1.085\ \text{mol}} = 0.997$$

$$p_{\text{toluene}} = x_{\text{toluene}} p^{*}_{\text{toluene}}\ [6.7] = 0.997 \times 5.00\ \text{kPa} = \boxed{4.99\ \text{kPa}}$$

E6.8 For the sake of convenience we make computations for 1.000 dm^3 of seawater. Taking this liberty is valid because it quickly gives us the mole fraction (an intensive property that is independent of the volume of solution used in the calculation) of water in the solution. Furthermore, being a dilute solution, we assume that 1.000 dm^3 of seawater contains roughly 1000 g of water.

$$n_{\text{water}} = (m/M)_{\text{water}} = (1000\text{ g})/(18.02\text{ g mol}^{-1}) = 55.5\text{ mol}$$

$$n_{\text{ions}} = 2Vc_{\text{NaCl}} = 2\times(1.000\text{ dm}^3)(0.50\text{ mol dm}^{-3}) = 1.0\text{ mol}$$

$$x_{\text{water}} = \frac{n_{\text{water}}}{n_{\text{water}} + n_{\text{ions}}} = \frac{55.5}{56.5} = 0.982$$

$$p_{\text{water}} = x_{\text{water}}p^*_{\text{water}}\ [6.7] = 0.982\times 2.338\text{ kPa} = \boxed{2.30\text{ kPa}}$$

E6.9 Check whether $p_{\text{HCl}}/x_{\text{HCl}}$ [6.11] is equal to a constant (K_{HCl}).

Show that sol. obey Henry.

x_{HCl}	0.005	0.012	0.019
p_{HCl} / kPa	32.0	76.9	121.8
$(p_{\text{HCl}}/\text{kPa})/x$	6.4×10^3	6.4×10^3	6.4×10^3

Hence, $K_{\text{HCl}} \approx \boxed{6.4\times10^3\text{ kPa.}}$

E6.10

$$K_{\text{CO}_2/\text{lipid}} = (8.6\times10^4\text{ Torr})\times(101.325\text{ kPa}/760\text{ Torr}) = 1.1\bar{5}\times10^4\text{ kPa}$$

$$x_{\text{CO}_2} = p_{\text{CO}_2}/K_{\text{CO}_2/\text{lipid}}\ [6.11] = (55\text{ kPa})/(1.1\bar{5}\times10^4\text{ kPa}) = \boxed{4.8\times10^{-3}}$$

E6.11

$$p_{\text{H}_2} = [\text{H}_2]K_{\text{H}_2}\ [6.12]$$

$$= \left(\frac{1.0\times10^{-3}\text{ mol}}{1\times10^{-3}\text{ m}^3}\right)\times(128\text{ kPa m}^3\text{ mol}^{-1})\ [\text{Table 6.1}] = \boxed{128\text{ kPa}}$$

E6.12

$$[\text{CO}_2] = p_{\text{CO}_2}/K_{\text{CO}_2}\ [6.12],\qquad K_{\text{CO}_2} = 2.937\text{ kPa m}^3\text{ mol}^{-1}$$

(a) $$[\text{CO}_2] = \frac{3.8\text{ kPa}}{2.937\text{ kPa m}^3\text{ mol}^{-1}} = 1.3\text{ mol m}^{-3} = \boxed{1.3\text{ mmol dm}^{-3}}$$

(b) $$[\text{CO}_2] = \frac{50.0\text{ kPa}}{2.937\text{ kPa m}^3\text{ mol}^{-1}} = 17.0\text{ mol m}^{-3} = \boxed{17.0\text{ mmol dm}^{-3}}$$

E6.13

$$[\text{J}] = p_{\text{J}}/K_{\text{J}}\ [6.12] = x_{\text{J}}(\text{gas})\times p/K_{\text{J}}\ [1.7]$$

We assume that $p = p^{\ominus} = 1.00\text{ bar} = 100\text{ kPa}$.

$$[\text{N}_2] = \frac{0.78\times(100\text{ kPa})}{156\text{ kPa m}^3\text{ mol}^{-1}}\ [\text{Table 6.1}] = 0.50\text{ mol m}^{-3} = 0.50\text{ mmol dm}^{-3}$$

$$[\text{O}_2] = \frac{0.21\times(100\text{ kPa})}{79.2\text{ kPa m}^3\text{ mol}^{-1}}\ [\text{Table 6.1}] = 0.27\text{ mol m}^{-3} = 0.27\text{ mmol dm}^{-3}$$

The magnitudes of molarity and molality concentrations are equal in very dilute solutions such as these. Consequently, $\boxed{b_{\text{N}_2} = 0.50\text{ mmol kg}^{-1}}$ and $b_{\text{O}_2} = \boxed{0.27\text{ mmol kg}^{-1}}$.

E6.14

$$p_{\text{CO}_2} = 1.0\text{ atm} = 101.325\text{ kPa and } K_{\text{CO}_2} = 2.937\text{ kPa m}^3\text{ mol}^{-1}\ [\text{Table 6.1}]$$

$$[\text{CO}_2] = p_{\text{CO}_2}/K_{\text{CO}_2}\ [6.12],\ K_{\text{CO}_2} = 2.937\text{ kPa m}^3\text{ mol}^{-1}$$

$$[\text{CO}_2] = \frac{p_{\text{CO}_2}}{K_{\text{CO}_2}}\ [6.12] = \frac{101.325\text{ kPa}}{2.937\text{ kPa m}^3\text{ mol}^{-1}} = 34.5\text{ mol m}^{-3} = \boxed{34.5\text{ mmol dm}^{-3}}$$

E6.15 $p = p_A + p_B = x_A p_A^* + x_B p_B^*$ [6.7] $= x p_A^* + (1 - x_A) p_B^*$ where the solution has the composition x_J Solving for x_A gives

$$x_A = \frac{p - p_B^*}{p_A^* - p_B^*}.$$

When boiling at 0.50 atm pressure (50.7 kPa), the combined vapor pressure must be 0.50 atm, hence if A = toluene and B = *o*-xylene

$$x_A = \frac{(50.7 - 20)\,\text{kPa}}{(53 - 20)\,\text{kPa}} = \boxed{0.93} \quad \text{and} \quad x_B = 1 - x_A = \boxed{0.07}.$$

The composition of the vapour (as the mole fractions y_J) is given by

$$y_A = \frac{p_A}{p}\ [1.7] = \frac{x_A p_A^*}{p} = \frac{0.93 \times (53\ \text{kPa})}{50.7\ \text{kPa}} = \boxed{0.97}$$

and $y_B = 1 - y_A = 1 - 0.97 = \boxed{0.03}$.

E6.16 A plot of the equations for the partial pressures of A and B is shown in Figure 6.1 along with the equation slopes in both the limit of the pure component (Raoult's limit) and the limit of very little component (Henry's limit). The equations are:

$$p_A/\text{Torr} = 68x_A - 12x_A^2 + 643x_A^3 - 283x_A^4 \quad \text{and} \quad p_B/\text{Torr} = 780 - 440x_A - 401x_A^2 + 92x_A^3.$$

We begin by checking that the eqns have the properties of providing an increasing partial pressure as the component mole fraction increases and reporting zero partial pressure when the component mole fraction equals zero. Figure 6.1 reveals that only one of these criteria are not satisfied. The eqn for component B does not give $p_B = 0$ at $x_B = 0$ ($x_A = 1$); it gives $p_B = 31$ Torr. Consequently, we conclude that the eqn for component B does not correctly reflect experimental data. In spite of this, we continue with the analysis.

Is there a quick visual check for Raoult's law? Yes. Examine the curve slopes in Raoult's limit ($x_J = 1$) and ask whether the slope is along the imaginary line between the partial pressures at $x_J = 1$(i.e., $p_J = p_J^*$) and $x_J = 0$ (i.e., $p_J = 0$). If the answer is no, the component does not conform to Raoult's law. **Neither A nor B satisfy Raoult's law**. A mathematical analysis of the equations follows.

What are the vapour pressures of the pure substances? These are the values when $x_A = 1$ and $x_A = 0$ in the respective eqns.

$$p_A^* = (68 - 12 + 643 - 283)\,\text{Torr} = 416\ \text{Torr} \quad \text{and} \quad p_B^* = 780\ \text{Torr}$$

Do components A and B conform to Raoult's law ($p_J = x_J p_J^*$ [6.7])? A substance that conforms to Raoult's law exhibits a slope at $x_J = 1$ (Raoult's limit in a p_J against x_J plot) that equals the vapour pressure of the pure substance; that is, $dp_J/dx_J = p_J^*$ in the limit as $x_J \to 1$. Evaluation of the slope for component A is straight forward.

$$dp_A/dx_A = (68 - 2\times 12x_A + 3\times 643x_A^2 - 4\times 283x_A^3)\,\text{Torr}$$

$$[dp_A/dx_A]_{x_A=1} = (68 - 2\times 12 + 3\times 643 - 4\times 283)\,\text{Torr} = 841\ \text{Torr}$$

Since 841 Torr is not equal to the value of p_A^*, we conclude that component **A does not conform to Raoult's law**. Evaluation of the slope for component B is somewhat more difficult.

$$\frac{dp_B}{dx_B} = \left(\frac{dx_A}{dx_B}\right)\times\left(\frac{dp_B}{dx_A}\right) = \left(\frac{d(1-x_B)}{dx_B}\right)\times\left(\frac{dp_B}{dx_A}\right) = -\frac{dp_B}{dx_A}$$

$$= -(-440 - 2\times 401x_A + 3\times 92x_A^2)\,\text{Torr} = (440 + 2\times 401x_A - 3\times 92x_A^2)\,\text{Torr}$$

$$\left[\frac{dp_B}{dx_B}\right]_{x_B=1} = \left[\frac{dp_B}{dx_B}\right]_{x_A=0} = 440\ \text{Torr}$$

Since 440 Torr is not equal to the value of p_B^*, we conclude that component B does not conform to Raoult's law.

What are the values of Henry's law constants K_J ($p_J = x_J K_J$ [6.11])? Henry's law exhibits a slope at $x_J = 0$ (Henry's limit in a p_J against x_J plot) that equals K_J; that is $dp_J/dx_J = K_J$ in the limit as $x_J \to 0$. We take the slopes found above and evaluate them in this limit.

$$K_A = [dp_A / dx_A]_{x_A=0} = (68 - 2\times12\times0 + 3\times643\times0 - 4\times283\times0)\,\text{Torr} = \boxed{68\ \text{Torr}}$$

$$K_B = [dp_B / dx_B]_{x_B=0} = -[dp_B / dx_A]_{x_A=1} = (440 + 2\times401\times1 - 3\times92\times1)\,\text{Torr} = \boxed{966\ \text{Torr}}$$

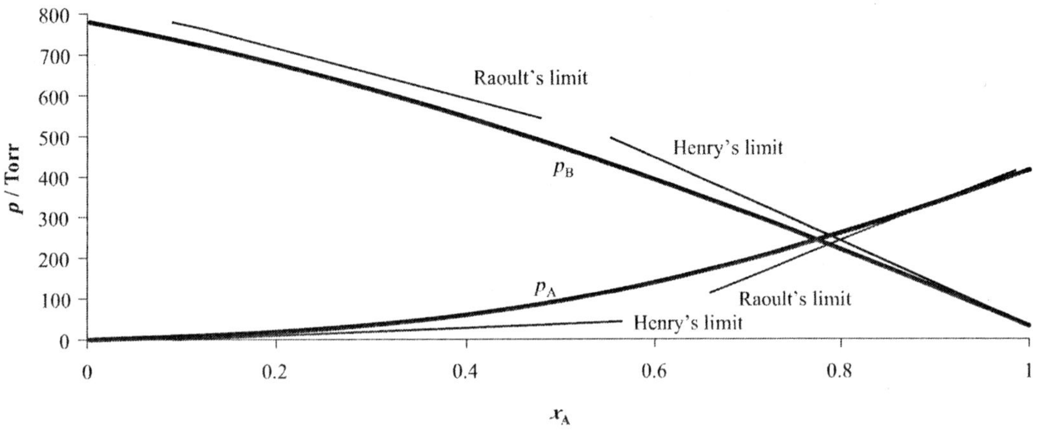

Figure 6.1

E6.17

x_A	0	0.20	0.40	0.60	0.80	1
p_A/Torr	0	127	246	357	457	539
p_B/Torr	701	631	526	394	234	0

A vapour pressure plot is presented in Figure 6.2 along with polynomial fits of the curves.

What are the vapour pressures of the pure substances as reported by the polynomial fits? These are the values when $x_A = 1$ and $x_A = 0$ in the respective eqns for A and B.

$$p_A^* = (-91 + 114 - 144 + 660)\,\text{Torr} = 539\ \text{Torr} \qquad \text{and} \qquad p_B^* = 701\ \text{Torr}$$

Do components A and B conform to Raoult's law ($p_J = x_J p_J^*$ [6.7])? Examine the curve slopes in Raoult's limit ($x_J = 1$) and ask whether the slope is along the imaginary line between the partial pressures at $x_J = 1$ (i.e., $p_J = p_J^*$) and $x_J = 0$ (i.e., $p_J = 0$). If the answer is no, the component does not conform to Raoult's law. Neither A nor B satisfy Raoult's law. A mathematical analysis that provides this same answer follows. A substance that conforms to Raoult's law exhibits a slope at $x_J = 1$ (Raoult's limit in a p_J against x_J plot) that equals the vapour pressure of the pure substance; that is, $dp_J / dx_J = p_J^*$ in the limit as $x_J \to 1$. Evaluation of the slope for component A is straight forward.

$$dp_A / dx_A = (660 - 2\times144 x_A + 3\times114 x_A^2 - 4\times91 x_A^3)\,\text{Torr}$$

$$[dp_A / dx_A]_{x_A=1} = (660 - 2\times144 + 3\times114 - 4\times91)\,\text{Torr} = 350\ \text{Torr}$$

Since 350 Torr is not equal to the value of p_A^*, we conclude that component $\boxed{\text{A does not conform to Raoult's law}}$. Evaluation of the slope for component B is somewhat more difficult.

$$\frac{dp_B}{dx_B} = \left(\frac{dx_A}{dx_B}\right)\times\left(\frac{dp_B}{dx_A}\right) = \left(\frac{d(1-x_B)}{dx_B}\right)\times\left(\frac{dp_B}{dx_A}\right) = -\frac{dp_B}{dx_A}$$

$$= -\left(-190 - 2\times984x_A + 3\times1177x_A^2 - 4\times703x_A^3\right)\text{Torr}$$

$$= \left(190 + 2\times984x_A - 3\times1177x_A^2 + 4\times703x_A^3\right)\text{Torr}$$

$$\left[\frac{dp_B}{dx_B}\right]_{x_B=1} = \left[\frac{dp_B}{dx_B}\right]_{x_A=0} = 190\ \text{Torr}$$

Since 190 Torr is not equal to the value of p_B^*, we conclude that component $\boxed{\text{B does not conform to Raoult's law}}$.

What are the values of Henry's law constants K_J ($p_J = x_J K_J$ [6.11])? Henry's law exhibits a slope at $x_J = 0$ (Henry's limit in a p_J against x_J plot) that equals K_J; that is $dp_J / dx_J = K_J$ in the limit as $x_J \to 0$. We take the slopes found above and evaluate them in this limit.

$$K_A = [dp_A / dx_A]_{x_A=0} = (660 - 2\times144\times0 + 3\times114\times0)\text{Torr} = \boxed{660\ \text{Torr}}$$

$$K_B = [dp_B / dx_B]_{x_B=0} = -[dp_B / dx_A]_{x_A=1}$$

$$= (190 + 2\times984\times1 - 3\times1177\times1 + 4\times703\times1)\text{Torr} = \boxed{1439\ \text{Torr}}$$

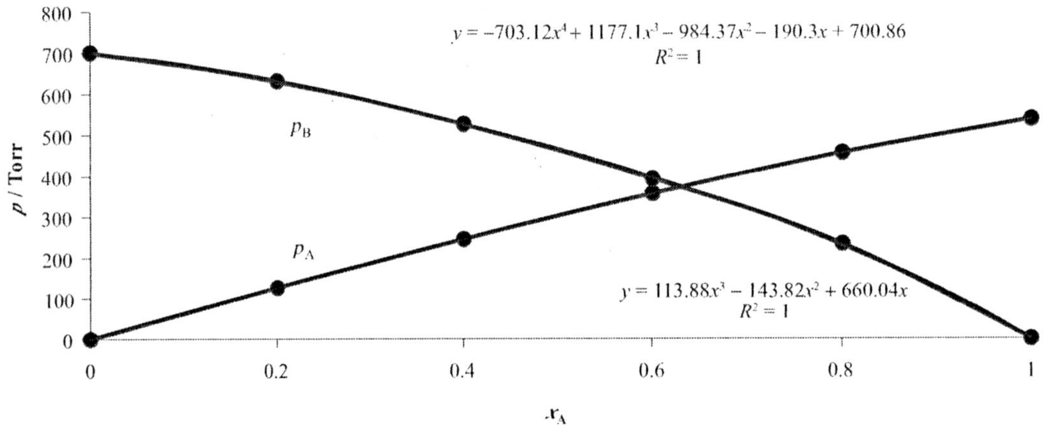

Figure 6.2

E6.18 $$\mu_{glu}^{\ominus} - \mu_{glu} = -RT\ln\left([\text{glu}]/c^{\ominus}\right)\ [6.14a]$$

$$= -(8.3145\ \text{J K}^{-1}\ \text{mol}^{-1})\times(293\ \text{K})\ln\left(0.10\ \text{mol dm}^{-3}/1.0\ \text{mol dm}^{-3}\right)$$

$$= \boxed{-5.6\ \text{kJ mol}^{-1}}$$

E6.19 Let B denote benzene and A the solute, then

$$p_B = x_B p_B^* \quad \text{and} \quad x_B = \frac{n_B}{n_A + n_B}.$$

Hence $p_B = \dfrac{n_B p_B^*}{n_A + n_B}$

which solves to

$$n_A = \frac{n_B(p_B^* - p_B)}{p_B}.$$

Then, because $n_A = m_A/M_A$, where m_A is the mass of A present,

$$M_A = \frac{m_A p_B}{n_B(p_B^* - p_B)} = \frac{m_A M_B p_B}{m_B(p_B^* - p_B)}$$

$$M_A = \frac{(0.133\ \text{g})\times(78.11\ \text{g mol}^{-1})\times(51.2\ \text{kPa})}{5.00\ \text{g}\times(53.0-51.2)\ \text{kPa}} = \boxed{59.1\ \text{g mol}^{-1}}.$$

E6.20 Assume 200 cm^3 of water has a mass of 0.200 kg.

$$\Delta T = K_f b_B\ [6.17] = 1.86\ \text{K kg mol}^{-1} \times \frac{2.5\ \text{g}}{342.3\ \text{g mol}^{-1}\times 0.200\ \text{kg}} = 0.068\ \text{K}$$

The freezing point will be approximately $\boxed{-0.068°\text{C}}$.

E6.21 Assume 200 cm^3 of water has a mass of 0.200 kg.

$$\Delta T = K_f b_B\ [6.17] = 1.86\ \text{K kg mol}^{-1} \times \frac{2.5\ \text{g}}{58.44\ \text{g mol}^{-1}\times 0.200\ \text{kg}} = 0.40\ \text{K}$$

The freezing point will be approximately $\boxed{-0.40°\text{C}}$.

E6.22 $\Delta T = K_f\, b_B$ [6.17] where b_B is the molality of B and is given by

$$b_B = \frac{n_B}{\text{mass of CCl}_4\ \text{in kg}} = \frac{28.0\ \text{g}}{M\times 0.750\ \text{kg}}$$
$$= \frac{37.3\ \text{g/kg}}{M} = \frac{\Delta T}{K_f}.$$

Solve for M,

$$M = \frac{37.3\ \text{g/kg}\times 30\ \text{K kg mol}^{-1}}{5.40\ \text{K}} = \boxed{207\ \text{g mol}^{-1}}.$$

E6.23 $K = \dfrac{[A_2]}{[A]^2}$ and let n denote the initial number of moles A.

At equilibrium $n_{A_2} = fn$, $n_A = (1-2f)n$, and the total amount of solute is $(1-f)n$. Therefore, if the volume is V,

$$K = \frac{fnV}{(1-2f)^2 n^2} = \frac{f}{(1-2f)^2 c} \qquad \text{where } c = n/V.$$

Vapor pressure, p is $p = x_{\text{solvent}}\, p^*$.

$$p = x_{\text{solvent}}\, p^* = \frac{n_{\text{solvent}}\, p^*}{n_A + n_{A_2} + n_{\text{solvent}}} = \frac{n_{\text{solvent}}\, p^*}{(1-f)n + n_{\text{solvent}}}$$

$n_{\text{solvent}} = Vr$ with $r = \rho/M$ and ρ = density of solvent.

$$p = \frac{rp^*}{(1-f)c + r} \quad \text{rearranging } f = 1 - \frac{r(p^*-p)}{cp} \quad \text{and, finally}$$

$$\boxed{K = \frac{1-\dfrac{r(p^*-p)}{cp}}{c\left(1-\dfrac{2r(p^*-p}{cp}\right)^2}}.$$

E6.24 For very dilute aqueous solutions: $[B] \approx \rho b_B$ where $\rho \approx 1.00 \text{ kg dm}^{-3}$.

$$\Pi = [B]RT\,[6.18b] = \rho b_B RT = \rho RT \times (\Delta T_f/K_f)\,[6.17]$$

$$\Delta T_f = K_f \Pi / (\rho RT)$$

Therefore, with $K_f = 1.86 \text{ K kg mol}^{-1}$ (Table 6.3)

$$\Delta T = \frac{(1.86 \text{ K kg mol}^{-1}) \times (150 \text{ kPa})}{(8.3145 \text{ kPa dm}^3 \text{ K}^{-1} \text{ mol}^{-1}) \times (300 \text{ K}) \times (1.00 \text{ kg dm}^{-3})} = 0.112 \text{ K}.$$

Therefore, the solution will freeze at about $\boxed{-0.11°\text{C}}$.

E6.25 Our strategy is to avoid assuming that these solutions behave as ideal-dilute solutions and to analyze the data as illustrated in Example 6.4. The method of analysis is suggested by the equation:

$$\frac{\Pi}{c} = \frac{RT}{M} + \left(\frac{RTB}{M^2}\right)c \qquad \text{where } c = m/V.$$

This says that a plot of Π/c against c has an intercept equal to RT/M and a slope equal to RTB/M^2 where B is osmotic virial coefficient. We draw up a table to calculate Π/c values, prepare a plot to check linearity, and perform a linear regression analysis of the plot with a scientific calculator (see Figure 6.3). The molar mass is given by $M = RT/intercept$. The virial coefficient is given by $B = slope \times M^2 / RT$.

$c/(\text{g dm}^{-3})$	2.042	6.613	9.521	12.602
Π/Pa	58.3	188.2	270.8	354.6
$\Pi/c/(\text{Pa g}^{-1} \text{ dm}^3)$	28.55	28.46	28.44	28.14

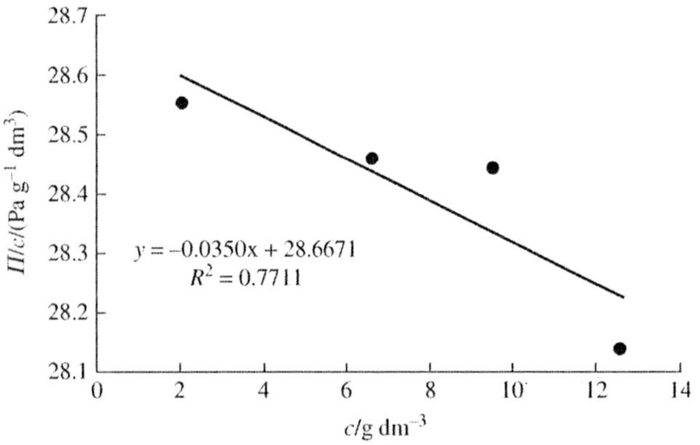

Figure 6.3

$$M = \left(\frac{(8.3145 \text{ J mol}^{-1} \text{ K}^{-1})(298 \text{ K})}{28.67 \text{ Pa g}^{-1} \text{ dm}^3}\right) \times \left(\frac{\text{dm}^3}{10^{-3} \text{ m}^3}\right) \times \left(\frac{10^{-3} \text{ kg}}{\text{g}}\right) = \boxed{86.4 \text{ kg mol}^{-1}}$$

$$B = \frac{(-0.0350 \text{ Pa g}^{-2} \text{ dm}^6)(8.64 \times 10^4 \text{ g mol}^{-1})^2}{(8.3145 \text{ J mol}^{-1} \text{ K}^{-1})(298 \text{ K})}\left(\frac{10^{-3} \text{ m}^3}{\text{dm}^3}\right) = -105 \text{ mol}^{-1} \text{ dm}^3$$

Note: The low value of the linear regression correlation coefficient (R^2) indicates that the apparent linearity is an artifact of experimental uncertainty of the individual data points. If this is confirmed, the data would be interpreted as indicating that B is too small to be confidently resolved with the above analysis and that the polymer solutions are behaving as ideal-dilute solutions within the limits of experimental data. In this case, the data should be analyzed with the van't Hoff equation [6.18b] in the form: $M = RTc/\Pi$. That is, the molar mass equals the mean of RTc/Π for the data set. This analysis gives $(c/\Pi)_{\text{mean}} = 3.522\times10^{-2}$ g dm^{-3} Pa^{-1} and, therefore,

$$M = (8.3145\ \text{J K}^{-1}\ \text{mol}^{-1})\times(298.15\ \text{K})\times(3.522\times10^{-2}\ \text{g dm}^{-3}\ \text{Pa}^{-1})\times\left(\frac{1\ \text{dm}^3}{10^{-3}\ \text{m}^3}\right)\times\left(\frac{10^{-3}\ \text{kg}}{\text{g}}\right)$$

$$= 87.\overline{31}\ \text{kg mol}^{-1}$$

E6.26 Our strategy is to avoid assuming that these solutions behave as ideal-dilute solutions and to analyze the data as illustrated in Example 6.4. The method of analysis is suggested by the equation:

$$\frac{\Pi}{c} = \frac{RT}{M} + \left(\frac{RTB}{M^2}\right)c \qquad \text{where } c = m/V.$$

$\Pi = \rho gh$ [hydrostatic pressure] so

$$\frac{h}{c} = \frac{\Pi}{\rho gc} = \left(\frac{RT}{\rho gM}\right) + \left(\frac{RTB}{\rho gM^2}\right)c.$$

This says that a plot of h/c against c has an intercept equal to $RT/\rho gM$ and a slope equal to $RTB/\rho gM^2$ where B is osmotic virial coefficient. We draw up a table to calculate h/c values

c/(mg cm^{-3})	3.221	4.618	5.112	6.722
h / cm	5.746	8.238	9.119	11.990
h/c / (mg^{-1} cm^4)	1.784	1.783	1.784	1.784

Inspection of the h/c values reveals that they are a constant for this experimental set. This implies that the virial coefficient equals zero and that these enzyme solutions are behaving as ideal-dilute solutions. The last term in the above equation vanishes giving the van't Hoff equation [6.18b]. Solving for M (assuming a density of 1.000 g cm^{-3}):

$$M = \frac{RT}{\rho g\times(h/c)}$$

$$= \frac{(8.3145\ \text{J K}^{-1}\ \text{mol}^{-1})\times(293.15\ \text{K})}{(1.000\ \text{g cm}^{-3})\times(9.807\ \text{m s}^{-2})\times(1.784\times10^{3}\ \text{g}^{-1}\ \text{cm}^{4})}\times\left(\frac{1\ \text{cm}}{10^{-2}\ \text{m}}\right)$$

$$= \boxed{13.9\overline{3}\ \text{kg mol}^{-1}}$$

E6.27 The data are plotted in Figure 6.4. From tie line (a) on the graph, the vapor in equilibrium with liquid of composition $x_T = 0.250$ has $y_T = \boxed{0.36}$. From tie line (b), for $x_O = 0.250$, $x_T = 0.750$, $y_T = \boxed{0.81}$.

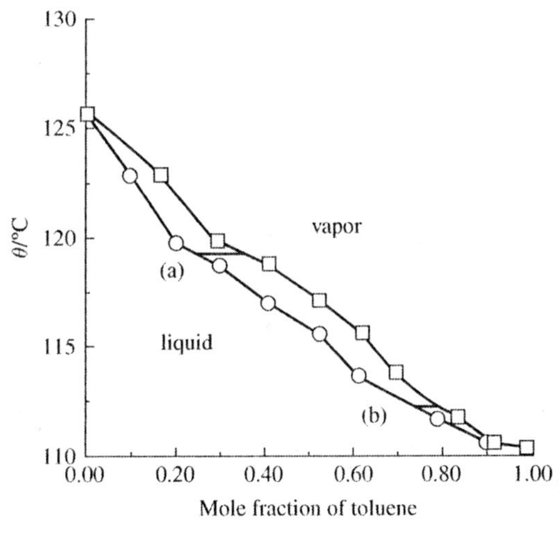

Figure 6.4

E6.28 The phase diagram of the NH_3/N_2H_4 system is sketched in Figure 6.5.

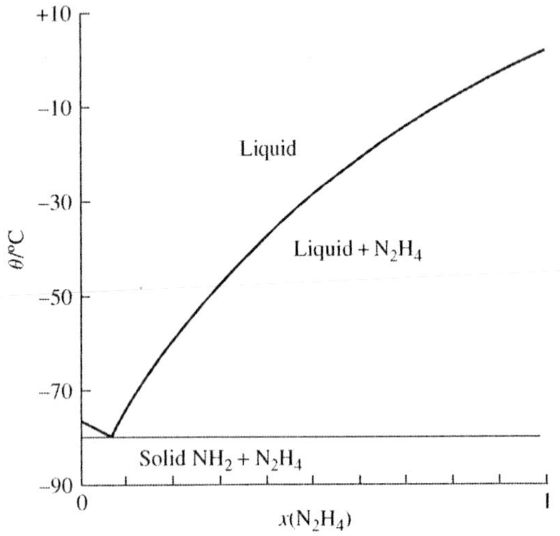

Figure 6.5

E6.29 Refer to Figure 6.37 of the text. At b_3 there are two partially miscible liquids at equilibrium with compositions $x_B = 0.34$ and $x_B = 0.62$; their abundances are in the ratio 1.5 [lever rule]. Because $C = 2$ and $P = 2$ we have $F = 2$ (such as p and x). On heating, the phases merge, and the single-phase liquid region is encountered near an upper critical point. Then $F = 3$ (such as p, T, and x). The liquid comes into equilibrium with its vapor (i.e., boils) when the isopleth cuts the phase line at b_2; the vapour composition is $x_B = 0.29$ and subsequent condensation and vaporization quickly leads to the low-boiling azeotrope at $x_B = 0.21$.

If heating begins in the single liquid phase region at point a_1 ($x_B = 0.73$), boiling occurs at a_2 with a vapour phase composition of $x_B = 0.51$. Once again, subsequent condensation and vaporization (in a fractionating column) leads to the low-boiling azeotrope at $x_B = 0.21$.

E6.30 The phase diagrams and cooling curves are shown in Figure 6.6.

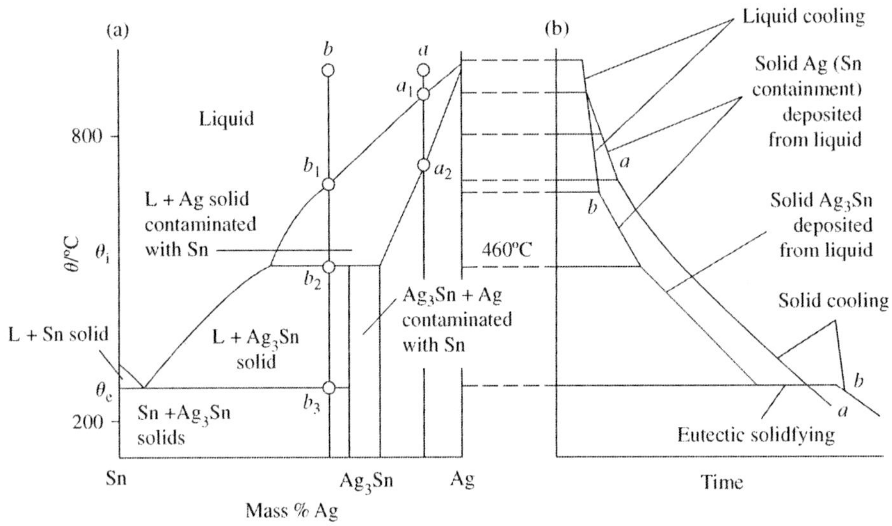

Figure 6.6

(*a*) Solid silver with dissolved tin begins to precipitate at a_1, and the sample solidifies completely at a_2.
(*b*) Solid silver with dissolved tin begins to precipitate at b_1, and the liquid becomes richer in Sn. The peritectic reaction occurs at b_2, and as cooling continues Ag_3 Sn is precipitated and the liquid becomes richer in tin. At b_3 the system has its eutectic composition (e) and freezes without further change.

E6.31 The curves are shown in part (b) of the figure in the solution for 6.30. Note the eutectic halt for the isopleth b.

E6.32 (a) Reading the silver composition from the 800°C solid phase line of text Figure 6.38, we find the solid to be 95% silver by mass. Thus, the solubility is 5% tin by mass. Furthermore, at this temperature the solid is in equilibrium with liquid that is 83% silver.

(b) Reading the silver composition from the 460°C solid phase line, we find the solid to be 82% silver by mass. Furthermore, at this temperature the solid is in equilibrium with liquid that is 62% silver. There is no Ag_3Sn in the solid at this temperature because the compound Ag_3Sn decomposes at this temperature.

(c) At 300°C the solid is 80% silver. Thus, the solubility is 20% Ag_3Sn by mass.

COMMENT. It is interesting to explore the equilibrium phases upon cooling from the compositions of points *a* and *b* of text Figure 3.38. Upon cooling from point *a*, a single liquid is observed until a solid of about 89% silver condensates at about 610°C. $Ag_3Sn(s)$ begins to appear in the liquid at about 430°C. Upon cooling from point *b*, a single liquid is observed until a solid of about 97% silver condensates at about 860°C.

E6.33 Follow the vertical line from point *a* to point *d* in the Figure 6.7 phase diagram of an alloy of copper and aluminium.

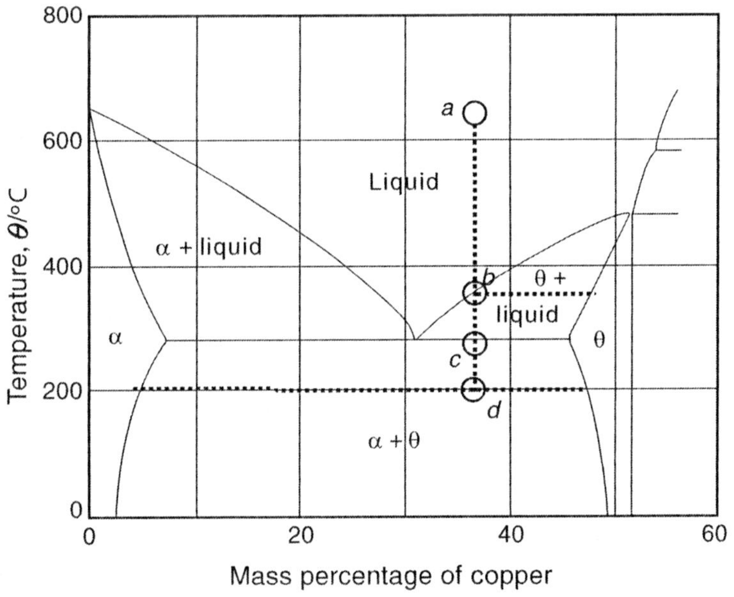

Figure 6.7

When lowering the temperature from point a toward point *b*, a two-component liquid of about 37% Cu is present. Upon reaching point b (~350°C) the solid θ, which is a 48% Cu alloy at this temperature, begins to come out of solution and the remaining liquid becomes richer in aluminium.

Upon reaching the temperature of point *c* (~300°C) the liquid composition has dropped to about 32% copper and the composition of solid θ has also dropped slightly to about 45% Cu. A second solid phase begins to come out of solution at this temperature. This is solid α, an aluminium alloy that is about 7% Cu.

Lowering the temperature below point *c* completely freezes the liquid into a heterogeneous mixture of θ and α phases. At point d (~200°C) the lever rule indicates that the mole ratio n_θ/n_α is about 3.4.

E6.34 Follow the vertical line from point *a* to point *e* in the Figure 6.8 phase diagram of a simple steel.

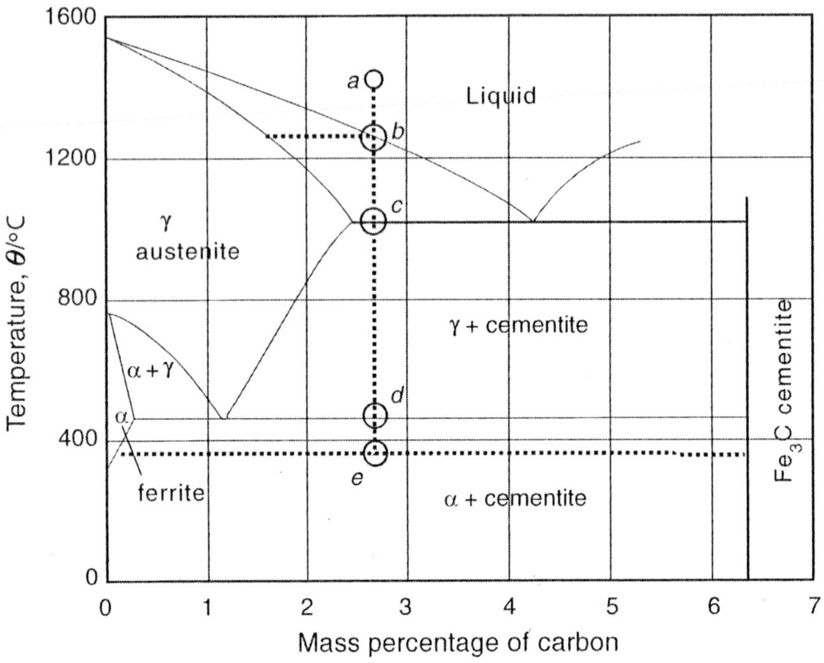

Figure 6.8

When lowering the temperature from point a toward point *b*, a two-component liquid of about 2.7% C and 97.3% Fe is present. Upon reaching point *b* (~1300°C) the solid γ (the mineral austenite), which is a 1.6% C alloy at this temperature, begins to come out of solution and the remaining liquid becomes richer in carbon.

Just before reaching the temperature of point *c* (~1000°C) the liquid composition has increased to about 4.2% C and the composition of solid γ has increased to about 2.3% C. Upon reaching point *c*, a second solid phase begins to come out of solution. This is solid Fe_3C (the mineral cementine), an iron alloy that is about 6.3% C. The liquid completely freezes below point *c*. At temperatures between points *c* and *d*, there is a heterogeneous mixture of γ and Fe_3C phase with a predominance of the γ phase.

Solid α (the mineral ferrite) appears in the equilibrium mixture at point *d* (~500°C). Below point *d*, γ is not present in the heterogeneous mixture and at room temperature (point *e*) only α and Fe_3C are observed.

E6.35 The temperature–composition diagram for hexane and perfluorohexane is sketched in Figure 6.9. (a) The mixture has a single liquid phase at all compositions above 22.7°C. (b) Upon adding perfluorohexane to hexane at 22.0°C, the perfluorohexane dissolves in the hexane until the mole fraction of perfluorohexane reaches 0.24. When the composition reaches $x(C_6F_{10})$ = 0.24, the mixture separates into two liquid phases of composition $x(C_6F_{10})$ = 0.24 and 0.48. The relative amounts of the two phases change as more perfluorohexane is added until the composition reaches $x(C_6F_{10})$ = 0.48. At all mole fractions greater than 0.48 in C_6F_{14} the mixture forms a single liquid phase.

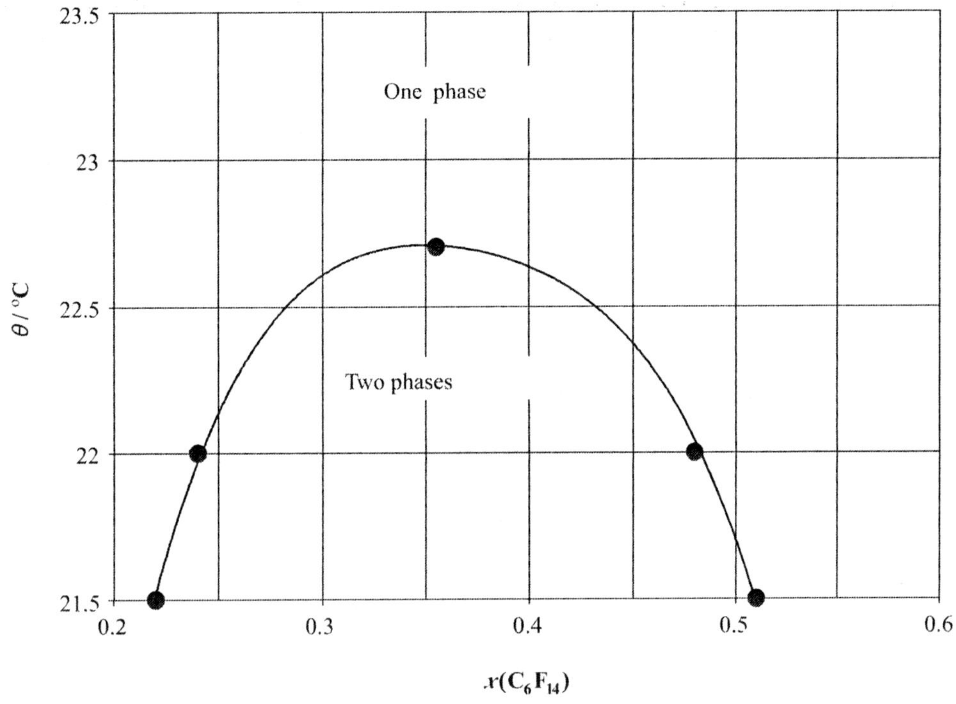

Figure 6.9

E6.36 (a) No, the region of stability of the molten-globule form does not extend below 0.1 concentration of denaturant.

(b) The native form converts to the molten-globule form at $T \approx 0.65$ and finally to the unfolded form at $T \approx 0.85$.

E6.37 At roughly 28°C, solid begins to form as the state point enters the two-phase region. Within the two-phase region, the proportion of liquid and solid can be determined by the lever rule. As the temperature is lowered through the two-phase region, the proportion of solid increases until, at roughly 23°C, the system becomes totally solid.

E6.38 The **Nernst distribution law** [6.21] tells us that the ratio of mole fractions of aspirin in the two immiscible liquids equals a constant, K.

$$K = \frac{x_{\text{aspirin}}(2)}{x_{\text{aspirin}}(1)} = \frac{0.18}{0.11}$$

Thus, when the amounts are changed so that $x_{\text{aspirin}}(1) = 0.15$, the mole fraction in liquid 2 becomes:

$$x_{\text{aspirin}}(2) = Kx_{\text{aspirin}}(1) = \left(\frac{0.18}{0.11}\right) \times (0.15) = \boxed{0.25}.$$

Solutions to projects

P6.39 (a) Figure 6.10 is a plot of the partial molar volume of ethanol solutions at 25°C.

$V_{\text{ethanol}}/(\text{cm}^3\ \text{mol}^{-1}) = 54.6664 - 0.72788\,b + 0.084768\,b^2$ where b is the magnitude of molality

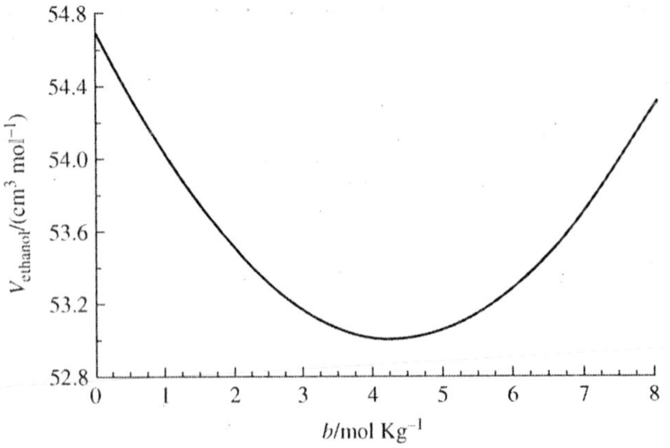

Figure 6.10

Examination of the plot shows that the minimum occurs at

$$b \cong \boxed{4.3\ \text{mol kg}^{-1}} \qquad \text{with } V_{\text{ethanol}} = \boxed{53.0\ \text{cm}^3\ \text{mol}^{-1}}.$$

To convert from molality (b) to mole fraction we proceed by considering a 1 kg mass of solvent. If the solvent is water,

$$b_{\text{water}} = \frac{1000\ \text{g/kg}}{18.02\ \text{g mol}^{-1}} = 55.5\ \text{mol kg}^{-1}$$

$$x_{\text{ethanol}} = \frac{n_{\text{ethanol}}}{n_{\text{ethanol}} + n_{\text{water}}} = \frac{b_{\text{ethanol}}}{b_{\text{ethanol}} + b_{\text{water}}} = \frac{4.3\ \text{mol kg}^{-1}}{(4.3 + 55.5)\ \text{mol kg}^{-1}} = \boxed{0.072}$$

(b) $$V_{\text{ethanol}} = c_0 + c_1 b + c_2 b^2$$

$$\frac{\text{d}V_{\text{ethanol}}}{\text{d}b} = c_1 + 2c_2 b$$

The minimum occurs at $b = b_{\text{min}}$ where $\dfrac{\text{d}V_{\text{ethanol}}(b_{\text{min}})}{\text{d}b} = c_1 + 2c_2 b_{\text{min}} = 0$

$$b_{\text{min}} = \frac{-c_1}{2c_2} = \frac{-(-0.72788\ \text{cm}^3\ \text{kg mol}^{-2})}{2(0.084768\ \text{cm}^3\ \text{kg}^2\ \text{mol}^{-3})} = \boxed{4.2934\ \text{mol kg}^{-1}}$$

$$V_{\text{min}} = 54.6664\ \text{cm}^3\ \text{mol}^{-1} - 0.72788\ \text{cm}^3\ \text{kg mol}^{-2}(4.2934\ \text{mol kg}^{-1})$$
$$+ 0.084768\ \text{cm}^3\ \text{kg}^2\ \text{mol}^{-3}(4.2934\ \text{mol kg}^{-1})^2$$
$$= \boxed{53.1039\ \text{cm}^3\ \text{mol}^{-1}}$$

P6.40 (a) $V/(1\ \text{kg water}) = (n_{\text{ethanol}}V_{\text{ethanol}} + n_{\text{water}}V_{\text{water}})/(1\ \text{kg water})$ [6.1]

$= b_{\text{ethanol}}V_{\text{ethanol}} + b_{\text{water}}V_{\text{water}}$ where $b_J = n_J/(1\ \text{kg water})$

$b_{\text{water}}V_{\text{water}}/\text{cm}^3 = V/(1\ \text{kg water}) - b_{\text{ethanol}}V_{\text{ethanol}} = V/(1\ \text{kg water}) - bV_{\text{ethanol}}$ where $b = b_{\text{ethanol}}$

$= \{1002.9 + 54.6664b - 0.36394b^2 + 0.028256b^3\}$

$-\{b\}\times\{54.6664 - 0.72788b + 0.084768b^2\}$ [provided in exercise 6.40 and 6.39]

$= 1002.9 + 0.36394b^2 - 0.056512b^3$

Since $b_{\text{water}} = (1000\ \text{g})/(18.01528\ \text{g mol}^{-1}) = 55.50844$ mol per kg water,

$$\boxed{V_{\text{water}}/\text{cm}^3 = 18.068 + 6.5565\times10^{-3}b^2 - 1.0181\times10^{-3}b^3}.$$

From an examination of Figure 6.11, we see that the maximum occurs at $b \approx 4.3$ mol kg^{-1} in agreement with the minimum in V_{ethanol}.

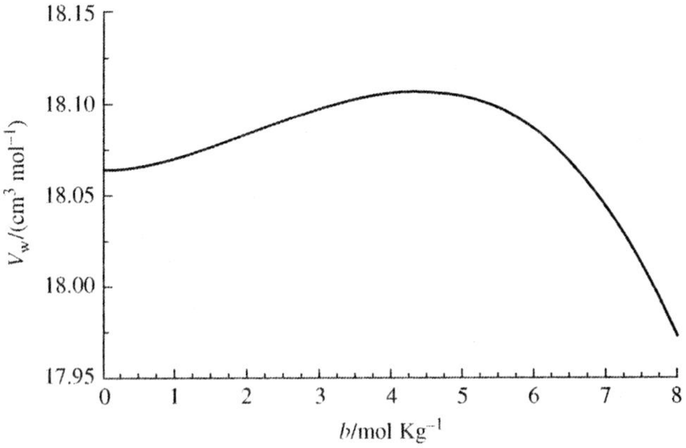

Figure 6.11

(b) The moles of ethanol (EtOH) in 1 kg of water, n_{EtOH}, is related to the molality b by the expression $b = \dfrac{n_{\text{EtOH}}}{1\ \text{kg}}$. Consequently, $\dfrac{db}{dn_{\text{EtOH}}} = \dfrac{1}{1\ \text{kg}} = 1\ \text{kg}^{-1}$.

$$V = c_0 + c_1 b + c_2 b^2 + c_3 b^3 \qquad \text{where } c_0 = 1002.93\ \text{cm}^3,\quad c_1 = 54.6664\ \text{cm}^3\ \text{mol}^{-1}\ \text{kg}$$

$$c_2 = -0.36394\ \text{cm}^3\ \text{mol}^{-2}\ \text{kg}^2,\quad c_3 = 0.028256\ \text{cm}^3\ \text{mol}^{-3}\ \text{kg}^3$$

$$V_{\text{EtOH}} = \frac{dV}{dn_{\text{EtOH}}} = \frac{dV}{db} \times \frac{db}{dn_{\text{EtOH}}} = \frac{dV}{db} \times (1\ \text{kg}^{-1})$$

$$= \frac{d(c_0 + c_1 b + c_2 b^2 + c_3 b^3)}{db} \times (1\ \text{kg}^{-1}) = (c_1 + 2c_2 b + 3c_3 b^2) \times (1\ \text{kg}^{-1})$$

This equation is used to calculate the points for a plot of V_{EtOH} as a function of b. Alternatively, a plot of V_{EtOH} as a function of x_{EtOH} may be prepared by calculating V_{EtOH} over a range of b values with the above equation and calculating corresponding value of x_{EtOH} at each b value with

$$x_{\text{EtOH}} = \frac{b \times (1\ \text{kg})}{b \times (1\ \text{kg}) + (10^3/18.02)\text{mol}}$$

Sample points for a $V_{\text{EtOH}}(x_{\text{EtOH}})$ plot are in the table and a complete plot appears in Figure 6.12.

b/m	x_{EtOH}	V_{EtOH} / cm^3 mol^{-1}
0	0	54.6664
1	0.017701	54.02312
2	0.034786	53.54904

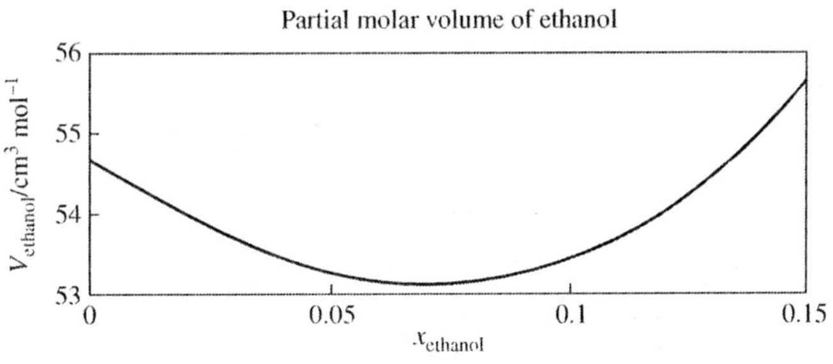

Figure 6.12

P6.41 (a) The 97 percent saturated haemoglobin (Hb) in the lungs releases oxygen in the capillary until the haemoglobin is 75 percent saturated.

100 cm^3 (= 0.100 dm^3) of blood in the lung containing 150 g dm^{-3} of Hb at 97 percent saturated with O_2 binds

$$0.97\times\left(\frac{1.34\ \text{cm}^3\ \text{O}_2}{\text{g Hb}}\right)\times\left(\frac{150\ \text{g Hb}}{\text{dm}^3}\right)\times\left(0.100\ \text{dm}^3\right)=20\ \text{cm}^3\ \text{O}_2.$$

The same 100 cm^3 of blood in the arteries would contain

$$20\ \text{cm}^3\ \text{O}_2\times\frac{75\%}{97\%}=15\ \text{cm}^3\ \text{O}_2.$$

Therefore, about (20 – 15) cm^3 or $\boxed{5\ \text{cm}^3}$ of O_2 is given up in the capillaries to body tissue.

(b) In this case, we write the Henry's law expression as

$$\text{mass of N}_2=p_{\text{N}_2}\times\text{mass of H}_2\text{O}\times K_{\text{N}_2}.$$

$$\text{At } p_{\text{N}_2}=0.78\times 4.0\ \text{atm}=3.1\ \text{atm},$$

$$\text{mass of N}_2=3.1\ \text{atm}\times 100\ \text{g H}_2\text{O}\times 1.8\times10^{-4}\ \text{mg N}_2/(\text{g H}_2\text{O atm})$$
$$=\boxed{0.056\ \text{mg N}_2}.$$

$$\text{At } p_{\text{N}_2}=0.78\ \text{atm},\qquad \text{mass of N}_2=\boxed{0.014\ \text{mg N}_2}.$$

(c) In fatty tissue the increase in N_2 concentration from 1 atm to 4 atm is

$$4\times(0.056-0.014)\ \text{mg N}_2=\boxed{0.17\ \text{mg N}_2}.$$

7 Chemical equilibrium: the principles

Answers to discussion questions

D7.1 The position of equilibrium is always determined by the condition that the reaction Gibbs energy equal zero at equilibrium:

$$\Delta_r G = 0 \text{ at equilibrium and, therefore, } \Delta_r G^\ominus = -RT\ln Q_{eq} = -RT\ln K\,[7.8].$$

If the mixing of reactants and products gives $\Delta_r G < 0$, reactant activities will spontaneously diminish to increase product activities until $\Delta_r G = 0$. If the mixing gives $\Delta_r G > 0$, product activities will spontaneously diminish to increase reactant activities until $\Delta_r G = 0$. For the general reaction

$$a\,A + b\,B \rightleftharpoons c\,C + d\,D \qquad Q = a_C^c a_D^d / a_A^a a_B^b \quad [7.5]$$

we say that, when the mixing gives $\Delta_r G < 0$, the reaction proceeds spontaneously to the right (*forward* reaction) until equilibrium is achieved but, when $\Delta_r G > 0$, the reaction proceeds spontaneously to the left (*reverse* direction). If reactant or product is added to an equilibrium mixture, the reaction spontaneously shifts in the direction that lowers the Gibbs energy of the reaction mixture (see text Figures 7.1–7.3, remember that $\Delta_r G$ is the *slope* of G plotted against composition). The reaction spontaneously shifts to the right upon addition of reactant to an equilibrium mixture; left upon addition of product. We must also remember that thermodynamics says nothing about the rate at which the reaction occurs or shifts. A spontaneous reaction may occur very rapidly, infinitely slowly, or at any intermediate speed.

D7.2 A non-spontaneous, endergonic reaction ($\Delta_r G > 0$) may be driven forward by a spontaneous, exergonic reaction ($\Delta_r G < 0$) that can supply the requisite reaction Gibbs energy. The total Gibbs energy change for this reaction coupling must be exergonic and the coupling is accomplished in many enzyme-catalyzed biochemical reactions where the enzyme serves as the transfer agent for the Gibbs energy. The exergonic reaction gives up a portion of its Gibbs energy not as heat but to the conversion of a low potential biochemical intermediate to a high potential one. The high potential species carries the energy to the endergonic reaction (on the enzyme surface for example) and in the process of releasing its chemical energy to the endergonic reaction it returns to its low potential form. The energy carrying intermediate effectively "couples" the two reactions. Adenosine diphosphate (ADP) and adenosine triphosphate (ADP) is an example set of coupling intermediates:

$$\text{ATP(aq)} + H_2O(l) \rightarrow \text{ADP(aq)} + P_i^-(aq) + H^+(aq) \qquad \Delta_r G^\oplus = -31\ \text{kJ mol}^{-1}.$$

D7.3 **Le Chatelier's principle**, an empirical rule: When a system at equilibrium is subjected to a disturbance, the composition of the system adjusts so as to tend to minimize the effect of the disturbance.

Thermodynamics provides both understanding of the rule's origin and relations that quantify catalytic, temperature change, and compression (pressure change) effects.

(1) Response to the presence of a catalyst. Neither the quantity $\Delta_r G^\oplus$ nor the equilibrium constant K is affected by a catalyst so the presence of a catalyst does not illicit a reaction response within a chemical mixture at equilibrium. The catalyst increases both the reaction rate to the left and the reaction rate to the left but these rates remain equal at dynamic equilibrium.

(2) Response to change in temperature. Eqn 7.15 shows that K decreases with increasing temperature when the reaction is exothermic ($\Delta_r H^\ominus > 0$); thus the reaction shifts to the left, the opposite occurs in endothermic reactions ($\Delta_r H^\ominus < 0$).

$$\ln K' - \ln K = \frac{\Delta_r H^\ominus}{R}\left(\frac{1}{T} - \frac{1}{T'}\right) \quad [7.15]$$

The reaction in which reactants and products exhibit identical standard enthalpies ($\Delta_r H^\ominus \simeq 0$) has an equilibrium constant that is independent of temperature ($K' = K$). A simple example is provided by mixing to form an ideal solution.

(3) Response to change in pressure. The relation $K = e^{-\Delta_r G^\ominus/RT}$ [7.8] indicates that the equilibrium constant for a gas phase reaction has no dependence upon pressure because the right side of the relation depends upon T alone and, therefore, the left side must depend upon T alone. Nonetheless, individual partial pressures and mole fractions can change as the total pressure changes. This will happen when there is a difference, $\Delta\nu_{gas}$, between the sums of the number of moles of gases on the product and reactant sides of the chemical equation. The requirement of an unchanged equilibrium constant implies that the side with the smaller number of moles of gas be favored as pressure increases.

These statements are based upon the thermodynamic analysis of the general gas-phase reaction equation:

$$a\,\text{A} + b\,\text{B} \rightleftharpoons c\,\text{C} + d\,\text{D} \qquad Q = a_C^c a_D^d / a_A^a a_B^b \quad [7.5]$$

Using the equilibrium (eq) definition $K_c = [\text{C}]_{eq}^c [\text{D}]_{eq}^d (c^\ominus)^{-\Delta\nu_{gas}} / [\text{A}]_{eq}^a [\text{B}]_{eq}^b$, the relevant thermodynamic relation is

$$K = K_c \times \left(\frac{c^\ominus RT}{p^\ominus}\right)^{\Delta\nu_{gas}} = K_c \times \left(\frac{T}{12.027\ \text{K}}\right)^{\Delta\nu_{gas}} \quad [7.13a, b]$$

$$\text{where } \Delta\nu_{gas} = \sum_{J=\text{product gases}} \nu_J - \sum_{J=\text{reactant gases}} |\nu_J|.$$

This equation clearly shows that, when a larger pressure is isothermally applied to an equilibrium mixture, the value of K_c remains unchanged because neither K nor the factor $T^{\Delta\nu_{gas}} / 12.027\ \text{K}$ change with changes in p. To find the Le Chatelier response to an increase in pressure, we now take a careful look at K_c in terms of mole fractions. For perfect gas component J we substitute the relation $[\text{J}]_{eq} = x_J p/RT$ into the expression for K_c. This gives

$$K_c = K_x \times (c^\ominus RT/p)^{-\Delta\nu_{gas}} \qquad \text{where } K_x = x_C^c x_D^d / x_A^a x_B^b.$$

For an isothermal compression this simplifies to

$$K_x \propto 1/p^{\Delta\nu_{gas}},$$

which says that the mole fraction reaction quotient depends upon pressure! If $\Delta\nu_{gas} > 0$, K_x is diminished by an increase in p. This is a decrease in the mole fractions of reaction products while the mole fractions of reactants increase – a shift to the left, the side with fewer moles of gas in the balanced reaction equation. For the same reason, the shift is to the right when $\Delta\nu_{gas} < 0$. If $\Delta\nu_{gas} = 0$, there will be no reaction shift when the pressure is changed.

Failure of Le Chatelier's principle is very unusual. The expected response to pressure may prove wrong should the gas components be non-perfect, real gases. Should the gases of the reaction side that is favored by Le Chatelier's principle strongly repel while the gas molecules of the other side strongly attract, a compression may shift the reaction in the direction opposite to that expected for perfect gases.

D7.4 Activities and activity coefficients, presented in Sections 6.5 and 7.2, are used to address questions that concern real, non-ideal mixtures. It is well worth remembering several useful activity forms. Eqn 6.15 provides the general definition of the activity for species J, a_J:

$$\mu_J = \mu_J^\ominus + RT \ln a_J \quad [6.15 \text{ and } 7.3]$$

where $\mu_J^\ominus$ is the value of the chemical potential of J in the standard state for which $a_J = 1$. The dimensionless activity coefficients, γ_J, of non-ideal mixtures are defined by eqns that have the general form $a_J = \gamma_J \times$ (concentration of J).

Perfect Gas:	$a_J = p_J / p^\ominus$	($\mu_J^\ominus$ depends upon T alone.)
Real Gas:	$a_J = \gamma_J p_J / p^\ominus$	($\mu_J^\ominus$ depends upon T alone.)
Ideal solutions:	$a_J = x_J$	
Ideal-dilute solutions:	$a_B = [B]/c^\ominus$ where $c^\ominus = 1$ mol dm^{-3}	
Solvent A of a non-ideal solution:	$a_A = \gamma_A x_A$	
Solute B of a non-ideal solution:	$a_B = \gamma_B [B]/c^\ominus$ where $c^\ominus = 1$ mol dm^{-3}	

Also, the activity of a pure solid or a pure liquid at 1 bar always equals 1 as these are standard states.

The significance of non-ideality is appreciated by inclusion of a non-unity activity coefficient. Consider the general real gas-phase reaction equation:

$$a\,\text{A} + b\,\text{B} \rightleftharpoons c\,\text{C} + d\,\text{D} \qquad Q = a_C^c a_D^d / a_A^a a_B^b \quad [7.5]$$

The equilibrium constant is given by

$$K = \left(a_C^c a_D^d / a_A^a a_B^b\right)_{\text{equilibrium}} \quad [7.7] = K_\gamma K_p$$

where the activity coefficient quotient is $K_\gamma = \gamma_C^c \gamma_D^d / \gamma_A^a \gamma_B^b$ and $K_p = p_C^c p_D^d / p_A^a p_B^b \times (p^\ominus)^{\Delta\nu_{\text{gas}}}$.

In discussion question 7.3 it is shown that K for a gas mixture depends upon T alone. Thus, for a perfect gas mixture (for which $K_\gamma = 1$) the reaction quotient K_p equals K and we conclude that K_p also depends upon T alone. For a real gas mixture, K_γ does not equal a constant; it is a complicated function of both T and p. This means that K_p must also be a complicated function of both T and p for a real gas mixture even though the function $K_\gamma K_p$ depends upon T alone. In the limit of very low pressures all gas mixtures behave ideally so in this case there is no difference between K and K_p. At higher pressures non-ideality and K_γ become important; the balance between molecular attractions and repulsions cause K_p to be either greater or smaller than K depending upon the value of the pressure.

Side note: The activity coefficient of a real gas gives an indication of dominant intermolecular forces. Coefficients greater than 1 are observed when repulsions dominate; coefficients less than one dominate when attractions dominate. In the limit of zero pressure all gases behave as perfect gases with activity coefficients equal to 1 and, therefore, $a_{J(\text{gas})} = \gamma_J p_J/p^\ominus = p_J/p^\ominus$ in the perfect gas case only. Thus, we conclude that intermolecular forces cause the thermodynamic equilibrium constant K to respond differently to changes in pressure from the equilibrium constant expressed in terms of partial pressures, K_p.

D7.5 The van't Hoff equation, written as $\ln K' - \ln K = \dfrac{\Delta_r H^\ominus}{R}\left(\dfrac{1}{T} - \dfrac{1}{T'}\right)$ [7.15], is valid over small temperature ranges in which neither $\Delta_r H^\ominus$ nor $\Delta_r S^\ominus$ vary much with temperature (see Derivation 7.2). Kirchhoff's law [3.6] indicates that the former criteria is often satisfied because $\Delta_r C_p^\ominus$ is often small. The latter criteria is justified by the relation $\Delta_r S^\ominus = \Delta_r H^\ominus / T$ for a constant T and p process [4.1]; the numerator does not change much with T and the large magnitude of T means the $1/T$ does not change much either so there is little variation of $\Delta_r S^\ominus$ over a small temperature range.

Solutions to exercises

E7.1 The general reaction equation and corresponding reaction quotient are:

$$a\,\mathrm{A} + b\,\mathrm{B} \rightleftharpoons c\,\mathrm{C} + d\,\mathrm{D} \qquad Q = a_\mathrm{C}^c a_\mathrm{D}^d / a_\mathrm{A}^a a_\mathrm{B}^b \ \ [7.5].$$

The activities of pure solids and liquids are equal to 1. Substitute $a_\text{solute} = \gamma_\text{solute}[\text{solute}]/c^\ominus \approx [\text{solute}]/c^\ominus$ for solute activities; assume perfect gas behavior with the substitution $a_\text{gas} = p_\text{gas}/p^\ominus$. $c^\ominus$ and $p^\ominus$ are omitted for convenience and must be replaced during computations. The concentrations and partial pressures of reaction quotient change in time as the reaction occurs until equilibrium is achieved.

(a) $Q = \dfrac{p_{\mathrm{CO}_2}^6}{[\mathrm{CH_3COCOOH}]^2\, p_{\mathrm{O}_2}^5}$ (b) $Q = \dfrac{[\mathrm{FeSO_4}]}{[\mathrm{PbSO_4}]}$

(c) $K = \dfrac{[\mathrm{HCl}]^2}{p_{\mathrm{H}_2}}$ (d) $Q = \dfrac{[\mathrm{CuCl_2}]}{[\mathrm{CuCl}]^2}$

E7.2 The general reaction equation and corresponding equilibrium are:

$$a\,\mathrm{A} + b\,\mathrm{B} \rightleftharpoons c\,\mathrm{C} + d\,\mathrm{D} \qquad K = \left(a_\mathrm{C}^c a_\mathrm{D}^d / a_\mathrm{A}^a a_\mathrm{B}^b\right)_\text{equilibrium}.$$

The activities of pure solids and liquids are equal to 1. Substitute $a_\text{solute} = \gamma_\text{solute}[\text{solute}]/c^\ominus \approx [\text{solute}]/c^\ominus$ for solute activities; assume perfect gas behavior with the substitution $a_\text{gas} = p_\text{gas}/p^\ominus$. $c^\ominus$ and $p^\ominus$ are omitted for convenience and must be replaced during computations. The concentrations and partial pressures of these expressions are the equilibrium values.

(a) $K = \dfrac{p_\mathrm{COCl}\, p_\mathrm{Cl}}{p_\mathrm{CO}\, p_{\mathrm{Cl}_2}}$ (b) $K = \dfrac{p_{\mathrm{SO}_3}^2}{p_{\mathrm{SO}_2}^2\, p_{\mathrm{O}_2}}$

(c) $K = \dfrac{p_\mathrm{HBr}^2}{p_{\mathrm{H}_2}\, p_{\mathrm{Br}_2}}$ (d) $K = \dfrac{p_{\mathrm{O}_2}^3}{p_{\mathrm{O}_3}^2}$

E7.3 $\tfrac{1}{2}\mathrm{N_2(g)} + \tfrac{3}{2}\mathrm{H_2(g)} \rightarrow \mathrm{NH_3(g)} \qquad \Delta_\mathrm{r}G = \Delta_\mathrm{r}G^\ominus + RT\ln Q \ \ [7.6]$

$$Q = \frac{p_{\mathrm{NH}_3}}{p_{\mathrm{N}_2}^{1/2}\, p_{\mathrm{H}_2}^{3/2}} = \frac{4.0}{(3.0)^{1/2}(1.0)^{3/2}} = \frac{4.0}{\sqrt{3.0}}$$

Therefore,

$$\Delta_\mathrm{r}G = -16.5\ \mathrm{kJ\ mol^{-1}} + RT\ln\left(\frac{4.0}{\sqrt{3.0}}\right) = -16.5\ \mathrm{kJ\ mol^{-1}} + 2.07\ \mathrm{kJ\ mol^{-1}} = \boxed{-14.4\ \mathrm{kJ\ mol^{-1}}}$$

Because $\Delta_\mathrm{r}G < 0$, the spontaneous direction of the reaction is toward the products.

E7.4 $K = \dfrac{[\mathrm{C}]}{[\mathrm{A}][\mathrm{B}]} = 0.432$ for the reaction $\mathrm{A} + \mathrm{B} \rightleftharpoons \mathrm{C}$.

The equilibrium constant for the reverse reaction is

$$K_\text{reverse reaction} = \frac{[\mathrm{A}][\mathrm{B}]}{[\mathrm{C}]} = \frac{1}{K} = \frac{1}{0.432} = \boxed{2.31}.$$

E7.5 $$K = \frac{[\mathrm{C}]^2}{[\mathrm{A}][\mathrm{B}]} = 7.2\times10^5$$

(a) $$K' = \frac{[\mathrm{C}]^4}{[\mathrm{A}]^2[\mathrm{B}]^2} = K^2 = \left(7.2\times10^5\right)^2 = \boxed{5.2\times10^{11}}$$

(b) $$K'' = \frac{[\mathrm{C}]}{[\mathrm{A}]^{1/2}[\mathrm{B}]^{1/2}} = K^{1/2} = \left(7.2\times10^5\right)^{1/2} = \boxed{8.5\times10^2}$$

E7.6 $$\Delta_r G^\ominus = -RT \ln K \;[7.8]$$

$$= -8.315 \text{ J K}^{-1} \text{ mol}^{-1} \times 400 \text{ K} \times \ln 2.07$$

$$= -2.42\times10^3 \text{ J mol}^{-1} = \boxed{-2.42 \text{ kJ mol}^{-1}}$$

E7.7 $$\Delta_r G^\ominus = -RT \ln K \;\; [7.8]$$

Therefore, $K = \mathrm{e}^{-\Delta G^\ominus/RT} = \mathrm{e}^{+3.67\times10^3 \text{ J mol}^{-1}/8.3145 \text{ J K}^{-1} \text{ mol}^{-1}\times 400 \text{ K}} = \boxed{3.01}$.

E7.8 $$\Delta_r G^\ominus = -RT \ln K \;[7.8] \qquad \text{or} \qquad K = \mathrm{e}^{-\Delta_r G^\ominus/RT}$$

The ratio of two equilibrium constants is given by $K_{r1}/K_{r2} = \mathrm{e}^{-\Delta_{r1}G^\ominus/RT} / \mathrm{e}^{-\Delta_{r2}G^\ominus/RT} = \mathrm{e}^{-\left(\Delta_{r1}G^\ominus - \Delta_{r2}G^\ominus\right)/RT}$.

For $\Delta_{r1}G^\ominus = -320 \text{ kJ mol}^{-1}$ and $\Delta_{r2}G^\ominus = -55 \text{ kJ mol}^{-1}$ the ratio at 300 K is

$$K_{r1}/K_{r2} = \mathrm{e}^{-\left(-320\times10^3 \text{ J mol}^{-1} - \left(-55\times10^3 \text{ J mol}^{-1}\right)\right)/\left(8.3145 \text{ J K}^{-1} \text{ mol}^{-1}\right)(300 \text{ K})}$$

$$= \mathrm{e}^{\left(265\times10^3 \text{ J mol}^{-1}\right)/\left(8.3145 \text{ J K}^{-1} \text{ mol}^{-1}\right)(300 \text{ K})}$$

$$= \boxed{1.38\times10^{46}}.$$

COMMENT. This is an enormous difference. Because of the exponential relation between K and $\Delta_r G^\ominus$ even small differences in $\Delta_r G^\ominus$ can make large differences in K.

E7.9 $$K_1 = 8.4 \times K_2$$

$$\Delta_{r2}G^\ominus = -RT \ln(K_1/8.4) = -RT \ln K_1 + RT \ln(8.4)$$

$$= -250 \text{ kJ mol}^{-1} + (2.478 \text{ kJ mol}^{-1})(2.1\overline{3}) \;\; \text{[assuming a temperature of 298.15 K]}$$

$$= \boxed{-245 \text{ kJ mol}^{-1}}$$

E7.10 $\Delta_r G^\ominus = -RT \ln K = 0$ so $\ln K = 0$ and $\boxed{K = 1}$.

E7.11 Let glucose-1-phosphate = G1P, glucose-6-phosphate = G6P, and glucose-3-phosphate = G3P.

$\ln K = -\Delta_r G^\oplus / RT$ or $K = \mathrm{e}^{-\Delta_r G^\oplus/RT}$ (The biological standard state, ⊕, has pH = 7.)

At 37°C, $RT = 8.3145 \text{ J K}^{-1} \text{ mol}^{-1} \times 310 \text{ K} = 2.577 \text{ kJ mol}^{-1}$.

$$K(\mathrm{G1P}) = \exp\left(\frac{21 \text{ kJ mol}^{-1}}{2.577 \text{ kJ mol}^{-1}}\right) = \boxed{3.5\times10^3}$$

$$K(\mathrm{G6P}) = \exp\left(\frac{14 \text{ kJ mol}^{-1}}{2.577 \text{ kJ mol}^{-1}}\right) = \boxed{2.3\times10^2}$$

$$K(\mathrm{G3P}) = \exp\left(\frac{9.2 \text{ kJ mol}^{-1}}{2.577 \text{ kJ mol}^{-1}}\right) = \boxed{36}$$

E7.12 $\text{ATP(aq)} + \text{H}_2\text{O(l)} \rightarrow \text{ADP(aq)} + \text{P}_\text{i}^-(\text{aq}) + \text{H}^+(\text{aq}) \qquad \Delta_\text{r}G^\oplus = -30.5 \text{ kJ mol}^{-1}$

The Gibbs energy quoted applies to biological standard state where $a_{\text{H}^+} = 10^{-7}$ (pH = 7) and all other activities are 1. The chemical standard state is defined with $a_J = 1$ for all reactants and products including $a_{\text{H}^+} = 1$ (pH = 0). The relation between the two standard states is provided by eqn 7.6:

$$\Delta_\text{r}G = \Delta_\text{r}G^\ominus + RT\ln Q \ [7.6] \qquad \text{and} \qquad \Delta_\text{r}G^\oplus = \Delta_\text{r}G^\ominus + RT\ln Q^\oplus .$$

Thus, when using the biological standard state, we may replace eqn 7.6 with the relation:

$$\Delta_\text{r}G = \Delta_\text{r}G^\oplus - RT\ln Q^\oplus + RT\ln Q.$$

$$\boxed{\Delta_\text{r}G = \Delta_\text{r}G^\oplus + RT\ln\left(Q/Q^\oplus\right)}$$

When discussing a solution with the hydrogen ion activity fixed at pH = 7, the reaction quotient ratio $Q/Q^\oplus$ effectively cancels out a_{H^+} factors. If a_{H^+} does not appear in Q (because H^+(aq) does not appear in the reaction equation), $Q^\oplus = 1$ and $\Delta_\text{r}G^\oplus = \Delta_\text{r}G^\ominus$.

For the above reaction

$$\Delta_\text{r}G = \Delta_\text{r}G^\oplus + RT\ln\left(\frac{a_{\text{ADP}}a_{\text{P}_\text{i}^-}}{a_{\text{ATP}}}\right) = \Delta_\text{r}G^\oplus + RT\ln\left(\frac{[\text{ADP}]\times[\text{P}_\text{i}^-]}{[\text{ATP}]c^\ominus}\right).$$

At 37°C, $RT = 8.3145 \text{ J K}^{-1}\text{ mol}^{-1} \times 310 \text{ K} = 2.577 \text{ kJ mol}^{-1}$.

(a) $\Delta_\text{r}G = -30.5 \text{ kJ mol}^{-1} + 2.577 \text{ kJ mol}^{-1} \times \ln(1.0\times10^{-3}) = \boxed{-48.3 \text{ kJ mol}^{-1}}$

(b) $\Delta_\text{r}G = -30.5 \text{ kJ mol}^{-1} + 2.577 \text{ kJ mol}^{-1} \times \ln\left(1.0\times10^{-6}\right) = \boxed{-66.1 \text{ kJ mol}^{-1}}$

E7.13 $\text{ATP(aq)} + \text{H}_2\text{O(l)} \rightarrow \text{ADP(aq)} + \text{P}_\text{i}^-(\text{aq}) + \text{H}^+(\text{aq}) \qquad \Delta_\text{r}G^\ominus = 10 \text{ kJ mol}^{-1}$

It is shown in Exercise 7.12 that $\Delta_\text{r}G^\oplus = \Delta_\text{r}G^\ominus + RT\ln Q^\oplus$ so

$$\begin{aligned}\Delta_\text{r}G^\oplus &= 10 \text{ kJ mol}^{-1} + \left(8.3145 \text{ J K}^{-1}\text{ mol}^{-1}\right)\times\left(298 \text{ K}\right)\times\ln a_{\text{H}^+} \\ &= 10 \text{ kJ mol}^{-1} + \left(2.478 \text{ kJ mol}^{-1}\right)\times\ln 10^{-7} \quad [a_{\text{H}^+} = 10^{-7} \text{ for the biological standard state}] \\ &= \boxed{-30 \text{ kJ mol}^{-1}}.\end{aligned}$$

E7.14 $\text{C}_6\text{H}_{12}\text{O}_6\text{(aq)} + 2\ \text{NAD}^+\text{(aq)} + 2\ \text{ADP(aq)} + 2\ \text{P}^-\text{(aq)} + 2\ \text{H}_2\text{O(l)}$

$\rightarrow 2\ \text{CH}_3\text{COCO}_2^-\text{(aq)} + 2\ \text{NADH(aq)} + 2\ \text{ATP(aq)} + 2\ \text{H}_3\text{O}^+\text{(aq)} \quad \Delta_\text{r}G^\oplus = -80.6 \text{ kJ mol}^{-1}$

It is shown in Exercise 7.12 that $\Delta_\text{r}G^\oplus = \Delta_\text{r}G^\ominus + RT\ln Q^\oplus$ and for this reaction $Q^\oplus = a_{\text{H}^+}^2 = \left(10^{-7}\right)^2 = 10^{-14}$.

Thus,

$$\begin{aligned}\Delta_\text{r}G^\ominus &= \Delta_\text{r}G^\oplus - RT\ln Q^\oplus \\ &= -80.6 \text{ kJ mol}^{-1} - \left(8.3145 \text{ J K}^{-1}\text{ mol}^{-1}\right)\times\left(298 \text{ K}\right)\times\ln\left(10^{-14}\right) = \boxed{-0.7 \text{ kJ mol}^{-1}}.\end{aligned}$$

E7.15 $\mu_\text{J} = \mu_\text{J}^\ominus + RT\ln a_\text{J} \ [7.3]$

Assume the ideal-dilute activity: $a_{\text{Na}^+} = [\text{Na}^+]$. Then,

$$\begin{aligned}\Delta G &= \mu_{\text{Na}^+}(\text{outside}) - \mu_{\text{Na}^+}(\text{inside}) \\ &= RT\ln(140) - RT\ln(10) = RT\ln\left(\frac{140}{10}\right) \\ &= 8.3145 \text{ J K}^{-1}\text{ mol}^{-1}\times 310 \text{ K}\times\ln 14 \\ &= 6.8\times10^3 \text{ J mol}^{-1} = \boxed{6.8 \text{ kJ mol}^{-1}}.\end{aligned}$$

E7.16 As the temperature of the decomposition reaction increases there comes a temperature T for which the reaction becomes spontaneous. At this temperature, $\Delta_r G^\ominus = \Delta_r H^\ominus - T\Delta_r S^\ominus = 0$ and $T = \Delta_r H^\ominus / \Delta_r S^\ominus$.

Of course values of $\Delta_r H^\ominus$ and $\Delta_r S^\ominus$ used in this last equation should be those observed at the temperature T but we make the estimate that these values are about the same as those computed from 298 K data found in the text *Data section*.

(a) $CaCO_3(s) \rightarrow CaO(s) + CO_2(g)$

$$\Delta_r H^\ominus = [-635.09 - 393.51 - (-1206.9)] \text{ kJ mol}^{-1} = \boxed{+178.3 \text{ kJ mol}^{-1}}$$

$$\Delta_r S^\ominus = [39.75 + 213.74 - 92.9] \text{ J K}^{-1} \text{ mol}^{-1} = \boxed{+160.6 \text{ J K}^{-1}\text{mol}^{-1}}$$

$$T = \frac{178.3\times 10^3 \text{ J mol}^{-1}}{160.6 \text{ J K}^{-1} \text{ mol}^{-1}} = \boxed{1110 \text{ K } (837°\text{C})}$$

(b) $CuSO_4 \cdot 5H_2O(s) \rightleftharpoons CuSO_4(s) + 5\ H_2O\ (g)$

$$\Delta_r H^\ominus = [-771.36 + 5\times(-241.82) - (-2279.7)] \text{ kJ mol}^{-1} = \boxed{+299.2 \text{ kJmol}^{-1}}$$

$$\Delta_r S^\ominus = [109 + (5\times 188.83) - 300.4] \text{ J K}^{-1} \text{ mol}^{-1} = \boxed{753 \text{ J K}^{-1}\text{mol}^{-1}}$$

$$T = \frac{299.2\times 10^3 \text{ J mol}^{-1}}{753 \text{ J K}^{-1} \text{ mol}^{-1}} = \boxed{397 \text{ K } (124°\text{C})}$$

E7.17 $Zn(s) + H_2O(g) \rightarrow ZnO(s) + H_2(g)$

$\Delta_r H^\ominus \simeq +224 \text{ kJ mol}^{-1}$ when $920 \text{ K} \le T \le 1280 \text{ K}$; $\Delta_r G^\ominus \simeq +33 \text{ kJ mol}^{-1}$ at 1280 K.

$\Delta_r G^\ominus$ is expected to change with temperature but we expect both $\Delta_r H^\ominus$ and $\Delta_r S^\ominus$ is display greater constancy over a reasonable temperature range. We can compute the value of $\Delta_r S^\ominus$ at these high temperatures with the relation $\Delta_r G^\ominus = \Delta_r H^\ominus - T\Delta_r S^\ominus$. Solving for $\Delta_r S^\ominus$ gives

$$\Delta_r S^\ominus = (\Delta_r H^\ominus - \Delta_r G^\ominus)/T = (224 \text{ kJ mol}^{-1} - 33 \text{ kJ mol}^{-1})/1280 \text{ K} = 0.149 \text{ kJ mol}^{-1}.$$

As the temperature of the redox reaction increases there comes a temperature T for which the reaction becomes spontaneous. At this temperature, $K = 1$ and $\Delta_r G^\ominus = \Delta_r H^\ominus - T\Delta_r S^\ominus = 0$. Consequently the temperature T is given by

$$\begin{aligned} T &= \Delta_r H^\ominus / \Delta_r S^\ominus \\ &= (224 \text{ kJ mol}^{-1})/(0.149 \text{ kJ mol}^{-1}) \\ &= \boxed{1.50\times 10^3 \text{ K}} \end{aligned}$$

E7.18 $I_2(g) \rightarrow 2\ I(g) \qquad \Delta\nu_{gas} = 1 \qquad K = 0.26$ at 1000 K

$$K_c = K\times\left(\frac{c^\ominus RT}{p^\ominus}\right)^{-\Delta\nu_{gas}} [7.13\text{a}] = 0.26\times\left(\frac{(1 \text{ mol dm}^{-3})\times(0.083145 \text{ dm}^3 \text{ bar K}^{-1} \text{ mol}^{-1})\times(1000 \text{ K})}{1 \text{ bar}}\right)^{-1}$$

$$= \boxed{0.0031}$$

E7.19 $G6P \rightarrow F6P \qquad Q = [F6P]/[G6P] \qquad \Delta_r G^\ominus = +1.7 \text{ kJ mol}^{-1}$ [Example 7.2]

$$f = \frac{[F6P]}{[F6P]+[G6P]} \text{ [Example 7.2]} = \frac{1}{1+\frac{[G6P]}{[F6P]}} = \frac{1}{1+\frac{1}{Q}}$$

Solving for Q gives

$$Q = \frac{f}{1-f}.$$

$$\Delta_r G = \Delta_r G^{\ominus} + RT\ln Q = \Delta_r G^{\ominus} + RT\ln\left(\frac{f}{1-f}\right) \qquad \text{where } T = 298K$$

$$= 1.7 \text{ kJ mol}^{-1} + \left(2.479 \text{ kJ mol}^{-1}\right)\times\ln\left(\frac{f}{1-f}\right)$$

Figure 7.1 gives a plot of $\Delta_r G$ against f. When $\Delta_r G < 0$, the reaction proceeds spontaneously to the right until $\Delta_r G = 0$ at the equilibrium value of f, i.e., $f_{eq} = 0.33$ (Example 7.2). When $\Delta_r G > 0$, the reaction proceeds spontaneously to the left until $\Delta_r G = 0$ at the equilibrium value of f.

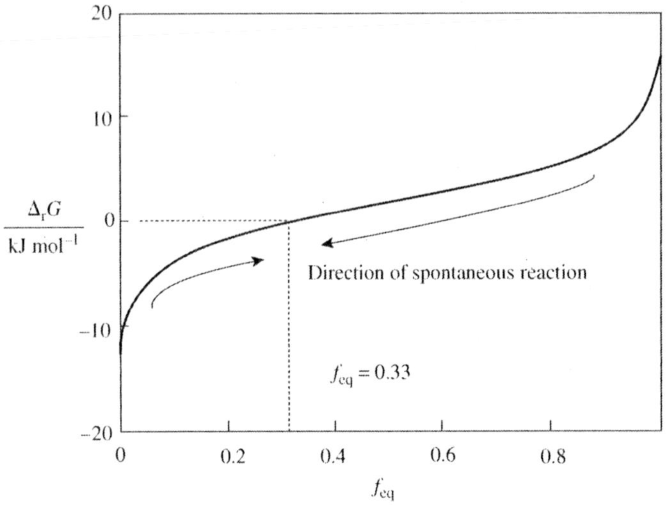

Figure 7.1

E7.20 We look up $\Delta_f G^{\ominus}$ for each compound and note the sign.

(a) –, exergonic (b) +, endergonic (c) +, endergonic (d) –, exergonic

E7.21 In each case, calculate $\Delta_r S^{\ominus}$ and $\Delta_r H^{\ominus}$ from information found in the text *Data section* and then calculate $\Delta_r G^{\ominus}$ from $\Delta_r G^{\ominus} = \Delta_r H^{\ominus} - T\Delta_r S^{\ominus}$.

(a) $\Delta_r S^{\ominus} = (94.6 - 186.91 - 192.45) \text{ J K}^{-1} \text{ mol}^{-1} = -284.8 \text{ J K}^{-1}\text{mol}^{-1}$

$\Delta_r H^{\ominus} = (-314.43 + 92.31 + 46.11) \text{ kJ mol}^{-1} = -176.01 \text{ kJ mol}^{-1}$

$\Delta_r G^{\ominus} = [-176.01 - 298\times(-0.2848)] \text{ kJ mol}^{-1} = \boxed{-91.14 \text{ kJ mol}^{-1}}$

(b) $\Delta_r S^{\ominus} = (4\times 28.33 + 3\times 41.84 - 2\times 50.92 - 3\times 18.83) \text{ J K}^{-1}\text{mol}^{-1}$

$= +80.51 \text{ J K}^{-1} \text{ mol}^{-1}$

$\Delta_r H^{\ominus} = [3\times(-910.93) - 2\times(-1675.7)] \text{ kJ mol}^{-1} = +618.6 \text{ kJ mol}^{-1}$

$\Delta_r G^{\ominus} = [+618.6 - 298\times(0.08051)] \text{ kJ mol}^{-1} = \boxed{+594.6 \text{ kJ mol}^{-1}}$

(c) $\Delta_r S^{\ominus} = (60.29 + 130.684 - 27.28 - 205.79) \text{ J K}^{-1}\text{mol}^{-1} = -42.10 \text{ J K}^{-1}\text{mol}^{-1}$

$\Delta_r H^{\ominus} = [-100.0 - (-20.63)] \text{ kJ mol}^{-1} = -79.4 \text{ kJ mol}^{-1}$

$\Delta_r G^{\ominus} = [-79.47 - 298\times(-0.04210)] \text{ kJ mol}^{-1} = \boxed{-66.8 \text{ kJ mol}^{-1}}$

(d) $\Delta_r S^{\ominus} = (27.28 + 2\times 205.79 - 52.93 - 2\times 130.684) \text{ J K}^{-1}\text{m}$

$= +124.56 \text{ J K}^{-1} \text{ mol}^{-1}$

$\Delta_r H^{\ominus} = [2\times(-20.63) - (-178.2)] \text{ kJ mol}^{-1} = +136.9 \text{ kJ mol}^{-1}$

$\Delta_r G^{\ominus} = (+136.9 - 298\times 0.12456) \text{ kJ mol}^{-1} = \boxed{+99.8 \text{ kJ mol}^{-1}}$

(e) $\Delta_r S^\ominus = (156.9 + 2\times130.684 - 2\times109.6 - 205.79)\ \mathrm{J\ K^{-1}\ mol^{-1}} = -6.7\ \mathrm{J\ K^{-1}\ mol^{-1}}$

$$\Delta_r H^\ominus = [-813.99 - 2\times(-187.78) - (-20.63)]\ \mathrm{kJ\ mol^{-1}} = -417.80\ \mathrm{kJ\ mol^{-1}}$$

$$\Delta_r G^\ominus = [-417.80 - 298\times(-6.7\times10^{-3})]\ \mathrm{kJ\ mol^{-1}} = \boxed{-415.80\ \mathrm{kJ\ mol^{-1}}}$$

E7.22 In each case, calculate $\Delta_r G^\ominus$ from the values of $\Delta_f G^\ominus$ found in the text *Data section.*

Then, from $\Delta_r G^\ominus = -RT\ln K$ decide which reactions have $K > 1$. If $\ln K > 0$, then $K > 1$ and $\Delta_r G^\ominus < 0$. If $\ln K < 0$, then $K < 1$ and $\Delta_r G^\ominus > 0$.

(a) $\Delta_r G^\ominus = 2\,\Delta_f G^\ominus(\mathrm{CH_3COOH, l}) - 2\,\Delta_f G^\ominus(\mathrm{CH_3CHO, g})$

$$= [2\times(-389.9) - 2\times(-128.86)\ \mathrm{kJ\ mol^{-1}}$$

$$= \boxed{-522.1\ \mathrm{kJ\ mol^{-1}},\ K > 1}$$

(b) $\Delta_r G^\ominus = 2\,\Delta_f G^\ominus(\mathrm{AgBr, s}) - 2\,\Delta_f G^\ominus(\mathrm{AgCl, s})$

$$= [2\times(-96.90) - 2\times(-109.79)]\ \mathrm{kJ\ mol^{-1}}$$

$$= \boxed{+25.78\ \mathrm{kJ\ mol^{-1}},\ K < 1}$$

(c) $\Delta_r G^\ominus = \Delta_f G^\ominus(\mathrm{HgCl_2, s}) = \boxed{-178.6\ \mathrm{kJ\ mol^{-1}},\ K > 1}$

(d) $\Delta_r G^\ominus = \Delta_f G^\ominus(\mathrm{Zn^{2+}, aq}) - \Delta_f G^\ominus(\mathrm{Cu^{2+}, aq})$

$$= [-147.06 - 65.49]\ \mathrm{kJ\ mol^{-1}} = \boxed{-212.55\ \mathrm{kJ\ mol^{-1}},\ K > 1}$$

(e) $\Delta_r G^\ominus = 12\,\Delta_f G^\ominus(\mathrm{CO_2, g}) + 11\,\Delta_f G^\ominus(\mathrm{H_2O, l}) - \Delta_f G^\ominus(\mathrm{C_{12}H_{22}O_{11}, s})$

$$= [12\times(-394.36) + 11\times(-237.13) - (-1543)]\ \mathrm{kJ\ mol^{-1}}$$

$$= \boxed{-5798\ \mathrm{kJ\ mol^{-1}},\ K > 1}$$

E7.23 $\mathrm{CH_4(g) + 2\,O_2(g) \rightarrow CO_2(g) + 2\,H_2O(l)}$

$\Delta_c H^\ominus$ and $\Delta_c G^\ominus$ are calculated from standard formation values found in the text *Data section.*

$$\Delta_c H^\ominus = 2\times\Delta_f H^\ominus(\mathrm{H_2O, l}) + \Delta_f H^\ominus(\mathrm{CO_2, g}) - \Delta_f H^\ominus(\mathrm{CH_4, g})$$

$$= [2\times(-285.83) - 393.51 + 74.81]\ \mathrm{kJ\ mol^{-1}}$$

$$= -890.21\ \mathrm{kJ\ mol^{-1}}$$

$$\Delta_c G^\ominus = 2\times\Delta_f G^\ominus(\mathrm{H_2O, l}) + \Delta_f G^\ominus(\mathrm{CO_2, g}) - \Delta_f G^\ominus(\mathrm{CH_4, g})$$

$$= [2\times(-237.13) - 394.36 + 50.72]\ \mathrm{kJ\ mol^{-1}}$$

$$= -817.90\ \mathrm{kJ\ mol^{-1}}$$

(a) heat given off by combustion $= -n(\mathrm{CH_4})\times\Delta_c H^\ominus$

$$= -\left(\frac{2.0\times10^3\ \mathrm{g}}{16.04\ \mathrm{g\ mol^{-1}}}\right)\times(-890.21\ \mathrm{kJ\ mol^{-1}}) = \boxed{1.1\times10^5\ \mathrm{kJ}}$$

(b) maximum non-expansion work done by combustion $= -n(\mathrm{CH_4})\times\Delta_c G^\ominus$

$$= -\left(\frac{2.0\times10^3\ \mathrm{g}}{16.04\ \mathrm{g\ mol^{-1}}}\right)\times(-817.90\ \mathrm{kJ\ mol^{-1}}) = \boxed{1.0\times10^5\ \mathrm{kJ}}$$

E7.24 (a) $C_6H_{12}O_6(s) + 12\ O_2(g) \rightarrow 6\ CO_2(g) + 6\ H_2O(l) \qquad \Delta_c H^{\ominus} = -2808\ \text{kJ mol}^{-1}$ (*Data section*)

The reaction of interest has water in the gas phase:

$$C_6H_{12}O_6(s) + 12\ O_2(g) \rightarrow 6\ CO_2(g) + 6\ H_2O(g).$$

$$\Delta_r H^{\ominus} = \Delta_c H^{\ominus} + \Delta_{vap} H^{\ominus} = \Delta_c H^{\ominus} + 6\times\{\Delta_f H^{\ominus}(H_2O,\ g) - \Delta_f H^{\ominus}(H_2O,\ l)\}$$

$$= (-2808 + 6\times\{-241.82 - (-285.83)\})\ \text{kJ mol}^{-1} \ \ (\textit{Data section})$$

$$= -2544\ \text{kJ mol}^{-1}$$

$$\Delta_r G^{\ominus} = 6\times\Delta_f G^{\ominus}(H_2O,\ g) + 6\times\Delta_f G^{\ominus}(CO_2,\ g) - \Delta_f G^{\ominus}(C_6H_{12}O_6, s)$$

$$= [6\times(-228.57) + 6\times(-394.36) + 910]\ \text{kJ mol}^{-1}$$

$$= -2828\ \text{kJ mol}^{-1}$$

(b) heat given off by combustion $= -n(C_6H_{12}O_6)\times\Delta_r H^{\ominus}$

$$= -\left(\frac{2.0\times10^3\ \text{g}}{180.16\ \text{g mol}^{-1}}\right)\times(-2544\ \text{kJ mol}^{-1}) = \boxed{2.8\times10^4\ \text{kJ}}$$

(c) maximum non-expansion work done by combustion $= -n(C_6H_{12}O_6)\times\Delta_r G^{\ominus}$

$$= -\left(\frac{2.0\times10^3\ \text{g}}{180.16\ \text{g mol}^{-1}}\right)\times(-2828\ \text{kJ mol}^{-1}) = \boxed{3.1\times10^4\ \text{kJ}}$$

E7.25 Questions about the energy effectiveness of fuels are discussed in Box 3.1. First, we note that there are no differences in the extent to which glucose ($C_6H_{12}O_6$, 180.16 g mol^{-1}, $\Delta_c H^{\ominus} = -2808\ \text{kJ mol}^{-1}$) and sucrose ($C_{12}H_{22}O_{11}$, 342.30 g mol^{-1}, $\Delta_c H^{\ominus} = -5645\ \text{kJ mol}^{-1}$) are partially oxidized so this does not distinguish their energy effectiveness. Sucrose does have a greater number of carbon atoms so its molar combustion enthalpy is expected to be larger and issues of energy effectiveness are customarily addressed by comparing specific combustion enthalpies. They are -15.6 kJ g^{-1} for glucose and -16.5 kJ g^{-1} for sucrose. These values are nearly identical but $\boxed{\text{sucrose has a slight advantage as an effective fuel}}$. The commercial advantage of sucrose is with a sweetness that is about 1.4 × that of glucose.

(a) Sucrose combustion with water vapor as product: $C_{12}H_{22}O_{11}(s) + 12\ O_2(g) \rightarrow 12\ CO_2(g) + 11\ H_2O(g)$.

$$\Delta_s G^{\ominus} = \{12\times\Delta_f G^{\ominus}(CO_2,\ g) + 11\times\Delta_f G^{\ominus}(H_2O,\ g) - \Delta_f G^{\ominus}(\text{sucrose})\}/M_{\text{sucrose}}$$

$$= \{[12\times(-394.36) + 11\times(-228.57) - (-1543)]\ \text{kJ mol}^{-1}\}/\{342.30\ \text{g mol}^{-1}\}$$

$$= -16.66\ \text{kJ g}^{-1}$$

$$w_{\text{non-expansion,s}} = \Delta_s G^{\ominus} = -16.66\ \text{kJ g}^{-1}$$

The specific expansion work is given by the constant p_{ex} expression:

$$w_{\text{expansion,s}} = -p_{ex}\Delta V/M_{\text{sucrose}} = -p^{\ominus}\Delta V/M_{\text{sucrose}}$$

where $\Delta V = V_{\text{products}} - V_{\text{reactants}} = V_{\text{gaseous products}} - V_{\text{gaseous reactants}}$

$$= \Delta\nu_{\text{gas}} RT/p^{\ominus} \quad \text{[volumes of solids and liquids are negligibly small].}$$

Thus,

$$w_{\text{expansion,s}} = -\Delta\nu_{\text{gas}} RT/M_{\text{sucrose}} = -(23-12)\times RT/M_{\text{sucrose}} = -11RT/M_{\text{sucrose}}$$

$$= -11\times(8.3145\ \text{J K}^{-1}\ \text{mol}^{-1})\times(298.15\ \text{K})/(342.30\ \text{g mol}^{-1})$$

$$= -79.7\ \text{J g}^{-1}.$$

$$w_{\text{total,s}} = w_{\text{expansion,s}} + w_{\text{non-expansion,s}} = -(0.0797 + 16.66)\ \text{kJ g}^{-1}$$

$$= -16.7\ \text{kJ g}^{-1}$$

The quantities for the combustion of 2 kg sucrose are:

Max. non-expansion work done by combustion $= -m_{\text{sucrose}} w_{\text{non-expansion,s}}$

$$= -(2000\ \text{g}) \times (-16.66\ \text{kJ g}^{-1})$$

$$= \boxed{3.33 \times 10^4\ \text{kJ}}$$

Expansion work done by combustion $= -m_{\text{sucrose}} w_{\text{expansion,s}}$

$$= -(2000\ \text{g}) \times (-79.7\ \text{J g}^{-1})$$

$$= \boxed{159\ \text{kJ}}$$

Total work done by combustion $= 3.33 \times 10^4\ \text{kJ} + 159\ \text{kJ} = \boxed{3.35 \times 10^4\ \text{kJ}}$

(b) Sucrose combustion with liquid water product: $C_{12}H_{22}O_{11}(s) + 12\ O_2(g) \rightarrow 12\ CO_2(g) + 11\ H_2O(l)$.

$$\Delta_s G^{\ominus} = \{12 \times \Delta_f G^{\ominus}(\text{CO}_2,\ \text{g}) + 11 \times \Delta_f G^{\ominus}(\text{H}_2\text{O},\ \text{g}) - \Delta_f G^{\ominus}(\text{sucrose})\} / M_{\text{sucrose}}$$

$$= \{[12 \times (-394.36) + 11 \times (-237.13) - (-1543)]\ \text{kJ mol}^{-1}\} / \{342.30\ \text{g mol}^{-1}\}$$

$$= -16.94\ \text{kJ g}^{-1}$$

$$w_{\text{non-expansion,s}} = \Delta_s G^{\ominus} = -16.94\ \text{kJ g}^{-1}$$

The specific expansion work is given by the constant p_{ex} expression:

$$w_{\text{expansion,s}} = -p_{\text{ex}} \Delta V / M_{\text{sucrose}} = -p^{\ominus} \Delta V / M_{\text{sucrose}}$$

where $\Delta V = V_{\text{products}} - V_{\text{reactants}} = V_{\text{gaseous products}} - V_{\text{gaseous reactants}}$

$$= \Delta \nu_{\text{gas}} RT / p^{\ominus}$$ [volumes of solids and liquids are negligibly small].

Thus,

$$w_{\text{expansion,s}} = -\Delta \nu_{\text{gas}} RT / M_{\text{sucrose}} = -(12 - 12) \times RT / M_{\text{sucrose}} = 0$$

$$w_{\text{total,s}} = w_{\text{expansion,s}} + w_{\text{non-expansion,s}} = (0 - 16.94)\ \text{kJ g}^{-1}$$

$$= -16.94\ \text{kJ g}^{-1}$$

The quantities for the combustion of 2 kg sucrose are:

Max. non-expansion work done by combustion $= -m_{\text{sucrose}} w_{\text{non-expansion,s}}$

$$= -(2000\ \text{g}) \times (-16.94\ \text{kJ g}^{-1})$$

$$= \boxed{3.39 \times 10^4\ \text{kJ}}$$

Expansion work done by combustion $= -m_{\text{sucrose}} w_{\text{expansion,s}} = \boxed{0}$

Total work done by combustion $= 3.39 \times 10^4\ \text{kJ} + 0 = \boxed{3.39 \times 10^4\ \text{kJ}}$

E7.26 To calculate the standard free energy of formation, we need to calculate the entropy of formation and the enthalpy of formation because $\Delta_f G^{\ominus} = \Delta_f H^{\ominus} - T\Delta_f S^{\ominus}$. The formation reaction is

$$6\,\text{C(s)} + 3\,\text{H}_2\text{(g)} + \tfrac{1}{2}\text{O}_2\text{(g)} \rightarrow \text{C}_6\text{H}_5\text{OH(s)}.$$

$$\Delta_f S^{\ominus} = S_m^{\ominus}(C_6H_5OH, s) - 6\, S_m^{\ominus}(C, s) - 3\, S_m^{\ominus}(H_2, g) - \tfrac{1}{2} S_m^{\ominus}(O_2, g)$$

$$= (144.0 - 6\times 5.740 - 3\times 130.68 - \tfrac{1}{2}\times 205.14)\ \text{J K}^{-1}\ \text{mol}^{-1}$$

$$= -385.05\ \text{J K}^{-1}\ \text{mol}^{-1}$$

The combustion reaction is $C_6H_5OH(s) + 7O_2(g) \rightarrow 6\ CO_2(g) + 3H_2O(l)$.

$$\Delta_c H^{\ominus} = 6\ \Delta_f H^{\ominus}(CO_2, g) + 3\ \Delta_f H^{\ominus}(H_2O, l) - \Delta_f H^{\ominus}(C_6H_5OH, s)$$

Solving for $\Delta_f H^{\ominus}(C_6H_5OH, s)$ gives

$$\Delta_f H^{\ominus}(C_6H_5OH, s) = 6\ \Delta_f H^{\ominus}(CO_2, g) + 3\ \Delta_f H^{\ominus}(H_2O, l) - \Delta_c H^{\ominus}$$

$$= [6\times(-393.51) + 3(-285.83) - (-3054)]\ \text{kJ mol}^{-1}$$

$$= -164.55\ \text{kJ mol}^{-1}.$$

Therefore,

$$\Delta_f G^{\ominus} = -164.55\ \text{kJ mol}^{-1} - 298.15\ \text{K} \times (-0.38505\ \text{kJ K}^{-1}\ \text{mol}^{-1}) = \boxed{-49.8\ \text{kJ mol}^{-1}}.$$

E7.27 $CH_4(g) + 2\ O_2(g) \rightarrow CO_2(g) + 2\ H_2O(l)$

$$w'_{max} = \Delta_r G\ [4.18]$$

Assume that $\Delta_r G = \Delta_r G^{\ominus}$ and calculate $\Delta_r G$ in the usual manner from $\Delta_f G^{\ominus}$ values found in the *Data section*. We obtain $\Delta_r G^{\ominus} = -817.90\ \text{kJ mol}^{-1}$.

Therefore, the maximum non-expansion work that can be obtained is $\boxed{817.90\ \text{kJ mol}^{-1}}$.

E7.28 $\text{Pyruvate}^{-1}(aq) + NADH(aq) + H^+(aq) \rightarrow \text{lactate}^{-1}(aq) + NAD^+(aq)\quad \Delta_r G^{\ominus} = -66.6\ \text{kJ mol}^{-1}$ at 37°C

Since $\Delta_r G = \Delta_r G^{\ominus} + RT\ln Q$ [7.6], we immediately have

$$\Delta_r G^{\oplus} = \Delta_r G^{\ominus} + RT\ln Q^{\oplus}$$

where $Q^{\oplus}$ is the reaction quotient for the biological standard state defined with $a_{H^+} = 10^{-7}$ (pH = 7) and all other activities are 1.

$$\Delta_r G^{\oplus} = \Delta_r G^{\ominus} + RT\ln\left(\frac{1}{a_{H^+}}\right) = (-66.6\ \text{kJ mol}^{-1}) + (8.3145\ \text{J K}^{-1}\ \text{mol}^{-1}) \times (310\ \text{K}) \times \ln\left(\frac{1}{10^{-7}}\right)$$

$$= \boxed{-25.1\ \text{kJ mol}^{-1}}$$

E7.29 The reaction is $AMP(aq) + H_2O(l) \rightarrow A(aq) + P_i^-(aq) + H^+(aq)\quad \Delta_r G^{\oplus} = -14\ \text{kJ mol}^{-1}$ at 298 K.

Since $\Delta_r G = \Delta_r G^{\ominus} + RT\ln Q$ [7.6], we immediately have

$$\Delta_r G^{\oplus} = \Delta_r G^{\ominus} + RT\ln Q^{\oplus}$$

where $Q^{\oplus}$ is the reaction quotient for the biological standard state defined with $a_{H^+} = 10^{-7}$ (pH = 7) and all other activities are 1. Therefore,

$$\Delta_r G^{\ominus} = \Delta_r G^{\oplus} - RT\ln Q^{\oplus} = \Delta_r G^{\oplus} - RT\ln(a_{H^+})$$

$$= (-14\ \text{kJ mol}^{-1}) - (8.3145\ \text{J K}^{-1}\ \text{mol}^{-1}) \times (298\ \text{K}) \times \ln(10^{-7})$$

$$= \boxed{26\ \text{kJ mol}^{-1}}$$

E7.30 $$\ln K = -\frac{\Delta_r G^{\ominus}}{RT} = -\frac{\Delta_r H^{\ominus\ominus}}{RT} + \frac{\Delta_r S^{\ominus}}{R} \text{ [7.8, 7.11, Derivation 7.2]}$$

For many reactions both $\Delta_r H^{\ominus}$ and $\Delta_r S^{\ominus}$ are weakly dependent upon temperature over a small temperature range and a data plot of ln K against $1/T$ is approximately linear. In this case, the equation indicates that a linear regression fit of the plot will yield a slope equal to $-\Delta_r H^{\ominus}/R$ and an intercept equal to $\Delta_r S^{\ominus}/R$. Consequently, $\Delta_r H^{\ominus}$ and $\Delta_r S^{\ominus}$ may often be determined by (i) measuring K over a range of temperatures, (ii) checking linearity of the ln$K$ against $1/T$ plot, and (iii) performing a linear regression fit of the plot. The thermodynamic properties are determined with the expressions

$$\Delta_r H^{\ominus} = -R \times \text{slope} \quad \text{and} \quad \Delta_r S^{\ominus} = R \times \text{intercept}.$$

E7.31 See Exercise 7.30 for the methodology for finding both $\Delta_r H^{\ominus}$ and $\Delta_r S^{\ominus}$ using the available data. Then, fill out a table that includes T and K data columns as well as $1/T$ and ln K transformations. Prepare a ln K against $1/T$ plot and visually check its linearity.

T / K	K	1000 K / T	ln K
300	4.00E+31	3.33E+00	72.77
500	4.00E+18	2.00E+00	42.83
1000	5.10E+08	1.00E+00	20.05

The plot (Figure 7.2) is linear with a slope equal to 22.6×10^3 K and an intercept equal to −2.46.

$$\Delta_r H^{\ominus} = -R \times slope = -(8.3145 \text{ J K}^{-1}\text{ mol}^{-1}) \times (22.6 \times 10^3 \text{ K}) = \boxed{-188 \text{ kJ mol}^{-1}}$$
$$\Delta_r S^{\ominus} = R \times intercept = (8.3145 \text{ J K}^{-1}\text{ mol}^{-1}) \times (-2.46) = \boxed{-20.5 \text{ J K}^{-1}\text{ mol}^{-1}}$$

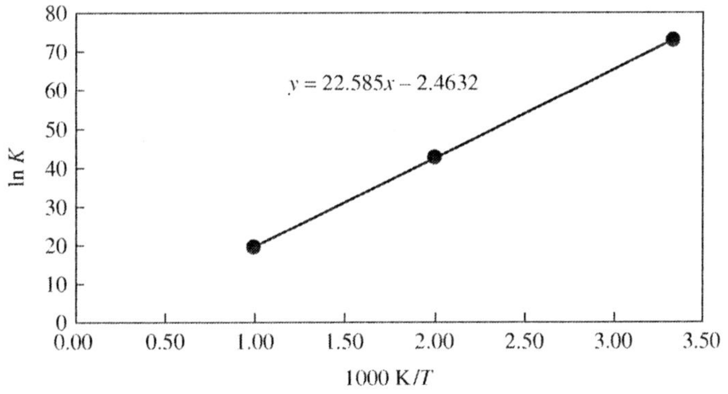

Figure 7.2

E7.32 First substitute in and solve for K at 390 K and 410 K from the equation given, thus

$$\ln K = -1.04 - \frac{1088 \text{ K}}{390} + \frac{1.51 \times 10^5 \text{ K}^2}{(390)^2} = -2.84.$$

Solving for K' where $T' = 410$ K, ln $K' = -2.80$.

Substitution of these values into eqn 7.15 gives

$$\ln K' - \ln K = -\frac{\Delta_r H^{\ominus}}{R}\left(\frac{1}{T'} - \frac{1}{T}\right)$$

$$0.04 = -\frac{\Delta_r H^{\ominus}}{8.3145 \text{ J K}^{-1}\text{ mol}^{-1}}\left(\frac{1}{410 \text{ K}} - \frac{1}{390 \text{ K}}\right).$$

Solving for $\Delta_r H^{\ominus}$ gives

$$\Delta_r H^{\ominus} = \boxed{2.66 \text{ kJ mol}^{-1}}.$$

$$\Delta_r G^\ominus = -RT \ln K \ [7.8] = RT \times \left(1.04 + \frac{1088}{T} - \frac{1.51 \times 10^5}{T^2}\right)$$

$$= (8.3145 \text{ J K}^{-1} \text{ mol}^{-1}) \times (400 \text{ K}) \times \left(1.04 + \frac{1088}{400} - \frac{1.51 \times 10^5}{(400)^2}\right)$$

$$= +9.36 \text{ kJ mol}^{-1}$$

Since $\Delta_r G^\ominus = \Delta_r H^\ominus - T\Delta_r S^\ominus$ [7.11],

$$\Delta_r S^\ominus = \frac{\Delta_r H^\ominus - \Delta_r G^\ominus}{T} = \frac{2.66 \text{ kJ mol}^{-1} - 9.36 \text{ kJ mol}^{-1}}{400 \text{ K}}$$

$$= \boxed{-16.8 \text{ J K}^{-1} \text{ mol}^{-1}}.$$

E7.33 Let B and I denote borneol and isoborneol, respectively. The reaction is:

$$\text{B(g)} \rightarrow \text{I(g)} \qquad \Delta_r G^\ominus = +9.4 \text{ kJ mol}^{-1} \text{ at } 503 \text{ K}.$$

$$\Delta_r G = \Delta_r G^\ominus + RT \ln Q$$

$$= \Delta_r G^\ominus + RT \ln\left(\frac{p_I}{p_B}\right) = \Delta_r G^\ominus + RT \ln\left(\frac{x_I p}{x_B p}\right) = \Delta_r G^\ominus + RT \ln\left(\frac{n_I}{n_B}\right)$$

$$= +9.4 \text{ kJ mol}^{-1} + (8.3145 \text{ J K}^{-1} \text{ mol}^{-1}) \times (503 \text{ K}) \times \ln\left(\frac{0.30}{0.15}\right)$$

$$= \boxed{+12.3 \text{ kJ mol}^{-1}}$$

E7.34 Let B and I denote borneol ($154.25 \text{ g mol}^{-1}$) and isoborneol, respectively. The reaction is:

$$\text{B(g)} \rightarrow \text{I(g)} \qquad K = 0.106 \text{ at } 503 \text{ K}.$$

$$K = \left(\frac{p_I}{p_B}\right)_{eq} = \left(\frac{x_I p}{x_B p}\right)_{eq} = \left(\frac{x_I}{x_B}\right)_{eq} = \left(\frac{x_I}{1 - x_I}\right)_{eq}$$

Solving for $x_{I,eq}$ gives

$$x_{I,eq} = \frac{K}{1+K} = \frac{0.106}{1.106}$$

$$= \boxed{0.0958}.$$

$$x_{B,eq} = 1 - x_{I,eq} = 1 - 0.0958 = \boxed{0.904}$$

The information provided in the exercise also permits the computation of the amounts of each at equilibrium. Because of the simple 1:1 stoichiometry of the isomerization reaction, the total number of moles does not change during the reaction. Thus,

$$n_{total} = (m_B + m_I)/M_B = (6.70 \text{ g} + 12.5 \text{ g})/(154.25 \text{ g mol}^{-1}) = 0.1245 \text{ mol}.$$

$$n_{I,eq} = x_{I,eq} n_{total} = (0.0958) \times (0.1245 \text{ mol}) = 0.0119 \text{ mol}$$

$$n_{B,eq} = x_{Beq} n_{total} = (0.9042) \times (0.1245 \text{ mol}) = 0.1126 \text{ mol}$$

E7.35

$K = 89.8$	$N_2(g)$	+	$3\ H_2(g)$	$\rightleftharpoons$	$2\ NH_3(g)$
Initial partial pressure / bar	1.00		4.00		0
Change / bar	$-x$		$-3x$		$+2x$
Equilibrium partial pressure / bar	$1.00 - x$		$4.00 - 3x$		$2x$

$$K=\left(\frac{p_{NH_3}^{\ 2}}{p_{N_2}p_{H_2}^{\ 3}}\right)_{eq}=\frac{(2x)^2}{(1.00-x)\times(4.00-3x)^3} \quad \text{or} \quad \frac{(2x)^2}{(1.00-x)\times(4.00-3x)^3}=89.8$$

It is not immediately obvious that the above 4th order polynomial can be algebraically solved for x. But since $K >> 1$ we can estimate that the limiting reagent completely reacts. This means that according to the reaction stoichiometry $x_{est} \sim 1.00$ bar of nitrogen consumes 3.00 bar of hydrogen to produce 2.00 bar ammonia. This numeric estimate leaves $(1.00 - x_{est})$ bar, about 0, nitrogen at equilibrium and the working equation becomes

$$\frac{2^2}{(1.00-x_{est})\times(4.00-3)^3}=89.8 \quad \text{or} \quad x_{est}=1-\frac{2^2}{89.8\times(4.00-3)^3}=0.9555\,.$$

We can now iterate to a better numeric estimate by using the above value of x_{est} in the hydrogen and ammonia factors of the equilibrium expression. This procedure is repeated until two successive iterations agree to within the desired precision (our equilibrium constant is reported to a precision of about 1%). The working equation becomes

$$\frac{(2\times\{0.9555\})^2}{(1.00-x_{est\,2})\times(4.00-3\times\{0.9555\})^3}=89.8 \quad \text{or} \quad x_{est\,2}=1-\frac{(2\times\{0.9555\})^2}{89.8\times(4.00-3\times\{0.9555\})^3}=0.9721.$$

$$x_{est\,3}=1-\frac{(2\times\{0.9721\})^2}{89.8\times(4.00-3\times\{0.9721\})^3}=0.9669$$

The last two estimates agree to within 1% so we conclude that $x = 0.97$. We can now calculate the equilibrium partial pressures.

$$p_{N_2}=(1.00-0.97)\text{ bar}=\boxed{0.03\text{ bar}}$$

$$p_{H_2}=(4.00-3\times0.97)\text{ bar}=\boxed{1.09\text{ bar}}$$

$$p_{NH_3}=2\times0.97\text{ bar}=\boxed{1.94\text{ bar}}$$

E7.36

$$SbCl_5(g) \rightleftharpoons SbCl_3(g)+Cl_2(g)$$

$$K=\frac{p_{SbCl_3}p_{Cl_2}}{p_{SbCl_5}}=3.5\times10^{-4}$$

$$=\frac{(0.22)\times p_{Cl_2}}{(0.17)}=3.5\times10^{-4}$$

$$p_{Cl_2}=\boxed{2.7\times10^{-4}\text{bar}}$$

E7.37

$$PCl_5(g) \rightleftharpoons PCl_3(g)+Cl_2(g) \qquad K=0.36,\ T=400\text{ K},\ V=0.250\text{ dm}^3$$

$$[PCl_5]_{initial}=\frac{(1.5\text{ g})/(208.24\text{ g mol}^{-1})}{0.250\text{ dm}^3}=0.028\overline{81}\text{ mol dm}^{-3}\,.$$

$$K_c=K\times\left(\frac{c^{\ominus}RT}{p^{\ominus}}\right)^{-\Delta\nu_{gas}} \quad [7.13a]$$

$$=(0.36)\times\left\{\frac{(1\text{ mol dm}^{-3})\times(0.0831447\text{ dm}^3\text{ bar K}^{-1}\text{ mol}^{-1})\times(400\text{ K})}{1\text{ bar}}\right\}^{-1}$$

$$=0.010\overline{8}$$

(a) Let x be the amount of PCl_5 that decomposes. Then, because of the reaction stoichiometry, the molar concentrations of both PCl_3 and Cl_2 equal x.

$$K_c = \left(\frac{[PCl_3][Cl_2]}{[PCl_5]c^{\ominus}}\right)_{eq} = \frac{x^2}{([PCl_5]_{initial} - x)c^{\ominus}}$$

$$x^2 + K_c c^{\ominus} x - K_c c^{\ominus}[PCl_5]_{initial} = 0$$

$$x/c^{\ominus} = \frac{-K_c \pm \sqrt{K_c^{\,2} + 4K_c[PCl_5]_{initial}/c^{\ominus}}}{2}$$

$$= \frac{-K_c + \sqrt{K_c^{\,2} + 4K_c[PCl_5]_{initial}/c^{\ominus}}}{2} \quad \text{(Selection of the + assures that } x > 0\text{)}$$

$$= \frac{-0.010\bar{8} + \sqrt{(0.010\bar{8})^2 + 4\times(0.010\bar{8})\times(0.028\bar{8}\bar{1})}}{2} = 0.013$$

At equilibrium,

$[PCl_3] = [Cl_2] = \boxed{0.013\ \text{mol dm}^{-3}}$

$[PCl_5] = (0.029 - 0.013)\ \text{mol dm}^{-3} = \boxed{0.016\ \text{mol dm}^{-3}}$.

(b) $$\%\ \text{decomposition} = \frac{\text{amount decomposed}}{\text{initial amount}}\times 100\% = \frac{1.3\times10^{-2}\ \text{mol dm}^{-3}}{2.9\times10^{-2}\ \text{mol dm}^{-3}}\times 100\% = \boxed{45\%}$$

E7.38

$K = 0.036$	$N_2(g)$	+	$3\,H_2(g)$	$\rightleftharpoons$	$2\,NH_3(g)$
Initial partial pressure / bar	0.020		0.020		0
Change / bar	$-x$		$-3x$		$+2x$
Equilibrium partial pressure / bar	$0.020 - x$		$0.020 - 3x$		$2x$

$$K = \frac{p^2_{NH_3}}{p_{N_2}p^3_{H_2}} = \frac{(2x)^2}{(0.020-x)\times(0.020-3x)^3} \quad \text{or} \quad \frac{(2x)^2}{(0.020-x)\times(0.020-3x)^3} = 0.036$$

It is not immediately obvious that the above 4th order polynomial can be algebraically solved for x. Consequently, we use a numerical method. The equilibrium is very small ($K \ll 1$) so we estimate that at equilibrium the partial pressures of both nitrogen and hydrogen are virtually unchanged from their initial values. The working eqn becomes:

$$\frac{(2x_{est})^2}{(0.020)\times(0.020)^3} = 0.036 \quad \text{or} \quad x_{est} = \tfrac{1}{2}\sqrt{(0.036)\times(0.020)\times(0.020)^3} = 3.795\times10^{-5}.$$

We can now iterate to a better numeric estimate by using the above value of x_{est} in the nitrogen and hydrogen factors of the equilibrium expression. This procedure is repeated until two successive iterations agree to within the desired precision (our equilibrium constant is reported to a precision of about 3%). The working equation becomes

$$x_{est\,2} = \tfrac{1}{2}\sqrt{(0.036)\times(0.020-3.795\times10^{-5})\times(0.020-3\times\{3.795\times10^{-5}\})^3} = 3.759\times10^{-5}$$

$$x_{est\,3} = \tfrac{1}{2}\sqrt{(0.036)\times(0.020-3.759\times10^{-5})\times(0.020-3\times\{3.759\times10^{-5}\})^3} = 3.759\times10^{-5}$$

The last two iterations agree to within 3% so we conclude that $x = 3.8 \times 10^{-5}$.

$$p_{N_2} = (0.020 - 3.8\times10^{-5})\ \text{bar} = \boxed{0.020\ \text{bar}}$$

$$p_{H_2} = (0.020 - 3\times3.8\times10^{-5})\ \text{bar} = \boxed{0.020\ \text{bar}}$$

$$p_{NH_3} = 2\times3.8\times10^{-5}\ \text{bar} = \boxed{7.6\times10^{-5}\ \text{bar}}$$

E7.39 Let α be the fraction of N_2O_4 molecules that dissociate as equilibrium is established.

	$N_2O_4(g)$	$\rightleftharpoons$	$2\ NO_2(g)$
Initial condition	n_i		0
Equilibrium amount	$(1-\alpha)n_i$		$2\alpha n_i$
$n_{\text{total at eq}} = (1+\alpha)n_i$			
Mole fraction, x_J	$(1-\alpha)/(1+\alpha)$		$2\alpha/(1+\alpha)$

$$K = \left(\frac{p_{NO_2}^2}{p_{N_2O_4}}\right)_{eq} = \frac{(x_{NO_2}p)^2}{x_{N_2O_4}p} = \frac{x_{NO_2}{}^2 p}{x_{N_2O_4}} = \frac{(2\alpha/\{1+\alpha\})^2}{(\{1-\alpha\}/\{1+\alpha\})}p = \left(\frac{4\alpha^2}{1-\alpha^2}\right)p$$

In the case for which $\alpha \ll 1$, the denominator factor $1-\alpha^2$ reduces to unity. Thus,

$$K = 4\alpha^2 p \qquad \text{or} \qquad \alpha = \tfrac{1}{2}K^{1/2}p^{-1/2}.$$

Since K is independent of pressure, we conclude that $\boxed{\alpha \propto p^{-1/2}}$ when $\alpha \ll 1$.

E7.40 $$\text{U(s)} + \tfrac{3}{2}\ \text{H}_2\text{(g)} \rightleftharpoons \text{UH}_3\text{(s)} \qquad K = (p/p^\ominus)^{-3/2} \qquad \text{where } p = p_{H_2} = 1.04\ \text{Torr}$$

$$p/p^\ominus = (1.04\ \text{Torr})\times(1.333\times10^{-3}\ \text{bar Torr}^{-1})/(1\ \text{bar}) = 1.39\times10^{-3}\ \text{at equilibrium.}$$

$$\Delta_r G^\ominus = -RT\ln K\ [7.8] = -RT\ln(p/p^\ominus)^{-3/2} = \tfrac{3}{2}RT\ln(p/p^\ominus)$$

$$= \tfrac{3}{2}\times(8.3145\ \text{J K}^{-1}\ \text{mol}^{-1})\times(500\ \text{K})\times\ln(1.39\times10^{-3})$$

$$= \boxed{-41.0\ \text{kJ mol}^{-1}}$$

E7.41 $$\ln\frac{K'}{K} = \frac{\Delta_r H^\ominus}{R}\left(\frac{1}{T}-\frac{1}{T'}\right)\ [7.15] \qquad \text{or} \qquad \Delta_r H^\ominus = R\left(\frac{1}{T}-\frac{1}{T'}\right)^{-1}\ln\frac{K'}{K}$$

$T = 298$ K and $T' = 308$ K, hence

$$\Delta_r H^\ominus = (8.3145\ \text{J K}^{-1}\ \text{mol}^{-1})\times\left(\frac{1}{298\ \text{K}}-\frac{1}{308\ \text{K}}\right)^{-1}\ln\frac{K'}{K} = (76.3\bar{1}\ \text{kJ mol}^{-1})\times\ln\frac{K'}{K}.$$

(a) $K'/K = 2,\ \Delta_r H^\ominus = (76.3\bar{1}\ \text{kJ mol}^{-1})\times\ln 2 = \boxed{+52.9\ \text{kJ mol}^{-1}}$

(b) $K'/K = \tfrac{1}{2},\ \Delta_r H^\ominus = (76.3\bar{1}\ \text{kJ mol}^{-1})\times\ln(\tfrac{1}{2}) = \boxed{-52.9\ \text{kJ mol}^{-1}}$

E7.42 $$\text{NH}_4\text{Cl(s)} \rightleftharpoons \text{NH}_3\text{(g)} + \text{HCl(g)}$$

$$p = p_{NH_3} + p_{HCl} = 2p_{NH_3}\ \text{[all } p\text{'s are relative to } p^\ominus = 1\ \text{bar]}$$

(a) $K = p_{NH_3} p_{HCl} = p_{NH_3}^2 = \frac{1}{4} p^2$

$$\text{At } 427°\text{C (700 K)}, \ K = \frac{1}{4}\left(\frac{608 \text{ kPa}}{100 \text{ k Pa bar}^{-1}}\right)^2 = \boxed{9.24}$$

$$\text{At } 459°\text{C (732 K)}, \ K = \frac{1}{4}\left(\frac{1115 \text{ kPa}}{100 \text{ kPa bar}^{-1}}\right)^2 = \boxed{31.08}$$

(b) $\Delta_r G^{\ominus} = -RT \ln K$ [7.8]

$$\text{At } 427°\text{C (700 K)}, \ \Delta_r G^{\ominus} = -(8.3145 \text{ J K}^{-1}\text{mol}^{-1}) \times (700 \text{ K}) \times \ln(9.24) = \boxed{-12.9 \text{ kJ mol}^{-1}}$$

$$\text{At } 459°\text{C (732 K)}, \ \Delta_r G^{\ominus} = -(8.3145 \text{ J K}^{-1}\text{mol}^{-1}) \times (732 \text{ K}) \times \ln(31.08) = \boxed{-20.9 \text{ kJ mol}^{-1}}$$

(c) $\ln \dfrac{K'}{K} = \dfrac{\Delta_r H^{\ominus}}{R}\left(\dfrac{1}{T} - \dfrac{1}{T'}\right)$ [7.15] or $\Delta_r H^{\ominus} = R\left(\dfrac{1}{T} - \dfrac{1}{T'}\right)^{-1} \ln \dfrac{K'}{K}$

$$\Delta_r H^{\ominus} = (8.3145 \text{ J K}^{-1}\text{mol}^{-1}) \times \left(\frac{1}{700 \text{ K}} - \frac{1}{732 \text{ K}}\right)^{-1} \times \ln(31.08/9.24) = \boxed{+161 \text{ kJ mol}^{-1}}$$

(d) $$\Delta_r S^{\ominus} = \frac{\Delta_r H^{\ominus} - \Delta_r G^{\ominus}}{T} \ [7.11] = \frac{161 \text{ KJ mol}^{-1} - (-12.9 \text{ kJ mol}^{-1})}{700 \text{ K}} = \boxed{+248 \text{ J K}^{-1} \text{ mol}^{-1}}$$

Answers to projects

P7.43 (a)

$M_I = 126.9$ g mol^{-1}, $T = 1000$ K	$I_2(g)$	$\rightleftharpoons$	2 I(g)
Initial mass / g	1.00		0
Initial amount / mmol	3.94		0
Equilibrium mass / g	0.830		1.00 – 0.830 = 0.17
Equilibrium amount / mmol	3.27		1.34
$n_{total, eq} = 4.61$ mmol			
Mole fraction at equilibrium, x_J	0.709		0.291

$$p = \frac{n_{total}RT}{V} = \frac{(4.61 \times 10^{-3} \text{ mol}) \times (0.831447 \text{ dm}^3 \text{ bar K}^{-1} \text{ mol}^{-1}) \times (1000 \text{ K})}{1.00 \text{ dm}^3} = 0.383 \text{ bar}$$

$$K = \left(\frac{p_I^2}{p_{I_2} p^{\ominus}}\right)_{eq} = \frac{(x_I p)^2}{x_{I_2} p p^{\ominus}} = \frac{x_I^2 p}{x_{I_2} p^{\ominus}} = \frac{(0.291)^2 \times (0.383)}{0.709} = \boxed{0.0457} \text{ at 1000 K}$$

(b) A series of experiments like that of part (a) in a range of temperatures around 1000 K finds that $\ln K = a + bT + cT^2$ where $a = -36.386$, $b = 0.0533 \text{ K}^{-1}$, $c = -2 \times 10^{-5} \text{ K}^{-2}$, and $950 \text{ K} \le T \le 1150 \text{ K}$.

Solving the van't Hoff equation for the standard reaction enthalpy, followed by substitution of the above relation for ln K yields an expression for the temperature dependence of $\Delta_r H^{\ominus}$.

$$\Delta_r H^{\ominus} = RT^2 \frac{d(\ln K)}{dT} = RT^2 \frac{d}{dT}(a + bT + cT^2)$$
$$= RT^2 (b + 2cT)$$

$\boxed{\Delta_r H^{\ominus} = bRT^2 + 2cRT^3}$ or $\Delta_r H^{\ominus} / \text{kJ mol}^{-1} = (4.43 \times 10^{-4}) \times (T/\text{K})^2 - (3.33 \times 10^{-7}) \times (T/\text{K})^3$

This relation is used to prepare the $\Delta_r H^\ominus$ against T plot of Figure 7.3.

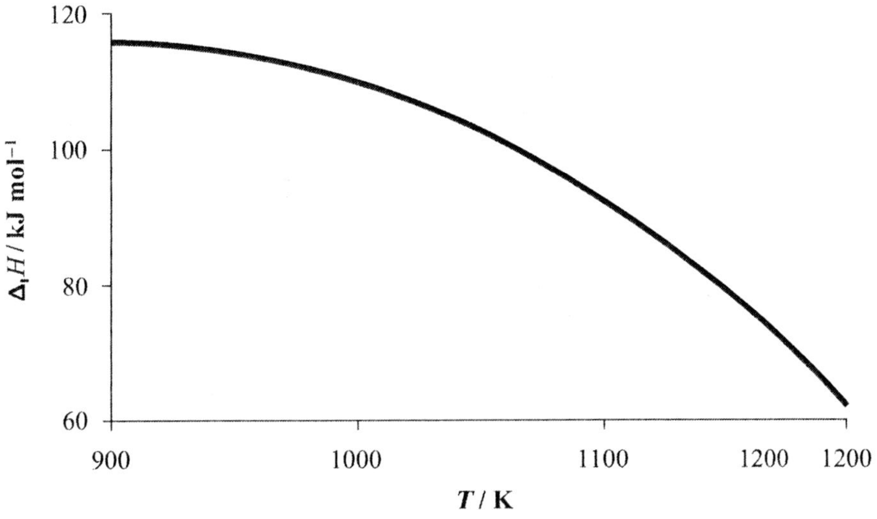

Figure 7.3

(c) $$K_c = K\left(\frac{c^\ominus RT}{p^\ominus}\right)^{-\Delta\nu_{\text{gas}}} \quad [7.13a]$$

$$\ln K_c = \ln K - \Delta\nu_{\text{gas}} \ln T - \Delta\nu_{\text{gas}} \ln\left(\frac{c^\ominus R}{p^\ominus}\right)$$

$$\frac{\mathrm{d}\ln K_c}{\mathrm{d}T} = \frac{\mathrm{d}\ln K}{\mathrm{d}T} - \Delta\nu_{\text{gas}}\frac{\mathrm{d}\ln T}{\mathrm{d}T} = \frac{\Delta_r H^\ominus}{RT^2} - \frac{\Delta\nu_{\text{gas}}}{T}$$

$$\boxed{\frac{\mathrm{d}\ln K_c}{\mathrm{d}T} = \frac{\Delta_r H^\ominus}{RT^2} - \frac{\Delta\nu_{\text{gas}}}{T}}$$

P7.44 (a) In order to calculate the fractional saturation at the requested pressure we need to know the value of K in the Hill equation. K can be determined approximately from the figure shown in the Box. For $s = 0.5$, $p = p_{50}$ where p_{50} is the pressure of O_2 at 50% saturation.

$$\log\left(\frac{0.5}{1-0.5}\right) = \nu \log p_{50} - \nu \log K$$

Therefore, $K = p_{50}$.

From the graph we estimate that $\boxed{\text{for hemoglobin, } p_{50} = K = 30 \text{ Torr; for myoglobin, } p_{50} = K = 5 \text{ Torr}}$. Solving the Hill equation for s we obtain

$$\frac{s}{1-s} = \left(\frac{p_{O_2}}{K}\right)^\nu \quad \text{and} \quad s = \frac{1}{(K/p_{O_2})^\nu + 1}.$$

Then for Hb we construct the following table.

p_{O_2} / Torr	5	10	20	30	60
s	9.8×10^{-3}	0.064	0.32	0.60	0.91

For Mb

p_{O_2} / Torr	5	10	20	30	60
s	0.50	0.66	0.80	0.86	0.92

The values of s match well the values read from the graph.

(b) Again we use $s = \dfrac{1}{(K/p_{O_2})^{\nu}+1}$ and construct the following tables.

For Hb

p_{O_2}/Torr	5	10	20	30	60
s	1.4×10^{-3}	0.021	0.26	0.64	0.996

For Mb

p_{O_2}/Torr	5	10	20	30	60
s	0.50	0.94	0.996	0.9992	0.99995

For $\nu = 4$, the curves lose their "S", sigmoid shape.

8 Chemical equilibrium: equilibria in solution

Answers to discussion questions

D8.1 (a) Figure 8.4 of the text illustrates the typical pH curve for the titrations of a weak acid with a strong base. Prior to reaching the stoichiometric point, when titrant volumes are small, the slope of the curve is positive but of small magnitude and an important inflection point is observed, which characterizes the pK_a of the weak acid. pH changes are small in this region because HA can both neutralize titrant and dissociate to maintain pH. Near the stoichiometric point the curve slope is very large because each addition of even a small drop of base causes a large pH change. The remaining amount of weak acid is not sufficient to maintain pH constancy through dissociation. At the stoichiometric point, $[A^-] = [HA]_{initial}V_{initial}/V$. After the stoichiometric point the curve levels off as the mixture approaches the pH of the titrant. If the acid is extremely weak, there is no large increase in pH near the stoichiometric point.

(b) Figure 8.5 of the text illustrates the typical pH curve for the titrations of a weak base with a strong acid. Prior to reaching the stoichiometric point, when titrant volumes are small, the slope of the curve is negative but of small magnitude and an important inflection point is observed, which characterizes the pK_b of the weak base. pH changes are small in this region because the base can both neutralize titrant and dissociate to maintain pH. Near the stoichiometric point the curve slope is very steep because each addition of even a small drop of acid causes a large pH change. The remaining amount of weak base is not sufficient to maintain pH constancy through dissociation. After the stoichiometric point the curve levels off as the mixture approaches the pH of the titrant. If the base is extremely weak, there is no large drop in pH near the stoichiometric point.

D8.2 An acid buffer is a solution of approximately equal concentrations of a weak acid and its salt. This solution maintains some constancy in the acidic pH range through its ability to neutralize both a small amount of strong base and a small amount of strong acid. A base buffer, a solution of equal concentrations of a weak base and its salt, does much the same thing but in the base pH range.

Indicators are weak acids which in their undissociated acid form have one colour, and in their dissociated anion form, another. In acidic solution, the indicator exists in the predominantly acid form (one colour), in basic solution in the predominantly anion form (the other colour). The ratio of the two forms is very pH sensitive because of the small value of pK_a of the indicator, so the colour change can occur very rapidly with change in pH. The indicator dye for an acid/base titration is chosen with care to match the pH at which colour change occurs with the stoichiometric point.

D8.3 At the **stoichiometric point** (also called the **equivalence point**) of an acid-base titration a stoichiometrically equivalent amount of acid has been added to a given amount of base. In a strong acid-strong base titration the pH changes sharply through several pH units around the stoichiometric point. The colour change of an indicator dye is often used to estimate the point at which the volume of titrant has provided acid-base equivalency. The titration point at which the indicator dye colour changes is called the titration **end point**. With a well-chosen indicator, the end point of the indicator coincides with the stoichiometric point of the titration. Other methods for detecting the end point include monitoring the electrical conductance of the solution and monitoring light absorption with a visible spectrophotometer.

D8.4 Phosphoric acid is a convenient example of a triprotic acid. Like all polyprotic acids its successive acid constants are progressively smaller because of the increased charge on the dissociating anion: $pK_{a1} = 2.12$, $pK_{a2} = 7.21$, $pK_{a3} = 12.67$. The fractional composition of the protonated and deprotonated forms of a triprotic acid are illustrated in text Figure 8.2 using phosphoric acid as the example. At the lowest pH values the fraction of the fully protonated species (H_3PO_4) is 1 but this fraction decreases with increasing pH. The Henderson–Hasselbalch eqn [8.13] estimates that when $pH = pK_{a1}$ the fraction of the fully protonated species equals its conjugate base (diprotic $H_2PO_4^-$). As the pH changes to greater values the fraction of the diprotic species increases to a maximum at $pH = \frac{1}{2}(pK_{a1} + pK_{a2})$ [8.12]. At yet larger pH the diprotic species fraction diminishes while its conjugate base (monoprotic HPO_4^{2-}) fraction increases; the fractions of these two are equal at $pH = pK_{a2}$. The monoprotic species fraction is a maximum at $pH = \frac{1}{2}(pK_{a2} + pK_{a3})$ after which it decreases until its fraction equals the fraction of its conjugate base (PO_4^{3-}) when $pH = pK_{a3}$. At the highest pH values the fully deprotonated species is present alone.

The fraction of a species at any chosen pH is computed with eqns that are analogous to eqn 8.11a–8.11c for a diprotic acid. Let D be the function: $D = [H_3O^+]^3 + [H_3O^+]^2 K_{a1} + [H_3O^+]K_{a1}K_{a2} + K_{a1}K_{a2}K_{a3}$. Then for a triprotic acid:

$$f(H_3PO_4) = \left[H_3O^+\right]^3 / D,$$

$$f(H_2PO_4^{1-}) = \left[H_3O^+\right]^2 K_{a1} / D,$$

$$f(HPO_4^{2-}) = \left[H_3O^+\right] K_{a1}K_{a2} / D, \text{ and finally}$$

$$f(PO_4^{3-}) = K_{a1}K_{a2}K_{a3} / D.$$

D8.5 The pH of the solution of an amphiprotic species is estimated with the relation $pH = \frac{1}{2}(pK_{a1} + pK_{a2})$ [8.12]. This relation is valid when the **formal concentration** F (the concentration as prepared) of the salt MHA, where HA^- is an amphiprotic anion, satisfies the condition $F/c^\ominus >> K_w/K_{a2}$ and $F/c^\ominus >> K_{a1}$ (see Derivation 8.1). Under these conditions eqn 8.12 indicates that the formal concentration does not determine the pH because F does not appear in the equation (i.e., the pH is constant over a considerable range of formal concentration). These conditions also cause $[H_2A] \approx [A^-]$.

D8.6 The Henderson-Hasselbalch equation, $pH = pK_a - \log([\text{acid}]/[\text{base}])$ [8.13], is limited to the condition that the weak acid solution has a **formal concentration** F (the concentration as prepared) that is dilute to the extent that it is valid to replace activities with concentrations in equilibrium expressions. Other conditions become more important when the equation is used in the form $pH \approx pK_a - \log(c_{HA}/c_{MA})$ where c_{HA} and c_{MA} are the formal concentrations of a solution prepared from a weak acid HA and a salt of its conjugate base A^-. The reversible acid dissociation reaction causes some dynamic adjustment in the concentration of both the acid and the conjugate base so we draw an equilibrium table to examine the changes closely.

	HA	⇌	A^-	H_3O^+
Initial molar concentration / mol dm^{-3}	c_{HA}		c_{MA}	0
Change to reach equilibrium / mol dm^{-3}	$-x$		$+x$	$+x$
Equilibrium concentration / mol dm^{-3}	$c_{HA} - x$		$c_{MA} + x$	x

$$K_a = \frac{a_{H_3O^+} a_{A^-}}{a_{HA}} \approx \frac{\left[H_3O^+\right]\left[A^-\right]}{[HA]} \approx \frac{\left[H_3O^+\right](c_{MA} + x)}{(c_{HA} - x)}$$

$$K_a \approx \left[H_3O^+\right]\left(\frac{c_{MA}}{c_{HA}}\right) \quad \text{and} \quad pH \approx pK_a - \log\left(\frac{c_{HA}}{c_{MA}}\right) \text{ provided that } |x| \ll c_{HA}, c_{MA}.$$

The condition $|x| = c_{HA}, c_{MA}$ is somewhat satisfied in practice. For example, the common 0.025 molar phosphate buffer is prepared with $c_{KH_2PO_4} = 25.0$ mmol dm^{-3} and $c_{Na_2HPO_4} = 25.0$ mmol dm^{-3} and has pH = 6.86 at 25°C but the eqn $pH \approx pK_a - \log(c_{HA}/c_{MA})$ predicts that $pH \approx pK_{a,H_2PO_4^{-1}} = 7.21$. The discrepancy between the actual pH and the prediction of the Henderson-Hasselbalch eqn has two origins: the activity coefficients do not equal 1 exactly and the condition $|x| = c_{\text{HA}}, c_{\text{MA}}$ is not entirely satisfied. To see the latter, solve the above eqn for x; if $|x|/c_{\text{HA}} < 0.01$, the condition $|x| = c_{\text{HA}}, c_{\text{MA}}$ is satisfied.

$$x = \frac{c_{\text{HA}}K_{\text{a}} - c_{\text{MA}}\left[\text{H}^+\right]}{K_{\text{a}} + \left[\text{H}^+\right]} = \frac{\left(25\ \text{mmol dm}^{-3}\right)\times\left(10^{-7.21} - 10^{-6.86}\right)}{\left(10^{-7.21} + 10^{-6.86}\right)} = -9.6\ \text{mmol dm}^{-3}$$

$$\frac{|x|}{c_{\text{HA}}} = \frac{9.6}{25} = 0.38 \text{ or } 38\%$$

We see that although $|x| < c_{\text{HA}}$, since the condition $|x|/c_{\text{HA}} < 0.01$ is not satisfied, the condition $|x| = c_{\text{HA}}, c_{\text{MA}}$ is not satisfied. Nevertheless, the Henderson-Hasselbalch eqn in the form $\text{pH} \approx \text{p}K_{\text{a}} - \log(c_{\text{HA}} / c_{\text{MA}})$ provides very useful approximation when working with a solution of a weak acid and its conjugate base.

D8.7 The common-ion effect is the phenomena in which the solubility of a sparingly soluble salt is reduced by a second salt when the two share a common-ion. For example, the solubility of lead(II) sulfate is reduced by sodium sulfate. This is expected from Le Chatelier's principle: when the concentration of the common ion is increased, the equilibrium shifts to minimize that increase. As a result, the solubility of the original salt can be expected to decrease.

Solutions to exercises

E8.1 (a)

conjugate (acid$_1$ ↔ base$_1$)

$$\underset{\text{acid}_1}{H_2SO_4} + \underset{\text{base}_2}{H_2O} \rightleftharpoons \underset{\text{acid}_2}{H_3O^+} + \underset{\text{base}_1}{HSO_4^-}$$

conjugate (base$_2$ ↔ acid$_2$)

(b)

conjugate (acid$_1$ ↔ base$_1$)

$$\underset{\text{acid}_1}{HF} + \underset{\text{base}_2}{H_2O} \rightleftharpoons \underset{\text{acid}_2}{H_3O^+} + \underset{\text{base}_1}{F^-}$$

conjugate (base$_2$ ↔ acid$_2$)

(c)

conjugate (acid$_1$ ↔ base$_1$)

$$\underset{\text{acid}_1}{C_6H_5NH_3^+} + \underset{\text{base}_2}{H_2O} \rightleftharpoons \underset{\text{acid}_2}{H_3O^+} + \underset{\text{base}_1}{C_6H_5NH_2}$$

conjugate (base$_2$ ↔ acid$_2$)

(d)

conjugate (acid$_1$ ↔ base$_1$)

$$\underset{\text{acid}_1}{H_2PO_4^-} + \underset{\text{base}_2}{H_2O} \rightleftharpoons \underset{\text{acid}_2}{H_3O^+} + \underset{\text{base}_1}{HPO_4^{2-}}$$

conjugate (base$_2$ ↔ acid$_2$)

(e)

conjugate (acid$_1$ ↔ base$_1$)

$$\underset{\text{acid}_1}{HCOOH} + \underset{\text{base}_2}{H_2O} \rightleftharpoons \underset{\text{acid}_2}{H_3O^+} + \underset{\text{base}_1}{HCO_2^-}$$

conjugate (base$_2$ ↔ acid$_2$)

(f)

conjugate (acid$_1$ ↔ base$_1$)

$$\underset{\text{acid}_1}{NH_2NH_3^+} + \underset{\text{base}_2}{H_2O} \rightleftharpoons \underset{\text{acid}_2}{H_3O^+} + \underset{\text{base}_1}{NH_2NH_2}$$

conjugate (base$_2$ ↔ acid$_2$)

E8.2 (a) $CH_3CH(OH)COOH + H_2O \rightleftharpoons CH_3CH(OH)CO_2^- + H_3O^+$

(b) $HOOC(CH_2)_2CH(NH_2)COOH + H_2O \rightleftharpoons HOOC(CH_2)_2CH(NH_2)CO_2^- + H_3O^+$

$HOOC(CH_2)_2CH(NH_2)CO_2^- + H_2O \rightleftharpoons {}^-O_2C(CH_2)_2CH(NH_2)CO_2^- + H_3O^+$

These are the carboxylic acid dissociation reactions. The amine group ($R{-}NH_2$) is a base so proton transfer reactions can also be written that show the proton transfer from water to the amine for each of the above four species. Example:

$$HOOC(CH_2)_2CH(NH_2)COOH + H_2O \rightleftharpoons HOOC(CH_2)_2CH(NH_3^+)COOH + OH^-$$

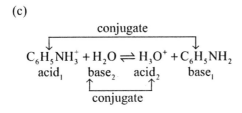

(c) $NH_2CH_2COOH + H_2O \rightleftharpoons NH_2CH_2CO_2^- + H_3O^+$

$NH_2CH_2COOH + H_2O \rightleftharpoons {}^+NH_3CH_2COOH + OH^-$

$NH_2CH_2CO_2^- + H_2O \rightleftharpoons {}^+NH_3CH_2CO_2^- + OH^-$

(d) $HOOCCOOH + H_2O \rightleftharpoons HOOCCO_2^- + H_3O^+$

$HOOCCO_2^- + H_2O \rightleftharpoons {}^-O_2CCO_2^- + H_3O^+$

E8.3 (a) $2\,H_2O \rightleftharpoons [H_3O^+] + [OH^-]$ and $[OH^-] = [H_3O^+]$ in pure water.

$K_w = 2.5 \times 10^{-14} = [H_3O^+][OH^-] = [H_3O^+]^2$

$[H_3O^+] = K_w^{1/2} = \sqrt{2.5\times10^{-14}} = \boxed{1.6\times10^{-7}\text{ mol dm}^{-3}}$

$pH = -\log[H_3O^+] = \boxed{6.80}$

(b) $[OH^-] = [H_3O^+] = \boxed{1.6 \times 10^{-7}\text{ mol dm}^{-3}}$

$pOH = -\log[OH^-] = \boxed{6.80}$

E8.4 (a) $2\,D_2O \rightleftharpoons D_3O^+ + OD^-$ and $[OD^-] = [D_3O^+]$ in pure dideuterium oxide.

(b) $K_{w,D} = [D_3O^+][OD^-] = 1.35 \times 10^{-15}$ $\quad pK_{w,D} = -\log K_{w,D} = \boxed{14.870}$

(c) $K_{w,D} = 1.35 \times 10^{-15} = [D_3O^+][OD^-] = [D_3O^+]^2$

$[D_3O^+] = K_{w,D}^{1/2} = \sqrt{1.35\times10^{-15}} = \boxed{3.67\times10^{-8}\text{ mol dm}^{-3}}$

$[OD^-] = [D_3O^+] = \boxed{3.67\times10^{-8}\text{ mol dm}^{-3}}$

(d) $pD = -\log(3.67 \times 10^{-8}) = \boxed{7.45} = pOD$

(e) $pD + pOD = pK_{w,D} = \boxed{14.870}$

E8.5 $pH = -\log a_{H_3O^+} = -\log\left(\gamma_{H_3O^+}[H_3O^+]\right)$ where $\gamma_{H_3O^+}$ is the activity coefficient.

Being a strong acid, a 0.50 molar hydrochloric acid solution is expected to be 100% dissociated with $[H^+] = 0.50\text{ mol dm}^{-3}$. Thus, with the usual approximation that $\gamma_{H_3O^+} \approx 1$ we find

$$pH_{approx} \approx -\log\left([H_3O^+]\right) = -\log(0.50) = \boxed{0.30}.$$

With the inclusion of the mean activity coefficient $\gamma_{H_3O^+} = 0.769$ we find

$$pH = -\log(0.769\times0.50) = \boxed{0.42}.$$

The difference of around 0.1 pH units between pH_{approx} and the actual pH is often negligible; however, the activity coefficient must be included in many exacting research situations.

E8.6 $$\ln K' - \ln K = \frac{\Delta_r H^\ominus}{R}\left(\frac{1}{T} - \frac{1}{T'}\right)$$ van't Hoff equation [7.15]

Use the relationship $\ln(x) = \ln(10) \times \log(x) = 2.303\log(x)$ in order to introduce log terms and subsequently pK terms into the van't Hoff equation.

$$\log K' - \log K = \frac{\Delta_r H^\ominus}{\ln 10\times R}\left(\frac{1}{T} - \frac{1}{T'}\right)$$

$$-\log K - (-\log K') = \frac{\Delta_r H^\ominus}{\ln 10\times R}\left(\frac{1}{T} - \frac{1}{T'}\right)$$

$$pK - pK' = \frac{\Delta_r H^\ominus}{\ln 10\times R}\left(\frac{1}{T} - \frac{1}{T'}\right)$$

K' can be considered as a reference equilibrium constant at the specific temperature T' while K is the equilibrium constant at the variable temperature T. This places the working equation in the linear form:

$$\text{p}K = \text{slope} \times \frac{1}{T} + \text{intercept}$$

where the slope equals $\boxed{\dfrac{\Delta_r H^\ominus}{\ln 10 \times R}}$ and the intercept equals $\text{p}K' - \dfrac{\Delta_r H^\ominus}{\ln 10 \times RT'}$

E8.7 As shown in Exercise 8.6 a plot of $\text{p}K_w$ against $1/T$ (see Figure 8.1) is linear with a slope equal to $\Delta_w H^\ominus / (\ln 10 \times R)$. The slope is determined with a linear least squares regression fit to the plotted data and we find that

$$\begin{aligned}\Delta_w H^\ominus &= \ln 10 \times R \times slope \\ &= \ln 10 \times (8.3145 \text{ J K}^{-1} \text{ mol}^{-1}) \times (2.985 \times 10^3 \text{ K}) \\ &= \boxed{57.1 \text{ kJ mol}^{-1}}\end{aligned}$$

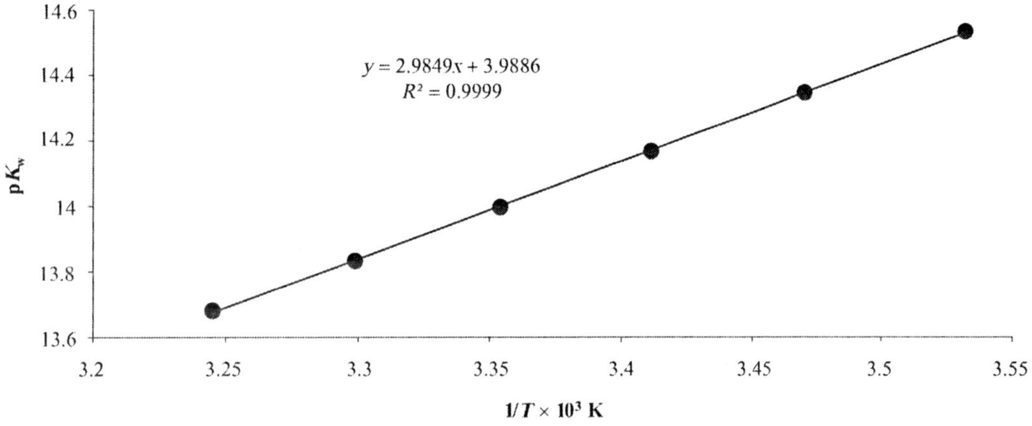

Figure 8.1

E8.8 As shown in Exercise 8.6 a plot of $\text{p}K_b$ against $1/T$ (see Figure 8.2) for ammonia is linear with a slope equal to $\Delta_b H^\ominus / (\ln 10 \times R)$. The slope is determined with a linear least squares regression fit to the plotted data and we find that

$$\begin{aligned}\Delta_b H^\ominus &= \ln 10 \times R \times slope \\ &= \ln 10 \times (8.3145 \text{ J K}^{-1} \text{ mol}^{-1}) \times (0.249 \times 10^3 \text{ K}) \\ &= \boxed{4.8 \text{ kJ mol}^{-1}}\end{aligned}$$

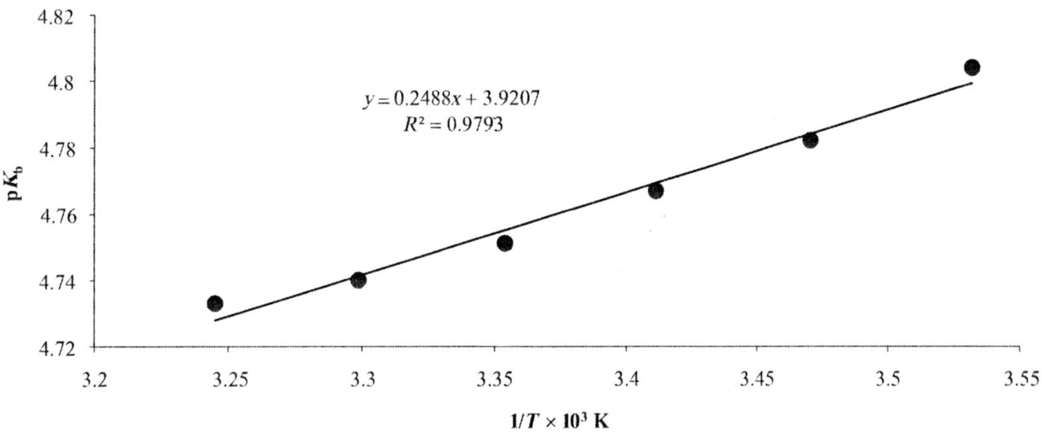

Figure 8.2

Furthermore, the pK_a of NH_4^+(aq), the conjugate acid of ammonia, over this temperature range is given by

$$pK_a = pK_w - pK_b \text{ [8.8b; the temperature dependence of } pK_w \text{ is found in Exercise 8.7]}$$

$$= (2.985 \times 10^3 \text{ K}) \times 1/T + 3.99 - [(0.249 \times 10^3 \text{ K}) \times 1/T + 3.92]$$

$$\boxed{pK_a = (2.736 \times 10^3 \text{ K}) \times 1/T + 0.07}.$$

$$\Delta_a H^\ominus = \ln 10 \times R \times \text{slope}$$

$$= \ln 10 \times (8.3145 \text{ J K}^{-1} \text{ mol}^{-1}) \times (2.736 \times 10^3 \text{ K})$$

$$= \boxed{52.4 \text{ kJ mol}^{-1}}$$

E8.9 Nicotine (Nic) protonation reaction: $Nic(aq) + H_2O \rightleftharpoons NicH^+(aq) + OH^-(aq)$ $pK_b = 5.98$

Deprotonation of conjugate acid: $NicH^+(aq) + H_2O \rightleftharpoons Nic(aq) + H_3O^+$ $pK_a = ?$

$$pK_a = pK_w - pK_b \text{ [8.8b]} = 14.00 - 5.98$$

$$= \boxed{8.02}$$

E8.10 $\text{pH} = -\log a_{H_3O^+} \ [8.3] \approx -\log([H_3O^+])$ and $\text{pOH} = pK_w - \text{pH}\ [8.9] = 14.00 - \text{pH}$

	$[H_3O^+]$/mol dm^{-3}	pH	pOH
(a)	1.50×10^{-5}	4.82	9.18
(b)	1.50×10^{-3}	2.82	11.18
(c)	5.1×10^{-14}	13.29	0.71
(d)	5.01×10^{-5}	4.30	9.70

E8.11 (a) amount (moles) $H_3O^+ = (0.0250 \text{ dm}^3) \times (0.144 \text{ mol dm}^{-3}) = 3.60 \times 10^{-3}$ mol

amount (moles) $OH^- = (0.0250 \text{ dm}^3) \times (0.125 \text{ mol dm}^{-3}) = 3.12 \times 10^{-3}$ mol

excess $H_3O^+ = (3.60 \times 10^{-3} - 3.12 \times 10^{-3}) = 0.48 \times 10^{-3}$ mol H_3O^+

$$[H_3O^+] = \frac{4.8 \times 10^{-4} \text{ mol}}{0.0500 \text{ dm}^3} = \boxed{9.60 \times 10^{-3} \text{ mol dm}^{-3}}$$

$$\text{pH} = -\log(9.60 \times 10^{-3}) = \boxed{2.02}$$

(b) amount of $H_3O^+ = (0.0250 \text{ dm}^3) \times (0.15 \text{ mol dm}^{-3}) = 3.75 \times 10^{-3}$ mol H_3O^+

amount of $OH^- = (0.0350 \text{ dm}^3) \times (0.15 \text{ mol dm}^{-3}) = 5.25 \times 10^{-3}$ mol OH^-

excess $OH^- = (5.25 \times 10^{-3} - 3.75 \times 10^{-3}) = 1.50 \times 10^{-3}$ mol OH^-

$$[OH^-] = \frac{1.50 \times 10^{-3} \text{ mol}}{0{,}060 \text{ dm}^3} = \boxed{0.025 \text{ mol dm}^{-3}}$$

$$\text{pOH} = -\log(0.025) = 1.60$$

$$\text{pH} = 14.00 - 1.30 = \boxed{12.40}$$

(c) amount of $H_3O^+ = (0.0212 \text{ dm}^3) \times (0.22 \text{ mol dm}^{-3}) = 4.7 \times 10^{-3}$ mol H_3O^+

amount of $OH^- = (0.0100 \text{ dm}^3) \times (0.30 \text{ mol dm}^{-3}) = 3.0 \times 10^{-3}$ mol OH^-

$$\text{concentration of excess } H_3O^+ = \frac{1.7 \times 10^{-3} \text{ mol}}{0.031 \text{ dm}^3} = 5.5 \times 10^{-2} \text{ M}$$

$$\text{pH} = \boxed{1.26}$$

E8.12 General acidity/basicity rules predict that an aqueous solution containing a/an

(1) salt of a strong acid and strong base is neutral;

(2) acid or a conjugate acid of a weak base is acidic;

(3) base or a conjugate base of a weak acid is basic;

(4) small, highly charged metal cation is acidic. *Example:* The solution $FeCl_3(aq)$ is acidic because Fe^{3+} acts as a Lewis acid in water: $[Fe(H_2O)_6]^{3+}(aq) + H_2O(l) \rightleftharpoons [Fe(H_2O)_5OH]^{2+}(aq) + H_3O^+(aq)$.

(a) acidic; $NH_4^+(aq) + H_2O(l) \rightleftharpoons H_3O^+(aq) + NH_3(aq)$

(b) basic; $H_2O(l) + CO_3^{2-}(aq) \rightleftharpoons HCO_3^-(aq) + OH^-(aq)$

(c) basic; $H_2O(l) + F^-(aq) \rightleftharpoons HF(aq) + OH^-(aq)$

(d) neutral

(e) acidic; $[Al(H_2O)_6]^{3+}(aq) + H_2O(l) \rightleftharpoons [Al(H_2O)_5OH]^{2+}(aq) + H_3O^+(aq)$

(f) acidic; $[Co(H_2O)_6]^{2+}(aq) + H_2O(l) \rightleftharpoons [Co(H_2O)_5OH]^{+}(aq) + H_3O^+(aq)$

E8.13

$$c_{NaCH_3CO_2} = n/V = m/M/V = (7.4\text{ g})/(82.03\text{ g mol}^{-1})/(0.250\text{ dm}^3) = 0.36\text{ mol dm}^{-3}$$

$pK_b = 9.25$	$CH_3CO_2^-(aq)$	+	$H_2O(l)$	$\rightleftharpoons$	$CH_3COOH(aq)$	+	$OH^-(aq)$
Formal conc./mol dm^{-3}	$c_{CH_3CO_2^-} = 0.36$						
Change/mol dm^{-3}	$-x$				$+x$		$+x$
Equil. conc./mol dm^{-3}	$c_{CH_3CO_2^-} - x \cong c_{CH_3CO_2^-}$				x		x

$$K_b = \left(\frac{[CH_3COOH][OH^-]}{[CH_3CO_2]}\right)_{eq} = \frac{x^2}{c_{CH_3CO_2^-}}$$

$$x = \sqrt{c_{CH_3CO_2^-}K_b} = \sqrt{(0.36)\times(10^{-9.25})}\text{ mol dm}^{-3} = 1.4\times10^{-5}\text{ mol dm}^{-3}$$

Note that we can now justify the approximation $c_{CH_3CO_2^-} - x \cong c_{CH_3CO_2^-}$ with the observation that the calculated value of x is much, much smaller than the value of $c_{CH_3CO_2^-}$.

$$pOH = -\log[OH^-] = -\log(1.4\times10^{-5}) = 4.8$$

$$pH = 14.0 - pOH = 14.0 - 4.8 = \boxed{9.2}$$

E8.14

$$c_{NH_4Cl} = n/V = m/M/V = (2.75\text{ g})/(53.49\text{ g mol}^{-1})/(0.100\text{ dm}^3) = 0.514\text{ mol dm}^{-3}$$

$pK_a = 9.25$	$NH_4^+(aq)$	+	$H_2O(l)$	$\rightleftharpoons$	$NH_3(aq)$	+	$H_3O^+(aq)$
Formal conc./mol dm^{-3}	$c_{NH_4^+} = 0.514$						
Change/mol dm^{-3}	$-x$				$+x$		$+x$
Equil. conc./mol dm^{-3}	$c_{NH_4^+} - x \cong c_{NH_4Cl}$				x		x

$$K_a = \left(\frac{[NH_3][H_3O^+]}{[NH_4^+]}\right)_{eq} = \frac{x^2}{c_{NH_4^+}}$$

$$x = \sqrt{c_{NH_4^+}K_a} = \sqrt{(0.514)\times(10^{-9.25})}\text{ mol dm}^{-3} = 1.70\times10^{-5}\text{ mol dm}^{-3}$$

Note that we can now justify the approximation $c_{NH_4^+} - x \cong c_{NH_4^+}$ with the observation that the calculated value of x is much, much smaller than the value of $c_{NH_4^+}$.

$$\text{pH} = -\log\left[H_3O^+\right] = -\log\left(1.70\times10^{-5}\right)$$
$$= \boxed{4.77}$$

E8.15 KBr is the salt of the strong acid HBr. Therefore, $\boxed{\text{none of the } Br^- \text{ is protonated}}$.

E8.16 (a) Let HL denote lactic acid: $HL(aq) + H_2O(l) \rightleftharpoons L^-(aq) + H_3O^+(l)$.

If $[L^-] = [HL]$, then $[H_3O^+] = 10^{-pH} = 10^{-3.08}$ mol dm^{-3} = 8.32×10^{-4} mol dm^{-3} and $K_a = \dfrac{[H_3O^+][L^-]}{[HL]} =$

$[H_3O^+] = \boxed{8.32\times10^{-4}}$. Additionally, p$K_a$ = pH = 3.08.

(b) $[HL] = 2\,[L^-]$

$$K_a = \frac{[H_3O^+][L^-]}{[HL]} = \frac{[H_3O^+][L^-]}{2\,[L^-]} = \frac{[H_3O^+]}{2}$$

$$[H_3O^+] = 2K_a = 2\times\left(8.32\times10^{-4}\right)\text{mol dm}^{-3} = 1.66\times10^{-3}\text{ mol dm}^{-3}$$

$$\text{pH} = -\log\left(1.66\times10^{-3}\right) = \boxed{2.78}$$

E8.17 Figure 8.3 is the titration of a strong base ($Ba(OH)_2$) with a strong acid (HCl). The titration curve will look roughly like Figure 8.3 of the text turned upside down with a pH equal to 7 at the stoichiometric point. We quantify two points on the curve: the initial pH (titrant volume is zero) and the titrant volume at the stoichiometric point.

Initial pH

$$[OH^-]_{\text{initial}} = 2\,c_{Ba(OH)_2} = 2 \times (0.15\text{ mol dm}^{-3}) = 0.30\text{ mol dm}^{-3}$$

$$\text{pOH}_{\text{initial}} = -\log(0.30) = 0.53$$

$$\text{pH}_{\text{initial}} = 14.00 - 0.53 = \boxed{13.47}$$

Stoichiometric point

$$Ba(OH)_2 + 2\,HCl(aq) \rightleftharpoons BaCl_2(aq) + 2\,H_2O(l)$$

$$\text{amount of HCl needed, } n_{HCl} = 2\times\left(Vc\right)_{\text{initial, }Ba(OH)_2} = 2\times\left(0.0250\text{ dm}^3\right)\times\left(0.15\text{ mol dm}^{-3}\right)$$
$$= 7.5\text{ mmol}$$

$$\text{volume of HCl needed} = \left(n/c\right)_{HCl} = \left(7.5\times10^{-3}\text{ mol}\right)/\left(0.22\text{ mol dm}^{-3}\right) = \boxed{34\text{ cm}^3}$$

Let's find one additional point on the curve. What's the pH when 50 cm^3 of titrant has been added?

$$\text{amount of unreacted HCl} = \left(Vc\right)_{HCl} - 2\times\left(Vc\right)_{\text{initial, }Ba(OH)_2}$$
$$= \left(0.0500\text{ dm}^3\right)\times\left(0.22\text{ mol dm}^{-3}\right) - 7.5\times10^{-3}\text{ mol}$$
$$= 3.5\times10^{-3}\text{ mol}$$

$$\text{total volume} = 0.0250\text{ dm}^3 + 0.0500\text{ dm}^3 = 0.0750\text{ dm}^3$$

$$\left[H_3O^-\right] = \left(\text{amount of unreacted HCl}\right)/\left(\text{total volume}\right)$$
$$= \left(3.5\times10^{-3}\text{ mol}\right)/\left(0.0750\text{ dm}^3\right) = 0.046\overline{7}\text{ mol dm}^{-3}$$

$$\text{pH} = -\log\left(\left[H_3O^+\right]\right) = -\log\left(0.046\overline{7}\right)$$
$$= \boxed{1.3}$$

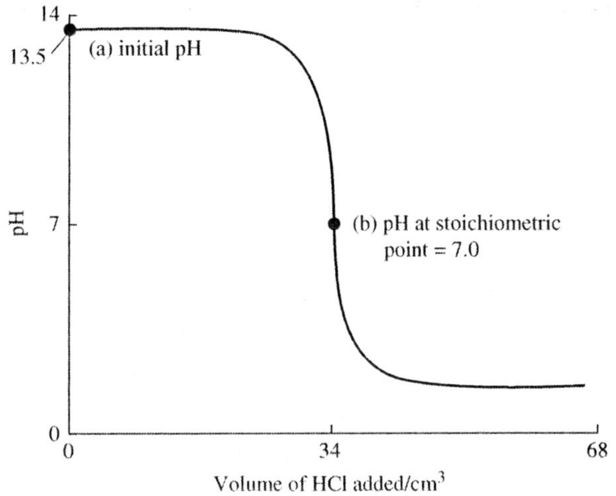

Figure 8.3

E8.18 The setups for the follow calculations assume the approximation that the formal concentration of the dissociating species is much, much larger than x, the equilibrium molarity of the species ion. In each case, the value of x is small enough to justify the approximation.

(a)

$K_a = 6.5 \times 10^{-5}$	$C_6H_5COOH(aq)$	+	$H_2O(l)$	$\rightleftharpoons$	$C_6H_5COO^-(aq)$	+	$H_3O^+(aq)$
Formal conc./mol dm^{-3}	0.250						
Change/mol dm^{-3}	$-x$				$+x$		$+x$
Equil. conc./mol dm^{-3}	$0.250 - x \cong 0.250$				x		x

$$K_a = \frac{x^2}{0.250} = 6.5\times10^{-5}$$

$x = 4.0 \times 10^{-3}$

$$\text{percentage deprotonated} = \frac{4.0\times10^{-3}}{0.250}\times100\% = \boxed{1.6\%}$$

(b)

$K_b = 1.7 \times 10^{-6}$	$NH_2NH_2(aq)$	+	$H_2O(l)$	$\rightleftharpoons$	$NH_2NH_3^+(aq)$	+	$OH^-(aq)$
Formal conc. / mol dm^{-3}	0.150						
Change / mol dm^{-3}	$-x$				$+x$		$+x$
Equil. conc. / mol dm^{-3}	$0.150 - x \cong 0.250$				x		x

$$K_b = \frac{[NH_2NH_3^+][OH^-]}{[NH_2NH_2]} \approx \frac{x^2}{0.150} = 1.7\times10^{-6}$$

$x = 5.0 \times 10^{-4}$

$$\text{percentage protonated} = \frac{5.0\times10^{-4}}{0.150}\times100\% = \boxed{0.33\%}$$

(c)

$K_b = 6.5 \times 10^{-5}$	$(CH_3)_3N(aq)$	+	$H_2O(l)$	$\rightleftharpoons$	$(CH_3)_3NH^+(aq)$	+	$OH^-(aq)$
Formal conc./mol dm^{-3}	0.112						
Change/mol dm^{-3}	$-x$				$+x$		$+x$
Equil. conc./mol dm^{-3}	$0.112 - x \cong 0.112$				x		x

$$K_b = \frac{[(CH_3)_3NH^+][OH^-]}{[(CH_3)_3N]} \approx \frac{x^2}{0.112} = 6.5\times10^{-5}$$

$x = 2.7 \times 10^{-3}$

$$\text{percentage protonated} = \frac{2.7\times10^{3}}{0.112}\times100\% = \boxed{2.4\%}$$

E8.19 See the solutions to previous exercises for detailed information on setting up the equilibrium calculations for weak acids.

(a) $$K_a = 8.4\times10^{-4} = \frac{[H_3O^+][CH_3CH(OH)CO_2^-]}{[CH_3CH(OH)COOH]} = \frac{x^2}{0.150-x} \approx \frac{x^2}{0.150}$$

$x = [H_3O^+] = 0.011$ M

$\text{pH} = -\log(0.011) = \boxed{2.0}$

$\text{pOH} = 14.00 - 2.0 = \boxed{12.0}$

$$\text{Fraction deprotonated} = \frac{0.011}{0.150} = \boxed{0.073}$$

Without the approximation $0.150 - x \sim 0.150$, the quadratic equation must be solved. With two significant figures the result is identical to that given by the above approximation.

(b) $$K_a = 8.4\times10^{-4} = \frac{[H_3O^+][CH_3CH(OH)CO_2^-]}{[CH_3CH(OH)COOH]} = \frac{x^2}{2.4\times10^{-4}-x}$$

The formal concentration of lactic acid is so small that we cannot estimate that x is negligibly small compared to it. The equation must be rearranged as a polynomial and solved as a quadratic equation.

$$x^2 + (8.4\times10^{-4})x - 2.016\times10^{-7} = 0$$

$$x = \tfrac{1}{2}\left\{-8.4\times10^{-4} + \sqrt{(8.4\times10^{-4})^2 + 4\times(2.016\times10^{-7})}\right\}$$

$$= 1.9\times10^{-4}$$

$\text{pH} = -\log(1.9\times10^{-4}) = \boxed{3.7}$

$\text{pOH} = 14.00 - 3.7 = \boxed{10.3}$

$$\text{Fraction deprotonated} = \frac{1.9\times10^{-4}}{2.4\times10^{-4}} = \boxed{0.79}$$

(c) $$K_a = \frac{[H_3O^+][C_6H_5SO_3^-]}{[C_6H_5SO_3H]} = \frac{x^2}{(0.25)-x} = 0.20$$

$$x^2 + 0.20x - 0.050 = 0$$

$$x = \tfrac{1}{2}\left\{-(0.20) + \sqrt{(0.20)^2 - 4\times(-0.05)}\right\} = 0.14\bar{5}$$

$\text{pH} = -\log(0.145) = \boxed{0.8}$

$\text{pOH} = 14.0 - 0.8 = \boxed{13.2}$

$$\text{Fraction deprotonated} = \frac{0.14\bar{5}}{0.25} = \boxed{0.58}$$

E8.20 The glycine (Gly) deprotonation reactions are:

$$H_2Gly^+(aq) + H_2O(l) \rightleftharpoons H_3O^+(aq) + HGly(aq) \quad pK_{a1} = 2.35 \text{ (Consult } \textit{CRC Handbook of Chemistry and Physics.)}$$

$$HGly + H_2O(l) \rightleftharpoons H_3O^+(aq) + Gly^- \quad pK_{a2} = 9.60$$

The fraction of a species at any chosen pH is computed with eqns that are analogous to eqn 8.11a–8.11c for a diprotic acid. Let D be the function: $D = [H_3O^+]^2 + [H_3O^+]K_{a1} + K_{a1}K_{a2}$.

$$f_1 = f(H_2Gly^+) = [H_3O^+]^2/D,$$
$$f_2 = f(HGly) = [H_3O^+]K_{a1}/D, \text{ and finally}$$
$$f_3 = f(Gly^-) = K_{a1}K_{a2}/D.$$

These fractions are plotted against pH in Figure 8.4. The molarity of a species at a specified pH is calculated with the relation: [species] = F(glycine) × f(species). For example, if pH = 9 and F(glycine) = 20 mmol dm^{-3}, then we find from the plot that f(HGly) = 0.80 and $f(Gly^-)$ = 0.20; therefore,

$$[HGly] = (20 \text{ mmol dm}^{-3}) \times (0.80) = 16 \text{ mmol dm}^{-3}$$

and

$$[Gly^-] = (20 \text{ mmol dm}^{-3}) \times (0.20) = 4 \text{ mmol dm}^{-3}.$$

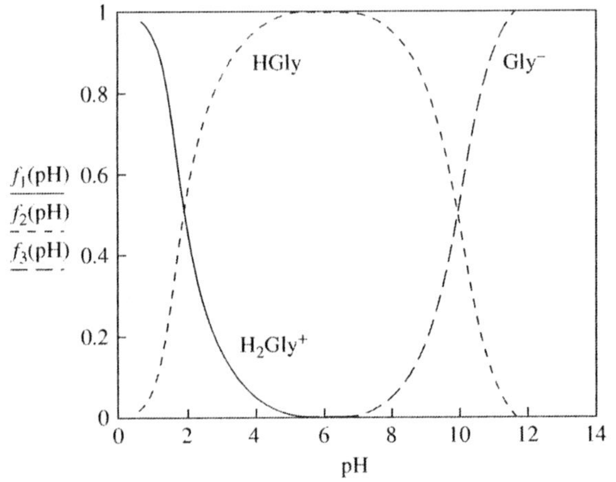

Figure 8.4

E8.21 The tyrosine (Tyr) deprotonation reactions are:

$$H_3Tyr^+(aq) + H_2O(l) \rightleftharpoons H_3O^+(aq) + H_2Tyr(aq) \quad pK_{a1} = 2.20$$

$$H_2Tyr(aq) + H_2O(l) \rightleftharpoons H_3O^+(aq) + HTyr^-(aq) \quad pK_{a2} = 9.11$$

$$HTyr^- + H_2O(l) \rightleftharpoons H_3O^+(aq) + Tyr^{2-} \quad pK_{a3} = 10.07$$

Equations that are analogous to those of Example 8.4 (and presented in Discussion question 8.4 for phosphoric acid) are used to calculate the fractional composition of each deprotonated species as a function of pH. Let D be the function: $D = [H_3O^+]^3 + [H_3O^+]^2 K_{a1} + [H_3O^+]K_{a1}K_{a2} + K_{a1}K_{a2}K_{a3}$. Then by generalizing the symmetry of the equations in Example 8.4 we find that

$$f_1 = f(H_3Tyr^+) = [H_3O^+]^3/D,$$
$$f_2 = f(H_2Tyr) = [H_3O^+]^2 K_{a1}/D,$$
$$f_3 = f(HTyr^-) = [H_3O^+]K_{a1}K_{a2}/D, \text{ and finally}$$
$$f_4 = f(Tyr^{2-}) = K_{a1}K_{a2}K_{a3}/D.$$

These fractions are plotted against pH in Figure 8.5. The molarity of a species at a specified pH is calculated with the relation: [species] = F(tyrosine) × f(species). For example, if pH = 9 and F(tyrosine) = 30 mmol dm^{-3}, then we find from the plot that $f(H_2Tyr) = 0.55$, $f(HTyr^-) = 0.40$, and $f(Tyr^{2-}) = 0.05$; therefore,

$$[H_2Tyr] = (30\ \text{mmol dm}^{-3}) \times (0.55) = 17\ \text{mmol dm}^{-3},$$

$$[HTyr^-] = (30\ \text{mmol dm}^{-3}) \times (0.40) = 12\ \text{mmol dm}^{-3},$$

and

$$[Tyr^{2-}] = (30\ \text{mmol dm}^{-3}) \times (0.05) = 1\ \text{mmol dm}^{-3}.$$

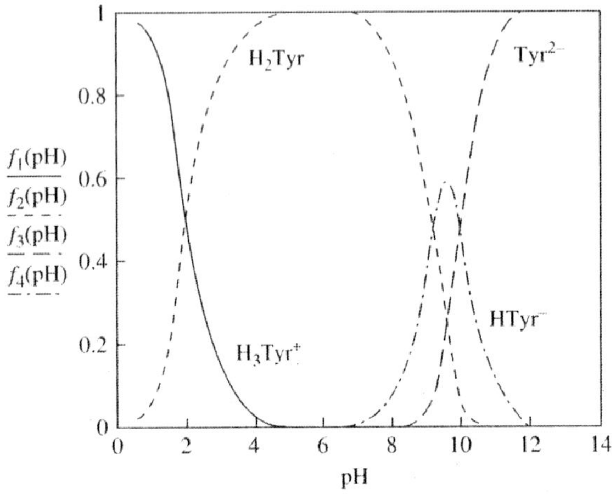

Figure 8.5

E8.22 For the case of an amphiprotic salt we use the relation $\text{pH} = \frac{1}{2}(\text{p}K_{a1} + \text{p}K_{a2})$ [8.12] and, as indicated by the eqn, it is not necessary to specify the concentration of the salt. For oxalic acid [Table 8.2], pK_{a1} = 1.23 and pK_{a2} = 4.19. Therefore,

$$\text{pH} = \tfrac{1}{2}(1.23 + 4.19) = \boxed{2.71}.$$

This relation is valid when the **formal concentration** F of the salt MHA, where HA^- is an amphiprotic anion, satisfies the condition $F/c^{\ominus} >> K_w/K_{a2}$ (the ratio is 1.5×10^{-10} in this exercise) and $F/c^{\ominus} >> K_{a1}$ (the constant is 0.059 in this exercise). These conditions are presented in Section 8.4 and under these conditions eqn 8.12 indicates that the formal concentration does not determine the pH because F does not appear in the equation (i.e., the pH is constant over a considerable range of formal concentration). These conditions also cause $[H_2A] \approx [A^-]$.

E8.23 Before justifying the various approximations that will be analyzed in this exercise, let us find the criterion that justifies the approximation [acid] = $c - x \simeq c$ where c is the formal molarity of a weak acid that has acid constant K and x is the conjugate base molarity. (Or, if c is the formal molarity of a weak base that has base constant K and x is the conjugate acid molarity, the approximation is [base] = $c - x \simeq c$.) The equilibrium constant has the general form

$$K = \frac{x^2}{c-x} \quad \text{or, writing the expression as a polynomial,} \quad x^2 + Kx - Kc = 0.$$

The quadratic equation solution for x is

$$x = \tfrac{1}{2}\left\{-K + \sqrt{K^2 + 4Kc}\right\}.$$

This expression for x can be used to calculate the value of x with given values of c and K. However, in many instances $4c >> K$, a case that causes the above expression to simplify into a very attractive computational equation: $x = \sqrt{Kc}$. In fact this is the equation we get by simply using the approximation the approximation [acid] = $c - x \simeq c$ in our equilibrium tables and expressions for equilibrium constant. The bottom line is: compare the value of $4c$ to K; if $4c >> K$, immediately apply the approximation [acid] = $c - x \simeq c$.

(a) Boric acid solution with $c_{B(OH)_3} = 1.0 \times 10^{-4}$ mol dm^{-3} and $K_a = 7.24 \times 10^{-10}$. Since $4\,c_{B(OH)_3} >> K_a$, we immediately apply the approximation $[B(OH)_3] = c_{B(OH)_3} - x \simeq c_{B(OH)_3}$.

$K_a = 7.24 \times 10^{-10}$	$B(OH)_3(aq)$	+	$2\,H_2O(l)$	$\rightleftharpoons$	$B(OH)_4^-(aq)$	+	$H_3O^+(aq)$
Formal conc./mol dm^{-3}	$c_{B(OH)_3} = 1.0\times10^{-4}$						
Change/mol dm^{-3}	$-x$				$+x$		$+x$
Equil. conc./mol dm^{-3}	$c_{B(OH)_3} - x \simeq c_{B(OH)_3}$				x		x

$$K_a = \frac{[H_3O^+][B(OH)_4^-]}{[B(OH)_3]} = \frac{x^2}{c_{B(OH)_3}}$$

$$x \approx \sqrt{K_a c_{B(OH)_3}} = \sqrt{(7.24\times10^{-10})\times(1.0\times10^{-4})}\ \text{mol dm}^{-3} = 2.69\times10^{-7}\ \text{mol dm}^{-3}$$

$$[H_3O^+] = 2.69 \times 10^{-7}\ \text{mol dm}^{-3}$$

$$\text{pH} = -\log(2.69 \times 10^{-7}) = 6.57$$

This value of $[H_3O^+]$ is not much different from the value for pure water, 1.0×10^{-7} mol dm^{-3}; hence, it is at the lower limit of safely ignoring the contribution to $[H_3O^+]$ from the autoprotolysis of water. Autoprotolysis contributes to the hydronium ion concentration so that $[H_3O^+]$ is not equal to $[B(OH)_4^-]$, nor is it equal to $[OH^-]$, as in pure water. However, the principle of electroneutrality specifies that all solutions must be electrically neutral so the sum of anion charge concentrations must equal the sum of cation charge concentrations. For our single-charge ions: $[B(OH)_4^-] + [OH^-] = [H_3O^+]$. Substituting $[OH^-] = K_w/[H_3O^+]$, letting $[H_3O^+] = x$, and solving for $[B(OH)_4^-]$ gives $[B(OH)_4^-] = x - K_w/x$. This is now substituted into the equilibrium constant expression to give an equation for x, the hydronium ion molarity.

$$K_a = \frac{[H_3O^+][B(OH)_4^-]}{[B(OH)_3]} = \frac{x\times(x - K_w/x)}{c_{B(OH)_3}} = \frac{x^2 - K_w}{c_{B(OH)_3}} \quad \text{or} \quad x = \sqrt{K_w + K_a c_{B(OH)_3}}$$

$$x = \sqrt{1.00\times10^{-14} + (7.24\times10^{-10})\times(1.0\times10^{-4})} = 2.9\times10^{-7}\ \text{mol dm}^{-3}$$

$$\text{pH} = -\log(2.9 \times 10^{-7}) = \boxed{6.54}$$

This value is slightly, but measurably, different from the value 6.57 obtained by ignoring the contribution to $[H_3O^+]$ from water. This last equation reduces to the approximate equation used above, $x \simeq \sqrt{K_a c_{B(OH)_3}}$, when $K_a c_{B(OH)_3} >> K_w$. This condition is not entirely satisfied in this exercise because $K_a c_{B(OH)_3} = 7.24\times10^{-14}$ is only 7.24 times K_w. and it must be, say, no less than 10 or more times K_w before the condition is satisfied.

(b) Phosphoric acid solution with $c_{H_3PO_4} = 0.015$ mol dm^{-3}; $K_{a1} = 7.59\times10^{-3}$, $K_{a2} = 6.17\times10^{-8}$, and $K_{a3} = 2.14\times10^{-13}$. Since $4\,c_{H_3PO_4}$ is not much, much greater than K_{a1}, we should not immediately apply the approximation $[H_3PO_4] = c_{H_3PO_4} - x \simeq c_{H_3PO_4}$. Furthermore, as the first acid constant is much, much larger than that of boric acid (see part (a) of this exercise) we expect that the autoprotolysis of water contributes negligibly to the hydronium ion concentration and we ignore autoprotolysis. We can check that the criterion, $K_{a1}c_{H_3PO_4} >> K_w$, derived in part (a) quantifies this expectation. It does because $K_{a1}c_{H_3PO_4} = 1.14\times10^{-4} >> K_w$ so we ignore autoprotolysis.

We also suspect that, since $K_{a2} << K_{a1}$, the pH contribution of the second acid constant is negligibly small and can be ignored. But how do we quantify this suspicion? The principle of electroneutrality (see part (a)) in an H_2A solution provides a useful relation after making substitutions that replace $[HA^-]$ and $[A^{2-}]$ with equilibrium constants and $[H_2A]$, which must be approximately equal to c_{H_2A} for a weak acid.

$$\left[\mathrm{H}^+\right]=\left[\mathrm{HA}^-\right]+2\left[\mathrm{A}^{2-}\right]+\left[\mathrm{OH}^-\right]$$

$$=\left[\mathrm{HA}^-\right]+2\left[\mathrm{A}^{2-}\right]\quad\left(\left[\mathrm{OH}^-\right]\text{ is negligible in all but the most weakly acid solutions}\right)$$

$$=\frac{K_{a1}\left[\mathrm{H_2A}\right]}{\left[\mathrm{H}^+\right]}+\frac{2K_{a2}\left[\mathrm{HA}^-\right]}{\left[\mathrm{H}^+\right]}$$

$$\left[\mathrm{H}^+\right]^2=K_{a1}\left[\mathrm{H_2A}\right]+2K_{a2}\left[\mathrm{HA}^-\right]=K_{a1}\left[\mathrm{H_2A}\right]+\frac{2K_{a1}K_{a2}\left[\mathrm{H_2A}\right]}{\left[\mathrm{H}^+\right]}$$

$$=K_{a1}\left[\mathrm{H_2A}\right]\left(1+\frac{2K_{a2}}{\left[\mathrm{H}^+\right]}\right)$$

Substitute the weak acid estimate that $\left[\mathrm{H_2A}\right]\simeq c_{\mathrm{H_2A}}$ and apply the approximation that $2K_{a2}/\left[\mathrm{H}^+\right]<<1$ to get the usual approximation that $\left[\mathrm{H}^+\right]\simeq\sqrt{K_{a1}c_{\mathrm{H_2A}}}$. Substitute the approximations $\left[\mathrm{H_2A}\right]\simeq c_{\mathrm{H_2A}}$ and $\left[\mathrm{H}^+\right]\simeq\sqrt{K_{a1}c_{\mathrm{H_2A}}}$ into the above working equations to find the improved approximation that $\left[\mathrm{H}^+\right]^2\simeq K_{a1}c_{\mathrm{H_2A}}\left(1+2K_{a2}/\sqrt{K_{a1}c_{\mathrm{H_2A}}}\right)$. This tells us that, if $2K_{a2}/\sqrt{K_{a1}c_{\mathrm{H_2A}}}<<1$, the second acid constant contributes negligibly to the pH. In this exercise $2K_{a2}/\sqrt{K_{a1}c_{\mathrm{H_2A}}}=1.16\times10^{-5}$ so the criteria for neglect of K_{a2} is satisfied. We have justified neglect of autoprotolysis and K_{a2} and proceed with these common approximations.

$K_{a1}=7.59\times10^{-3}$	$H_3PO_4(aq)$	+	$H_2O(l)$	$\rightleftharpoons$	$H_2PO_4^-(aq)$	+	$H_3O^+(aq)$
Formal conc./mol dm^{-3}	$c_{\mathrm{H_3PO_4}}=0.0150$						
Change/mol dm^{-3}	$-x$				$+x$		$+x$
Equil. conc./mol dm^{-3}	$c_{\mathrm{H_3PO_4}}-x$				x		x

$$K_{a1}=\frac{x^2}{c_{\mathrm{H_3PO_4}}-x}$$

$$x^2+K_{a1}x-K_{a1}c_{\mathrm{H_3PO_4}}=0$$

$$x=\tfrac{1}{2}\left\{-K_{a1}+\sqrt{K_{a1}^{\ 2}+4K_{a1}c_{\mathrm{H_3PO_4}}}\right\}$$

$$=\tfrac{1}{2}\left\{-\left(7.59\times10^{-3}\right)+\sqrt{\left(7.59\times10^{-3}\right)^2+4\left(7.59\times10^{-3}\right)\times\left(0.015\right)}\right\}$$

$$=7.53\times10^{-3}\text{ mol dm}^{-3}$$

$$\mathrm{pH}=-\log(7.53\times10^{-3})=\boxed{2.12}$$

(c) Sulfurous acid solution with $c_{\mathrm{H_2SO_3}}=0.10\text{ mol dm}^{-3}$: $K_{a1}=1.55\times10^{-2}$ and $K_{a2}=1.23\times10^{-7}$. Since $4c_{\mathrm{H_2SO_3}}$ is not much, much greater than K_{a1}, we will not apply the approximation $[H_2SO_3]$ = $c_{\mathrm{H_2SO_3}}-x\simeq c_{\mathrm{H_2SO_3}}$. Furthermore, as the first acid constant is much, much larger than that of boric acid (see part (a) of this exercise) we expect that the autoprotolysis of water contributes negligibly to the hydronium ion concentration and we ignore autoprotolysis. We can check that the criterion, $K_{a1}c_{\mathrm{H_2SO_3}}>>K_w$, derived in part (a) to quantify this expectation. Since $K_{a1}c_{\mathrm{H_2SO_3}}=1.55\times10^{-3}>>K_w$, we can ignore autoprotolysis. We also suspect that, since $K_{a2}<<K_{a1}$, the pH contribution of the second acid constant is negligibly small and can be ignored. We check this suspicion with the criterion $2K_{a2}/\sqrt{K_{a1}c_{\mathrm{H_2A}}}<<1$. In this exercise $2K_{a2}/\sqrt{K_{a1}c_{\mathrm{H_2A}}}=6.25\times10^{-6}$ so the criterion for neglect of K_{a2} is satisfied. We have justified neglect of autoprotolysis and K_{a2}.

$K_{a1} = 1.55 \times 10^{-2}$	$H_2SO_3(aq)$	+	$H_2O(l)$	$\rightleftharpoons$	$HSO_3^-(aq)$	+	$H_3O^+(aq)$
Formal conc./mol dm^{-3}	$c_{H_2SO_3} = 0.10$						
Change/mol dm^{-3}	$-x$				$+x$		$+x$
Equil. conc./mol dm^{-3}	$c_{H_2SO_3} - x$				x		x

$$K_{a1} = \frac{x^2}{c_{H_2SO_3} - x}$$

$$x^2 + K_{a1}x - K_{a1}c_{H_2SO_3} = 0$$

$$x = \tfrac{1}{2}\left\{-K_{a1} + \sqrt{K_{a1}^2 + 4K_{a1}c_{H_2SO_3}}\right\}$$

$$= \tfrac{1}{2}\left\{-(1.55\times10^{-2}) + \sqrt{(1.55\times10^{-2})^2 + 4(1.55\times10^{-2})\times(0.10)}\right\}$$

$$= 3.24\times10^{-2}\ \text{mol dm}^{-3}$$

$$\text{pH} = -\log(3.24\times10^{-2}) = \boxed{1.49}$$

E8.24 The Henderson-Hasselbalch equation, $\text{pH} = \text{p}K_a - \log([\text{acid}]/[\text{base}])$ [8.13], indicates that at half neutralization, pH = pK_a = 8.3. This is the pH at which the buffering action is best as solution components provide the capacity to neutralize small amounts of either base or acid that may be added without drastic effect on solution pH.

E8.25

$$\text{pH} = \text{p}K_a - \log\frac{[\text{acid}]}{[\text{base}]}\ [8.13] \quad \text{or} \quad \frac{[\text{acid}]}{[\text{base}]} = 10^{\text{p}K_a - \text{pH}}$$

(a) $$\frac{[\text{acid}]}{[\text{base}]} = 10^{2.20-7.00} = \boxed{1.59\times10^{-5}}$$

(b) $$\frac{[\text{acid}]}{[\text{base}]} = 10^{2.20-2.20} = \boxed{0}$$

(c) $$\frac{[\text{acid}]}{[\text{base}]} = 10^{2.20-1.50} = \boxed{5.01}$$

E8.26 (a) Oxalic acid solution with $c_{(COOH)_2} = 0.15$ mol dm^{-3}; $K_{a1} = 5.89\times10^{-2}$ and $K_{a2} = 6.46\times10^{-5}$. Since $4c_{(COOH)_2}$ is not much, much greater than K_{a1}, we will not apply the approximation $[(COOH)_2] = c_{(COOH)_2} - x \simeq c_{(COOH)_2}$. (See the first paragraph of Exercise 8.23 for discussion of this justification.) Furthermore, neither autoprotolysis nor the second acid constant provide a significant pH contribution (an assertion that can be checked with criteria summarized in Exercise 8.23 (c)) so common pH approximations are appropriate.

$K_{a1} = 5.89 \times 10^{-2}$	$(COOH)_2(aq)$	+	$H_2O(l)$	$\rightleftharpoons$	$HOOCO_2^-(aq)$	+	$H_3O^+(aq)$
Formal conc./mol dm^{-3}	$c_{(COOH)_2} = 0.15$						
Change/mol dm^{-3}	$-x$				$+x$		$+x$
Equil. conc./mol dm^{-3}	$c_{(COOH)_2} - x$				x		x

$$K_{a1} = \frac{x^2}{c_{(COOH)_2} - x}$$

$$x^2 + K_{a1}x - K_{a1}c_{(COOH)_2} = 0$$

$$x = \tfrac{1}{2}\left\{-K_{a1} + \sqrt{K_{a1}^2 + 4K_{a1}c_{(COOH)_2}}\right\}$$

$$= \tfrac{1}{2}\left\{-(5.89\times10^{-2}) + \sqrt{(5.89\times10^{-2})^2 + 4(5.89\times10^{-2})\times(0.15)}\right\} = 6.91\times10^{-2}\ \text{mol dm}^{-3}$$

$$\left[\mathrm{H_3O^+}\right]=\boxed{6.91\times10^{-2}\ \mathrm{mol\ dm^{-3}}} \quad \text{and} \quad \left[\mathrm{OH^-}\right]=K_w/\left[\mathrm{H_3O^+}\right]=\boxed{1.45\times10^{-13}\ \mathrm{mol\ dm^{-3}}}$$

$$\left[(\mathrm{COOH})_2\right]=c_{(\mathrm{COOH})_2}-x=\boxed{8.09\times10^{-2}\ \mathrm{mol\ dm^{-3}}}$$

The $HOOCO_2^-$(aq) molarity is very slightly diminished below $x = 0.0691$ mol dm^{-3} by the second acid constant. However, this does not contribute significantly to pH. It is the value of y in the following equilibrium table that we must find.

$K_{a2} = 6.46 \times 10^{-5}$	$HOOCO_2^-$(aq)	+	H_2O(l)	$\rightleftharpoons$	$^-O_2CO_2^-$(aq)	+	H_3O^+(aq)
First est./mol dm^{-3}	$x = 0.0691$				0		x
Change/mol dm^{-3}	$-y$				$+y$		0
Equil. conc./mol dm^{-3}	$x - y$				y		x

$$K_{a2}=\frac{xy}{x-y} \quad \text{or} \quad y=\frac{K_{a2}x}{x+K_{a2}}$$

But $x >> K_{a2}$ so the above eqn reduces to $y=\dfrac{K_{a2}x}{x}=K_{a2}$.

$$[^-O_2CO_2^-] = K_{a2} = \boxed{6.46 \times 10^{-5}\ \mathrm{mol\ dm^{-3}}}$$

$$[HOOCO_2^-] = x - K_{a2} = \boxed{0.0691\ \mathrm{mol\ dm^{-3}}}$$

(b) Hydrosulfuric acid solution with $c_{H_2S} = 0.065$ mol dm^{-3} ; $K_{a1} = 1.32\times10^{-7}$ and $K_{a2} = 7.08\times10^{-15}$. Since $4c_{H_2S}$ is much, much greater than K_{a1}, we will apply the approximation $[H_2S] = c_{H_2S} - x \simeq c_{H_2S}$. (See the first paragraph of Exercise 8.23 for discussion of this justification.) Neither autoprotolysis nor the second acid constant provide a significant pH contribution (an assertion that can be checked with criteria summarized in Exercise 8.23 (c)) so common pH approximations are appropriate.

$K_{a1} = 1.32 \times 10^{-7}$	H_2S(aq)	+	H_2O(l)	$\rightleftharpoons$	HS^-(aq)	+	H_3O^+(aq)
Formal conc./mol dm^{-3}	$c_{H_2S} = 0.065$						
Change/mol dm^{-3}	$-x$				$+x$		$+x$
Equil. conc./mol dm^{-3}	$c_{H_2S} - x \simeq c_{H_2S}$				x		x

$$K_{a1}=\frac{x^2}{c_{H_2S}}$$

$$x=\sqrt{K_{a1}c_{H_2S}}$$
$$=\sqrt{(1.32\times10^{-7})\times(0.065)}=9.26\times10^{-5}\ \mathrm{mol\ dm^{-3}}$$

$$\left[\mathrm{H_3O^+}\right]=\boxed{9.26\times10^{-5}\ \mathrm{mol\ dm^{-3}}} \quad \text{and} \quad \left[\mathrm{OH^-}\right]=K_w/\left[\mathrm{H_3O^+}\right]=\boxed{1.08\times10^{-10}\ \mathrm{mol\ dm^{-3}}}$$

$$\left[\mathrm{H_2S}\right]=c_{H_2S}=\boxed{0.065\ \mathrm{mol\ dm^{-3}}}$$

The HS^-(aq) molarity is very slightly diminished below $x = 9.26 \times 10^{-5}$ mol dm^{-3} by the second acid constant. However, this does not contribute significantly to pH. It is the value of y in the following equilibrium table that we must find. Since $4x >> K_{a2}$, we can use the approximation $x - y \simeq x$.

$K_{a2} = 7.08 \times 10^{-15}$	HS^-(aq)	+	H_2O(l)	$\rightleftharpoons$	S^{2-}(aq)	+	H_3O^+(aq)
First est./mol dm^{-3}	$x = 9.26 \times 10^{-5}$				0		x
Change/mol dm^{-3}	$-y$				$+y$		0
Equil. conc./mol dm^{-3}	$x - y \simeq x$				y		x

$$K_{a2} = \frac{xy}{x} \quad \text{or} \quad y = K_{a2}$$

$$[S^{2-}] = K_{a2} = \boxed{7.08 \times 10^{-15}\ \text{mol dm}^{-3}}$$

$$[HS^-] = x = \boxed{9.26 \times 10^{-5}\ \text{mol dm}^{-3}}$$

E8.27 (a) Acetic acid with $c_{CH_3COOH} = 0.10$ mol dm^{-3} and $K_a = 1.8 \times 10^{-5}$. Since $4c_{CH_3COOH}$ is much, much greater than K_a, we will apply the approximation $[CH_3COOH] = c_{CH_3COOH} - x \simeq c_{CH_3COOH}$. (See the first paragraph of Exercise 8.23 for discussion of this justification.)

$K_a = 1.8 \times 10^{-5}$	$CH_3COOH(aq)$	+	$H_2O(l)$	$\rightleftharpoons$	$CH_3CO_2^-(aq)$	+	$H_3O^+(aq)$
Formal conc./mol dm^{-3}	$c_{CH_3COOH} = 0.10$						
Change/mol dm^{-3}	$-x$				$+x$		$+x$
Equil. conc./mol dm^{-3}	$c_{CH_3COOH} - x \simeq c_{CH_3COOH}$				x		x

$$K_a = \frac{x^2}{c_{CH_3COOH}} \quad \text{or} \quad x = \sqrt{K_a c_{CH_3COOH}}$$

$$x = \sqrt{(1.8\times10^{-5})\times(0.10)} = 1.34\times10^{-3}\ \text{mol dm}^{-3}$$

$$\text{pH} = -\log(1.3\times10^{-3}) = \boxed{2.9}$$

(b) Let $\Delta c = (n_{CH_3COOH} - n_{NaOH})/V_{total}$ and let x be the molarity of $CH_3CO_2^-(aq)$ that originates from the acid dissociation of unneutralized CH_3COOH only; the hydronium ion molarity also equals x.

$K_a = 1.8 \times 10^{-5}$	$CH_3COOH(aq)$	+	$NaOH(l)$	$\rightleftharpoons$	$CH_3CO_2^-(aq) + H_2O(aq)$
Amount mixed	$n_{CH_3COOH} = (cV)_{CH_3COOH}$		$n_{NaOH} = (cV)_{NaOH}$		0
Amount changed	$-n_{NaOH} - xV_{total}$		$-n_{NaOH}$		$n_{NaOH} + xV_{total}$
Equil. molarity	$\Delta c - x \simeq \Delta c$				$n_{NaOH}/V_{total} + x \simeq n_{NaOH}/V_{total}$

In this exercise: $n_{CH_3COOH} = (0.10\ \text{mol dm}^{-3})\times(0.0250\ \text{dm}^{-3}) = 2.5\ \text{mmol}$,

$$n_{NaOH} = (0.10\ \text{mol dm}^{-3})\times(0.0100\ \text{dm}^{-3}) = 1.0\ \text{mmol},$$

$$V_{total} = (0.0250 + 0.0100)\ \text{dm}^3 = 0.0350\ \text{dm}^3,$$

$$\Delta c = (n_{CH_3COOH} - n_{NaOH})/V_{total} = 0.0429\ \text{mol dm}^{-3},$$

$$n_{NaOH}/V_{total} = 0.0286\ \text{mol dm}^{-3}.$$

$$K_a = \frac{[CH_3COO^-][H_3O^+]}{[CH_3COOH]} = \frac{(n_{NaOH}/V_{total})x}{\Delta c}$$

$$x = K_a\Delta c/(n_{NaOH}/V_{total})$$
$$= (1.8\times10^{-5})\times(0.0429)/(0.0286) = 2.7\times10^{-5}\ \text{mol dm}^{-3}$$

$$\text{pH} = -\log(2.7 \times 10^{-5}) = \boxed{4.6}$$

(c) Because the acetic acid and sodium hydroxide solutions have equal molarities, their volumes are equal at the stoichiometric point. Therefore, 25.0 cm^3 NaOH are required to reach the stoichiometric point and $\boxed{12.5\ \text{cm}^3\ \text{of}\ 0.10\ \text{M NaOH(aq)}}$ are required to reach halfway to the stoichiometric point.

(d) At the half stoichiometric point, pH = pK_a and pH = $\boxed{4.74}$.

(e) $\boxed{25.0\ \text{cm}^3}$; see part (c).

(f) The stoichiometric pH is that of 0.050 M $NaCH_3CO_2$.

$K_b = K_w/K_a = 5.6 \times 10^{-10}$	$CH_3CO_2^-(aq)$	+	$H_2O(l)$	$\rightleftharpoons$	$CH_3COOH(aq)$	+	$OH^-(aq)$
Formal conc./mol dm^{-3}	$c_{CH_3CO_2^-} = 0.050$						
Change/mol dm^{-3}	$-x$				$+x$		$+x$
Equil. conc./mol dm^{-3}	$c_{CH_3CO_2^-} - x \simeq c_{CH_3CO_2^-}$				x		x

$$K_b = \frac{x^2}{c_{CH_3CO_2^-}} \quad \text{or} \quad x = \sqrt{K_b c_{CH_3CO_2^-}}$$

$$x = \sqrt{(5.56\times10^{-10})\times(0.050)} = 5.27\times10^{-6} \text{ mol dm}^{-3}$$

$$\text{pOH} = -\log(5.27\times10^{-6}) = 5.28$$

$$\text{pH} = 14.00 - \text{pH} = \boxed{8.72}$$

E8.28 $$K_a = \frac{[H_3O^+][CH_3CO_2^-]}{[CH_3COOH]} \quad \text{and} \quad \text{pH} = \text{p}K_a + \log\frac{[CH_3CO_2^-]}{[CH_3COOH]} \quad [8.13] \text{ where p}K_a = 4.75$$

(a) $$\text{pH} = \text{p}K_a + \log\frac{[0.10]}{[0.10]} = \text{p}K_a = \boxed{4.75}$$

(b) We recognize that 3.3 mmol NaOH = 3.3×10^{-3} mol OH^- [strong base] produces 3.3×10^{-3} mol $CH_3CO_2^-$ from CH_3COOH. Since both the formal amount of acetic acid and the formal amount of acetate ion are equal to 0.10 mol dm^{-3} × 0.100 dm^3 = 1.0×10^{-2} mol before adding the base, the molarities after addition of the base are

$$[CH_3COOH] = \frac{1.0\times10^{-2} - 3.3\times10^{-3}}{0.10 \text{ dm}^3} = 6.7\times10^{-2} \text{ mol dm}^{-3}$$

$$\left[CH_3CO_2^-\right] = \frac{1.0\times10^{-2} + 3.3\times10^{-3}}{0.10 \text{ dm}^3} = 0.13 \text{ mol dm}^{-3}.$$

The Henderson-Hasselbalch equation can be used to calculate pH.

$$\text{pH} = 4.75 + \log\frac{0.13}{0.067} = \boxed{5.04}$$

(c) We recognize that 6.0 mmol HNO_3 = 6.0×10^{-3} mol H_3O^+ [strong acid] produces 6.0×10^{-3} mol CH_3COOH from $CH_3CO_2^-$. Since both the formal amount of acetic acid and the formal amount of acetate ion are equal to 0.10 mol dm^{-3} × 0.100 dm^3 = 1.0×10^{-2} mol before adding the base, the molarities after addition of the base are

$$[CH_3COOH] = \frac{1.0\times10^{-2} + 6.0\times10^{-3}}{0.10 \text{ dm}^3} = 0.16 \text{ mol dm}^{-3}$$

$$\left[CH_3CO_2^-\right] = \frac{1.0\times10^{-2} - 6.0\times10^{-3}}{0.10 \text{ dm}^3} = 0.040 \text{ mol dm}^{-3}.$$

The Henderson-Hasselbalch equation can be used to calculate pH.

$$\text{pH} = 4.75 + \log\frac{0.040}{0.16} = \boxed{4.15}$$

E8.29 The rule of thumb we use is that the effective range of a buffer is roughly within plus or minus one pH unit of the pK_a of the acid. Therefore,

(a) pK_a = 3.08; pH range, $\boxed{2\text{–}4}$

(b) pK_a = 4.19; pH range, $\boxed{3\text{–}5}$

(c) $pK_{a3} = 12.68$; pH range, $\boxed{11.5-13.5}$

(d) $pK_{a2} = 7.21$; pH range, $\boxed{6-8}$

(e) $pK_b = 7.97$, $pK_a = 6.03$; pH range, $\boxed{5-7}$

E8.30 At the halfway point the Henderson–Hasselbalch equation gives

$$\text{pH} = \text{p}K_a + \log\frac{[CH_3CO_2^-]}{[CH_3COOH]}\ [8.13] = \text{p}K_a + \log(1) = \text{p}K_a = \boxed{5.16} \text{ and, therefore,}$$

$$K_a = 10^{-\text{p}K_a} = 10^{-5.16} = \boxed{6.92\times10^{-6}}.$$

For a solution that has the formal concentration $c_{acid} = 0.025$ mol dm^{-3} and $K_a = 6.92 \times 10^{-6}$, $4c_{acid}$ is much, much greater than K_a so we will apply the approximation [acid] $= c_{acid} - x \simeq c_{acid}$. (See the first paragraph of Exercise 8.23 for discussion of this justification.)

$K_a = 6.92 \times 10^{-6}$	HA(aq)	+	H_2O(l)	$\rightleftharpoons$	A^-(aq)	+	H_3O^+(aq)
Formal conc./mol dm^{-3}	$c_{acid} = 0.025$						
Change/mol dm^{-3}	$-x$				$+x$		$+x$
Equil. conc./mol dm^{-3}	$c_{acid} - x \simeq c_{acid}$				x		x

$$K_a = \frac{x^2}{c_{acid}} \quad \text{or} \quad x = \sqrt{K_a c_{acid}}$$

$$x = \sqrt{(6.92\times10^{-6})\times(0.025)} = 4.1\overline{6}\times10^{-4} \text{ mol dm}^{-3}$$

$$\text{pH} = -\log(4.1\overline{6}\times10^{-4}) = \boxed{3.4}$$

E8.31 (a) Ammonium chloride, $[NH_4Cl]_{formal} = [NH_4^+] = [Cl^-] = 0.10$ mol dm^{-3}. Since $4c_{acid}$ is much, much greater than K_a, we will apply the approximation [acid] $= c_{acid} - x \simeq c_{acid}$. (See the first paragraph of Exercise 8.23 for discussion of this justification.)

$K_a = 5.6 \times 10^{-10}$	NH_4^+(aq)	+	H_2O(l)	$\rightleftharpoons$	NH_3(aq)	+	H_3O^+(aq)
Formal conc./mol dm^{-3}	$c_{acid} = 0.10$						
Change/mol dm^{-3}	$-x$				$+x$		$+x$
Equil. conc./mol dm^{-3}	$c_{acid} - x \simeq c_{acid}$				x		x

$$K_a = \frac{x^2}{c_{acid}} \quad \text{or} \quad x = \sqrt{K_a c_{acid}}$$

$$x = \sqrt{(5.6\times10^{-10})\times(0.10)} = 7.5\times10^{-6} \text{ mol dm}^{-3}$$

$$\text{pH} = -\log(7.5\times10^{-6}) = \boxed{5.1}$$

(b) Sodium acetate, $[CH_3CO_2^-] = 0.25$ mol dm^{-3}. Since $4c_{base}$ is much, much greater than K_b, we will apply the approximation [base] $= c_{base} - x \simeq c_{base}$. (See the first paragraph of Exercise 8.23 for discussion of this justification.)

$K_b = 5.6 \times 10^{-10}$	$CH_3CO_2^-$(aq)	+	H_2O(l)	$\rightleftharpoons$	CH_3COOH(aq)	+	OH^-(aq)
Formal conc./mol dm^{-3}	$c_{base} = 0.25$						
Change/mol dm^{-3}	$-x$				$+x$		$+x$
Equil. conc./mol dm^{-3}	$c_{base} - x \simeq c_{base}$				x		x

$$K_b = \frac{x^2}{c_{base}} \quad \text{or} \quad x = \sqrt{K_b c_{base}}$$

$$x = \sqrt{(5.6\times10^{-10})\times(0.25)} = 1.1\times10^{-5}\ \text{mol dm}^{-3}$$

$$\text{pOH} = -\log(1.1\times10^{-5}) = \boxed{5.0}$$

$$\text{pH} = 14.00 - \text{pOH} = \boxed{9.0}$$

(c) Acetic acid, [CH_3COOH] = 0.20 mol dm^{-3}. Since $4c_{acid}$ is much, much greater than K_a, we will apply the approximation [acid] = $c_{acid} - x \simeq c_{acid}$. (See the first paragraph of Exercise 8.23 for discussion of this justification.)

$K_a = 1.8 \times 10^{-5}$	$CH_3COOH(aq)$	+	$H_2O(l)$	$\rightleftharpoons$	$CH_3CO_2^-(aq)$	+	$H_3O^+(aq)$
Formal conc./mol dm^{-3}	$c_{acid} = 0.20$						
Change/mol dm^{-3}	$-x$				$+x$		$+x$
Equil. conc./mol dm^{-3}	$c_{acid} - x \simeq c_{acid}$				x		x

$$K_a = \frac{x^2}{c_{acid}} \quad \text{or} \quad x = \sqrt{K_a c_{acid}}$$

$$x = \sqrt{(1.8\times10^{-5})\times(0.20)} = 1.9\times10^{-3}\ \text{mol dm}^{-3}$$

$$\text{pH} = -\log(1.9\times10^{-3}) = \boxed{2.7}$$

E8.32 At the stoichiometric point the solution will consist of the lactate ion, which is a weak base, and Na^+ ions. To calculate the pH we first need to calculate the total volume of solution at the stoichiometric point.

amount (moles) lactate ion = 0.02500 dm^3 × 0.150 mol dm^{-3} = 3.75 × 10^{-3} mol

volume of base added = (3.75 × 10^{-3} mol)/(0.188 mol dm^{-3}) = 0.0199 dm^3

total volume = 0.02500 dm^3 + 0.0199 dm^3 = 0.0449 dm^3

[lactate ion] = (3.75 × 10^{-3} mol)/(0.0449 dm^3) = 0.0835 mol dm^{-3}

$$K_b = 1.2\times10^{-11} = \frac{[\text{lactic acid}][\text{OH}^-]}{[\text{lactate ion}]} = \frac{x^2}{0.0835 - x} \approx \frac{x^2}{0.0835}$$

x = [OH$^-$] = 1.00 × 10^{-6} M

pOH = −log(1.00 × 10^{-6}) = 6.00

pH = 14.00 − 6.00 = $\boxed{8.00}$

E8.33 The initial pH is calculated from

$$K_b = \frac{[\text{CH}_3\text{COOH}][\text{OH}^-]}{[\text{CH}_3\text{CO}_2^-]} \approx \frac{x^2}{0.10} = 5.6\times10^{-10}.$$

x = [OH$^-$] = 7.5 × 10^{-6}, pOH = 5.12, pH = 8.88

Use the Henderson-Hasselbalch equation to calculate the other pHs for specific concentrations of CH_3COOH.

$$\text{pH} = \text{p}K_a - \log\frac{[\text{acid}]}{[\text{base}]}\ [8.13] \text{ with p}K_a = 4.74 \text{ and [base]} = [\text{NaCH}_3\text{CO}_2] = 0.10 \text{ mol dm}^{-3}$$

$$= 4.74 - \log\frac{[\text{CH}_3\text{COOH}]}{0.10\ \text{mol dm}^{-3}}$$

Draw a table of calculated pH values over a range of acid concentrations and prepare the pH against $[CH_3COOH]$ as shown in Figure 8.6.

$[CH_3COOH]$/ mol dm^{-3}	0	2.5×10^{-5}	5×10^{-5}	6×10^{-5}	7×10^{-5}	8.5×10^{-5}	1×10^{-4}	1×10^{-3}	5×10^{-3}	1×10^{-2}
pH	8.88	8.34	8.04	7.96	7.89	7.81	7.74	6.74	6.04	5.74

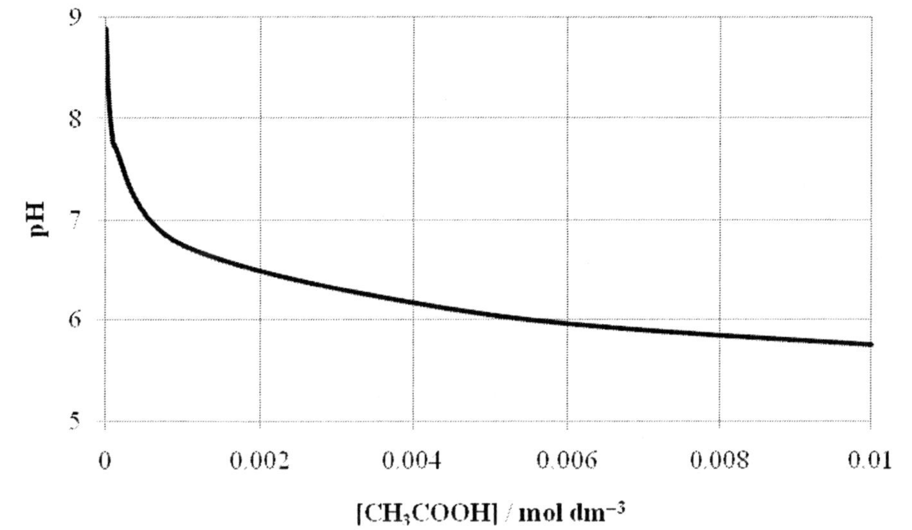

Figure 8.6

E8.34 Choose a buffer system in which the conjugate acid has a pK_a close to the desired pH. Therefore,

(a) H_3PO_4 and NaH_2PO_4

(b) NaH_2PO_4 and Na_2HPO_4, or $NaHSO_3$ and Na_2SO_3

E8.35 (a) $K_s = [Ag^+][I^-]$ (b) $K_s = [Hg_2^{+2}][S^{2-}]$

(c) $K_s = [Fe^{3+}][OH^-]^3$ (d) $K_s = [Ag^+]^2[CrO_4^{2-}]$

E8.36 (a) $K_s = [Ba^{2+}][SO_4^{2-}] = S^2 = 1.1\times10^{-10}$; $S = \boxed{1.0 \times 10^{-5} \text{ mol dm}^{-3}}$

(b) $K_s = [Ag^+]^2[CO_3^{2-}] = (2S)^2(S) = 4S^3 = 6.2\times10^{-12}$; $S = \boxed{1.2 \times 10^{-4} \text{ mol dm}^{-3}}$

(c) $K_s = [Fe^{3+}][OH^-]^3 = (S)(3S)^3 = 27S^4 = 2.0\times10^{-39}$; $S = \boxed{9.3 \times 10^{-11} \text{ mol dm}^{-3}}$

(d) $K_s = [Hg_2^{2+}][Cl^-]^2 = S(2S)^2 = 4S^3 = 1.3\times10^{-18}$; $S = \boxed{6.9 \times 10^{-7} \text{ mol dm}^{-3}}$

E8.37 Solubility (S) computations in the presence of a common ion give us pause to discover the condition under which the attractive approximation $c_{ion} + S \simeq c_{ion}$, where c_{ion} is the formal concentration of the common ion, is valid. For MX salts, the solubility constant has the form: $K_s = S \times (c_{ion} + S)$. Now, we write this expression as a polynomial, apply the quadratic equation solution, and expand the square root in a Taylor series.

$$S^2 + c_{ion}S - K_s = 0$$

$$S = \tfrac{1}{2}\left\{-c_{ion} + \sqrt{c_{ion}^2 + 4K_s}\right\} = \tfrac{1}{2}\left\{-c_{ion} + c_{ion}\sqrt{1 + 4K_s/c_{ion}^2}\right\} = \tfrac{c_{ion}}{2}\left\{-1 + \sqrt{1 + 4K_s/c_{ion}^2}\right\}$$

$$= \tfrac{c_{ion}}{2}\left\{-1 + \left[1 + \tfrac{1}{2}\left(4K_s/c_{ion}^2\right) - \tfrac{1}{8}\left(4K_s/c_{ion}^2\right)^2 + Order\left(4K_s/c_{ion}^2\right)^3\right]\right\}$$ (Taylor series expansion when $4K_s/c_{ion}^2 << 1$)

$$= \tfrac{c_{ion}}{2}\left\{-1 + \left[1 + \tfrac{1}{2}\left(4K_s/c_{ion}^2\right)\right]\right\}$$ (Successively higher powers are negligibly small when $4K_s/c_{ion}^2 << 1$ or $c_{ion} >> 2K_s^{1/2}$)

$$= K_s/c_{ion} \quad (\text{condition: } c_{ion} >> 2K_s^{1/2})$$

The above expression, $S = K_s/c_{ion}$, is the approximation we get from the original expression for the solubility constant by simply making the approximation $c_{ion} + S \simeq c_{ion}$, so we conclude that the approximation is justified when $c_{ion} >> 2K_s^{1/2}$.

(a) Solubility of silver bromide ($K_s = 7.7 \times 10^{-13}$) in 1.4×10^{-3} M NaBr(aq). Since $c_{ion} >> 2K_s^{1/2}$, the approximation $c_{Br^-} + S \simeq c_{Br^-}$ is justified.

$K_s = 7.7 \times 10^{-13}$	$AgBr(s)$	$\rightleftharpoons$	$Ag^+(aq)$	+	$Br^-(aq)$
Formal conc./mol dm^{-3}			0		$c_{Br^-} = 1.4 \times 10^{-3}$
Change/mol dm^{-3}			$+S$		$+S$
Equilibrium conc./mol dm^{-3}			S		$c_{Br^-} + S \simeq c_{Br^-}$

$$K_s = [Ag^+][Br^-] = S \times c_{Br^-}$$

$$S = K_s / c_{Br^-} = 7.7 \times 10^{-13} / 1.4 \times 10^{-3} = \boxed{5.5 \times 10^{-10} \text{ mol dm}^{-3}}.$$

(b) Solubility of magnesium carbonate ($K_s = 1.0 \times 10^{-5}$) in 1.1×10^{-5} M Na_2CO_3(aq). Since c_{ion} is not much, much larger than $2K_s^{1/2}$, we will not use the approximation $c_{Br^-} + S \simeq c_{Br^-}$.

$K_s = 1.0 \times 10^{-5}$	$MgCO_3(s)$	$\rightleftharpoons$	$Mg^{2+}(aq)$	+	$CO_3^{2-}(aq)$
Formal conc./mol dm^{-3}			0		$c_{CO_3^{2-}} = 1.1 \times 10^{-5}$
Change/ mol dm^{-3}			$+S$		$+S$
Equilibrium conc./mol dm^{-3}			S		$c_{CO_3^{2-}} + S$

$$K_s = [Mg^{2+}][CO_3^{2-}] = S \times (c_{CO_3^{2-}} + S)$$

$$S^2 + c_{CO_3^{2-}} S - K_s = 0$$

$$S = \tfrac{1}{2}\left\{-c_{CO_3^{2-}} + \sqrt{\left(c_{CO_3^{2-}}\right)^2 + 4K_s}\right\}$$

$$= \tfrac{1}{2}\left\{-\left(1.1\times10^{-5}\right) + \sqrt{\left(1.1\times10^{-5}\right)^2 + 4\left(1.0\times10^{-5}\right)}\right\}$$

$$= \boxed{3.2\times10^{-3} \text{ mol dm}^{-3}}$$

(c) Solubility of lead sulfate ($K_s = 1.6 \times 10^{-8}$) in 0.10 M $CaSO_4$(aq). Since $c_{ion} >> 2K_s^{1/2}$, the approximation $c_{SO_4^{2-}} + S \simeq c_{SO_4^{2-}}$ is justified.

$K_s = 1.6 \times 10^{-8}$	$PbSO_4(s)$	$\rightleftharpoons$	$Pb^{2+}(aq)$	+	$SO_4^{2-}(aq)$
Formal conc./mol dm^{-3}			0		$c_{SO_4^{2-}} = 0.10$
Change/mol dm^{-3}			$+S$		$+S$
Equilibrium conc./mol dm^{-3}			S		$c_{SO_4^{2-}} + S \simeq c_{SO_4^{2-}}$

$$K_s = [Pb^{2+}][SO_4^{2-}] = S \times c_{SO_4^{2-}}$$

$$S = K_s / c_{SO_4^{2-}} = 1.6 \times 10^{-8} / 0.10 = \boxed{1.6 \times 10^{-7} \text{ mol dm}^{-3}}.$$

(d) Solubility of nickel(II) hydroxide ($K_s = 6.5 \times 10^{-18}$) in 2.7×10^{-5} M $NiSO_4$(aq). In this example the solubility constant is extremely small so we will assume that $c_{Ni^{2+}} + S \simeq c_{Ni^{2+}}$. We check that validity of the assumption after calculation of S.

$K_s = 6.5 \times 10^{-18}$	$Ni(OH)_2(s)$	$\rightleftharpoons$	$Ni^{2+}(aq)$	+	$2\,OH^-(aq)$
Formal conc./mol dm^{-3}			$c_{Ni^{2+}} = 2.7 \times 10^{-5}$		0
Change/mol dm^{-3}			$+S$		$+2S$
Equilibrium conc./mol dm^{-3}			$c_{Ni^{2+}} + S \simeq c_{Ni^{2+}}$		$2S$

$$K_s = [Ni^{2+}]\,[OH^-]^2 = c_{Ni^{2+}} \times (2S)^2 = 4\,c_{Ni^{2+}}\,S^2$$

$$S = \tfrac{1}{2}\sqrt{K_s / c_{Ni^{2+}}} = \tfrac{1}{2}\sqrt{(6.5\times10^{-18})/(2.7\times10^{-5})} = \boxed{2.5\times10^{-7}\ \text{mol dm}^{-3}}$$

Since S is much, much smaller than $c_{Ni^{2+}}$, the approximation $c_{Ni^{2+}} + S \simeq c_{Ni^{2+}}$ is justified.

E8.38 $\quad Hg_2I_2(s) \rightleftharpoons Hg_2^{\ 2+}(aq) + 2\,I^-(aq)$

Given $S = 0.24$ nmol dm^{-3} (value calculated from information that can be found in the CRC *Handbook of Chemistry and Physics*), what is $\Delta_s G^{\ominus}$ at 25°C?

$$K_s = \left[Hg_2^{2+}\right]\left[I^-\right]^2 = S\times(2S)^2 = 4S^3$$

$$= 4\left(0.24\times10^{-9}\right)^3 = 5.5\times10^{-29}$$

$$\Delta_s G^{\ominus} = -RT\ln K_s \ [7.8]$$

$$= -\left(8.3145\ \text{J K}^{-1}\ \text{mol}^{-1}\right)\times\left(298.15\ \text{K}\right)\times\ln\left(5.5\times10^{-29}\right)$$

$$= \boxed{161\ \text{kJ mol}^{-1}}$$

E8.39 $\quad HgCl_2(s) \rightleftharpoons Hg^{2+}(aq) + 2Cl^-(aq)$

Calculation of $\Delta_s G^{\ominus}$:

$$\Delta_s G^{\ominus} = \Delta_f G^{\ominus}(Hg^{2+}) + 2\Delta_f G^{\ominus}(Cl^-) - \Delta_f G^{\ominus}(HgCl_2)$$

$$= +164.40 + 2\times(-131.23) - (-178.6)\ \text{kJ mol}^{-1} = 80.54\ \text{kJ mol}^{-1}$$

Calculation of K_s:

$$\ln K_s = \frac{-\Delta_s G^{\ominus}}{RT}\ [7.8] = \frac{-80.54\times10^3\ \text{J mol}^{-1}}{8.3145\ \text{J K}^{-1}\text{mol}^{-1}\times298.15\ \text{K}} = -32.49$$

$$K_s = e^{-32.49} = 7.76\times10^{-15}$$

Calculation of S:

$$K_s = \left[Hg^{2+}\right]\left[Cl^-\right]^2 = S\times(2S)^2 = 4S^3$$

$$S = \left(K_s/4\right)^{1/3}$$

$$= \left(7.76\times10^{-15}/4\right)^{1/3} = \boxed{1.25\times10^{-5}\ \text{mol dm}^{-3}}$$

E8.40 (a) $\quad AgCl(s) \rightleftharpoons Ag^+(aq) + Cl^-(aq)$

$$K_s = a_{Ag^+}a_{Cl^-} = S^2 \qquad \text{where } S = \left[Ag^+\right] = \left[Cl^-\right]$$

$$\ln K_s = \ln S^2 = 2\ln S \qquad \text{at temperature } T$$

$$\ln K_s' = \ln\left(S'\right)^2 = 2\ln S' \qquad \text{at temperature } T'$$

Subtracting the expression at temperature T from the expression at temperature T' gives:

$$\ln K_s' - \ln K_s = 2\ln S' - 2\ln S = 2\ln\left(\frac{S'}{S}\right) = \frac{\Delta_s H^{\ominus}}{R}\left(\frac{1}{T} - \frac{1}{T'}\right) \quad \text{van't Hoff equation [7.15]}$$

$$\ln\left(\frac{S'}{S}\right) = \frac{\Delta_s H^{\ominus}}{2R}\left(\frac{1}{T} - \frac{1}{T'}\right)$$

$$\boxed{\frac{S'}{S} = \mathrm{e}^{\frac{\Delta_s H^{\ominus}}{2R}\left(\frac{1}{T} - \frac{1}{T'}\right)}}$$

If $T' > T$ and $\Delta_s H^{\ominus} > 0$, then $S' > S$ because $\frac{1}{T} - \frac{1}{T'} > 0$ and e to a positive exponent is greater than 1; solubility increases. If $T' > T$ and $\Delta_s H^{\ominus} < 0$, then $S' < S$ because $\frac{1}{T} - \frac{1}{T'} > 0$ and e to a negative exponent is less than 1; solubility decreases.

(b) $\Delta_s H^{\ominus}(\mathrm{AgCl}) = \Delta_f H^{\ominus}(\mathrm{Ag^+(aq)}) + \Delta_f H^{\ominus}(\mathrm{Cl^-(aq)}) - \Delta_f H^{\ominus}(\mathrm{AgCl(s)})$

$$= \{105.58 + (-167.16) - (-127.07)\}\ \mathrm{kJ\ mol^{-1}} = 65.49\ \mathrm{kJ\ mol^{-1}}$$

Thus, because $\Delta_s H^{\ominus}(\mathrm{AgCl}) > 0$, the solubility of silver chloride $\boxed{\text{increases}}$ as the temperature increases.

Answers to projects

P8.41 Derive the fractional composition of each protonated species by analogy to those of oxalic in Example 8.4. Since oxalic acid is a diprotic acid and lysine may be triprotonated in very acid solutions, draw up a comparative table of the fractional composition numerators for each species. Begin with a deduction of the numerator of the fractional composition for the totally deprotonated species and proceed to deduction of the numerator of the fractional composition for the maximally protonated species. The deduction is an analogy to the equations that appear in Example 8.4. The denominator of the fractional compositions equals the sum of these numerators.

Species	Fractional composition numerator	Species	Fractional composition numerator
$C_2O_4^{2-}$	$K_{a1}K_{a2}$	Lys^-	$K_{a1}K_{a2}K_{a3}$
$HC_2O_4^-$	$[H_3O^+]\,K_{a1}$	HLys	$[H_3O^+]\,K_{a1}K_{a2}$
$H_2C_2O_4$	$[H_3O^+]^2$	H_2Lys^+	$[H_3O^+]^2K_{a1}$
		H_3Lys^{2+}	$[H_3O^+]^3$
Fractional composition denominator	$[H_3O^+]^2 + [H_3O^+]\,K_{a1} + K_{a1}K_{a2}$		$[H_3O^+]^3 + [H_3O^+]^2K_{a1} + [H_3O^+]K_{a1}K_{a2} + K_{a1}K_{a2}K_{a3}$

The table gives formulas for the fractional composition for every species of a lysine solution. For example,

$$f_{\mathrm{HLys}} = \frac{[\mathrm{H_3O^+}]K_{a1}K_{a2}}{[\mathrm{H_3O^+}]^3 + [\mathrm{H_3O^+}]^2 K_{a1} + [\mathrm{H_3O^+}]K_{a1}K_{a2} + K_{a1}K_{a2}K_{a3}}$$

A speciation diagram (Figure 8.7) is prepared over the pH range of 0 to 14 by calculating the hydronium ion concentration at each point in the range with the expression $[H_3O^+] = 10^{-pH}$. The data editor of scientific graphing calculator serves as an efficient calculation environment.

Alternatively, software such as Microsoft Excel or Mathcad are especially effective for preparing the calculations and graph. A Mathcad worksheet is shown below.

$K_{a1} := 10^{-2.18}$ $\quad K_{a2} := 10^{-8.95}$ $\quad K_{a3} := 10^{-10.53}$

$i := 0...1000$ $\quad pH_i := \dfrac{i}{1000}\cdot 14$ $\quad H_i := 10^{-pH_i}$

$$D_i := (H_i)^3 + (H_i)^2\cdot K_{a1} + H_i\cdot K_{a1}\cdot K_{a2} + K_{a1}\cdot K_{a2}\cdot K_{a3}$$

$$f_{H3Lys_i} := \frac{(H_i)^3}{D_i} \qquad f_{H2Lys_i} := \frac{(H_i)^2\cdot K_{a1}}{D_i}$$

$$f_{HLys_i} := \frac{H_i\cdot K_{a1}\cdot K_{a2}}{D_i} \qquad f_{Lys_i} := \frac{K_{a1}\cdot K_{a2}\cdot K_{a3}}{D_i}$$

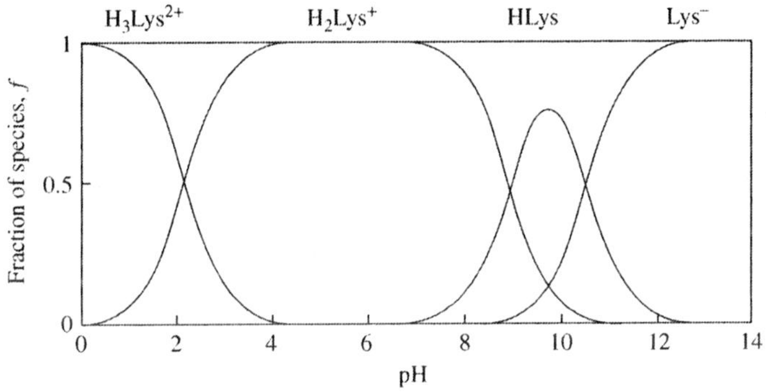

Figure 8.7

P8.42

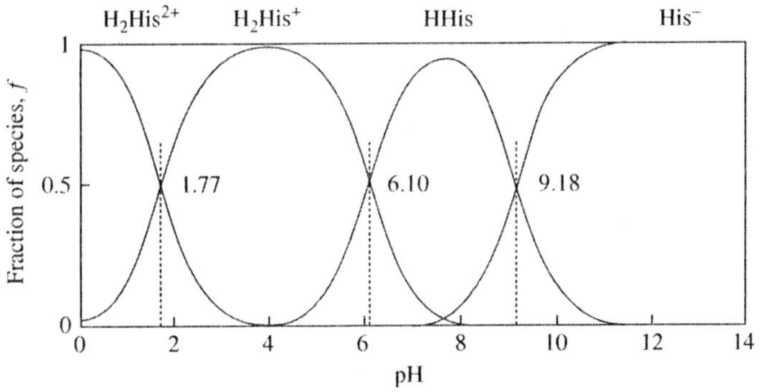

Figure 8.8

P8.43 (a) $H_2CO_3(aq) \rightleftharpoons H^+(aq) + HCO_3^-(aq)$

$$K_{a1} = \frac{[H^+][HCO_3^-]}{[H_2CO_3]} = 7.9\times10^{-7} \text{ at the physiological temperature of 37°C}$$

$$\frac{[HCO_3^-]}{[H_2CO_3]} = \frac{7.9\times10^{-7}}{[H^+]} = \frac{7.9\times10^{-7}}{10^{-pH}}$$

Normal ratio (pH = 7.40): $\dfrac{[HCO_3^-]}{[H_2CO_3]} = \dfrac{7.9\times10^{-7}}{10^{-7.40}} = 20.\bar{0}$

Ratio at onset of acidosis (pH = 7.35): $\dfrac{[HCO_3^-]}{[H_2CO_3]} = \dfrac{7.9\times10^{-7}}{10^{-7.35}} = \boxed{17.\bar{7}}$

Ratio at onset of alkalosis (pH = 7.45): $\dfrac{[HCO_3^-]}{[H_2CO_3]} = \dfrac{7.9\times10^{-7}}{10^{-7.45}} = \boxed{22.\bar{3}}$

(b) The Bohr effect observes that haemoglobin binds O_2 strongly when it is deprotonated at slightly higher blood pH. That is, the Hb saturation s, which is proportional to bound O_2, increases. The mathematical model

$$\log\left(\frac{s}{1-s}\right) = \nu \log p_{O_2} - \nu \log K \qquad \text{where } \nu \text{ is the Hill coefficient}$$

defines the Hill coefficient. At high partial pressures, the last term is not the major influence. Ignoring it, we see that the left side of the equation increases as s increases and, consequently, the Hill coefficient must increase should this occur at constant pressure. We conclude that the Hill coefficient increases with a slight increase in pH.

9 Chemical equilibrium: electrochemistry

Answers to discussion questions

D9.1 The **Debye–Hückel theory** of very dilute ionic solutions yields a method for calculating the **mean activity coefficient, $\gamma_\pm$**, of an electrolyte. The theory emphasizes the long range Coulombic (electrostatic) interaction between ions that brings anions and cations into energetically favorable proximity. Averaged over time, **counter ions** (ions of opposite charge) are more likely to be found near any given ion. This time-averaged, spherical haze around the central ion, in which counter ions outnumber ions of the same charge as the central ion, has a net charge equal in magnitude but opposite in sign to that on the central ion, and is called the **ionic atmosphere**. The energy, and therefore the chemical potential, of any given central ion is lowered below its ideal value as a result of its net electrostatic attraction with its ionic atmosphere. The ideal value of both an activity coefficient and a mean activity coefficient is equal to 1 but the attraction of the ionic atmosphere causes $\gamma_\pm$ to be less than 1. The Debye–Hückel theory is correct in the limit of infinitely dilute electrolyte molality, but a rule-of-thumb accepts its application when the total ion concentration is less than around 10^{-3} mol kg^{-1}. When interested in property trends alone, we often accept, within limits, the estimate that activity coefficients are approximately equal to 1 and activities equal concentration.

The Debye–Hückel limiting theory of activity coefficients fails in more concentrated electrolyte solutions because ionic size becomes important and the effective hydration spheres of ions is reduced by competitive attraction by ions for available water molecules. These affect the ionic atmosphere stabilization effect of the Debye–Hückel limiting theory. First, direct association of oppositely charged ions yields ion pairs of zero net charge. The ion pairs have a dipole but a reduced attraction to the ionic atmosphere. Secondly, strong repulsions between either cations or anions diminish the validity of the hazy ionic atmosphere model. The charge imbalance of the ionic atmosphere may decrease or ions may begin to align in very small, localized patterns that resemble an ionic crystal lattice when time-averaged. Consequently, as concentration approaches moderate levels the activity coefficient stops the decline of the Debye–Hückel limiting theory and begins to grow toward a value of 1 as the ionic atmosphere stabilization effect is lost. At higher concentrations the activity coefficient may even become larger than 1.

D9.2 The migration of aqueous protons is mechanistically very different from the migration of other ions. In a simplified view a proton on one water molecule (H_3O^+) migrates to a neighbour water molecule, a proton on that water molecule then migrates to its neighbour, and so on along a chain of water molecules. The motion of protons is therefore an *effective* motion of a proton, not the actual motion of a single proton. This causes proton migration to be far more rapid than the migration of other ions, which must move as a single, individual unit from one position to another. (Compare the ionic mobilities of Table 9.2.)

According to the detailed **Grotthus mechanism**, there is an effective motion of a proton that involves the rearrangement of bonds in a group of water molecules. However, the actual mechanism is still highly contentious. Attention focuses on the $H_9O_4^+$ unit in which the nearly trigonal planar H_3O^+ ion is linked to three strongly solvating H_2O molecules. This cluster of atoms is itself hydrated, but the hydrogen bonds in the secondary sphere are weaker than in the primary sphere. It is envisaged that the rate-determining step is the cleavage of one of the weaker hydrogen bonds of this secondary sphere. After this bond cleavage has taken place, and the released molecule has rotated through a few degrees (a process that takes about 1 ps), there is a rapid adjustment of bond lengths and angles in the remaining cluster, to form an $H_5O_2^+$ cation of structure $H_2O...H^+...OH_2$. Shortly after this reorganization has occurred, a new $H_9O_4^+$ cluster forms as other molecules rotate into a position where they can become members of a secondary hydration sphere, but now the positive charge is located one molecule to the right of its initial location. According to this model, there is no coordinated motion of a proton along a chain of molecules, simply a very rapid hopping between neighbouring sites, with low activation energy. The model is consistent with the observation that the molar conductivity of protons increases as the pressure is raised, for increasing pressure ruptures the hydrogen bonds in water.

D9.3 A **galvanic cell** uses a spontaneous chemical reaction to generate a potential difference and deliver an electric current to an external device. An **electrolytic cell** uses an external potential difference to drive a chemical reaction in the cell that is by itself non-spontaneous. In their essential features, these two kinds of cells can be considered opposites of each other, in the sense that an electrolytic cell can be thought of as a galvanic cell operating in the reverse direction. For some electrochemical cells, this is easy to accomplish. We say they are rechargeable. The most common example is the lead-acid battery used in automobiles. For many other cells, however, this kind of reversibility cannot be achieved. A **fuel cell**, like the galvanic cell, uses a spontaneous chemical reaction to generate a potential difference and deliver an electric current to an external device. Unlike the galvanic cell, the fuel cell must receive reactants from an external storage tank.

A **salt bridge** connecting two half-cells compartments is usually a U-tube filled with potassium chloride in agar jelly. It provides the mobile electrolyte for completing the circuit of an electrochemical cell. In its absence, the cell cannot generate an electrical current through the single wire that connects the two electrodes and the circuit is said to be "open". No electron can leave, or enter, either half-cell, because this act would cause the net electronic charge of the half-cell to be non-zero. The strong electrostatic force prevents this from happening and causes macroscopic objects to normally have a zero net electrical charge. However, a salt bridge provides an anion to the anodic half-cell for every electron that leaves while simultaneously providing a cation to the cathodic half-cell for every electron that enters. This is a "closed" electrical circuit in which the net charge of each half-cell remains zero but an electric current can be generated. The salt bridge uses KCl specifically because the mobilities of the K^+ and Cl^- ions are very similar, which minimizes the **liquid junction potential** across the interface of the two electrolyte compartments.

D9.4 The **electrochemical series** lists metallic elements and hydrogen in the order of their reducing power as measured by their standard potentials in aqueous solution (Table 9.3). It is used to quickly determine whether one metal can spontaneously displace another from solution at 298 K. The application rules is: a couple with a low standard potential has a thermodynamic tendency to reduce a couple with a high standard potential.

Couple	$E^\ominus$/V at 25°C
Ag^+/Ag	+0.80
Cu^{2+}/Cu	+0.34
H^+/H_2	0, by definition
Zn^{2+}/Zn	−0.76

For example, zinc is lower than hydrogen in the series. So zinc will spontaneously react with the hydronium cation, $H^+(aq)$, to form the zinc cation and hydrogen gas: $Zn(s) + 2\,H^+(aq) \rightarrow Zn^{2+}(aq) + H_2(g)$. Silver is higher in the series so it will not displace the hydronium ion.

D9.5 The table of standard potentials provides the starting point for the determination of thermodynamic properties of a redox reaction. To calculate the standard reaction potential we take the difference of the difference of the right (R, cathode) and left (L, anode) standard potential of the half-reactions.

$$E^\ominus_{\text{cell}} = E^\ominus_{\text{R}} - E^\ominus_{\text{L}} \quad [9.17]$$

The standard reaction Gibbs energy and equilibrium constant can be calculated with the relations

$$\Delta_r G^{\ominus} = -\nu F E^{\ominus}_{\text{cell}} \quad [9.15] \qquad \text{and} \qquad K = e^{-\Delta_r G^{\ominus}/RT} = e^{\nu F E^{\ominus}_{\text{cell}}/RT} \quad [9.16]$$

where is the amount of electrons transferred in the reaction. With $E^{\ominus}_{\text{cell}}$ measured at temperature T and $E^{\ominus\prime}_{\text{cell}}$ measured at temperature T' the standard reaction entropy is given by

$$\Delta_r S^{\ominus} = \frac{\nu F\left(E^{\ominus}_{\text{cell}} - E^{\ominus\prime}_{\text{cell}}\right)}{T - T'} \quad [9.18].$$

Finally, the standard reaction enthalpy is calculated with $\Delta_r H^{\ominus} = \Delta_r G^{\ominus} + T\Delta_r S^{\ominus}$.

A few words can also be said about the electrochemical measurement method. The cell potential E_{cell} is measured with a high impedance voltmeter under zero current conditions. When using SHE as a reference electrode, E_{cell} is the desired half-reaction potential [9.17]. Should the redox couple have one or more electroactive species (i) that are solvated with concentration b_i, E must be measured over a range of b_i values. The Nernst equation [9.14], with Q being the cell reaction quotient, is the starting point for analysis of the $E(b_i)$ data.

$$E_{\text{cell}} = E^{\ominus}_{\text{cell}} - \frac{RT}{\nu F}\ln Q \quad [9.14]$$

It would seem that substitution of E and Q values would allow the computation of the standard redox potential $E^{\ominus}_{\text{cell}}$ for the couple. However, a problem arises because the calculation of Q requires not only knowledge of the concentrations of the species involved in the cell reaction but also of their activity coefficients. These coefficients are not usually available, so the calculation cannot be directly completed. However, at very low concentrations, the Debye–Hückel limiting law for the coefficients holds. The procedure then is to substitute the Debye–Hückel law for the activity coefficients into the specific form of the Nernst equation for the cell under investigation and carefully examine the equation to determine what kind of plot to make of the $E(b_i)$ data so that extrapolation of the plot to zero concentration, where the Debye-Hückel law is valid, gives a plot intercept that equals $E^{\ominus}_{\text{cell}}$. For example, the reaction of the Harned cell, ½ H_2(g) + AgCl(s) → HCl(aq) + Ag(s), is analyzed with a plot of $E_{\text{cell}} + (2RT/F)\ln b$ against $b^{1/2}$ for it can be shown that the plot intercept equals $E^{\ominus}_{\text{cell}}$ (see Project 9.43 of the text exercises).

Solutions to exercises

For notational simplicity, we have used both the molality concentration expression $a_J = \gamma_J b_J / b^{\ominus}$ where $b^{\ominus} = 1\ \text{mol kg}^{-1}$ [9.1a] and $a_J = \gamma_J b_J$ [9.1b] where b_J is the unitless magnitude of molality. The convention of Eq. 9.1b is most often used in calculations of ionic strength while the convention of Eq. 9.1a appears in Nernst equation computations.

E9.1 $$I = I_{\text{KCl}} + I_{\text{CuSO}_4} = \tfrac{1}{2}\left(z_+^2 b_+ + z_-^2 b_-\right)_{\text{KCl}} + \tfrac{1}{2}\left(z_+^2 b_+ + z_-^2 b_-\right)_{\text{CuSO}_4} \quad [9.5]$$

Let the preparation molality, the formal concentration, be b. Determination of solution ionic strength requires the deduction of z_+^2, b_+, z_-^2, and b_- for each ionic compound. These values are substituted into the above equation.

$$\text{KCl(s)} \xrightarrow{\text{water}} \text{K}^+\text{(aq)} + \text{Cl}^-\text{(aq)}$$

$$z_+^2 = (+1)^2 = 1 \qquad b_+ = b_{\text{KCl}} \qquad z_+^2 b_+ = b_{\text{KCl}}$$

$$z_-^2 = (-1)^2 = 1 \qquad b_- = b_{\text{KCl}} \qquad z_-^2 b_- = b_{\text{KCl}}$$

$$I_{\text{KCl}} = \tfrac{1}{2}(b+b)_{\text{KCl}} = b_{\text{KCl}} = 0.15$$

$$\text{CuSO}_4\text{(s)} \xrightarrow{\text{water}} \text{Cu}^{2+}\text{(aq)} + \text{SO}_4^{2-}\text{(aq)}$$

$$z_+^2 = (+2)^2 = 4 \qquad b_+ = b_{\text{CuSO}_4} \qquad z_+^2 b_+ = 4b_{\text{CuSO}_4}$$

$$z_-^2 = (-2)^2 = 4 \qquad b_- = b_{\text{CuSO}_4} \qquad z_-^2 b_- = 4b_{\text{CuSO}_4}$$

$$I_{\text{CuSO}_4} = \tfrac{1}{2}(4b+4b)_{\text{CuSO}_4} = 4b_{\text{CuSO}_4} = 4\times(0.30) = 1.20$$

$$I = I_{\text{KCl}} + I_{\text{CuSO}_4} = 0.15 + 1.20 = \boxed{1.35}$$

COMMENT. Note that the ionic strength of a solution of more than one electrolyte may be calculated by summing the ionic strengths of each electrolyte considered as a separate solution, as in the solution to this exercise, or by summing the product $\tfrac{1}{2}b_J z_J^2$ for each individual ion, as in the definition of I [9.5].

E9.2 $$I_{KNO_3} = b_{KNO_3} = 0.150$$

Therefore, the ionic strengths of the added salts must be 0.100 to result in a total of 0.250.

(a) $$I_{Ca(NO_3)_2} = \tfrac{1}{2}(2^2 + 2)\times b = 3b$$

Therefore, the solution should be made $\tfrac{1}{3}(0.100\ \text{mol kg}^{-1}) = 0.0333\ \text{mol kg}^{-1}$ in $Ca(NO_3)_2$.

The mass that should be added to 500 g of the KNO_3 solution is therefore

$$(0.500\ \text{kg})\times(0.0333\ \text{mol kg}^{-1})\times(164\ \text{g mol}^{-1}) = \boxed{2.73\ \text{g}}$$

(b) $I_{NaCl} = b$; therefore, with $b = 0.100\ \text{mol kg}^{-1}$

$$(0.500\ \text{kg})\times(0.100\ \text{mol kg}^{-1})\times(58.4\ \text{g mol}^{-1}) = \boxed{2.92\ \text{g}}$$

E9.3 For the salt M_pX_q: $\gamma_\pm = (\gamma_+^p\gamma_-^q)^{1/s}$ where $s = p + q$ [9.3b]

For MgF_2: $p = 1$, $q = 2$, $s = 3$, $\boxed{\gamma_\pm = (\gamma_+\gamma_-^2)^{1/3}}$.

E9.4 The concentrations $b_{MgF_2}/b^\ominus = 0.015$ and $b_{NaCl}/b^\ominus = 0.025$ are sufficiently dilute for the Debye-Hückel limiting law to give a reasonable estimate of the mean ionic activity coefficients.

$$I = \tfrac{1}{2}\sum_i z_i^2 b_i\ [9.5b] = I_{MgF_2} + I_{NaCl}$$

$$= \tfrac{1}{2}\{(4\times0.015) + (1\times0.030)\} + \tfrac{1}{2}\{(1\times0.025) + (1\times0.025)\} = \boxed{0.070}$$

For MgF_2(aq):

$$\log(\gamma_\pm)_{MgF_2} = -A|z_+z_-|I^{1/2}\ [9.4] = -0.509\times|2\times(-1)|\times(0.070)^{1/2} = -0.27$$

$$(\gamma_\pm)_{MgF_2} = 10^{-0.27} = 0.54 = \boxed{0.54}$$

$$a_{Mg^{2+}} = (\gamma_\pm)_{MgF_2} b_{Mg^{2+}} = (0.54)\times(.015) = \boxed{0.0081}$$

$$a_{F^-} = (\gamma_\pm)_{MgF_2} b_{F^-} = (0.54)\times(.030) = \boxed{0.016}$$

For NaCl(aq):

$$\log(\gamma_\pm)_{NaCl} = -A|z_+z_-|I^{1/2}\ [9.4] = -0.509\times|1\times(-1)|\times(0.070)^{1/2} = -0.13\overline{5}$$

$$(\gamma_\pm)_{NaCl} = 10^{-0.13\overline{5}} = 0.73 = \boxed{0.73}$$

$$a_{Na^+} = (\gamma_\pm)_{NaCl} b_{Na^+} = (0.73)\times(.025) = \boxed{0.018}$$

$$a_{Cl^-} = (\gamma_\pm)_{NaCl} b_{Cl^-} = (0.73)\times(.025) = \boxed{0.018}$$

E9.5 $$\log\gamma_\pm = -\frac{A|z_+z_-|I^{1/2}}{1 + BI^{1/2}}\qquad [9.6\ \text{with}\ C = 0]$$

Solving for B,

$$B = -\left(\frac{1}{I^{1/2}} + \frac{A|z_+z_-|}{\log\gamma_\pm}\right).$$

Recognizing that for HBr we know that $|z_+z_-| = 1$ and $I = \tfrac{1}{2}(b_{H^+} + b_{Br^{-1}}) = b_{HBr} = b$, the equation for B simplifies to

$$B = -\left(\frac{1}{b^{1/2}} + \frac{0.509}{\log \gamma_\pm}\right).$$

We do a table calculation of the right side of this equation for each solution and average the results.

$b/b^\ominus$	0.0050	0.0100	0.0200
$\gamma_\pm$	0.930	0.907	0.879
B	2.01	2.01	2.02

The constancy of B indicates that the mean activity coefficient of HBr obeys the extended Debye-Hückel law very well with $\boxed{B = 2.01}$.

E9.6 The basis for the solution is Kohlrausch's law of independent migration of ions [9.8 and especially 9.9]. Switching counterions does not affect the mobility of the other ion at infinite dilution.

$$\Lambda_m^\circ(\text{KCl}) = \lambda(\text{K}^+) + \lambda(\text{Cl}^-)\ [9.9] = 14.99\ \text{mS m}^2\ \text{mol}^{-1}$$

$$\Lambda_m^\circ(\text{KNO}_3) = \lambda(\text{K}^+) + \lambda(\text{NO}_3^-) = 14.50\ \text{mS m}^2\ \text{mol}^{-1}$$

$$\Lambda_m^\circ(\text{AgNO}_3) = \lambda(\text{Ag}^+) + \lambda(\text{NO}_3^-) = 13.34\ \text{mS m}^2\ \text{mol}^{-1}$$

Hence,

$$\Lambda_m^\circ(\text{AgCl}) = \Lambda_m^\circ(\text{AgNO}_3) + \Lambda_m^\circ(\text{KCl}) - \Lambda_m^\circ(\text{KNO}_3)$$
$$= (13.34 + 14.99 - 14.50)\ \text{mS m}^2\ \text{mol}^{-1} = \boxed{13.83\ \text{mS m}^2\ \text{mol}^{-1}}.$$

E9.7 Molar ionic conductivity λ is proportional to mobility u. The constant of proportionality must include a factor for the charge carried by the ion and a factor for the Coulomb charge per mole of charge, which is Faraday's constant F, so we conclude that $\lambda = |z|Fu$.

$$\lambda_{\text{Cl}^-} = |z|Fu$$
$$= |-1| \times (96485\ \text{C mol}^{-1}) \times (7.91\times10^{-8}\ \text{m}^2\ \text{s}^{-1}\ \text{V}^{-1})$$
$$= 7.63\times10^{-3}\ \text{C V}^{-1}\ \text{s}^{-1}\ \text{m}^2\ \text{mol}^{-1}$$ (Recognize that the siemens unit, S, is defined by $1\ \text{S} = \Omega^{-1} = \text{C V}^{-1}\ \text{s}^{-1}$.)

$$= \boxed{7.63\ \text{mS m}^2\ \text{mol}^{-1}}$$

E9.8 The drift velocity s and the electric field $\mathcal{E}$ created by the potential difference $\Delta\phi$ that is applied between two electrodes separated by the very small distance l is given by the relation:

$$s = u\mathcal{E}\ [9.10] = u\Delta\phi / l.$$

Therefore,

$$s = (7.92\times10^{-8}\ \text{m}^2\,\text{s}^{-1}\,\text{V}^{-1}) \times (35.0\,\text{V}) / (8.00\times10^{-3}\,\text{m}) = 3.47\times10^{-4}\,\text{m s}^{-1} = \boxed{347\ \mu\text{m s}^{-1}}.$$

E9.9 $$\Lambda_m = \Lambda_m^\circ - \mathcal{K}c^{1/2}\ [9.8], \qquad \Lambda_m = \frac{\kappa}{c}[9.7] = \frac{C}{cR} = \frac{0.2063\ \text{cm}^{-1}}{cR}$$

(The cell constant C is determined by measuring the conductivity and resistance of a standard solution. Then, $C = (\kappa R)_{\text{std. soln.}}$, which is 0.2063 cm^{-1} in this example.)

We draw up the following table using 1 M = 1 mol dm^{-3}.

c/M	0.0005	0.001	0.005	0.010	0.020	0.050
$(c/\mathrm{M})^{1/2}$	0.224	0.032	0.071	0.100	0.141	0.224
R/Ω	3314	1669	342.1	174.1	89.08	37.14
$\Lambda_{\rm m}/(\mathrm{mS\ m^2\ mol^{-1}})$	12.45	12.36	12.06	11.85	11.58	11.11

The values of $\Lambda_{\rm m}$ are plotted against $c^{1/2}$ in Figure 9.1.

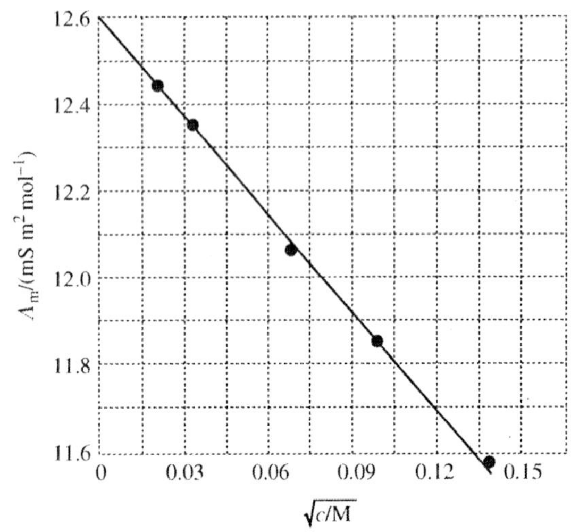

Figure 9.1

(a) The plot of $_{\rm m}$ against $c^{1/2}$ is linear. Hence, Kohlrausch's law is obeyed. The limiting value is $\Lambda_{\rm m}^{\rm o} = \boxed{12.6\,\mathrm{mS\ m^2\ mol^{-1}}}$.

(b) The slope is –7.30; hence $\mathcal{K} = \boxed{7.30\ \mathrm{mS\ m^2\ mol^{-1}\ M^{-1/2}}}$.

(c) $\Lambda_{\rm m}^{\rm o}(\mathrm{NaI}) = \lambda(\mathrm{Na^+}) + \lambda(\mathrm{I^-})$ [9.9]

$= (5.01 + 7.68)\ \mathrm{mS\ m^2\ mol^{-1}} = 12.69\ \mathrm{mS\ m^2\ mol^{-1}}$

For a 0.010 M NaI(aq):

(i) $\Lambda_{\rm m} = \Lambda_{\rm m}^{\rm o} - \mathcal{K}c^{1/2}$ [9.8]

$= 12.69\ \mathrm{mS\ m^2\ mol^{-1}} - (7.30\ \mathrm{mS\ m^2\ mol^{-1}}) \times (0.010)^{1/2}$

$= \boxed{11.96\,\mathrm{mS\ m^2\ mol^{-1}}}$

(ii) $\kappa = c\Lambda_{\rm m} = (10\ \mathrm{mol\ m^{-3}}) \times (11.96\ \mathrm{mS\ m^2 mol^{-1}}) = 119.6\,\mathrm{mS\ m^2\ m^{-3}} = \boxed{119.6\ \mathrm{mS\ m^{-1}}}$

(iii) $R = C/\kappa = (20.63\mathrm{m^{-1}})/(119.6\ \mathrm{mS\ m^{-1}}) = \boxed{172.5\ \Omega}$

E9.10 $c = \dfrac{\kappa}{\Lambda_{\rm m}}$ [9.7] $\approx \dfrac{\kappa}{\Lambda_{\rm m}^{\rm o}}$ [c small, conductivity of water allowed for in the data]

$c \approx \dfrac{1.887 \times 10^{-6}\,\mathrm{S cm^{-1}}}{138.3\mathrm{S cm^2\ mol^{-1}}}$ [Exercise 9.6 gives the limiting molar conductivity of AgCl]

$\approx 1.36 \times 10^{-8}\ \mathrm{mol cm^{-3}} = \text{solubility} = \boxed{1.36 \times 10^{-5}\ \mathrm{M}}$

E9.11 The fraction α of HCOOH molecules present as ions is given by its relation to the molar conductivity.

$$\alpha = \frac{\Lambda_{\rm m}}{\Lambda_{\rm m}^{\rm o}} = \frac{\Lambda_{\rm m}}{\lambda_+ + \lambda_-}[9.9] = \frac{3.83\ \mathrm{mS\ m^2\ mol^{-1}}}{(34.96 + 5.46)\mathrm{mS\ m^2\ mol^{-1}}\,[\text{Table 9.1}]} = 0.0948$$

We now write the acid constant in terms of α and calculate both K_a and pK_a.

	HA	$\rightleftharpoons$	A^-	H_3O^+
Initial molar concentration / mol dm^{-3}	c_{HA}		0	0
Change to reach equilibrium / mol dm^{-3}	$-\alpha c_{HA}$		$+\alpha c_{HA}$	$+\alpha c_{HA}$
Equilibrium concentration / mol dm^{-3}	$c_{HA}(1-\alpha)$		αc_{HA}	αc_{HA}

$$K_a = \frac{[H_3O^+][HCOO^-]}{[HCOOH]} = \frac{(\alpha c_{HCOOH})^2}{c_{HCOOH}(1-\alpha)} = \frac{\alpha^2 c_{HCOOH}}{1-\alpha}$$

$$= \left\{\frac{(0.0948)^2}{1-0.0948}\right\}(0.020) = 1.99\times10^{-4}$$

$$pK_a = -\log K_a = -\log(1.99\times10^{-4}) = \boxed{3.70}$$

E9.12 We prepare a plot, shown in Figure 9.2, of pH against velocity and find the linear least squares regression fit to the points.

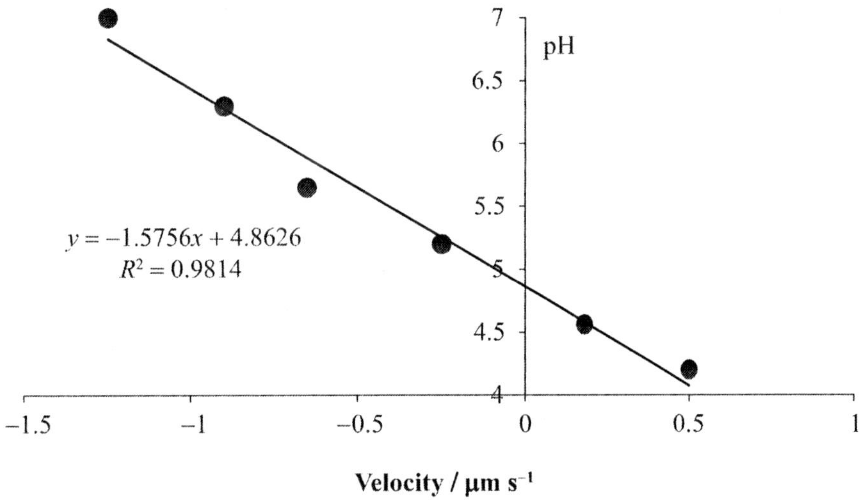

Figure 9.2

The plot and regression fit indicate a zero velocity when pH = $\boxed{4.9}$.

E9.13 L: cystine(aq) + 2 H^+(aq) + 2 e^- $\rightarrow$ 2 cysteine(aq)

R: O_2(g) + 4 H^+(aq) + 4 e^- $\rightarrow$ 2 H_2O

The overall reaction, which must not contain free electrons, is obtained as R − 2 × L and is: 4 cysteine(aq) + O_2(g) $\rightarrow$ 2 cystine(aq) + 2 H_2O(1).

E9.14 The half-reactions are

L: NAD^+(aq) + H^+(aq) + 2 e^- $\rightarrow$ NADH(aq)

R: O_2(g) + 4 H^+(aq) + 4 e^- $\rightarrow$ 2 H_2O(l)

The overall reaction, which must not contain free electrons, is obtained as R − 2 × L and is:

$$2\ NADH(aq) + O_2(g) + 2\ H^+(aq) \rightarrow 2\ NAD^+(aq) + 2\ H_2O(l).$$

The biological standard potential and Gibbs energy for the overall reaction are then

$$E^{\oplus} = E_R^{\oplus} - E_L^{\oplus} \ [9.17]$$
$$= +0.82 \text{ V} - (-0.32 \text{ V}) = +1.14 \text{ V}$$

$$\Delta_r G^{\oplus} = -\nu F E^{\oplus} \ [9.15]$$
$$= -4 \times (96{,}485 \text{ C mol}^{-1}) \times (+1.14 \text{ V})$$
$$= \boxed{-440 \text{ kJ mol}^{-1}}$$

E9.15 The half-reaction and standard potential for the electrode are: $2\,H^+(aq) + 2\,e^- \rightarrow H_2(g)$ and $E^{\ominus} = 0$.

The Nernst equation applies to half-cells as well as cells so at any hydrogen partial pressure and HBr(aq) concentration c we have the relation:

$$E = E^{\ominus} - \frac{RT}{\nu F} \ln Q \ [9.14]$$
$$= E^{\ominus} - \frac{25.693 \text{ mV}}{\nu} \ln Q$$
$$= E^{\ominus} - \frac{25.693 \text{ mV}}{2} \ln\left(p_{H_2} / a_{H^+}{}^2\right) = E^{\ominus} - (12.85 \text{ mV}) \times \ln\left(p_{H_2} / a_{H^+}{}^2\right)$$
$$= E^{\ominus} - (12.85 \text{ mV}) \times \ln\left(p_{H_2} / c^2\right) \quad \text{because } a_{H^+} = \gamma_{H^+} c \simeq c.$$

Thus, at constant p_{H_2} the difference between the potential when c_i = 5.0 mmol dm^{-3} and the potential when c_f = 15.0 mmol dm^{-3} is:

$$\Delta E = E_f - E_i$$
$$= \left\{E^{\ominus} - (12.85 \text{ mV}) \times \ln\left(p_{H_2} / c_f{}^2\right)\right\} - \left\{E^{\ominus} - (12.85 \text{ mV}) \times \ln\left(p_{H_2} / c_i{}^2\right)\right\}$$
$$= (12.85 \text{ mV}) \times \ln\left(c_f / c_i\right)^2 = (25.693 \text{ mV}) \times \ln\left(c_f / c_i\right)$$
$$= (25.693 \text{ mV}) \times \ln(15.0 / 5.0) = \boxed{28 \text{ mV}}.$$

E9.16 R: $Cl_2(g) + 2\,e^- \rightarrow 2\,Cl^-(aq)$ $\quad E^{\ominus} = +1.36$ V

L: $Mn^{2+}(aq) + 2\,e^- \rightarrow Mn(s)$ $\quad E^{\ominus} = ?$

The cell corresponding to these half reactions is

$$Mn \,|\, MnCl_2(aq) \,|\, Cl_2(g) \,|\, Pt \qquad E^{\ominus} = 2.54 \text{ V} = 1.36 \text{ V} - E^{\ominus}\left(Mn^{2+}/Mn\right).$$

Hence, $E^{\ominus}(Mn^{2+}/Mn) = 1.36 \text{ V} - 2.54 \text{ V} = \boxed{-1.18 \text{V}}$.

E9.17 (a) R: $Ag^+(aq, b_R) + e^- \rightarrow Ag(s)$

L: $Ag^+(aq, b_L) + e^- \rightarrow Ag(s)$

R–L: $Ag^+(aq, b_R) \rightarrow Ag^+(aq, b_L)$

(b) R: $2\,H^+(aq) + 2\,e^- \rightarrow H_2(g, p_R)$

L: $2\,H^+(aq) + 2\,e^- \rightarrow H_2(g, p_L)$

R–L: $H_2(g, p_L) \rightarrow H_2(g, p_R)$

(c) R: $MnO_2(s) + 4\,H^+(aq) + 2\,e^- \rightarrow Mn^{2+}(aq) + 2\,H_2O(l)$

L: $[Fe(CN)_6]^{3-}(aq) + e^- \rightarrow [Fe(CN)_6]^{4-}(aq)$

R–L: $MnO_2(s) + 4\,H^+(aq) + 2\,[Fe(CN)_6]^{4-}(aq) \rightarrow Mn^{2+}(aq) + 2\,[Fe(CN)_6]^{3-}(aq) + 2\,H_2O(l)$

(d) R: $Br_2(l) + 2\,e^- \rightarrow 2\,Br^-(aq)$

L: $Cl_2(g) + 2\,e^- \rightarrow 2\,Cl^-(aq)$

R–L: $Br_2(l) + 2\,Cl^-(aq) \rightarrow Cl_2(g) + 2\,Br^-(aq)$

(e) R: $Sn^{4+}(aq) + 2\,e^- \rightarrow Sn^{2+}(aq)$

L: $2\,Fe^{3+}(aq) + 2\,e^- \rightarrow 2\,Fe^{2+}(aq)$

R–L: $Sn^{4+}(aq) + 2\,Fe^{2+}(aq) \rightarrow Sn^{2+}(aq) + 2\,Fe^{3+}(aq)$

(f) R: $MnO_2(s) + 4\,H^+(aq) + 2e^- \rightarrow Mn^{2+}(aq) + 2\,H_2O(l)$

L: $Fe^{2+}(aq) + 2\,e^- \rightarrow Fe(s)$

R–L: $Fe(s) + MnO_2(s) + 4\,H^+(aq) \rightarrow Fe^{2+}(aq) + Mn^{2+}(aq) + 2\,H_2O(l)$

E9.18 (a) $$E = E^{\ominus} - \frac{RT}{F}\ln\frac{b_L}{b_R}$$

(b) $$E = E^{\ominus} - \frac{RT}{2F}\ln\frac{p_R}{p_L}$$

In the following Nernst equations involving ions in aqueous solution we have replaced activities with molar concentrations

(c) $$E = E^{\ominus} - \frac{RT}{2F}\ln\left(\frac{[Mn^{2+}][Fe(CN)_6^{3-}]^2}{[H^+]^4[Fe(CN)_6^{4-}]^2}\right)$$

(d) $$E = E^{\ominus} - \frac{RT}{2F}\ln\left(\frac{p_{Cl_2}[Br^-]^2}{[Cl^-]^2}\right)$$

(e) $$E = E^{\ominus} - \frac{RT}{2F}\ln\left(\frac{[Sn^{2+}][Fe^{3+}]^2}{[Sn^{4+}][Fe^{2+}]^2}\right)$$

(f) $$E = E^{\ominus} - \frac{RT}{2F}\ln\left(\frac{[Fe^{2+}][Mn^{2+}]}{[H^+]^4}\right)$$

E9.19 (a) R: $PbSO_4(s) + 2\,e^- \rightarrow Pb(s) + SO_4^{2-}(aq)$

L: $Fe^{2+}(aq) + 2\,e^- \rightarrow Fe(s)$

The cell is: $Fe(s)\,|\,FeSO_4(aq)\,|\,PbSO_4(s)\,|\,Pb(s)$ $\boxed{\nu = 2}$

(b) R: $Hg_2Cl_2(s) + 2\,e^- \rightarrow 2\,Hg(l) + 2\,Cl^-(aq)$

L: $2\,H^+(aq) + 2\,e^- \rightarrow H_2(g)$

The cell is: $Pt\,|\,H_2(g)\,|\,HCl(aq)\,|\,Hg_2Cl_2(s)\,|\,Hg(l)$ $\boxed{\nu = 2}$

(c) R: $O_2(g) + 4\,H^+(aq) + 4\,e^- \rightarrow 2\,H_2O(l)$

L: $4\,H^+(aq) + 4\,e^- \rightarrow 2\,H_2(g)$

The cell is: $Pt\,|\,H_2(g)\,|\,H^+(aq), H_2O\,|\,O_2(g)\,|\,Pt\,|$ $\boxed{\nu = 4}$

(d) R: $O_2(g) + 2\,H^+(aq) + 2\,e^- \rightarrow H_2O_2(aq)$

L: $2\,H^+(aq) + 2\,e^- \rightarrow H_2(g)$

The cell is: $Pt\,|\,H_2(g)\,|\,H^+(aq), H_2O_2(aq)\,|\,O_2(g)\,|\,Pt$ $\boxed{\nu = 2}$

(e) R: $I_2(s) + 2\,e^- \rightarrow 2\,I^-(aq)$

L: $2\,H^+(aq) + 2\,e^- \rightarrow H_2(g)$

The cell is: $Pt\,|\,H_2(g)\,|\,H^+(aq), I^-(aq)\,|\,I_2(s)\,|\,Pt$

or more simply: $Pt\,|\,H_2(g)\,|\,HI(aq)\,|\,I_2(s)\,|\,Pt$ $\boxed{\nu = 2}$

(f) R: $Cu^+(aq) + e^- \rightarrow Cu(s)$

L: $Cu^{2+}(aq) + e^- \rightarrow Cu^+(aq)$

The cell is: $Pt\,|\,CuCl_2(aq)\,||\,CuCl(aq)\,|\,Cu(s)$ $\boxed{\nu = 1}$

E9.20 See the half-reactions shown in Exercise 9.17 and identify the standard potentials in either Table 9.3 or the CRC *Handbook of Chemistry and Physics*.

(a) $E^\ominus_{cell} = E^\ominus_R - E^\ominus_L = \boxed{0}$ (Same electrode on right and left.)

(b) $E^\ominus_{cell} = E^\ominus_R - E^\ominus_L = \boxed{0}$ (Same electrode on right and left.)

(c) $E^\ominus_{cell} = E^\ominus_R - E^\ominus_L = 1.23\text{ V} - 0.36\text{ V} = \boxed{+0.87\text{ V}}$

(d) $E^\ominus_{cell} = E^\ominus_R - E^\ominus_L = 1.09\text{ V} - 1.36\text{ V} = \boxed{-0.27\text{ V}}$ This is a non-spontaneous cell reaction.

(e) $E^\ominus_{cell} = E^\ominus_R - E^\ominus_L = 0.15\text{ V} - 0.77\text{ V} = \boxed{-0.62\text{ V}}$ This is a non-spontaneous cell reaction.

(f) $E^\ominus_{cell} = E^\ominus_R - E^\ominus_L = 1.23\text{ V} - (-0.44\text{ V}) = \boxed{+1.67\text{ V}}$

E9.21 See the half-reactions shown in Exercise 9.19 and identify the standard potentials in either Table 9.3 or the CRC *Handbook of Chemistry and Physics*.

(a) $E^\ominus_{cell} = E^\ominus_R - E^\ominus_L = -0.36\text{ V} - (-0.44\text{ V}) = \boxed{0.08\text{ V}}$

(b) $E^\ominus_{cell} = E^\ominus_R - E^\ominus_L = 0.27\text{ V} - 0 = \boxed{+0.27}$ (Same electrode on right and left.)

(c) $E^\ominus_{cell} = E^\ominus_R - E^\ominus_L = 1.23\text{ V} - 0 = \boxed{+1.23\text{ V}}$

(d) $E^\ominus_{cell} = E^\ominus_R - E^\ominus_L = 0.695\text{ V} - 0 = \boxed{+0.695\text{ V}}$

(e) $E^\ominus_{cell} = E^\ominus_R - E^\ominus_L = 0.54\text{ V} - 0 = \boxed{+0.54\text{ V}}$

(f) $E^\ominus_{cell} = E^\ominus_R - E^\ominus_L = 0.52\text{ V} - 0.16\text{ V} = \boxed{+0.36\text{ V}}$

E9.22 The net chemical reactions of the hydrogen/oxygen fuel cell and the benzene/oxygen fuel cell, shown below, are identical to those of complete combustion. Yet, the mechanism by which the reaction occurs in a fuel cell is very different from the mechanism of complete combustion. Collisions occur in a mixture of reactant molecules and their fragments in a combustion reaction. Reactant molecules of a fuel cell, being in separate compartments, do not directly collide but electrons supplied by the reductant flow through an external circuit from the anode to the cathode where they are received by the oxidant. In spite of this remarkable difference, the laws of thermodynamics and electrochemistry emphasize that changes in properties like the Gibbs energy and chemical potential do not depend upon the process path or reaction mechanism. This is a very powerful feature and we use it in this exercise by noting that $\Delta_r G^\ominus$ for these fuel cells is identical to the Gibbs energy of a combustion reaction, $\Delta_c G^\ominus$. Thus, the $\Delta_r G^\ominus$ value can be directly found in tabulations of combustion thermodynamic values or it can be calculated with tabulated values of formation Gibbs energies. We calculate the standard cell potential with eqn 9.15: $E^\ominus = -\Delta_c G^\ominus / \nu F$.

(a) $H_2(g) + \frac{1}{2}\,O_2 \rightarrow H_2O(l)$ $\Delta_c G^\ominus = \Delta_f G^\ominus(H_2O, l) = -237.13\text{ kJ mol}^{-1}$

The oxidation number of the oxygen atom changes from 0 on the left to −2 on the right. Thus, $\nu = 2$.

$$E^\ominus = -\Delta_c G^\ominus / \nu F = -(-237.13\times10^3\text{ J mol}^{-1})/(2\times96485.3\text{ C mol}^{-1}) = \boxed{+1.23\text{ V}}$$

(b) $C_6H_6(l) + {}^{15}/_2\ O_2 \rightarrow 6\ CO_2(g) + 3\ H_2O(l)$

$$\begin{aligned}\Delta_c G^\ominus &= 6\Delta_f G^\ominus(CO_2,g) + 3\Delta_f G^\ominus(H_2O,l) - \Delta_f G^\ominus(C_6H_6,l)\\ &= [6(-394.36) + 3(-237.13) - (124.3)]\ \text{kJ mol}^{-1}\\ &= -3201.9\ \text{kJ mol}^{-1}\end{aligned}$$

The oxidation number of each oxygen atom changes from 0 on the left to –2 on the right. Thus, $\nu = 2$ for each oxygen atom and, since there are 15 oxygen atoms on each side, $\nu = 30$ for the reaction as written.

$$E^\ominus = -\Delta_c G^\ominus / \nu F = -(-3.2019\times10^6\ \text{J mol}^{-1})/(30\times96485.3\ \text{C mol}^{-1}) = \boxed{+1.11\ \text{V}}$$

E9.23 (a) In either a galvanic cell or a fuel cell, the cathode is the electrode that receives electrons spontaneously from an external source; it is the site of reduction. Furthermore, the cathode has a higher potential than the anode or, synonymously, the Gibbs energy of the cathode reaction is lower than the anode. In an effort to apply these principles to a fuel cell that in some sense involves oxidation of methane at both electrodes we first compare the complete combustion reaction of methane with incomplete combustion.

Complete combustion: $CH_4(g) + 2\ O_2(g) \rightarrow CO_2(g) + 2\ H_2O(l)$ $\quad \Delta_c G^\ominus = -817.90\ \text{kJ mol}^{-1}$

The oxidation number of carbon changes from –4 to +4 so $\nu = 8$.

Incomplete combustion: $CH_4(g) + {}^{3}/_2\ O_2(g) \rightarrow CO(g) + 2\ H_2O(l)$ $\quad \Delta_c G^\ominus = -560.71\ \text{kJ mol}^{-1}$

The oxidation number of carbon changes from –4 to +2 so $\nu = 6$.

It is apparent that complete combustion is far more exergonic than incomplete combustion, which leads us to expect that complete oxidation of methane has the strongest tendency to provide electrons. Complete oxidation of methane should occur at the anode while a half-reaction that involves **partial oxidation of methane occurs at the cathode**.

(b) Cathode: $CO(g) + 5\ H_2O(l) + 6\ e^- \rightarrow CH_4(g) + 6\ OH^-(aq)$

Anode: $CH_4(g) + 8\ OH^-(aq) \rightarrow CO_2(g) + 6\ H_2O(l) + 8\ e^-$

Net Reaction: $4\ CO(g) + 2\ H_2O(l) \rightarrow CH_4(g) + 3\ CO_2(g)$ $\quad \nu = 24$

We use standard formation Gibbs energies to calculate $\Delta_r G^\ominus$.

$$\Delta_r G^\ominus = \{1\times(-50.72) + 3\times(-394.36) - [4\times(-137.17) + 2(-237.13)]\}\ \text{kJ mol}^{-1} = -210.86\ \text{kJ mol}^{-1}$$

$$E^\ominus = -\frac{\Delta_r G^\ominus}{\nu F}\ [17.30] = -\frac{(-210.86\times10^3)\ \text{J mol}^{-1}}{24\times(9.6485\times10^4\ \text{C mol}^{-1})} = \boxed{0.09\ \text{V}}$$

The net reaction indicates that this fuel cell spontaneously produces a positive net amount of methane but it gives only a very small zero-current potential. It is unlikely that such a cell could be used to drive a commercial process but we wonder whether an industrial activity that produces waste carbon monoxide could be used to supply the cathode reactant to produce a valuable fuel, methane.

COMMENT. The simplest methane fuel cell oxidizes methane at the anode and reduces oxygen at the cathode.

Cathode: $O_2(g) + 2\ H_2O(l) + 4\ e^- \rightarrow 4\ OH^-(aq)$

Anode: $CH_4(g) + 8\ OH^-(aq) \rightarrow CO_2(g) + 6\ H_2O(l) + 8\ e^-$

Net Reaction: $CH_4(g) + 2\ O_2(g) \rightarrow CO_2(g) + 2\ H_2O(l)$ $\quad \nu = 8$

$$\Delta_r G^\ominus = \{1\times(-394.36) + 2\times(-237.13) - 1\times(-50.72)\}\ \text{kJ mol}^{-1} = -817.9\ \text{kJ mol}^{-1}$$

$$E^\ominus = -\frac{\Delta_r G^\ominus}{\nu F}\ [9.15] = -\frac{(-817.9\times10^3)\ \text{J mol}^{-1}}{8\times(9.6485\times10^4\ \text{C mol}^{-1})} = 1.06\ \text{V}$$

This cell consumes methane and produces a voltage large enough to drive a commercial process.

E9.24 $$MnO_4^- + 8\,H^+ + 5\,e^- \rightarrow Mn^{2+} + 4\,H_2O \qquad E^\ominus = 1.51\ V$$

(a) Using the Nernst equation to determine the reduction potential for pH 6.00, keeping $a_{MnO_4^-}$ and $a_{Mn^{2-}} = 1.00$.

$$E = E^\ominus - \frac{25.7\ mV}{v}\ln Q \qquad \text{where } \ln Q = \ln\frac{1}{a_{H^+}^8} = 8\times 2.303\times pH$$

Thus, $E = 1.51\ V - \dfrac{25.7\ mV\times 8\times 2.303\times 6.00}{5} = \boxed{+0.94\ V}$.

(b) In general, $\boxed{E = 1.51 - 0.0947\ pH.}$

E9.25 (a) E decreases, $E = E^\ominus - \dfrac{RT}{F}\ln\left(\dfrac{[Ag^+]_L}{[Ag^+]_R}\right)$

(b) E increases, $E = E^\ominus - \dfrac{RT}{2F}\ln\left(\dfrac{p_R}{p_L}\right)$

(c) E increases, $E = E^\ominus - \dfrac{RT}{2F}\ln\left(\dfrac{[Mn^{2+}][Fe(CN)_6^{3-}]^2}{[H^+]^4[Fe(CN)_6^{4-}]^2}\right)$

(d) E increases, $E = E^\ominus - \dfrac{RT}{2F}\ln\left(\dfrac{[Br^-]^2 p_{Cl_2}}{[Cl^-]^2}\right)$

(e) E decreases, $E = E^\ominus - \dfrac{RT}{2F}\ln\left(\dfrac{[Sn^{2+}][Fe^{3+}]^2}{[Sn^{4+}][Fe^{2+}]^2}\right)$

(f) E increases, $E = E^\ominus - \dfrac{RT}{2F}\ln\left(\dfrac{[Fe^{2+}][Mn^{2+}]}{[H^+]^4}\right)$

E9.26 (a) The cell reaction is $PbSO_4(s) + Fe(s) \rightarrow Pb(s) + Fe^{2+}(aq) + SO_4^{2-}(aq)$.

The Nernst equation is $E = E^\ominus - \dfrac{RT}{2F}\ln[Fe^{2+}][SO_4^{2-}]$. Increasing $[Fe^{2+}]$, $\boxed{\text{decreases } E}$.

(b) The cell reaction is $Hg_2Cl_2(s) + H_2(g) \rightarrow 2\,Hg(l) + 2\,H^+(aq) + 2\,Cl^-(aq)$.

The Nernst equation is $E = E^\ominus - \dfrac{RT}{2F}\ln\left(\dfrac{[H^+]^2[Cl^-]^2}{(p_{H_2})^4}\right)$. Increase $[H^+]$, $\boxed{\text{decreases } E}$.

(c) The cell reaction is $2\,H_2(g) + O_2(g) \rightarrow 2\,H_2O(l)$.

The Nernst equation is $E = E^\ominus - \dfrac{RT}{4F}\ln\left(\dfrac{1}{p_{H_2}^2 p_{O_2}}\right)$. Increasing $p\,(O_2)$, $\boxed{\text{increases } E}$.

(d) The cell reaction is $H_2(g) + O_2(g) \rightarrow H_2O_2(aq)$.

The Nernst equation is $E = E^\ominus - \dfrac{RT}{2F}\ln\dfrac{[H_2O_2]}{p_{H_2}p_{O_2}}$. Increasing $p(H_2)$, $\boxed{\text{increases } E}$.

(e) The cell reaction is $H_2(g) + I_2(s) \rightarrow 2\,H^+(aq) + 2\,I^-(aq)$.

The Nernst equation is $E = E^\ominus - \dfrac{RT}{2F}\ln\dfrac{[H^+]^2[I^-]^2}{p_{H_2}}$.

(i) Increasing $[H^+]$, $\boxed{\text{decreases } E}$.

(ii) Increasing both $[H^+]$ and $[I^-]$, $\boxed{\text{decreases } E}$.

(f) The cell reaction is $2\,Cu^+(aq) \rightarrow Cu(s) + Cu^{2+}(aq)$.

The Nernst equation is $E = E^{\ominus} - \frac{RT}{F}\ln\frac{[Cu^{2+}]}{[Cu^+]^2}$. Adding HCl(aq) has $\boxed{\text{no effect}}$ on E.

E9.27 R: $2\,Tl^+(aq) + 2\,e^- \rightarrow 2\,Tl(s)$ $\quad -0.34$ V

L: $Hg^{2+}(aq) + 2\,e^- \rightarrow Hg(l)$ $\quad +0.86$ V

(a) Combining for the cell potential $E^{\ominus} = E_R^{\ominus} - E_L^{\ominus} = \boxed{-1.20\text{ V}}$

(b) Overall: $2\,Tl^+(aq) + Hg(l) \rightarrow 2\,Tl(s) + Hg^{2+}(aq)$

Replacing activities by molar concentrations, we have $Q = \frac{[Hg^{2+}]}{[Tl^+]^2}$

$$E = E^{\ominus} - \frac{RT}{\nu F}\ln Q = E^{\ominus} - \frac{25.7\text{ mV}}{\nu}\ln\frac{[Hg^{2+}]}{[Tl^+]^2}$$
$$= -1.20\text{ V} - \frac{25.7\text{ mV}}{2}\times\ln\frac{0.230}{(0.720)^2}$$
$$= -1.20\text{ V} + 0.001\text{ V} = \boxed{-1.19\text{ V}}$$

E9.28 (a) $Zn^{2+}(aq) + 2\,Hg(l) \rightarrow Zn(s) + Hg_2^{2+}(aq)$

$$E_{cell}^{\ominus} = E_R^{\ominus} - E_L^{\ominus} = E_{Zn^{2+}/Zn}^{\ominus} - E_{Hg_2^{2+}/Hg}^{\ominus} = -0.76\text{ V} - (+0.79\text{ V}) = \boxed{-1.55\text{ V}}$$

Mercury does not displace zinc from solution because $E_{cell}^{\ominus} < 0$.

(b) R: $Cl_2(g) + 2\,e^- \rightarrow 2\,Cl^-(aq)$ $\quad E^{\ominus} = +1.36$ V

L: $O_2(g) + 4\,H^+(aq) + 4\,e^- \rightarrow 2\,H_2O(l)$ $\quad E^{\ominus} = +1.23$ V

R–L: $2\,Cl_2(g) + 2\,H_2O(l) \rightarrow 4\,Cl^-(aq) + O_2(g) + 4\,H^+(aq)$ $\quad E_{cell}^{\ominus} = +0.13$ V $\quad \nu = 4$

Since $E_{cell}^{\ominus} > 0$ we conclude that chlorine spontaneously oxidizes water to oxygen under standard conditions for which $a_{H^+} = 1$ and the solution is acidic. We could use the Nernst equation to check spontaneity under basic conditions where $a_{H^+} > 7$ but we chose to take a conceptual short cut by recognizing that according to Le Chatelier's principle addition of base to an equilibrium standard state shifts the net reaction to the right by removal of H^+ through acid-base neutralization. This means that the equilibrium constant is larger under basic conditions and that $E_{cell}^{\ominus} > +0.13$ V. Thus, $\boxed{\text{chlorine spontaneously oxidizes water to oxygen under both acidic and basic conditions.}}$ However, this prediction gives no indication as to whether the oxidation will occur slowly or rapidly.

E9.29 Each reaction has to be broken down into the two half–reactions from which it is formed. The standard potential for the cell is calculated from the standard electrode potentials in Table 9.3. The standard Gibbs energies of the reaction are calculated from $\Delta_r G^{\ominus} = -\nu F E^{\ominus}$.

(a) R: $2\,H_2O(l) + 2\,e^- \rightarrow H_2(g) + 2\,OH^-(aq)$ $\quad E_R^{\ominus} = -0.83$ V

L: $Ca^{2+}(aq) + 2\,e^- \rightarrow Ca(s)$ $\quad E_L^{\ominus} = -2.87$ V

R–L: $2\,H_2O(l) + Ca(s) \rightarrow Ca(OH)_2(aq) + H_2(g)$ $\quad E_{cell}^{\ominus} = E_R^{\ominus} - E_L^{\ominus} = +2.04$ V $\quad \nu = 2$

$$\Delta_r G^{\ominus} = -\nu F E^{\ominus} = -2\times(96.485\text{ kC mol}^{-1})\times(2.04\text{ V}) = \boxed{-394\text{ kJ mol}^{-1}}$$

(b) This net reaction is 2 times the net reaction of part (a). Therefore,

$$\Delta_r G^{\ominus} = 2\times(-394 \text{ kJ mol}^{-1}) = \boxed{-788 \text{ kJ mol}^{-1}}.$$

(c) R: $2\,H_2O(l) + 2\,e^- \rightarrow H_2(g) + 2\,OH^-(aq)$ $\quad E_R^{\ominus} = -0.83$ V

L: $Fe^{2+}(aq) + 2\,e^- \rightarrow Fe(s)$ $\quad E_L^{\ominus} = -0.44$ V

R–L: $2\,H_2O(l) + Fe(s) \rightarrow Fe(OH)_2(aq) + H_2(g)$ $\quad E_{cell}^{\ominus} = E_R^{\ominus} - E_L^{\ominus} = -0.39$ V $\quad \nu = 2$

$$\Delta_r G^{\ominus} = -\nu F E^{\ominus} = -2\times(96.485 \text{ kC mol}^{-1})\times(-0.39 \text{ V}) = \boxed{+75 \text{ kJ mol}^{-1}}$$

(d) R: $S_2O_8^{2-}(aq) + 2\,e^- \rightarrow 2\,SO_4^{2-}(aq)$ $\quad E_R^{\ominus} = +2.05$ V

L: $I_2(s) + 2\,e^- \rightarrow 2\,I^-(aq)$ $\quad E_L^{\ominus} = +0.54$ V

R–L: $S_2O_8^{2-}(aq) + 2\,I^-(aq) \rightarrow I_2(s) + 2\,SO_4^{2-}(aq)$ $\quad E_{cell}^{\ominus} = E_R^{\ominus} - E_L^{\ominus} = +1.51$ V $\quad \nu = 2$

$$\Delta_r G^{\ominus} = -\nu F E^{\ominus} = -2\times(96.485 \text{ kC mol}^{-1})\times(+1.51 \text{ V}) = \boxed{-291 \text{ kJ mol}^{-1}}$$

(e) Although the spectator ions of parts (d) and (e) differ, the net ionic equations and Gibbs energies are identical. Thus, $\Delta_r G^{\ominus} = \boxed{-291 \text{ kJ mol}^{-1}}$.

(f) R: $2\,Na^{2+}(aq) + 2\,e^- \rightarrow 2\,Na(s)$ $\quad E_R^{\ominus} = -2.71$ V

L: $Pb^{2+}(aq) + 2\,e^- \rightarrow Pb(s)$ $\quad E_L^{\ominus} = -0.13$ V

R–L: $Pb(s) + 2\,Na^{2+}(aq) + 2\,e^- \rightarrow Pb^{2+}(aq) + 2\,Na(s)$ $\quad E_{cell}^{\ominus} = E_R^{\ominus} - E_L^{\ominus} = -2.58$ V $\quad \nu = 2$

$$\Delta_r G^{\ominus} = -\nu F E^{\ominus} = -2\times(96.485 \text{ kC mol}^{-1})\times(-2.58 \text{ V}) = \boxed{+498 \text{ kJ mol}^{-1}}$$

E9.30 (a) $2\,NADH(aq) + O_2(g) + 2\,H^+(aq) \rightarrow 2\,NAD^+(aq) + 2\,H_2O(l)$ $\quad E^{\oplus} = +1.14$ V

The oxidation number of each oxygen atom changes from 0 on the left to −2 on the right. Since there are two oxygen atoms undergoing this change, we deduce that $\nu = 4$.

$$\Delta_r G^{\oplus} = -\nu F E^{\oplus} \; [9.15]$$
$$= -4\times(96{,}485 \text{ C mol}^{-1})\times(+1.14 \text{ V})$$
$$= \boxed{-440 \text{ kJ mol}^{-1}}$$

(b) Malate(aq) + NAD^+(aq) → oxaloacetate(aq) + NADH + H^+(aq) $\quad E^{\oplus} = -0.154$ V

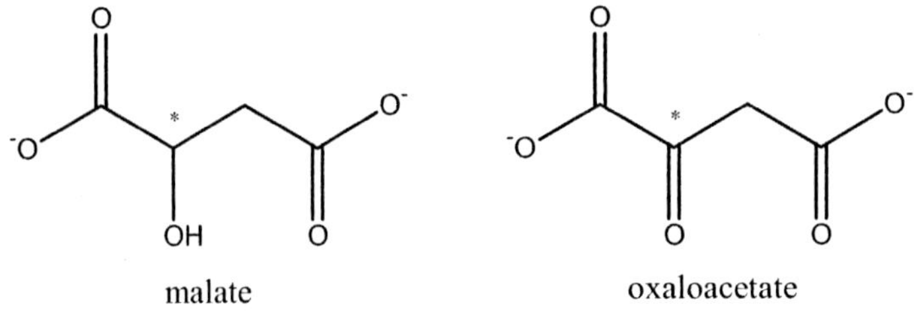

The starred carbon in the above structure of malate may be viewed as having an oxidation number of zero and, using the same counting method, the starred carbon in oxaloacetate has an oxidation number of +2. Thus, $\nu = 2$.

$$\Delta_r G^{\oplus} = -\nu F E^{\oplus} \; [9.15]$$
$$= -2\times(96{,}485 \text{ C mol}^{-1})\times(-0.154 \text{ V})$$
$$= \boxed{+29.7 \text{ kJ mol}^{-1}}$$

(c) $O_2(g) + 4\,H^+(aq) + 4\,e^- \rightarrow 2\,H_2O(l)$ $\quad E^{\ominus} = +0.81\ \text{V} \quad \nu = 4$

$$\Delta_r G^{\ominus} = -\nu F E^{\ominus} \ [9.15]$$
$$= -4 \times (96{,}485\ \text{C mol}^{-1}) \times (+0.81\ \text{V})$$
$$= \boxed{-313\ \text{kJ mol}^{-1}}$$

E9.31 $K_2CrO_4(aq) + 2\,Ag(s) + 2\,FeCl_3(aq) \rightarrow Ag_2CrO_4(s) + 2\,FeCl_2(aq) + 2\,KCl(aq)$

$$\Delta_r G^{\ominus} = -62.5\ \text{kJ mol}^{-1}$$

The oxidation number of each iron atom changes from +3 on the left to +2 on the right, a change of 1. Since there are 2 iron atoms, we deduce that $\nu = 2$.

(a) $$E^{\ominus} = \frac{-\Delta_r G^{\ominus}}{\nu F} = \frac{+62.5\ \text{kJ mol}^{-1}}{2 \times 96.485\ \text{kC mol}^{-1}} = \boxed{+0.324\ \text{V}}$$

(b) $$E^{\ominus} = E^{\ominus}(Fe^{3+}/Fe^{2+}) - E^{\ominus}(Ag_2CrO_4/Ag,\ CrO_4^{2-})$$

Therefore,

$$E^{\ominus}(Ag_2CrO_4/Ag,\ CrO_4^{2-}) = E^{\ominus}(Fe^{3+}/Fe^{2+}) - E^{\ominus}$$
$$= +0.77 - 0.324\ \text{V} = \boxed{+0.45\ \text{V}}.$$

E9.32 R: $Ag^+(aq) + e^- \rightarrow Ag(s)$ $\quad E^{\ominus}(Ag^+/Ag) = 0.80\ \text{V}$

L: $AgCl(s) + e^- \rightarrow Ag(s) + Cl^-(aq)$ $\quad E^{\ominus}(AgCl/Ag,\ Cl^-) = +0.22\ \text{V}$

R–L: $Ag^+(aq) + Cl^-(aq) \rightarrow AgCl(s)$ $\quad E^{\ominus} = E_R^{\ominus} - E_L^{\ominus} = +0.58\ \text{V} \quad \nu = 1$

The potential under non-standard conditions is calculated with the Nernst equation.

$$E_{\text{cell}} = E_R - E_L = E_R^{\ominus} - \frac{RT}{\nu F}\ln Q_R - \left\{E_L^{\ominus} - \frac{RT}{\nu F}\ln Q_L\right\} \ [9.14]$$
$$= E_R^{\ominus} - E_L^{\ominus} - \frac{RT}{\nu F}\ln\frac{Q_R}{Q_L}$$
$$= E_{\text{cell}}^{\ominus} - \frac{RT}{\nu F}\ln\left(\frac{1/b_{R,Ag^+}}{b_{L,Cl^-}}\right)$$
$$= E_{\text{cell}}^{\ominus} - \frac{RT}{\nu F}\ln\left(\frac{1}{b_{R,Ag^+}b_{L,Cl^-}}\right) = E_{\text{cell}}^{\ominus} + \frac{RT}{\nu F}\ln\left(b_{R,Ag^+}b_{L,Cl^-}\right)$$
$$= 0.58\ \text{V} + (0.025693\ \text{V}) \times \ln(\{0.01\} \times \{0.025\})$$
$$= \boxed{0.37\ \text{V}}$$

E9.33 (a) $2\,Ag(s) + Cu^{2+}(aq) \rightarrow 2\,Ag^+(aq) + Cu(s)$ $\quad \nu = 2$

$$E = E^{\ominus}(Cu^{2+}/Cu) - E^{\ominus}(Ag^+/Ag) = +0.34 - 0.80\ \text{V} = \boxed{-0.46\ \text{V}}$$

$$\Delta_r G^{\ominus} = 2\Delta_f G^{\ominus}(Ag^+,\ aq) - \Delta_f G^{\ominus}(Cu^{2+},\ aq)$$
$$= [(2 \times 77.1) - (65.49)]\ \text{kJ mol}^{-1} = \boxed{+88.7\ \text{kJ mol}^{-1}}$$

Alternatively,

$$\Delta_r G^{\ominus} = -\nu F E^{\ominus} = -(2) \times (96.485\ \text{kC mol}^{-1}) \times (-0.46\ \text{V}) = +88.8\ \text{kJ mol}^{-1}.$$

$$\Delta_r H^{\ominus} = 2\Delta_f H^{\ominus}(Ag^+,\ aq) - \Delta_f H^{\ominus}(Cu^{2+},\ aq)$$
$$= [(2 \times 105.58) - (64.77)]\ \text{kJ mol}^{-1} = \boxed{146\ \text{kJ mol}^{-1}}$$

$$\Delta_r S^\ominus = \frac{\Delta_r H^\ominus - \Delta_r G^\ominus}{T} \quad [\Delta_r G = \Delta_r H - T\Delta_r S]$$

$$= \boxed{188\ \text{J mol}^{-1}\ \text{K}^{-1}}$$

(b) Assuming that $\Delta_r S^\ominus$ and $\Delta_r H^\ominus$ do not vary greatly with temperature,

$$\Delta_r G(T) \approx \Delta_r H^\ominus - T\Delta_r S^\ominus$$

Therefore, $\Delta_r G^\ominus(308\ \text{K}) \approx [146 - 308 \times 0.188]\ \text{kJ mol}^{-1} \approx \boxed{+88\ \text{kJ mol}^{-1}}$.

E9.34 R: $O_2(g) + 4\,H^+(aq) + 4\,e^- \rightarrow 2\,H_2O$ $\qquad E_R^\ominus = +1.23\ \text{V}$

L: $\text{cystine(aq)} + 2\,H^+(aq) + 2\,e^- \rightarrow 2\ \text{cysteine(aq)}$ $\qquad E_L^\ominus = -0.34\ \text{V}$

R–L: $4\ \text{cysteine(aq)} + O_2(g) \rightarrow 2\ \text{cystine(aq)} + 2\,H_2O(l)$ $\qquad E_{\text{cell}}^\ominus = E_R^\ominus - E_L^\ominus = \boxed{+1.57\ \text{V}}$ $\quad \nu = 4$

(a) At 25°C, $\Delta_r G^\ominus = -\nu F E^\ominus = -4 \times (96.485\ \text{kC mol}^{-1}) \times (1.57\ \text{V}) = \boxed{-606\ \text{kJ mol}^{-1}}$

Assuming that the aqueous phase reaction entropy changes balance to a value that is negligibly small compared to the entropy change of gas consumption, the standard reaction entropy is given by

$$\begin{aligned}\Delta_r S^\ominus &= 2\,S_m^\ominus(\text{cystine, aq}) + 2\,S_m^\ominus(H_2O, l) - 4\,S_m^\ominus(\text{cysteine, aq}) - S_m^\ominus(O_2, g) \quad [4.13]\\ &\simeq -S_m^\ominus(O_2, g)\\ &\simeq -205\ \text{J K}^{-1}\ \text{mol}^{-1} \quad [\text{Table D1.2}]\end{aligned}$$

$$\begin{aligned}\Delta_r H^\ominus &= \Delta_r G^\ominus + T\Delta_r S^\ominus \quad [4.16]\\ &\simeq -606\ \text{kJ mol}^{-1} + (298.15\ \text{K}) \times (-205\ \text{J K}^{-1}\ \text{mol}^{-1}) \simeq \boxed{-667\ \text{kJ mol}^{-1}}\end{aligned}$$

(b) Upon a temperature increase from 25°C to 35°C, we do not expect $\Delta_r S^\ominus$ to change significantly so the change in $\Delta_r G^\ominus$ is given by

$$\begin{aligned}\text{Change in } \Delta_r G^\ominus &= -(\text{change in } T) \times \Delta_r S^\ominus \quad [7.14]\\ &= -(10\ \text{K}) \times (-205\ \text{J K}^{-1}\ \text{mol}^{-1}) = 2.05\ \text{kJ mol}^{-1}\end{aligned}$$

$$\begin{aligned}\Delta_r G^\ominus(308.15\ \text{K}) &= \Delta_r G^\ominus(298.15\ \text{K}) + \text{change in } \Delta_r G^\ominus\\ &\simeq (-606\ \text{kJ mol}^{-1}) + (2.05\ \text{kJ mol}^{-1}) \simeq \boxed{-604\ \text{kJ mol}^{-1}}\end{aligned}$$

E9.35 The pyruvic acid/lactic acid couple (HP/HL) is

$$CH_3COCOOH + 2\,H^+ + 2\,e^- \rightarrow CH_3CH(OH)COOH \qquad E^\oplus = -0.19\ \text{V} \quad \nu = 2.$$

The Nernst equation for the couple is

$$E = E^\ominus - \frac{RT}{2F}\ln\frac{a_{\text{HL}}}{a_{\text{HP}}a_{\text{H}^+}^2} = E^\ominus - \frac{RT}{2F}\ln\frac{a_{\text{HL}}}{a_{\text{HP}}} + \frac{RT}{F}\ln a_{\text{H}^+} \quad \text{or} \quad E^\ominus = E + \frac{RT}{2F}\ln\frac{a_{\text{HL}}}{a_{\text{HP}}} - \frac{RT}{F}\ln a_{\text{H}^+}.$$

At the biological standard state $a_{\text{HL}} = a_{\text{HP}} = 1$ and $a_{\text{H}^+} = 1\times10^{-7}$ (pH = 7) so evaluation of the Nernst eqn at the state $E = E^\oplus$ gives

$$\begin{aligned}E^\ominus &= E^\oplus + \frac{RT}{2F}\ln 1 - \frac{RT}{F}\ln(1\times10^{-7})\\ &= -0.19\ \text{V} - (0.025693\ \text{V})\ln(1\times10^{-7})\\ &= \boxed{+0.22\ \text{V}}.\end{aligned}$$

E9.36 (a) L: $HCO_3^-(aq) + e^- \rightarrow CO_3^{2-}(aq) + \frac{1}{2} H_2(g)$ $\quad \nu = 1$

$$\Delta_L G^\ominus = \Delta_f G^\ominus(CO_3^{2-},aq) + \Delta_f G^\ominus(H_2,g) - \Delta_f G^\ominus(HCO_3^-,aq)$$
$$= (-527.81 \text{ kJ mol}^{-1}) + 0 - (-586.77 \text{ kJ mol}^{-1})$$
$$= +58.96 \text{ kJ mol}^{-1}$$

$$E_L^\ominus = -\frac{\Delta_L G^\ominus}{\nu F} = -\frac{58.96 \text{ kJ mol}^{-1}}{1 \times (96.485 \text{ kC mol}^{-1})}$$
$$= \boxed{-0.6111 \text{ V}}$$

(b) We now wish to calculate the standard potential of a cell in which the cell net ionic reaction is

$$CO_3^{2-}(aq) + H_2O(l) \rightarrow HCO_3^-(aq) + OH^-(aq).$$

This is the **base dissociation reaction** of CO_3^{2-} and it is the net ionic reaction for combined L half-reaction presented in part (a) and the half-reaction

R: $H_2O(l) + e^- \rightarrow \frac{1}{2} H_2(g) + OH^-(aq)$ $\quad E_R^\ominus = -0.83 \text{ V} \quad \nu = 1.$

Thus, the cell standard potential is

$$E_{cell}^\ominus = E_R^\ominus - E_L^\ominus = -0.83 \text{ V} - (-0.61) = \boxed{-0.22 \text{ V}}.$$

Notice that we can now calculate $\Delta_{cell} G^\ominus$ and the basicity constant K_b of CO_3^{2-}.

$$\Delta_r G^\ominus = -\nu F E^\ominus$$
$$= -1 \times (96.485 \text{ kC mol}^{-1}) \times (-0.22 \text{ V})$$
$$= +21 \text{ kJ mol}^{-1}$$

$$K_b = e^{-\Delta_r G^\ominus / RT} = e^{-(21 \times 10^3 \text{ J mol}^{-1}) / \{(8.31447 \text{ J K}^{-1} \text{ mol}^{-1}) \times (298.15 \text{ K})\}} = 2.1 \times 10^{-4}$$

Also, $pK_b = -\log K_b = -\log(2.1 \times 10^{-4}) = 3.67.$

This value of pK_b is used in part (e) to calculate pK_a of HCO_3^-.

(c) The Nernst eqn for the cell of part (b) is

$$E_{cell} = E_{cell}^\ominus - \frac{RT}{\nu F} \ln Q \quad [9.14].$$

Thus,

$$\boxed{E_{cell} = E_{cell}^\ominus - \frac{RT}{F} \ln\left(\frac{a_{HCO_3^-} a_{OH^-}}{a_{CO_3^{2-}}}\right)}.$$

(d) At the biological standard state $a_{HCO_3^-} = a_{CO_3^{2-}} = 1$ and $a_{OH^-} = a_{H^+} = 1 \times 10^{-7}$ (i.e., pOH = pH = 7).

Thus, evaluation of the above Nernst eqn at $E_{cell} = E_{cell}^\oplus$ gives

$$E_{cell}^\oplus = E_{cell}^\ominus - \frac{RT}{F} \ln a_{H^+}$$

$$E_{cell}^\oplus - E_{cell}^\ominus = -\frac{RT}{F} \ln a_{H^+}$$
$$= -(0.025693 \text{ V}) \ln(1 \times 10^{-7})$$
$$= \boxed{+0.41}.$$

(e) Within the acid-base discussion of text Ch. 8 it is shown that for a conjugate acid-base pair: $pK_a + pK_b = pK_w$.

In part (b) it is shown that $pK_b = 3.67$ for the HCO_3^-/CO_3^{2-} conjugate pair. Thus, the pK_a of HCO_3^- is $pK_a = pK_w - pK_b = 14.00 - 3.67 = \boxed{10.33}$.

E9.37 In each case we use $\ln K = \nu FE^\ominus / RT$ [9.16] and $K = e^{\ln K}$.

(a) $Sn(s) + Sn^{4+}(aq) \rightleftharpoons 2Sn^{2+}(aq)$

$$\begin{array}{llr} \text{R:} & Sn^{4+}(aq) + 2e^- \rightarrow Sn^{2+}(aq) & +0.15\ \text{V} \\ \text{L:} & Sn^{2+}(aq) + 2e^- \rightarrow Sn(s) & -0.14\ \text{V} \\ & & E^\ominus = +0.29\ \text{V} \end{array}$$

$$\ln K = \frac{\nu FE^\ominus}{RT} = \frac{2\times 0.29\ \text{V}}{25.693\times 10^{-3}\ \text{V}} = 22.6, \qquad K = e^{22.6} = \boxed{6.5\times 10^{9}}$$

(b) $Sn(s) + 2\,AgBr(s) \rightleftharpoons SnBr_2(aq) + 2\,Ag(s)$

$$\begin{array}{llr} \text{R:} & AgBr(s) + e^- \rightarrow Ag(s) + Br^-(aq) & +0.07\ \text{V} \\ \text{L:} & Sn^{2+}(aq) + 2e^- \rightarrow Sn(s) & -0.14\ \text{V} \\ & & E^\ominus = +0.21\ \text{V} \end{array}$$

$$\ln K = \frac{2\times 0.21\ \text{V}}{25.693\times 10^{-3}\ \text{V}} = 16.3, \qquad K = e^{16.3} = \boxed{1.2\times 10^{7}}$$

(c) $Fe(s) + Hg(NO_3)_2(aq) \rightleftharpoons Hg(l) + Fe(NO_3)_2(aq)$

$$\begin{array}{llr} \text{R:} & Hg^{2+}(aq) + 2e^- \rightarrow Hg(l) & +0.86\ \text{V} \\ \text{L:} & Fe^{2+}(aq) + 2e^- \rightarrow Fe(s) & -0.44\ \text{V} \\ & & E^\ominus = +1.30\ \text{V} \end{array}$$

$$\ln K = \frac{2\times 1.30\ \text{V}}{25.693\times 10^{-3}\ \text{V}} = 160, \qquad \boxed{K = e^{101} = 7.3\times 10^{43}}$$

(d) $Cd(s) + CuSO_4(aq) \rightleftharpoons Cu(s) + CdSO_4(aq)$

$$\begin{array}{llr} \text{R:} & Cu^{2+}(aq) + 2e^- \rightarrow Cu(s) & +0.34\ \text{V} \\ \text{L:} & Cd^{2+}(aq) + 2e^- \rightarrow Cd(s) & -0.40\ \text{V} \\ & & E^\ominus = +0.74\ \text{V} \end{array}$$

$$\ln K = \frac{2\times 0.74\ \text{V}}{25.693\times 10^{-3}\ \text{V}} = 57.6, \qquad K = e^{57.6} = \boxed{1.0\times 10^{25}}$$

(e) $Cu^{2+}(aq) + Cu(s) \rightleftharpoons 2\,Cu^{+}(aq)$

$$\begin{array}{llr} \text{R:} & Cu^{2+}(aq) + e^- \rightarrow Cu^{+}(aq) & +0.16\ \text{V} \\ \text{L:} & Cu^{+}(aq) + e^- \rightarrow Cu(s) & +0.52\ \text{V} \\ & & E^\ominus = -0.36\ \text{V} \end{array}$$

$$\ln K = \frac{-0.36\ \text{V}}{25.693\times 10^{-3}\ \text{V}} = -14.0, \qquad K = e^{-14.0} = \boxed{8.3\times 10^{-7}}$$

(f) $3\,Au^{+}(aq) \rightleftharpoons 2\,Au(s) + Au^{3+}(aq)$

$$\begin{array}{llr} \text{R:} & Au^{+}(aq) + e^- \rightarrow Au(s) & 1.69\ \text{V} \\ \text{L:} & Au^{3+}(aq) + 3\,e^- \rightarrow Au(s) & 1.50\ \text{V} \\ & & E^\ominus = 0.19\ \text{V} \end{array}$$

$$\ln K = \frac{0.19\ \text{V}}{25.693\times 10^{-3}\ \text{V}} = 7.40, \qquad K = e^{7.40} = \boxed{1.6\times 10^{3}}$$

E9.38 The half-reaction is

$$Cr_2O_7^{2-}(aq) + 14\,H^{+}(aq) + 6\,e^- \rightarrow 2\,Cr^{3+}(aq) + 7\,H_2O(l).$$

The reaction quotient is

$$Q=\frac{a^{2}_{\mathrm{Cr^{3+}}}}{a_{\mathrm{Cr_2O_7{}^{2-}}}a^{14}_{\mathrm{H^+}}} \qquad \nu=6.$$

Hence,

$$E=E^{\ominus}-\frac{RT}{6F}\ln\left(\frac{a^{2}_{\mathrm{Cr^{3+}}}}{a_{\mathrm{Cr_2O_7^{2-}}}a^{14}_{\mathrm{H^+}}}\right).$$

E9.39 Assume that all activity coefficients are 1.

(1) $\mathrm{AgCl(s)} \rightleftharpoons \mathrm{Ag^+(aq)+Cl^-(aq)}$

Since all stoichiometric coefficients are 1, $S(\mathrm{AgCl}) = b(\mathrm{Ag^+}) = b(\mathrm{Cl^-})$. Hence,

$$K_{\mathrm{S}}=\frac{b(\mathrm{Ag^+})\times b(\mathrm{Cl^-})}{\left(b^{\ominus}\right)^2}=\frac{S^2}{\left(b^{\ominus}\right)^2}=\left(1.34\times10^{-5}\right)^2=\boxed{1.80\times10^{-10}}$$

(2) $\mathrm{BaSO_4(s)} \rightleftharpoons \mathrm{Ba^{2+}(aq)+SO_4^{2-}(aq)}$

Since all stoichiometric coefficients are 1, $S\left(\mathrm{BaSO_4}\right)=b(\mathrm{Ba^{2+}})=b(\mathrm{SO_4^{2-}})$. Hence,

$$K_{\mathrm{S}}=\frac{S^2}{\left(b^{\ominus}\right)^2}=\left(9.51\times10^{-4}\right)^2=\boxed{9.04\times10^{-7}}$$

E9.40 R: $\mathrm{2\,AgCl(s)+2\,e^- \rightarrow 2\,Ag(s)+2\,Cl^-(aq)}$ $\qquad E^{\ominus}_{\mathrm{R}}=+0.22\ \mathrm{V}$

L: $\mathrm{2\,H^+(aq)+2\,e^- \rightarrow H_2(g)}$ $\qquad E^{\ominus}_{\mathrm{L}}=0$

R–L: $\mathrm{2\,AgCl(s)+H_2(g) \rightarrow 2\,Ag(s)+2\,Cl^-(aq)+2\,H^+(aq)}$ $\qquad E^{\ominus}_{\mathrm{cell}}=E^{\ominus}_{\mathrm{R}}-E^{\ominus}_{\mathrm{L}}=+0.22\ \mathrm{V}$

For this cell $a_{\mathrm{H^+}}=a_{\mathrm{Cl^-}}$, ν = 2, and E = 0.312 V. The Nernst eqn can be used to calculate the pH under these conditions.

$$\begin{aligned}E&=E^{\ominus}-\frac{RT}{2F}\ln\left(a^2_{\mathrm{H^+}}a^2_{\mathrm{Cl^-}}\right)\\&=E^{\ominus}-\frac{RT}{2F}\ln a^4_{\mathrm{H^+}}=E^{\ominus}-\frac{2RT}{F}\ln a_{\mathrm{H^+}}=E^{\ominus}-\frac{2RT\ln10}{F}\log a_{\mathrm{H^+}}\\&=E^{\ominus}+\frac{2RT\ln10}{F}\ \mathrm{pH}\end{aligned}$$

Solving for pH gives

$$\begin{aligned}\mathrm{pH}&=\frac{F}{2RT\ln10}\times(E-E^{\ominus})=\frac{0.312\ \mathrm{V}-0.22\ \mathrm{V}}{0.1183\ \mathrm{V}}\\&=\boxed{0.78}.\end{aligned}$$

E9.41 R: $\mathrm{AgBr(s)+e^- \rightarrow Ag(s)+Br^-(aq)}$

L: $\mathrm{Ag^+(aq)+e^- \rightarrow Ag(s)}$

R–L: $\mathrm{AgBr(s) \rightarrow Ag^+(aq)+Br^-(aq)}$ $\qquad$ = 1

Given that the molar solubility of AgBr(s) is $S = 7.31\times10^{-7}\ \mathrm{mol\ dm^{-3}}$, the equilibrium constant for the net reaction, the solubility reaction, and the standard potential of this cell can be calculated.

$$K=a_{\mathrm{Ag^+}}a_{\mathrm{Cl^-}}=S^2=\left(7.31\times10^{-7}\right)^2=5.34\times10^{-13}$$

$$\begin{aligned}E^{\ominus}_{\mathrm{cell}}&=RT\ln K/\nu F\ [9.16]\\&=(0.025693\ \mathrm{V})\ln\left(5.34\times10^{-13}\right)\\&=\boxed{-0.73\ \mathrm{V}}\end{aligned}$$

E9.42 (a) Overall: $Ag^+(aq) + I^-(aq) \rightarrow AgI(s)$ $\quad \nu = 1 \quad E^\ominus = +0.9509$ V

$$E^\ominus_{cell} = \frac{RT}{\nu F}\ln K \ [9.16] = \frac{RT}{F}\ln\left(\frac{1}{a_{Ag^+}a_{I^-}}\right) = \frac{RT}{F}\ln\left(\frac{1}{S^2}\right)$$

$$= -\frac{2RT}{F}\ln S$$

Solving for the molar solubility S gives

$$S = e^{-FE^\ominus_{cell}/2RT} = e^{-(0.9509)/(0.051386)}$$

$$= \boxed{9.19\times10^{-9}\ \text{mol dm}^{-3}}.$$

(b) $K_S = S^2 = (9.19\times10^{-9})^2 = \boxed{8.45\times10^{-17}}$

Answers to projects

P9.43 Harned cell: $Pt(s)|H_2(g, 1\ bar)|HCl(aq, b)|AgCl(s)|Ag(s)$

The cell reaction is ½ $H_2(g) + AgCl(s) \rightarrow HCl(aq) + Ag(s)$ for which $\nu = 1$.

(a) $E = E^\ominus_{cell} - \frac{RT}{\nu F}\ln Q$ [9.14]

$$= E^\ominus_{cell} - \frac{RT}{F}\ln\left(\frac{a_{H^+}a_{Cl^-}}{a_{H_2}^{1/2}}\right)$$

$$= E^\ominus_{cell} - \frac{RT}{F}\ln\left(a_{H^+}a_{Cl^-}\right) \qquad \text{(i.e., } a_{H_2} = p_{H_2}/p^\ominus = 1\ \text{bar}/p^\ominus = 1\text{)}$$

$$= E^\ominus - \frac{RT}{F}\ln\left\{\left(\gamma_\pm b/b^\ominus\right)\times\left(\gamma_\pm b/b^\ominus\right)\right\} \qquad [9.1\text{a and } 9.3\text{a}];\ b = b_{HCl}$$

$$= E^\ominus - \frac{RT}{F}\ln\left(b/b^\ominus\right)^2 - \frac{RT}{F}\ln\left(\gamma_\pm\right)^2$$

$$= E^\ominus - \frac{2RT}{F}\ln\left(b/b^\ominus\right) - \frac{2RT}{F}\ln\left(\gamma_\pm\right)$$

From the Debye-Hückel limiting law [9.4] we know that

$$\log\gamma_\pm = -A|z_+z_-|I^{1/2} = -A|z_+z_-|\left\{\tfrac{1}{2}\left(z_+^2b_+ + z_-^2b_-\right)/b^\ominus\right\}^{1/2} \qquad [9.5];\ (z_+^2 + z_-^2)b = (1+1)b = 2b$$

$$\frac{\ln\gamma_\pm}{\ln(10)} = -A\left(b/b^\ominus\right)^{1/2}$$

$$\ln\gamma_\pm = -A\ln(10)\left(b/b^\ominus\right)^{1/2}.$$

Substitution into the working equation for E yields

$$E = E^\ominus - \frac{2RT}{F}\ln\left(b/b^\ominus\right) + \frac{2RTA\ln(10)}{F}\left(b/b^\ominus\right)^{1/2}$$

$$\boxed{E + \frac{2RT}{F}\ln\left(b/b^\ominus\right) = E^\ominus + \frac{2RTA\ln(10)}{F}\left(b/b^\ominus\right)^{1/2}}.$$

This equation indicates that a plot of $E + \frac{2RT}{F}\ln\left(b/b^\ominus\right)$ against $\left(b/b^\ominus\right)^{1/2}$ should be linear with an intercept equal to $E^\ominus$.

(b) We draw up a data table for preparation of the plot discussed above using the definitions

$$Y = E + \frac{2RT}{F}\ln\left(b/b^\ominus\right) = E + (0.051386\ \text{V})\ln\left(b/b^\ominus\right) \qquad \text{and} \qquad X = \left(b/b^\ominus\right)^{1/2}.$$

b / 10^{-3} mol dm^{-3}	3.215	5.619	9.138	25.63
E / V	0.52053	0.49257	0.46860	0.41824
X / 10^{-2}	5.670	7.496	9.559	16.01
Y / V	0.22558	0.22631	0.22733	0.22996

Y is plotted against X and from the linear regression fit, which is a box insert in the figure, we see an intercept of 0.2232 and conclude that $E^\ominus = \boxed{0.2232\text{ V}}$.

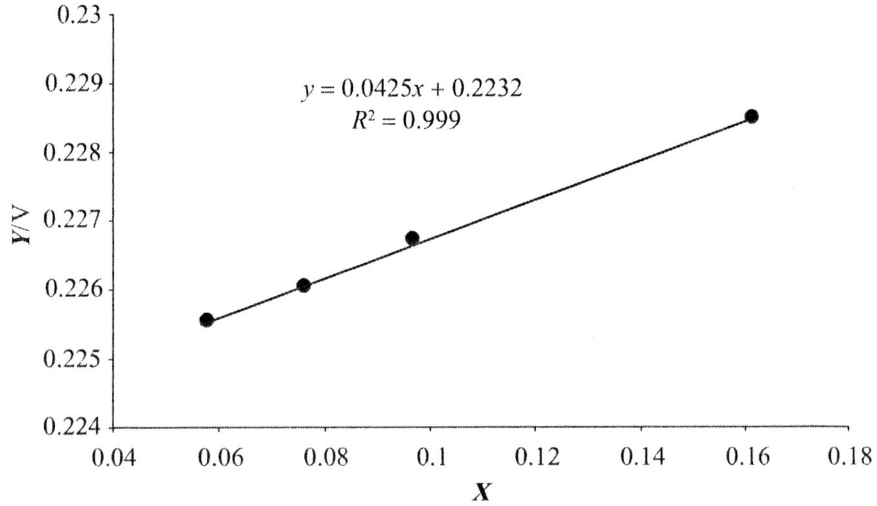

Figure 9.3

P9.44 The half-reactions involved are:

R: $\text{cyt}_{\text{ox}} + e^- \rightarrow \text{cyt}_{\text{red}}$ $\quad E^\ominus_{\text{cyt}}$

L: $\text{D}_{\text{ox}} + e^- \rightarrow \text{D}_{\text{red}}$ $\quad E^\ominus_{\text{D}}$

The overall cell reaction is:

R−L: $\text{cyt}_{\text{ox}} + \text{D}_{\text{red}} \rightleftharpoons \text{cyt}_{\text{red}} + \text{D}_{\text{ox}} \qquad E^\ominus = E^\ominus_{\text{cyt}} - E^\ominus_{\text{D}} \quad \nu = 1$

(a) The Nernst equation for the cell reaction is

$$E = E^\ominus - \frac{RT}{F}\ln\frac{[\text{cyt}_{\text{red}}][\text{D}_{\text{ox}}]}{[\text{cyt}_{\text{ox}}][\text{D}_{\text{red}}]}.$$

$E = 0$ at equilibrium. Therefore,

$$\ln\frac{[\text{cyt}_{\text{red}}]_{\text{eq}}[\text{D}_{\text{ox}}]_{\text{eq}}}{[\text{cyt}_{\text{ox}}]_{\text{eq}}[\text{D}_{\text{red}}]_{\text{eq}}} = \frac{F}{RT}(E^\ominus_{\text{cyt}} - E^\ominus_{\text{D}})$$

$$\ln\left(\frac{[\text{D}_{\text{ox}}]_{\text{eq}}}{[\text{D}_{\text{red}}]_{\text{eq}}}\right) = \ln\left(\frac{[\text{cyt}_{\text{ox}}]_{\text{eq}}}{[\text{cyt}_{\text{red}}]_{\text{eq}}}\right) + \frac{F}{RT}(E^\ominus_{\text{cyt}} - E^\ominus_{\text{D}}).$$

Therefore a plot of $\ln([\text{D}_{\text{ox}}]_{\text{eq}}/[\text{D}_{\text{red}}]_{\text{eq}})$ against $\ln([\text{cyt}_{\text{ox}}]_{\text{eq}}/[\text{cyt}_{\text{red}}]_{\text{eq}})$ is linear with a slope of one and an intercept of $F(E^\ominus_{\text{cyt}} - E^\ominus_{\text{D}})/RT$.

(b) Draw up the following table.

$\ln([\text{D}_{\text{ox}}]_{\text{eq}}/[\text{D}_{\text{red}}]_{\text{eq}})$	−5.882	−4.776	−3.661	−3.002	−2.593	−1.436	−0.6274
$\ln([\text{cyt}_{\text{ox}}]_{\text{eq}}/[\text{cyt}_{\text{red}}]_{\text{eq}})$	−4.547	−3.772	−2.415	−1.625	−1.094	−0.2120	−0.3293

The plot of $\ln([D_{ox}]_{eq}/[D_{red}]_{eq})$ against $\ln([cyt_{ox}]_{eq}/[cyt_{red}]_{eq})$ is shown in Figure 9.4. The figure insert reports the linear least-squares regression fit from which we see that the intercept is -1.2124. In part (a) it is shown that $F\left(E^{\ominus}_{cyt}-E^{\ominus}_{D}\right)/RT$ equals the intercept so, given that $E^{\ominus}_{D}=+0.237\text{ V}$, the intercept value allows the calculation of $E^{\ominus}_{cyt}$.

$$\begin{aligned} E^{\ominus}_{cyt} &= (RT/F)\times(\text{intercept})+E^{\ominus}_{D} \\ &= (0.025693\text{ V})\times(-1.2124)+0.237\text{ V} \\ &= \boxed{+0.206\text{ V}} \end{aligned}$$

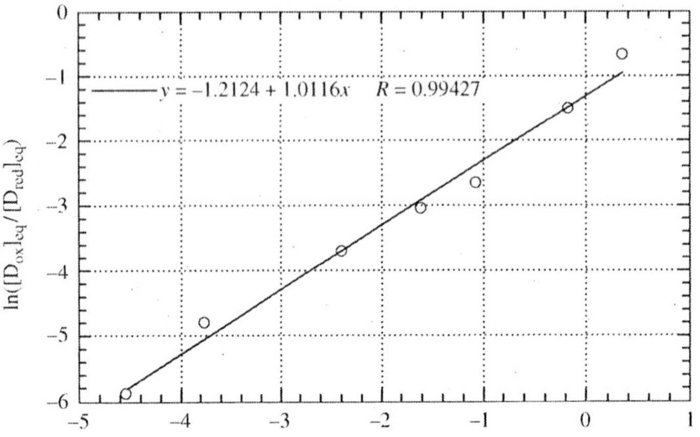

Figure 9.4

P9.45 (a) With the equilibrium approximation $\Delta G_m = 0$, the electrostatic potential difference for $[K^+]_{in}/[K^+]_{out} = 20$ is estimated by solving the following equation for $\Delta\phi$.

$$\Delta G_m = RT\ln\frac{[K^+]_{in}}{[K^+]_{out}} + zF\Delta\phi = 0 \qquad \text{or} \qquad \Delta\phi = -\frac{RT}{zF}\ln\frac{[K^+]_{in}}{[K^+]_{out}}$$

$$\Delta\phi = -\frac{(8.3145\text{ J K}^{-1}\text{ mol}^{-1})(298\text{ K})}{1\times(9.649\times10^4\text{ C mol}^{-1})}\ln 20 = \boxed{-76.9\text{ mV}} \qquad (\textit{Note:}\ 1\text{ J C}^{-1}=1\text{ V})$$

With the equilibrium approximation $\Delta G_m = 0$, the electrostatic potential difference for $[Na^+]_{in}/[Na^+]_{out} = 0.10$ is estimated by solving the following equation for $\Delta\phi$.

$$\Delta G_m = RT\ln\frac{[Na^+]_{in}}{[Na^+]_{out}} + zF\Delta\phi = 0 \qquad \text{or} \qquad \Delta\phi = -\frac{RT}{zF}\ln\frac{[Na^+]_{in}}{[Na^+]_{out}}$$

$$\Delta\phi = -\frac{(8.3145\text{ J K}^{-1}\text{ mol}^{-1})(298\text{ K})}{1\times(9.649\times10^4\text{ C mol}^{-1})}\ln 0.10 = \boxed{+59.1\text{ mV}} \qquad (\textit{Note:}\ 1\text{ J C}^{-1}=1\text{ V})$$

Only the $\Delta\phi$ value for K^+ is comparable to the observed resting potential of −62 mV. The $\Delta\phi$ value for Na^+ has the opposite sign. These computations are not expected to match the resting potential because the cell is not at equilibrium.

(b) $$\Delta\phi = \frac{RT}{F}\ln\left(\frac{\sum_i P_i[M_i^+]_{out}+\sum_j P_j[X_j^-]_{in}}{\sum_i P_i[M_i^+]_{in}+\sum_j P_j[X_j^-]_{out}}\right) \qquad \text{Goldman equation}$$

$$\Delta\phi = \frac{(8.3145\text{ J K}^{-1}\text{ mol}^{-1})(298\text{ K})}{9.649\times10^4\text{ C mol}^{-1}}\ln\left\{\frac{(0.04\times440)+(1.0\times20)+(0.45\times50)}{(0.04\times50)+(1.0\times400)+(0.45\times560)}\right\}$$

$$= \boxed{-61.\bar{3}\text{ mV}} \qquad (\textit{Note:}\ 1\text{ J C}^{-1}=1\text{ V})$$

The Goldman equation is in good agreement with the observed resting potential of −62 mV.

10 Chemical kinetics: the rates of reactions

Answers to discussion questions

D10.1 A wide range of **proton transfer rate constants**, and timescales, are observed in aqueous solution. This is especially illustrated by the reverse transfers in the following reactions at 25°C.

Reaction	$k_{forward}$ / mol^{-1} dm^3 s^{-1}	$k_{reverse}$ / mol^{-1} dm^3 s^{-1}
$H_3O^+ + OH^- \rightleftharpoons H_2O + H_2O$	1.4×10^{11}	2.5×10^{-5}
$H_3O^+ + SO_4^{2-} \rightleftharpoons H_2O + HSO_4^-$	1×10^{11}	7×10^{7}
$H_3O^+ + NH_3 \rightleftharpoons H_2O + NH_4^+$	4.3×10^{10}	8.4×10^{5}
$H_2O + C_2H_5O^- \rightleftharpoons OH^- + C_2H_5OH$	—	3×10^{6}

We form an idea of the timescale of a second-order reaction with the time constant τ. In analogy to eqn 10.17 we take it to be given by $\tau = 1/k_r[A]$ where [A] is the molar concentration of reactants. (This eqn is not strictly correct but it does provide an estimate of the reaction timescale.) For the first of the above reactions at pH = pOH = 7:

$$\tau_{H_3O^+/OH^-} = 1/\{(1.4\times10^{11}\ \text{mol}^{-1}\ \text{dm}^3\ \text{s}^{-1})\times(1.0\times10^{-7}\ \text{mol dm}^{-3})\} = 7.1\times10^{-5}\ \text{s} = 71\ \mu\text{s}.$$

For the reverse of this proton transfer:

$$\tau_{H_2O/H_2O} = 1/\{(2.5\times10^{-5}\ \text{mol}^{-1}\ \text{dm}^3\ \text{s}^{-1})\times(55.5\ \text{mol dm}^{-3})\} = 7.2\times10^{2}\ \text{s}.$$

Proton transfer between the hydronium and hydroxide ions is about 10 million times more rapid than transfer between water molecules! It is very interesting to compare these rate constants to the rate constant for a **diffusion-controlled limit** (also called the **encounter-controlled limit**) in which reactants immediately react upon diffusing into proximity. This topic is discussed in text Section 11.10 where it is shown that in the diffusion-controlled limit the rate constant is inversely proportional to the media viscosity and for aqueous solution at 25°C it equals 7.4×10^9 mol^{-1} dm^3 s^{-1}. The fact that the proton transfer reactions involving the hydronium ion have larger rate constants than the diffusion-limit is evidence that the proton-carrying species does not necessarily need to diffuse into proximity with a Brønsted base; the proton may rapidly migrate in a cooperative process along a network of hydrogen-bonded water molecules (see the discussion of the **Grotthus mechanism** in text Section 9.2).

A one electron transfer event between complex ions in aqueous solution often takes place by **outer-sphere electron transfer** in which no change occurs in the coordination sphere of the atom undergoing a change of oxidation number (i.e., the redox center). Transfer rates can be fast for this mechanism and logarithms of rates are often observed to be proportional to differences in standard potentials. Self-exchange rate constants of the type $M(L)_6^{3+} + M(L)_6^{2+} \rightarrow M(L)_6^{2+} + M(L)_6^{3+}$ exhibit outer-sphere electron transfer rates of 4 mol^{-1} dm^3 s^{-1} for the $Fe(H_2O)_6^{3+}/Fe(H_2O)_6^{2+}$ couple, 50 mol^{-1} dm^3 s^{-1} for the $Ru(H_2O)_6^{3+}/Ru(H_2O)_6^{2+}$ couple, 2.8×10^4 mol^{-1} dm^3 s^{-1} for the $Ru(NH_3)_6^{3+}/Ru(NH_3)_6^{2+}$ couple, and 8×10^{-6} mol^{-1} dm^3 s^{-1} for the $Co(NH_3)_6^{3+}/Co(NH_3)_6^{2+}$ couple. Only the $Co(NH_3)_6^{3+}/Co(NH_3)_6^{2+}$ couple occurs on a long timescale and this happens because the transfer requires an electron spin flip that is forbidden by the rules of quantum chemistry (see text Section 13.6 and 13.7). Taking the smallest and largest of these rate constants and assuming concentrations of the order of one mole per dm^3, we find a timescale range for electron transfer of about 4×10^{-5} s to 1×10^5 s.

A one electron transfer event can also occur by **inner-sphere electron transfer** in which a change does occur in the coordination sphere of the complex ion. An example is the reduction of pentaamminechlorocobalt(2+) by hexaaquochromiun(2+) for which the second-order rate constant is $k_r = 6 \times 10^5\ \text{mol}^{-1}\ \text{dm}^3\ \text{s}^{-1}$:

$$[\text{Co(NH}_3)_5\text{Cl}]^{2+} + \text{Cr(H}_2\text{O)}_6^{\ 2+} \rightarrow \text{Co(H}_2\text{O)}_6^{\ 2+} + [\text{Cr(H}_2\text{O)}_5\text{Cl}]^{2+} + 5\ \text{NH}_3.$$

This electron transfer occurs following the formation of a chloro bridge between Co(III) and Cr(II): $[(\text{NH}_3)_5\text{Co—Cl—Cr(H}_2\text{O)}_5]^{4+}$.

The **harpoon mechanism** of one electron transfer is between neutral molecular or atomic entities in which long-range electron transfer is followed by a considerable reduction of the distance between donor and acceptor sites as a result of the electrostatic attraction in the ion pair created. In the gas phase transfer of an electron from a potassium atom to a bromine molecule, the gaseous K^+ and Br_2^- ions are observed to form when their centers are no more than about 580 pm apart. The strong attractive Coulomb force brings the two ions together after which KBr forms with the ejection of a Br atom. We assume that (a) the transfer must occur before the potassium atom (235 pm radius) and bromine molecule (228 pm bond length and radius of about 228 pm) are spaced by 463 pm and (b) the relative approach speed is 437 m s^{-1} (the K atom average speed at 298 K), then the timescale for the transfer must be less than the time needed to close the reactants from 580 pm to 463 pm: (580 – 463)pm / (437 m s^{-1}) = 0.3 ps. Relative speed in crossed atom/molecular-beam measurements may be 1/10th this value so an estimated timescale of about 1 ps is reasonable.

There are two important timescales for **collisions in liquids**. The first involves the concept of collisions between a molecule or ion and members of its solvent cage, a situation for which an exact definition of collision may not be possible as the central molecule and its solvent cage are in continuous contact. However, it is reasonable to assume that something like one collision occurs in each vibrational cycle of the central molecule so the timescale for a collision has an order of magnitude similar to the inverse of the molecular vibrational frequency. With typical vibrational frequencies between 1.2×10^{13} Hz and 9.0×10^{13} Hz, which corresponds to infrared vibrational absorbances in the range 400 cm^{-1} and 3000 cm^{-1}, the range of timescales between collisions with the solvent cage is 0.01 ps to 0.1 ps.

The second timescale for collisions in liquids concerns the question of the average time between collisions of solute molecule A with solute molecule B. Equation 11.17 provides an answer by relating the rate constant for a diffusion controlled encounter to solvent viscosity according to $k_{r,d} = 8RT/3\eta = 7.4 \times 10^9\ \text{mol}^{-1}\ \text{dm}^3\ \text{s}^{-1}$ at 25°C. Assuming solute concentrations of the order of one mole per dm^3 gives a timescale of about 100 ps as the lower limit estimate for the time between collisions by a solute molecule with other solute molecules; lower concentrations have longer average times between encounter collisions.

There are many literature sources for this discussion question. You may enjoy a survey of the following articles and their references in your literature search.

S.V. Rosokha and J.K. Kochi, *Acc. Chem. Res., 41*(5), 641–653, 2008. Fresh Look at Electron–Transfer Mechanisms via the Donor/Acceptor Bindings in the Critical Encounter Complex.

R. Zhang and M. Newcomb, *Acc. Chem. Res., 41*(3), 468–477, 2008. Laser Flash Photolysis Generation of High–Valent Transition Metal–Oxo Species: Insights from Kinetic Studies in Real Time.

D.H. Evans, *Chem. Rev., 108*(7), 2113–2144, 2008. One–Electron and Two–Electron Transfers in Electrochemistry and Homogeneous Solution Reactions.

D10.2 The information provided by the determination of a reaction rate under different conditions of pressure, temperature, and the presence of a catalyst makes it possible to quickly predict reactant and product concentrations throughout the reaction progression toward equilibrium. With this knowledge we may be able to optimize the rate by the appropriate choice of conditions. Also, the study of reaction rates is a prerequisite to the understanding of the **mechanism of a reaction**, its analysis into a sequence of elementary steps.

D10.3 The determination of a rate law is simplified by the **isolation method** in which the concentrations of all the reactants except one are in large excess. If B is in large excess, for example, then to a good approximation its concentration is constant throughout the reaction. Although the true rate law might be rate = $k_r[A][B]$, we can approximate [B] by $[B]_0$ and write

$$\text{rate} = k_{\text{eff}}[A], \qquad \text{where } k_{\text{eff}} = k[B]_0$$

which has the form of a first-order rate law. Because the true rate law has been forced into first-order form by assuming that the concentration of B is constant, it is called a **pseudofirst-order** rate law. [A] has been isolated. The dependence of the rate on the concentration of each of the reactants may be found by isolating them in turn (by having all the other substances present in large excess), and so constructing a picture of the overall rate law.

In the **method of initial rates**, which is often used in conjunction with the isolation method, the rate is measured at the beginning of the reaction for several different initial concentrations of reactants. We shall suppose that the rate law for a reaction with A isolated is rate = $k_r[A]^a$; then its initial rate, rate_0, is given by the initial values of the concentration of A, and we write $\text{rate}_0 = k_r[A]_0^a$. Taking logarithms gives:

$$\log(\text{rate}_0) = \log k_r + a \log [A]_0 \quad [10.11]$$

For a series of initial concentrations, a plot of the logarithms of the initial rates against the logarithms of the initial concentrations of A should be a straight lime with slope a.

The method of initial rates might not reveal the full rate law, for the products may participate in the reaction and affect the rate. For example, products participate in the synthesis of HBr, where the full rate law depends on the concentration of HBr. To avoid this difficulty, the rate law should be fitted to the data throughout the reaction. The fitting may be done, in simple cases at least, by using a proposed rate law to predict the concentration of any component at any time, and comparing it with the data.

Because rate laws are differential equations, we must integrate them if we want to find the concentrations as a function of time. Even the most complex rate laws may be integrated numerically. However, in a number of simple cases analytical solutions are easily obtained, and prove to be very useful. A first-order rate law shows linearity when the logarithm of concentration is plotted against time while a second-order rate law exhibits linearity in a plot of inverse concentration against time. The slope of the linear plot equals the second-order rate constant in the latter case and the negative of the first-order rate constant in the former case.

D10.4 Consider a rate law of the form: rate = $k_r [A]^m [B]^n$ where the concentration orders m and n equal either zero or a positive integer. If the sum $m + n$ equals zero, the rate is zeroth-order and the rate is independent of species concentration. If the sum equals either 1 or 2, the rate is first-order or second-order, respectively. In the case for which $m = 1$ and $n \neq 0$, the reaction order will appear to be 1 if the concentration of B is a large excess and [B] remains basically unchanged during the course of reaction. This is the pseudofirst-order reaction rate for which

$$\text{rate} = k_r [A][B]^n = (k_r [B]^n)[A] = k_{\text{eff}} [A]$$

where $k_{\text{eff}} = k [B]^n$ = pseudofirst-order rate constant.

The apparent order of a reaction changes whenever a concentration varies to make one term in the rate law smaller, or larger, relative to another term. For example, a typical rate law for the action of an enzyme E on a substrate S is (see Chapter 11)

$$\text{rate} = \frac{k_r[E][S]}{[S]+K_M}$$

where K_M is a constant. This rate law does not have a definite order overall. If the substrate concentration is so low that $[S] << K_M$, the rate becomes

$$\text{rate} = \frac{k_r}{K_M}[E][S]$$

which is first-order in S, first-order in E, and second-order overall. If the substrate concentration is so large that [S] >> K_M, the rate becomes

$$\text{rate} = k_r[\text{E}]$$

which is zero-order in S, first-order in E, and first-order overall.

D10.5 The **Arrhenius equation,** $k_r = Ae^{-E_a/RT}$ [10.19] or $\ln k_r = \ln A - \frac{E_a}{RT}$ [10.18], provides for the variation of the reaction rate constant with temperature, $k_r(T)$. The constants A and E_a, which are determined through experimental effort, are called the **Arrhenius parameters**. More specifically, the parameter A is the **pre-exponential factor** and E_a is the **activation energy**. The pre-exponential factor is proportional to the rate at which reactant molecules collide while the activation energy is the minimum kinetic energy required for a collision to result in a reaction. Determination of the Arrhenius parameters involves preparation of a plot of experimentally determined values of ln $k_r(T)$ against $1/T$. The Arrhenius equation predicts that the plot will be linear with a slope equal to $-E_a/R$ and an extrapolated intercept equal to ln A. Thus, if the plot is linear, a linear least-square regression fit of the data plot yields values for the slope and intercept from which the Arrhenius parameters are calculated with the relations $E_a = -R \times \textit{slope}$ and $A = e^{\textit{intercept}}$.

The Arrhenius equation describes the variation of the reaction rate constant with temperature whenever the activation energy is temperature independent. There are instances for which the plot of $k_r(T)$ against $1/T$ is not Arrhenius-like, in the sense that a straight line is not obtained. However, it is still possible to define a general activation energy as

$$E_a = RT^2\left(\frac{d\ln k_r}{dT}\right)$$

This definition accounts for the temperature dependent activation energy. It reduces to the Arrhenius equation (as the slope of a straight line) for a temperature-independent activation energy. However, this definition is more general, because it allows E_a to be obtained from the slope (at the temperature of interest) of a plot of ln k_r against $1/T$ even if the Arrhenius plot is not a straight line. Non-Arrhenius behaviour is sometimes a sign that quantum mechanical tunnelling is playing a significant role in the reaction.

D10.6 The **quasi-steady state approximation** for the concentration of chemical species X recognizes a situation for which the rate of reaction of X is negligibly small so that [X] remains both constant and small throughout the reaction (except right at the beginning and right at the end). The approximation is often useful to the description of an intermediate concentration in a reaction mechanism that consists of multiple elementary steps. Analysis of the reaction mechanism is simplified by making the approximation and insight as to whether the mechanism is compatible to experimental observations, which are often unable to directly measure [X], becomes feasible. For example, suppose that the rate of X depends upon species A, B, and X in the form: d[X]/dt = f_1(A, B, X) + f_2(A, B, X) where f_1 and f_2 are different functions. The quasi-steady state approximation for X assumes that d[X]/dt ~ 0 and the mechanism analysis is simplified with the relation f_1(A, B, X) = $-f_2$(A, B, X). In addition to this and the discussions of Chapter 11 here's a relevent reference to this topic:

B. Li, Y. Shen, and B. Li, *J. Phys. Chem. A, 112*(11), 2311–2321, 2008. Quasi–Steady State Laws in Enzyme Kinetics.

D10.7 The Eyring equation, $k_{TS} = \kappa \times (kT/h) \times K^{\ddagger}$ [eqn 10.23], results from activated complex theory which is an attempt to account for the rate constants of bimolecular reactions of the form $A + B \rightleftharpoons C^{\ddagger} \rightarrow P$ in terms of the formation of an activated complex $C^{\ddagger}$. In the formulation of the theory, it is assumed that the activated complex and the reactants are in equilibrium, and the concentration of activated complex is calculated in terms of an equilibrium constant, which in turn is calculated from the partition functions (Ch. 22) of the reactants and a postulated form of the activated complex. It is further supposed that one normal vibrational mode of the activated complex, the one corresponding to displacement along the reaction coordinate, has a very low force constant and displacement along this normal mode leads to products provided that the complex enters a certain configuration of its atoms, which is known as the transition state.

Collision theory provides the computational equations for collision rates between gaseous hard spheres but activation energies, collision cross-sections for non-hard spheres, and steric factors remain as empirical parameters in the computation of a rate constant. Consequently, the empirical parameters are often deduced from measured values of a rate constant under a variety of conditions. The Eyring theory of reaction rates provides a computational framework based upon the fundamental principles of quantum chemistry and statistical thermodynamics so the Eyring theory, when computationally feasible, provides detailed insight into formation of the activated complex and movement of atoms to form products in either the solvated or gaseous state. Calculations of the equilibrium constant $K^{\ddagger}$ and transmission coefficient are very difficult but, when fundamental insight into a reaction mechanism is the goal, the Eyring theory is superior to collision theory. If the goal is acquisition of activation energies, collision cross-sections for non-hard spheres, and steric factor knowledge and a detailed computation of atomic movement in the activated complex is not feasible or not desired, collision theory is superior.

Solutions to exercises

E10.1

$$A = \log\left(\frac{I_0}{I}\right) = -\log\left(\frac{I}{I_0}\right) = -\log(0.398) = 0.400 \qquad [10.3]$$

$$[\text{cyt P450}] = \frac{A}{\varepsilon l} \qquad [10.4]$$

$$= \frac{0.400}{(291\ \text{dm}^3\ \text{mol}^{-1}\ \text{cm}^{-1})(0.65\ \text{cm})} = 2.1\times10^{-3}\ \text{mol dm}^{-3} = \boxed{2.1\ \text{mmol dm}^{-3}}$$

E10.2 Because the rate of formation of C is known, the reaction stoichiometry can be used to determine the rates of consumption and formation of the other participants in the reaction. The reaction is: $2\text{ A} + \text{B} \rightarrow 4\text{C} + 3\text{ D}$ and the rate of formation of C equals 3.2 mol dm^{-3} s^{-1}.

$$\text{rate of consumption of A} = \frac{2}{4}\times\text{rate of formation of C} = \frac{1}{2}\times(3.2\ \text{mol dm}^{-3}\ \text{s}^{-1})$$
$$= \boxed{1.6\ \text{mol dm}^{-3}\ \text{s}^{-1}}$$

$$\text{rate of consumption of B} = \frac{1}{4}\times\text{rate of formation of C}$$
$$= \boxed{0.80\ \text{mol dm}^{-3}\ \text{s}^{-1}}$$

$$\text{rate of formation of D} = \frac{3}{4}\times\text{rate of formation of C}$$
$$= \boxed{2.4\ \text{mol dm}^{-3}\ \text{s}^{-1}}$$

E10.3 The reaction rate v can be expressed in terms of any reactant or product according to $v = \dfrac{1}{\nu_J}\dfrac{\text{d}[\text{J}]}{\text{d}t}$ [10.5c]. Thus, for the reaction $2\text{ A} + \text{B} \rightarrow 4\text{C} + 3\text{ D}$

$$v = -\frac{1}{2}\frac{\text{d}[\text{A}]}{\text{d}t} = -\frac{\text{d}[\text{B}]}{\text{d}t} = \frac{1}{4}\frac{\text{d}[\text{C}]}{\text{d}t} = \frac{1}{3}\frac{\text{d}[\text{D}]}{\text{d}t}.$$

When the rate of formation of C equals 3.2 mol dm^{-3} s^{-1},

$$v = \frac{1}{4}\times(3.2\ \text{mol dm}^{-3}\ \text{s}^{-1}) = \boxed{0.80\ \text{mol dm}^{-3}\ \text{s}^{-1}}.$$

E10.4 $v = k_r[\mathrm{A}][\mathrm{B}][\mathrm{C}]$

The reaction rate has units of concentration per unit time (mol dm^{-3} s^{-1}), hence

$$\text{mol dm}^{-3}\ \text{s}^{-1} = [\text{units of } k] \times (\text{mol dm}^{-3})^3.$$

$$\text{Therefore, [units of } k] = \frac{\text{mol dm}^{-3}\ \text{s}^{-1}}{(\text{mol dm}^{-3})^3} = \boxed{\text{mol}^{-2}\ \text{dm}^6\ \text{s}^{-1}}.$$

E10.5 $v = k_{r1}[\mathrm{A}][\mathrm{B}]/(1 + k_{r2}[\mathrm{B}])$

Starting with the denominator on the right side of the rate expression, we see that, since it contains a unitless constant, it must be the sum of unitless terms. Thus, $k_{r2}[\mathrm{B}]$ must be unitless and $\boxed{\text{unit of } k_{r2} = \text{mol}^{-1}\ \text{dm}^3}$.

We now observe that the reaction rate and the numerator on the right side of the rate expression must have the same unit. The rate has units of concentration per unit time (mol dm^{-3} s^{-1}), hence

$$\text{mol dm}^{-3}\ \text{s}^{-1} = [\text{units of } k_{r1}] \times (\text{mol dm}^{-3})^2.$$

$$\text{Therefore, [units of } k_{r1}] = \frac{\text{mol dm}^{-3}\ \text{s}^{-1}}{(\text{mol dm}^{-3})^2} = \boxed{\text{mol}^{-1}\ \text{dm}^3\ \text{s}^{-1}}.$$

E10.6 $v = k_{r1} p_{\mathrm{A}}^{3/2} p_{\mathrm{B}} / (p_{\mathrm{A}} + k_{r2} p_{\mathrm{B}})$

Starting with the denominator on the right side of the rate expression, we see that, since the first term has the unit kPa, it must be the sum of terms of unit kPa. Thus, $k_{r2} p_{\mathrm{B}}$ has the unit kPa and $\boxed{k_{r2} \text{ is unitless}}$. Now, compare the units on the left side of the rate expression to those on the right.

$$\text{kPa s}^{-1} = [\text{units of } k_{r1}] \times (\text{kPa})^{5/2} / \text{kPa}.$$

$$\text{Therefore, [units of } k_{r1}] = \boxed{\text{kPa}^{-1/2}\ \text{s}^{-1}}.$$

E10.7 Multiplication by Avogadro's constant converts the 'per molecule' unit to 'per mol' and there is 0.1 dm per cm. Thus,

$$v = (6.2\times10^{-14}\ \text{cm}^3\ \text{molecule}^{-1}\ \text{s}^{-1}) \times (6.022\times10^{23}\ \text{molecules mol}^{-1}) \times (0.1\ \text{dm cm}^{-1})^3$$

$$= \boxed{3.7\times10^7\ \text{dm}^3\ \text{mol}^{-1}\ \text{s}^{-1}}.$$

E10.8 $v = k_{r1}[\mathrm{A}][\mathrm{B}]^{3/2} / (k_{r2}[\mathrm{A}] + k_{r3}[\mathrm{B}]^{1/2})$

This rate law does not have a definite order overall. However, there are two special cases for which the rate expression simplifies to a definite order.

Case 1. If $\boxed{k_{r2}[\mathrm{A}] >> k_{r3}[\mathrm{B}]^{1/2}}$, the rate law simplifies to $v = \frac{k_{r1}}{k_{r2}}[\mathrm{B}]^{3/2}$. This is zero order in A while having an order of $^3/_2$ in B and an overall order of $^3/_2$.

Case 2. If $\boxed{k_{r2}[\mathrm{A}] << k_{r3}[\mathrm{B}]^{1/2}}$, the rate law simplifies to $v = \frac{k_{r1}}{k_{r3}}[\mathrm{A}][\mathrm{B}]$. This is first order in A, first order in B, and second order overall.

E10.9 We suppose that the rate law for the reaction of isolated glucose (glu) with the enzyme hexokinase at 1.34 mmol dm^{-3} is $v = k_{r,\mathrm{eff}}[\mathrm{glu}]^a$. Evaluating this rate law at initial conditions and taking the logarithms gives eqn 10.11.

$$\log v_0 = \log k_{r,\mathrm{eff}} + a \log[\mathrm{glu}]_0 \quad [10.11]$$

Thus, if the supposition is correct, a plot of $\log v_0$ against $\log[\mathrm{glu}]_0$ with be linear with a slope equal to the reaction order a and an intercept equal to $\log k_{r,\mathrm{eff}}$. We draw the following table with the requisite logarithm transformations, prepare the plot (see Figure 10.1), and check whether the plot is linear.

$[glu]_0$/mmol dm^{-3}	1.00	1.54	3.12	4.02
v_0/mol dm^{-3} s^{-1}	5.0	7.6	15.5	20.0
log ($[glu]_0$/mmol dm^{-3})	0.00	0.188	0.494	0.604
log(v_0/mol dm^{-3} s^{-1})	0.699	0.881	1.19	1.30

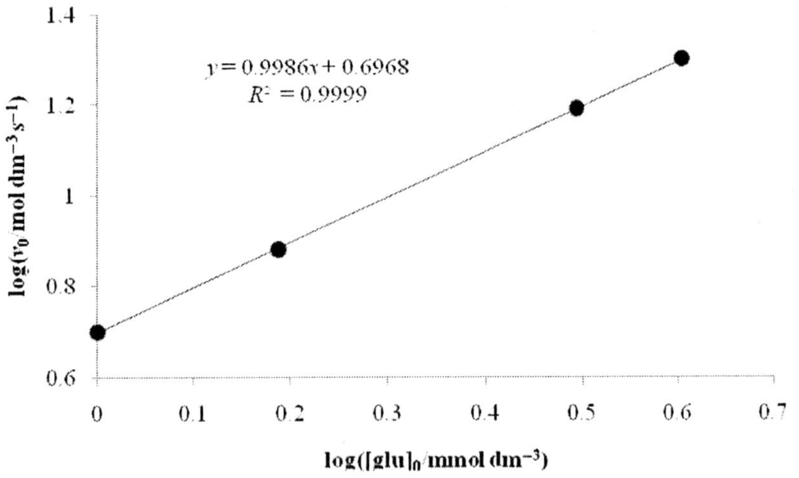

Figure 10.1

Inspection of the plot reveals that it is linear so we conclude that the supposed form of the rate law is valid and we perform the computation of the linear least squares regression fit of the data, which is shown in Figure 10.1.

(a) The plot slope is extremely close to 1.00 so we conclude that the reaction order w/r/t glucose is $\boxed{1}$.

(b) The regression intercept tells us that $\log\left(k_{r,eff} / \text{mol dm}^{-3}\ \text{s}^{-1}\right) = 0.6968$. Thus,

$$k_{r,eff} = 10^{0.6968}\ \text{mol dm}^{-3}\ \text{s}^{-1} = \boxed{5.0\ \text{mol dm}^{-3}\ \text{s}^{-1}}.$$

E10.10 In this exercise we follow the strategy of text Example 10.1. Begin with the supposition that the rate law has the form $v_0 = k_r[\text{complex}]_0^a[\text{Y}]_0^b$ or, taking logarithms,

$$\log(v_0) = \log(k_r) + a\log([\text{complex}]_0) + b\log([\text{Y}]_0) = \log(k_{r,eff}) + a\log([\text{complex}]_0)$$

where $\log(k_{r,eff}) = \log(k_r) + b\log([\text{Y}]_0)$.

Thus, if the supposition is correct, a plot of $\log(v_0)$ against $\log([\text{complex}]_0)$ at fixed $[\text{Y}]_0$ will be linear with a slope equal to a and an intercept equal to $\log(k_{r,eff})$. A subsequent plot of $\log(k_{r,eff})$ against $\log([\text{Y}]_0)$ will have a slope equal to b and an intercept equal to $\log(k_r)$. Plots of $\log(v_0)$ against $\log([\text{complex}]_0)$ for $[\text{Y}]_0 = 2.7$ mmol dm^{-3} and 6.1 mmol dm^{-3} are shown in Figure 10.2.

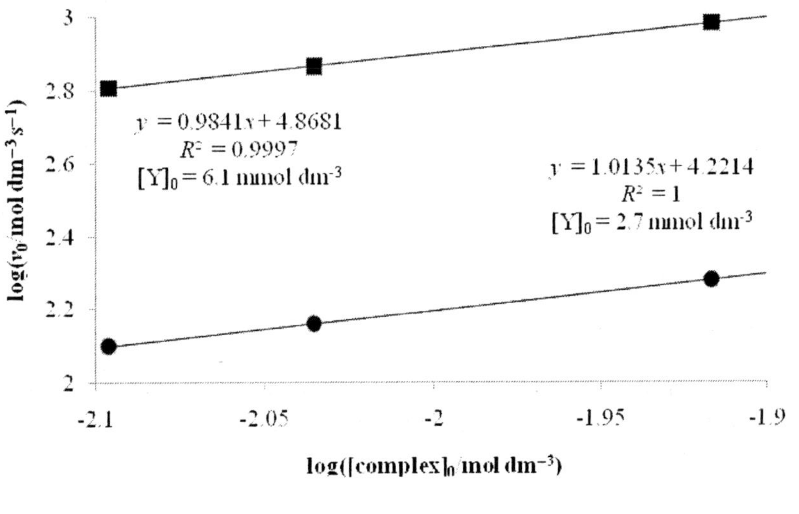

Figure 10.2

(a) The plots of Figure 10.2, being linear, confirm the postulated form of the rate law and the box inserts display the linear least squares regression fits of the data. Since both plots have slopes very close to 1 with small deviations due to experimental error, we conclude that the reaction order w/r/t the complex is $\boxed{1}$. Rather than preparing a two-point plot of $\log(k_{r,eff})$ against $\log([Y]_0)$ for the determination of b, we will simply use the mathematic definition of a slope and the intercepts of Figure 10.2 in a computation.

$$b = slope = \frac{d(\log k_{r,eff})}{d(\log[Y]_0)} = \frac{\Delta(\log k_{r,eff})}{\Delta(\log[Y]_0)} = \frac{\log k_{r,eff\,1} - \log k_{r,eff\,2}}{\log[Y]_{0,1} - \log[Y]_{0,2}} = \frac{\log k_{r,eff\,1} - \log k_{r,eff\,2}}{\log([Y]_{0,1}/[Y]_{0,2})}$$

$$= \frac{4.2214 - 4.8681}{\log(2.7/6.1)}$$

$$= 1.83$$

The computed value of b is close to 2 with deviation due to experimental error so we conclude that the reaction order w/r/t Y is $\boxed{2}$ and the overall reaction order is $\boxed{3}$.

(b) We have not prepared a two-point plot of $\log(k_{r,eff})$ against $\log([Y]_0)$ so we calculate k_r using the definition that $k_r = k_{r,eff}\,[Y]_0^{-b}$ where $b = 2$.

For the $[Y]_0 = 2.7$ mmol dm^{-3} data,

$$k_r = (10^{4.2214}\ s^{-1}) \times (2.7 \times 10^{-3}\ mol\ dm^{-3})^{-2} = 2.3 \times 10^9\ mol^{-2}\ dm^6\ s^{-1}.$$

For the $[Y]_0 = 6.1$ mmol dm^{-3} data,

$$k_r = (10^{4.8681}\ s^{-1}) \times (6.1 \times 10^{-3}\ mol\ dm^{-3})^{-2} = 2.0 \times 10^9\ mol^{-2}\ dm^6\ s^{-1}.$$

Since these values reasonably agree, we conclude that on the average $k_r = \boxed{2.2 \times 10^9\ mol^{-2}\ dm^6\ s^{-1}}$.

E10.11 $N_2O_5(g) \rightarrow 2\,NO_2(g) + \tfrac{1}{2}\,O_2(g)$ $d[N_2O_5]/dt = -k_r[N_2O_5]$ where $k_r = 3.38 \times 10^{-5}\ s^{-1}$ at 25°C

The integrated form of this rate expression is: $\ln([N_2O_5]/[N_2O_5]_0) = -k_rt$ (the integration is presented as a note at the end of the exercise). Thus, evaluation of the integrated expression at $t = t_{1/2}$, when $[N_2O_5] = [N_2O_5]_0/2$, gives

$$t_{1/2} = \ln([N_2O_5]_0/2\,[N_2O_5]_0)/(-k_r) = \ln(2)/(k_r)\ [10.16]$$

$$= \ln(2) / (3.38 \times 10^{-5}\ s^{-1})$$

$$= 2.05 \times 10^4\ s = \boxed{5.70\ h}.$$

Since partial pressure is proportional to concentration, the integrated rate law may be written in the form:

$$\ln\left(p_{N_2O_5} / p_{N_2O_5}(0)\right) = -k_rt \qquad \text{or} \qquad p_{N_2O_5} = p_{N_2O_5}(0)\,e^{-k_rt}$$

where $p_{N_2O_5}(0)$ is the initial partial pressure of $N_2O_5(g)$ and $p_{N_2O_5}$ is the partial pressure at time t. The decrease in $p_{N_2O_5}$ that occurs in time t is $p_{N_2O_5}(0) - p_{N_2O_5} = p_{N_2O_5}(0)\times(1-e^{-k_rt})$ and the reaction stoichiometry tells us that $p_{NO_2} = 2(p_{N_2O_5}(0) - p_{N_2O_5})$ and $p_{O_2} = \tfrac{1}{2}(p_{N_2O_5}(0) - p_{N_2O_5})$. Thus, the total pressure at time t is

$$\begin{aligned}p_{\text{total}} &= p_{N_2O_5} + p_{NO_2} + p_{O_2}\\ &= p_{N_2O_5}(0)\,e^{-k_rt} + 2\left(p_{N_2O_5}(0) - p_{N_2O_5}\right) + \tfrac{1}{2}\left(p_{N_2O_5}(0) - p_{N_2O_5}\right)\\ &= p_{N_2O_5}(0)\,e^{-k_rt} + 2p_{N_2O_5}(0)\times\left(1 - e^{-k_rt}\right) + \tfrac{1}{2}p_{N_2O_5}(0)\times\left(1 - e^{-k_rt}\right)\\ &= \tfrac{1}{2}p_{N_2O_5}(0)\left\{5 - 3e^{-k_rt}\right\} \qquad \text{where } p_{N_2O_5}(0) = 78.4\text{ kPa.}\end{aligned}$$

(a) When $t = 5.0$ s,

$$p_{\text{total}} = \tfrac{1}{2}(78.4\text{ kPa})\left\{5 - 3e^{-(3.38\times10^{-5}\text{ s}^{-1})\times(5.0\text{ s})}\right\} = \boxed{78.4\text{ kPa}}.$$

(b) When $t = 5.0$ min $= 300$ s,

$$p_{\text{total}} = \tfrac{1}{2}(78.4\text{ kPa})\left\{5 - 3e^{-(3.38\times10^{-5}\text{ s}^{-1})\times(300\text{ s})}\right\} = \boxed{79.6\text{ kPa}}.$$

Note: The integration of the expression $d[N_2O_5]/dt = -k_r[N_2O_5]$ proceeds by moving constants and time factors to the right of the equality while moving $[N_2O_5]$ factors to the left.

$$\frac{d[N_2O_5]}{[N_2O_5]} = -k_r dt$$

Integrate both sides of the expression from the initial conditions to the conditions at time t. Constants move outside of integrals.

$$\int_{[N_2O_5]_0}^{[N_2O_5]} \frac{d[N_2O_5]}{[N_2O_5]} = -k_r \int_0^t dt$$

Substituting the integrated forms of these standard integral expressions gives

$$\begin{aligned}\ln[N_2O_5]\Big|_{[N_2O_5]_0}^{[N_2O_5]} &= -k_rt\Big|_0^t\\ \ln[N_2O_5] - \ln[N_2O_5]_0 &= -k_r\times(t-0)\\ \ln\left([N_2O_5]/[N_2O_5]_0\right) &= -k_rt.\end{aligned}$$

E10.12 Use $\ln\dfrac{[A]_0}{[A]} = k_rt$ [10.13a] and solve for k_r.

$$k_r = \frac{\ln(220/56.0)}{1.22\times10^4\text{ s}} = \boxed{1.12\times10^{-4}\text{s}^{-1}}$$

E10.13 The reaction is $CH_3COCO_2^-(aq) + H^+(aq) \rightarrow CH_3CHO(aq) + CO_2(g)$.

Data provided: $V_{\text{gas}} = 250\text{ cm}^3$, $V_{\text{solution}} = 100\text{ cm}^3$, $T = 293$ K, $[P]_0 = 3.23\text{ mmol dm}^{-3}$, $p_{CO_2}(t) =$ 100 Pa at $t = 422$ s, and $p_{CO_2}(0) = 0$.

We can assume that the concentration of pyruvate (P) decreases in proportion to the increase in p_{CO_2}. For the CO_2(g) formed, we may write the perfect gas relation.

$$p_{CO_2}V_{\text{gas}} = n_{CO_2}RT$$

$$n_{CO_2} = \frac{V_{\text{gas}}}{RT}p_{CO_2} = \Delta n_{CO_2}$$

$$\Delta n_{CO_2} = \frac{250\times10^{-6}\text{ m}^3\times100\text{ Pa}}{8.3145\text{ J K}^{-1}\text{ mol}^{-1}\times293\text{ K}} = 0.0102\overline{6}\text{ mmol}$$

$$[\mathrm{P}]=[\mathrm{P}]_0-\frac{\Delta n_{\mathrm{CO_2}}}{V_{\text{solution}}}=[\mathrm{P}]_0-\frac{\Delta n_{\mathrm{CO_2}}}{0.100\ \mathrm{dm^3}}=3.23\ \mathrm{mmol\ dm^{-3}}-0.102\overline{6}\ \mathrm{mmol\ dm^{-3}}$$

$$=3.12\overline{7}\ \mathrm{mmol\ dm^{-3}}$$

For a first-order reaction:

$$[\mathrm{P}]=[\mathrm{P}]_0\mathrm{e}^{-k_r t}\quad[10.13\mathrm{c}]$$

$$k_r=\frac{\ln\left(\dfrac{[\mathrm{P}]_0}{[\mathrm{P}]}\right)}{t}$$

$$=\frac{\ln\left(\dfrac{3.23\ \mathrm{mmol\ dm^{-3}}}{3.12\overline{7}\ \mathrm{mmol\ dm^{-3}}}\right)}{422\ \mathrm{s}}$$

$$=\boxed{7.7\times10^{-5}\ \mathrm{s^{-1}}}.$$

Note: In the above solution we have focused on the particular values of $p_{\mathrm{CO_2}}$, V_{gas}, V_{solution}, $[\mathrm{P}]_0$, and T. However, an experimentalist may need a general expression for k_r because these are variables in many experiments. Can you derive the general expression? It is

$$k_r=-\frac{1}{t}\ln\left(1-\frac{p_{\mathrm{CO_2}}V_{\text{gas}}}{RT[\mathrm{P}]_0V_{\text{solution}}}\right).$$

E10.14

$$\frac{1}{[\mathrm{A}]}-\frac{1}{[\mathrm{A}]_0}=k_r t\quad[10.15\mathrm{a}]$$

$$k_r=\frac{1}{t}\left(\frac{1}{[\mathrm{A}]}-\frac{1}{[\mathrm{A}]_0}\right)$$

$$=\left(\frac{1}{1.22\times10^4\ \mathrm{s}}\right)\times\left(\frac{1}{56.0\ \mathrm{mmol\ dm^{-3}}}-\frac{1}{220\ \mathrm{mmol\ dm^{-3}}}\right)$$

$$=1.09\times10^{-6}\ \mathrm{dm^3\ mmol^{-1}\ s^{-1}}=\boxed{1.09\times10^{-3}\ \mathrm{dm^3\ mol^{-1}\ s^{-1}}}$$

E10.15

$$\mathrm{CO_2(aq)+H_2O\xrightarrow{\text{carbonic anhydrase}}H_2CO_3(aq)}$$

We assume that the reaction rate is pseudofirst-order in carbon dioxide as the enzyme concentration may appear in the general rate expression. Then,

$$\ln\frac{[\mathrm{CO_2}]_0}{[\mathrm{CO_2}]}=k_{\mathrm{r,eff}}t\quad[10.13\mathrm{a}]$$

$$k_{\mathrm{r,eff}}=\frac{1}{t}\ln\frac{[\mathrm{CO_2}]_0}{[\mathrm{CO_2}]}$$

$$=\left(\frac{1}{1.22\times10^4\ \mathrm{s}}\right)\times\ln\left(\frac{220}{56.0}\right)$$

$$=\boxed{1.12\times10^{-4}\ \mathrm{s^{-1}}}.$$

E10.16 We have the reaction $\mathrm{NO(g)+\tfrac{1}{2}\,Cl_2(g)\rightarrow NOCl(g)}$.

Data provided: $p_{\mathrm{NO}}(0)=300$ Pa, $p_{\mathrm{NOCl}}(522\ \mathrm{s})=100$ Pa, $p_{\mathrm{NOCl}}(0)=0$.

The reaction rate is stated to be a pseudo-second order reaction in NO. Thus, $\frac{dp_{NO}}{dt} = k_{r,eff}\, p_{NO}^2$ where $k_{r,eff}$ is the pseudo-second order rate constant and the integrated form of the rate law is

$$\frac{1}{p_{NO}} - \frac{1}{p_{NO}(0)} = k_{r,eff}t \ \ [10.15a].$$

Since the reaction stoichiometry indicates that $p_{NOCl} = p_{NO}(0) - p_{NO}$, we substitute $p_{NO} = p_{NO}(0) - p_{NOCl}$ and solve for $k_{r,eff}$.

$$k_{r,eff} = \frac{1}{t}\left(\frac{1}{p_{NO}(0) - p_{NOCl}} - \frac{1}{p_{NO}(0)}\right)$$

$$= \left(\frac{1}{522\ s}\right) \times \left(\frac{1}{(300-100)\ Pa} - \frac{1}{300\ Pa}\right)$$

$$= \boxed{3.19\times10^{-6}\ Pa^{-1}\ s^{-1}}$$

E10.17 Let p be the partial pressure of ammonia at time t. Then, for a reaction that is zero-order in ammonia:

$$-\frac{dp}{dt} = k_r$$

$$\int_{p_0}^{p} dp = -k_r \int_0^t dt$$

$$p - p_0 = -k_r t$$

(a) $$k_r = \frac{p_0 - p}{t} = \frac{(21-10)\ kPa}{770\ s} = \boxed{0.014\ kPa\ s^{-1}}$$

(b) $$t = \frac{p_0}{k_r} = \frac{21\ kPa}{0.014\ kPa\ s^{-1}} = \boxed{1.5\times10^3\ s}$$

E10.18 (a) $\boxed{v = k_r[ICl][H_2]}$

In order to deduce this rate law, compare experiments that have identical initial H_2 concentration then compare experiments that have identical initial ICl concentration. Experiments 1 and 2 have identical $[H_2]_0$ values but the 2nd has twice the $[ICl]_0$ value and an initial rate that is twice as large. The rate must be first-order in [ICl]. Similarly, Experiments 2 and 3 have identical $[ICl]_0$ values but the 3rd has thrice the $[H_2]_0$ value and an initial rate that is three times as large. Once again, the rate is proportional to the concentration so it must be first-order in $[H_2]$.

(b) $k_r = v/([ICl][H_2])$

$$= 3.7\times10^{-7}\ mol\ dm^{-3}\ s^{-1} / (1.5\times10^{-3}\ mol\ dm^{-3})^2 = \boxed{0.16\ dm^3\ mol^{-1}\ s^{-1}}$$

(c) $v_0 = (0.16\ dm^3\ mol^{-1}\ s^{-1})\times(4.7\times10^{-3}\ mol\ dm^{-3})\times(2.7\times10^{-3}\ mol\ dm^{-3})$

$$= \boxed{2.0\times10^{-6}\ mol\ dm^{-3}\ s^{-1}}$$

E10.19 $Hg(CH_3)_2(g) \rightarrow Hg(g) + 2\ CH_3(g)$

Since concentration is proportional to partial pressure, eqn. 10.13a for a first-order rate law can be written in the form $\ln(p_0/p) = k_r t$ where p is the partial pressure of mercury dimethyl at time t. Thus, should a plot of $\ln(p_0/p)$ against t be linear, we conclude that the rate is first-order and that the plot slope equals k_r. We draw a table with the requisite data transform and prepare the plot shown in Figure 10.3.

t/s	0	1.0	2.0	3.0	4.0
p/kPa	15.1	11.8	9.21	7.2	5.6
$\ln(p_0/p)$	0	0.247	0.494	0.74	0.99

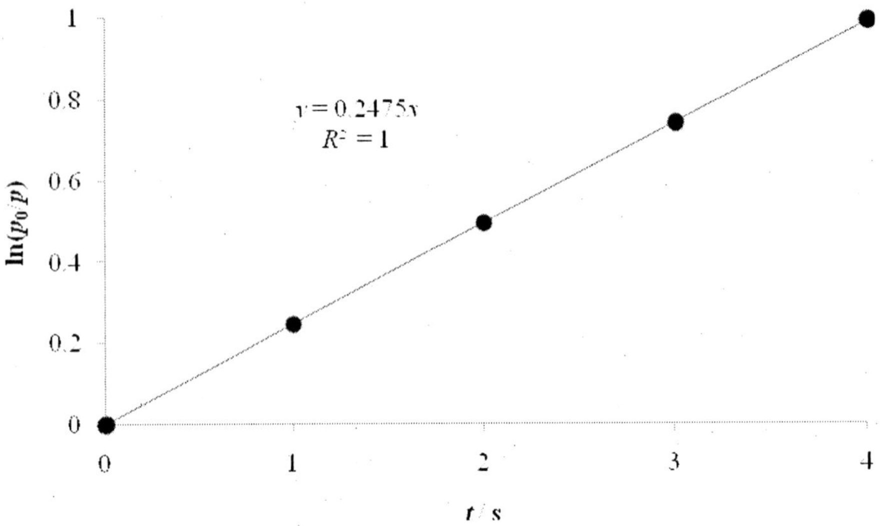

Figure 10.3

Figure 10.3 shows a linear plot so we conclude that the rate is first-order. The linear least squares regression fit to the plot is shown as an insert in the figure. It provides the slope which indicates that $k_r = \boxed{0.248\ \mathrm{s}^{-1}}$.

E10.20 $2\ \mathrm{HI(g)} \rightarrow \mathrm{H_2(g)} + \mathrm{I_2(g)}$

If the reaction is first-order in HI:

(i) The differential rate expression is

$$v = -\tfrac{1}{2}\mathrm{d}[\mathrm{HI}]\mathrm{d}t\ [10.5\mathrm{c}] = k_r[\mathrm{HI}] \quad \text{(Note the factor } \tfrac{1}{2}!\text{)}.$$

(ii) The integrated rate expression is $\ln([\mathrm{HI}]_0/[\mathrm{HI}]) = 2k_r t$.

(iii) A plot of $\ln([\mathrm{HI}]_0/[\mathrm{HI}])$ against t will be linear with slope equal to $2k_r$.

If the reaction is second-order in HI:

(i) The differential rate expression is

$$v = -\tfrac{1}{2}\mathrm{d}[\mathrm{HI}]\mathrm{d}t\ [10.5\mathrm{c}] = k_r[\mathrm{HI}]^2 \quad \text{(Note the factor } \tfrac{1}{2}!\text{)}.$$

(ii) The integrated rate expression is $\dfrac{1}{[\mathrm{HI}]} - \dfrac{1}{[\mathrm{HI}]_0} = 2k_r t\ [10.15\mathrm{a}]$.

(iii) A plot of $1/[\mathrm{HI}]$ against t will be linear with slope equal to $2k_r$.

We draw a table of data and data transformations needed to make plots to test both the first-order and second-order hypothesis. The plots are found in Figures 10.4 and 10.5.

t/s	0	1000	2000	3000	4000
$[\mathrm{HI}]/\mathrm{mol\ dm^{-3}}$	1.00	0.112	0.061	0.041	0.031
$\ln([\mathrm{HI}]_0/[\mathrm{HI}])$	0	2.19	2.80	3.19	3.47
$\mathrm{mol\ dm^{-3}}/[\mathrm{HI}]$	1.00	8.93	16.4	24.4	32.3

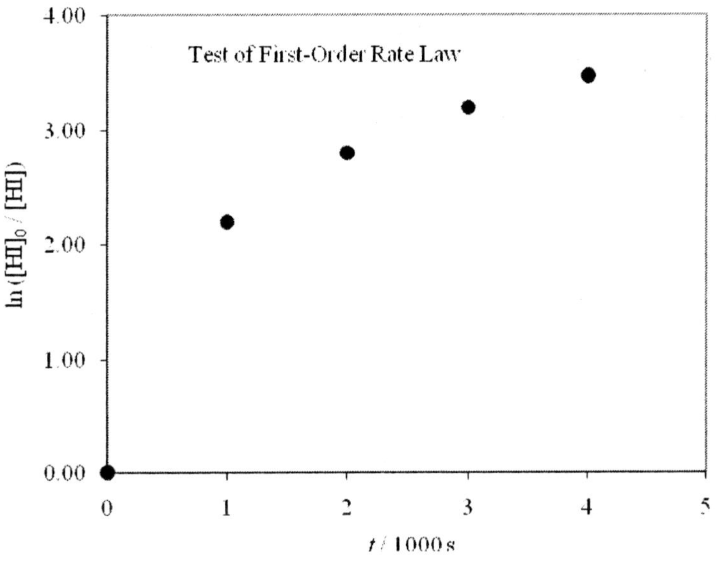

Figure 10.4

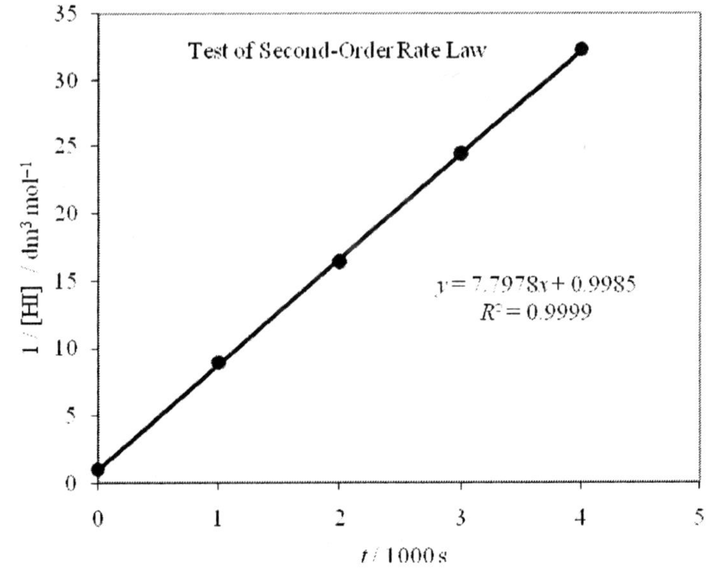

Figure 10.5

(a) Since the plot of Figure 10.4 is non-linear, we conclude that the reaction rate is not first-order in [HI]. The plot of Figure 10.5 is linear so we conclude that the reaction rate is second-order in [HI].

(b) The linear regression fit of the second-order rate is displayed in Figure 10.5. The reaction rate constant equals the regression slope divided by 2 because the stoichiometric coefficient of HI is 2. Thus,

$$k_r = slope/2 = \boxed{3.90 \times 10^{-3}\ \text{dm}^3\ \text{mol}^{-1}\ \text{s}^{-1}}\ .$$

E10.21 $2\ HI(g) \rightarrow H_2(g) + I_2(g)$

If the reaction is first-order in HI:

(i) The differential rate expression is

$$v = -\tfrac{1}{2}\text{d}[\text{HI}]\text{d}t\ \ [10.5\text{c}] = k_r[\text{HI}]\ \ \text{(Note the factor } \tfrac{1}{2}!\text{)}\ .$$

(ii) The integrated rate expression is $\ln([\text{HI}]_0/[\text{HI}]) = 2k_r t$.

(iii) A plot of $\ln([\text{HI}]_0/[\text{HI}])$ against t will be linear with slope equal to $2k_r$.

If the reaction is second-order in HI:

(i) The differential rate expression is

$$v = -\tfrac{1}{2}\mathrm{d}[\mathrm{HI}]\mathrm{d}t \ \ [10.5\mathrm{c}] = k_{\mathrm{r}}[\mathrm{HI}]^2 \quad \text{(Note the factor } \tfrac{1}{2}!).$$

(ii) The integrated rate expression is $\dfrac{1}{[\mathrm{HI}]} - \dfrac{1}{[\mathrm{HI}]_0} = 2k_{\mathrm{r}}t \ \ [10.15\mathrm{a}]$.

(iii) A plot of $1/[\mathrm{HI}]$ against t will be linear with slope equal to $2k_{\mathrm{r}}$.

We draw a table of data and data transformations needed to make plots to test both the first-order and second-order hypothesis. The plots are found in Figures 10.6 and 10.7.

t / s	0	1	2	3	4
[HI]/mol dm^{-3}	1.00	0.43	0.27	0.2	0.16
ln([HI]$_0$/[HI])	0	0.84	1.3	1.6	1.8
mol dm^{-3}/[HI]	1.00	2.3	3.7	5.0	6.3

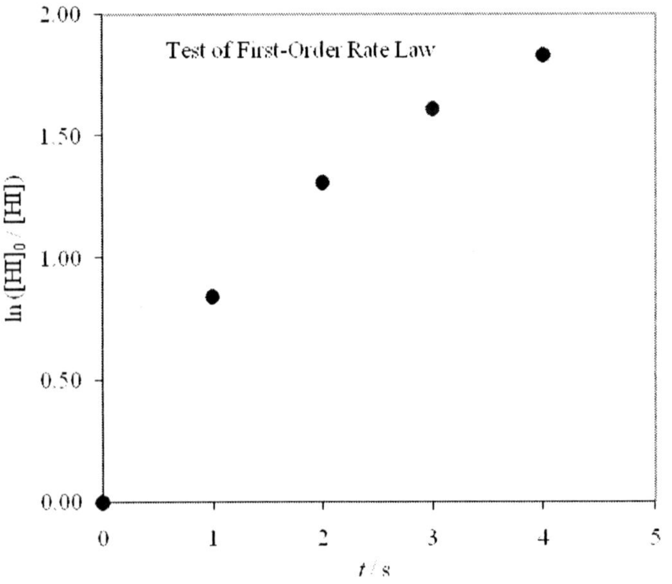

Figure 10.6

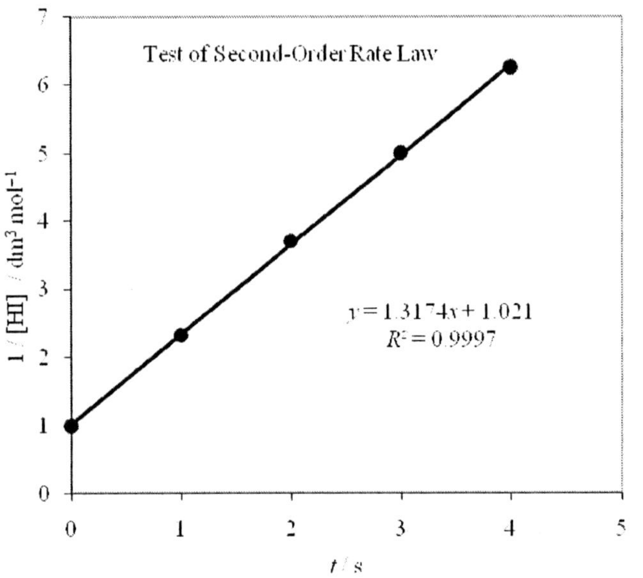

Figure 10.7

(a) Since the plot of Figure 10.6 is non-linear, we conclude that the reaction rate is not first-order in [HI]. The plot of Figure 10.7 is linear so we conclude that the reaction rate is $\boxed{\text{second-order}}$ in [HI].

(b) The linear regression fit of the second-order rate is displayed in Figure 10.7. The reaction rate constant equals the regression slope divided by 2 because the stoichiometric coefficient of HI is 2. Thus, the reaction rate constant for the above balanced reaction equation is

$$k_r = slope/2 = \boxed{0.66\ \text{dm}^3\ \text{mol}^{-1}\ \text{s}^{-1}}\,.$$

E10.22 $\text{Mb} + \text{CO} \rightarrow \text{MbCO} \qquad k_r = 5.8 \times 10^5\ \text{dm}^3\ \text{mol}^{-1}\ \text{s}^{-1}$

Since the carbon monoxide concentration (400 mmol dm^{-3}) is much larger than the initial myoglobin concentration (10 mmol dm^{-3}), we assume that Mb is isolated. Additionally, although the exercise suggests the rate to be pseudofirst-order in Mb, we assume that the reaction rate is first-order in the CO. Thus, $v = -\text{d[Mb]}/\text{d}t = k_r[\text{CO}][\text{Mb}] = k_{r,\text{eff}}[\text{Mb}]$ where

$$k_{r,\text{eff}} = (5.8 \times 10^5\ \text{dm}^3\ \text{mol}^{-1}\ \text{s}^{-1}) \times (400\ \text{mmol dm}^{-3}) = 2.3 \times 10^5\ \text{s}^{-1}.$$

The integrated form of this rate law is $\ln([\text{Mb}]_0/[\text{Mb}]) = k_{r,\text{eff}}t$ or $[\text{Mb}] = [\text{Mb}]_0\,\text{e}^{-k_{r,\text{eff}}t}$. The latter equation is used to prepare the plot of [Mb] against t shown in Figure 10.8. The time range for the computation is chosen by calculating the pseudofirst-order time constant.

$$\tau_{\text{eff}} = 1/k_{r,\text{eff}}\ [10.17] = 1/(2.3\times10^5\ \text{s}^{-1}) = 4.3\times10^{-6}\ \text{s} = 4.3\ \mu\text{s}$$

The range is chosen to be 0–10 μs or about 2.5 τ_{eff}.

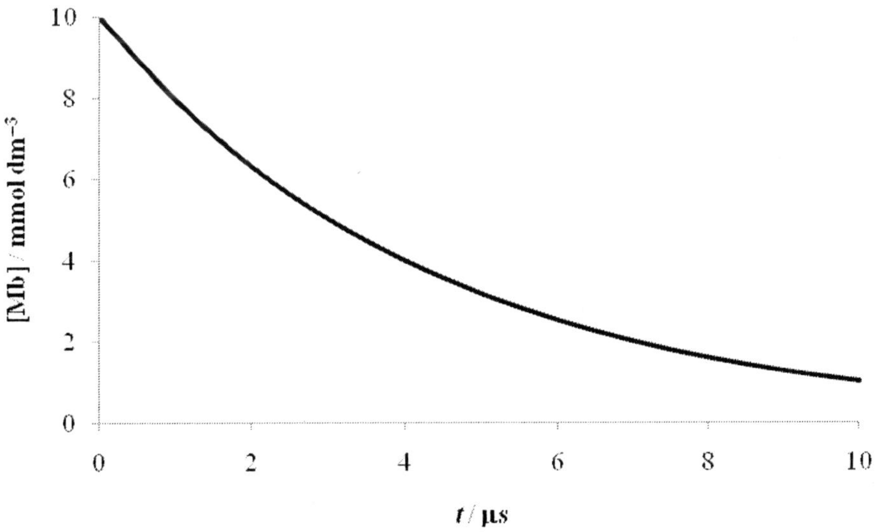

Figure 10.8

E10.23 $3\,\text{A} \rightarrow \text{B}$

The reaction stoichiometry indicates that

$$[\text{B}] = [\text{B}]_0 + \tfrac{1}{3}\{[\text{A}]_0 - [\text{A}]\}.$$

Substitution of the provided relation $[\text{A}] = [\text{A}]_0/(1 + k_r t[\text{A}]_0)$ and simplification gives

$$[\text{B}] = [\text{B}]_0 + \tfrac{1}{3}\{[\text{A}]_0 - [\text{A}]_0/(1 + k_r t[\text{A}]_0)\}$$

$$= \boxed{[\text{B}]_0 + \tfrac{1}{3}[\text{A}]_0\{k_r t[\text{A}]_0/(1 + k_r t[\text{A}]_0)\}}.$$

E10.24 For the reaction $2\text{A} \rightarrow \text{B}$, $[\text{A}] = [\text{A}]_0 - 2[\text{B}]$.

Hence, $[\text{A}]_0 = 2[\text{B}]_\infty = 2\times(0.500\ \text{mol dm}^{-3}) = 1.000\ \text{mol dm}^{-3}$.

We can therefore draw up the following table, which includes the original data along with the calculation of [A] and ln([A]$_0$/[A]). The calculation of ln([A]$_0$/[A]) is included in the table because a plot of ln([A]$_0$/[A]) against t provides a test of the hypothesis that the reaction rate is first-order in [A]; if the plot is linear, the hypothesis is valid. The test plot is shown in Figure 10.9. Clearly, the plot is nonlinear and we reject the first-order hypothesis. We add the row calculation of 1/[A] to the data table because a plot of 1/[A] against t provides a test of the hypothesis that the rate is second-order in [A]; if the plot is linear, the hypothesis is valid. The test plot is shown in Figure 10.10. This plot is linear so the reaction rate is second-order in [A].

t / min	0	10	20	30	40
[B] / mol dm^{-3}	0	0.372	0.426	0.448	0.460
[A] / mol dm^{-3}	1.000	0.256	0.148	0.104	0.080
ln([A]$_0$/[A])	0	1.36	1.91	2.26	2.53
mol dm^{-3}/[A]	1	3.91	6.76	9.62	12.5

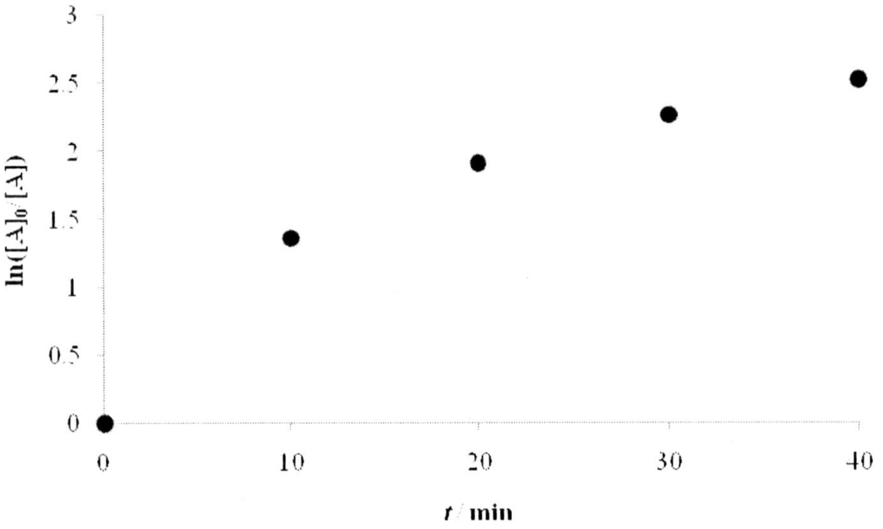

Figure 10.9

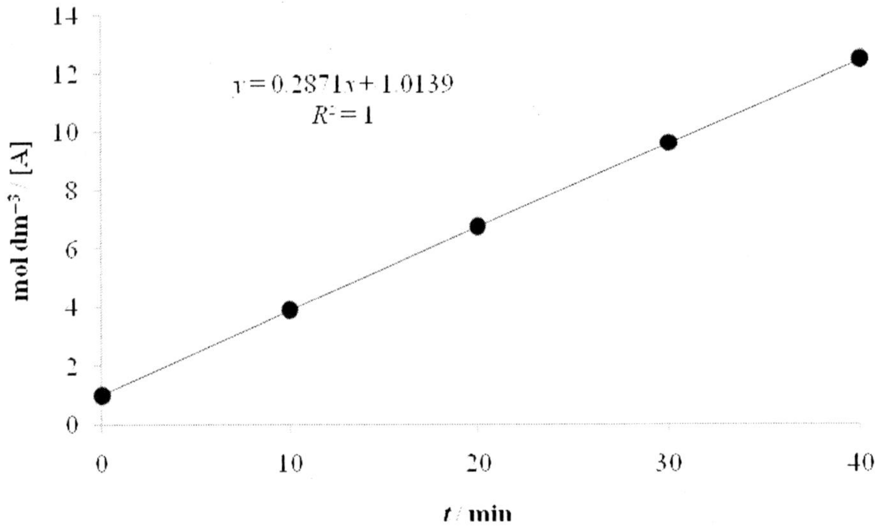

Figure 10.10

Since the reaction rate law is second-order in A, the differential rate expression is

$$v = -\tfrac{1}{2}\mathrm{d}[\mathrm{A}]\mathrm{d}t \ [10.5\mathrm{c}] = k_{\mathrm{r}}[\mathrm{A}]^2.$$

Note the factor $\tfrac{1}{2}$ that originates from the stoichiometric coefficient of A!

The integrated rate expression is $\dfrac{1}{[A]}-\dfrac{1}{[A]_0}=2k_rt$ [10.15a], which indicates that the linear plot of $1/[A]$ against t has a slope equal to $2k_r$. The least squares linear regression fit of the plot is shown in Figure 10.10 so we conclude that

$$k_r = slope/2 = (0.287\ \text{mol}^{-1}\ \text{dm}^3\ \text{min}^{-1})\ /\ 2 = \boxed{0.144\ \text{mol}^{-1}\ \text{dm}^3\ \text{min}^{-1}}.$$

E10.25 The reaction of text Example 10.2 is $CH_3N_2CH_3(g) \rightarrow CH_3CH_3(g) + N_2(g)$ and the integrated rate law (p_A is the partial pressure of azomethane) and rate constant are:

$$p_A = p_{A,0}e^{-k_rt} \qquad \text{where } p_{A,0} = 0.1020\ \text{Torr} = 13.60\ \text{Pa} \quad \text{and} \quad k_r = 4.04\times10^{-4}\ \text{s}^{-1}.$$

Let p_E and p_N be the partial pressures of ethane and dinitrogen. Then, according to the reaction stoichiometry, and assuming that the initial pressures of ethane and dinitrogen are zero, the total pressure at time t is given by

$$\begin{aligned} p_{\text{total}} &= p_A + p_E + p_N \\ &= p_A + (p_{A,0} - p_A) + (p_{A,0} - p_A) \\ &= p_{A,0}e^{-k_rt} + (p_{A,0} - p_{A,0}e^{-k_rt}) + (p_{A,0} - p_{A,0}e^{-k_rt}) \\ &= \boxed{p_{A,0}\{2 - e^{-k_rt}\}}. \end{aligned}$$

A plot of the total pressure is presented in Figure 10.11.

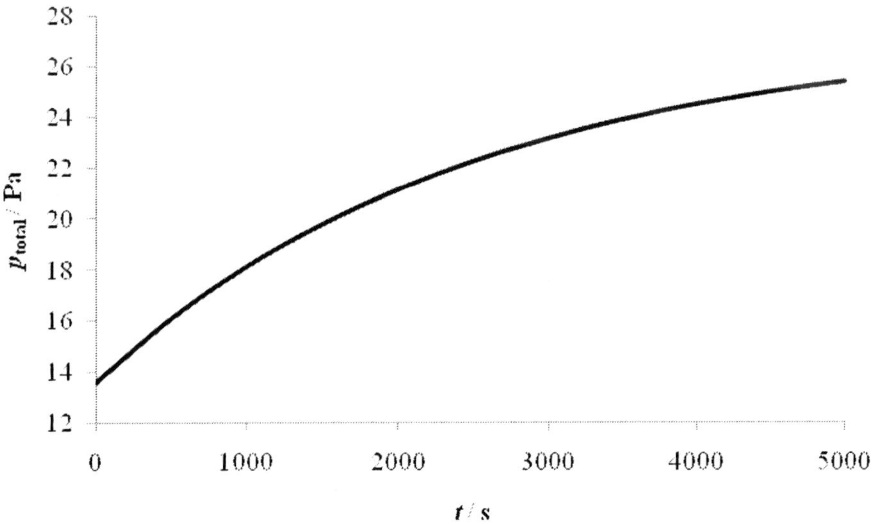

Figure 10.11

E10.26 Because $(1/2)^6 = 1/64$, 6 half-lives must pass for the concentration to fall to 1/64th of its initial values for this first-order reaction.

$$t = t_{1/2}\times6 = 221\ \text{s}\times6 = \boxed{1.33\times10^3\ \text{s}}$$

E10.27 For a first-order reaction: $k_r = \dfrac{\ln 2}{t_{1/2}}$ [10.16]. Therefore,

$$t = \frac{1}{k_r}\ln\frac{[^{14}C]_0}{[^{14}C]}\ [10.13a] = \frac{t_{1/2}}{\ln 2}\ln\frac{[^{14}C]_0}{[^{14}C]}$$

$$= \frac{(5730\ \text{a})}{\ln 2}\ln\frac{[^{14}C]_0}{0.69[^{14}C]_0} = \frac{(5730\ \text{a})}{\ln 2}\ln\frac{1}{0.69}$$

$$= \boxed{3067\ \text{a}} \pm 100\ \text{a}.$$

E10.28

$$m_{^{90}\mathrm{Sr}} = m_{^{90}\mathrm{Sr},0}\,\mathrm{e}^{-k_\mathrm{r}t}\ [10.13\mathrm{c}] = m_{^{90}\mathrm{Sr},0}\,\mathrm{e}^{-t\times(\ln 2)/t_{1/2}}\ [10.16] = (1.00\ \mu\mathrm{g})\times\mathrm{e}^{-t\times(\ln 2)/(28.1\ \mathrm{a})}$$

(a) When $t = 19$ a,

$$m_{^{90}\mathrm{Sr}} = (1.00\ \mu\mathrm{g})\times\mathrm{e}^{-(19\ \mathrm{a})\times(\ln 2)/(28.1\ \mathrm{a})} = \boxed{0.63\ \mu\mathrm{g}}$$

(b) When $t = 75$ a,

$$m_{^{90}\mathrm{Sr}} = (1.00\ \mu\mathrm{g})\times\mathrm{e}^{-(75\ \mathrm{a})\times(\ln 2)/(28.1\ \mathrm{a})} = \boxed{0.16\ \mu\mathrm{g}}$$

E10.29 The time constant τ is given by

$$\tau = 1/k_\mathrm{r}\ [10.17] = t_{1/2}/\ln 2\ [10.16]$$
$$= (439\ \mathrm{s})/\ln 2 = \boxed{633\ \mathrm{s}}.$$

E10.30 This is a reaction of the type $\mathrm{A} + \mathrm{B} \rightarrow \mathrm{P}$.

This integrated rate law is given in Table 10.4 and is

$$[\mathrm{P}] = \frac{[\mathrm{A}]_0[\mathrm{B}]_0\left[1-\exp\left[([\mathrm{B}]_0 - [\mathrm{A}]_0)k_\mathrm{r}t\right]\right]}{[\mathrm{A}]_0 - [\mathrm{B}]_0\exp\left[([\mathrm{B}]_0 - [\mathrm{A}]_0)k_\mathrm{r}t\right]}.$$

(a)
$$[\mathrm{P}] = \frac{(0.150)(0.055)\left[1-\exp\left[(0.055-0.150)\times 0.11\times 15\right]\right]}{0.150-0.055\ \exp\left[(0.055-0.150)\times 0.11\times 15\right]}\ \mathrm{mol\ dm^{-3}}$$
$$= 1.16\times 10^{-2}\ \mathrm{mol\ dm^{-3}}$$

$$[\mathrm{CH_3COOC_2H_5}] = 0.150\ \mathrm{mol\ dm^{-3}} - 0.012\ \mathrm{mol\ dm^{-3}} = \boxed{0.138\ \mathrm{mol\ dm^{-3}}}$$

(b) Perform the above calculation of [P] with $t = 15\ \mathrm{min} = 900$ s.

$$[\mathrm{P}] = 5.50\times 10^{-2}\ \mathrm{mol\ dm^{-3}}$$

$$[\mathrm{CH_3COOC_2H_5}] = 0.150\ \mathrm{mol\ dm^{-3}} - 0.055\ \mathrm{mol\ dm^{-3}} = \boxed{0.095\ \mathrm{mol\ dm^{-3}}}$$

E10.31

$$2\,\mathrm{A} \rightarrow \mathrm{P}$$

Since the rate law is second-order in A, the differential rate expression is

$$v = -\tfrac{1}{2}\mathrm{d}[\mathrm{A}]\mathrm{d}t\ [10.5\mathrm{c}] = k_\mathrm{r}[\mathrm{A}]^2.$$

Note the factor $\tfrac{1}{2}$ that originates from the stoichiometric coefficient of A!

The integrated rate expression is $\dfrac{1}{[\mathrm{A}]} - \dfrac{1}{[\mathrm{A}]_0} = 2k_\mathrm{r}t\ [10.15\mathrm{a}]$.

Solving for t gives

$$t = \frac{1}{2k_\mathrm{r}}\left(\frac{1}{[\mathrm{A}]} - \frac{1}{[\mathrm{A}]_0}\right)$$
$$= \frac{1}{2\times(1.44\ \mathrm{dm^3 mol^{-1}\ s^{-1}})}\left(\frac{1}{0.046\ \mathrm{mol\ dm^{-3}}} - \frac{1}{0.460\ \mathrm{mol\ dm^{-3}}}\right)$$
$$= 6.8\ \mathrm{s}.$$

E10.32 $t_{1/2}=\dfrac{\ln 2}{k_r}$ [10.16] $=\dfrac{\ln 2}{A\mathrm{e}^{-E_a/RT}}$ [10.19] where $A=10^{15.6}$ s and $E_a=261\times10^3$ J mol^{-1}

(a) When $T = 293$ K,

$$t_{1/2}=\frac{\ln 2}{\left(10^{15.6}\ \text{s}\right)\times \mathrm{e}^{-\left(261\times10^{3}\ \text{J mol}^{-1}\right)/\left(8.3145\ \text{J mol}^{-1}\ \text{K}^{-1}\times 293\ \text{K}\right)}}=5.88\times10^{30}\ \text{s}=\boxed{1.86\times10^{23}\ \text{a}}.$$

The combination of high activation energy and low temperature causes a basically infinite time for appreciable reaction.

(b) When $T = 793$ K,

$$t_{1/2}=\frac{\ln 2}{\left(10^{15.6}\ \text{s}\right)\times \mathrm{e}^{-\left(261\times10^{3}\ \text{J mol}^{-1}\right)/\left(8.3145\ \text{J mol}^{-1}\ \text{K}^{-1}\times 793\ \text{K}\right)}}=\boxed{27.1\ \text{s}}.$$

At high temperature the reaction occurs rapidly.

E10.33 $\ln\dfrac{k_r(T')}{k_r(T)}=\dfrac{E_a}{R}\left(\dfrac{1}{T}-\dfrac{1}{T'}\right)$ [10.20]

Solve the above equation for E_a.

$$E_a=\frac{R}{\left(\dfrac{1}{T}-\dfrac{1}{T'}\right)}\ln\frac{k_r(T')}{k_r(T)}$$

$$=\left(\frac{8.3145\ \text{J K}^{-1}\ \text{mol}^{-1}}{\dfrac{1}{292\ \text{K}}-\dfrac{1}{310\ \text{K}}}\right)\times\ln\left(\frac{3.38\times10^{-3}}{2.78\times10^{-4}}\right)$$

$$=\boxed{104\ \text{kJ mol}^{-1}}$$

For A, use

$$A=k_r(T)\times\mathrm{e}^{E_a/RT}\quad[10.19]$$

$$=2.78\times10^{-4}\ \text{mol dm}^{-3}\ \text{s}^{-1}\times\mathrm{e}^{104000/(8.3145\times292)}$$

$$=\boxed{1.12\times10^{15}\ \text{mol dm}^{-3}\ \text{s}^{-1}}.$$

E10.34 $\ln\dfrac{k_r(T')}{k_r(T)}=\dfrac{E_a}{R}\left(\dfrac{1}{T}-\dfrac{1}{T'}\right)$ [10.20]

Let T = 298.15 K and solve the above equation for T where $k_r(T')$ is 10% larger than $k_r(T)$. Consider all factors are exact, including the factor 1.1 that accounts for the 10% rate increase, so that basic concepts may be examined.

$$T' = \left[\frac{1}{T} - \frac{R}{E_a}\ln\frac{k_r(T')}{k_r(T)}\right]^{-1}$$

$$= \left[\frac{1}{298.15\ \text{K}} - \frac{8.3145\ \text{J mol}^{-1}\ \text{K}^{-1}}{99.1\times10^3\ \text{J mol}^{-1}}\ln\frac{1.1\times k_r(T)}{k_r(T)}\right]^{-1}$$

$$= \left[\frac{1}{298.15\ \text{K}} - \frac{8.3145\ \text{J mol}^{-1}\ \text{K}^{-1}}{99.1\times10^3\ \text{J mol}^{-1}}\ln(1.1)\right]^{-1}$$

$$= \boxed{298.86\ \text{K}}$$

The temperature need only be increased by 0.71 Kelvin to achieve a 10% rate increase when $E_a = 99.1\times10^3\ \text{J mol}^{-1}$! This suggests that the rate is very sensitive to temperature when the activation energy is large. To confirm this hypothesis, we repeat the calculation with a low activation energy of $E_a = 1\times10^3\ \text{J mol}^{-1}$.

$$T' = \left[\frac{1}{298.15\ \text{K}} - \frac{8.3145\ \text{J mol}^{-1}\ \text{K}^{-1}}{1\times10^3\ \text{J mol}^{-1}}\ln(1.1)\right]^{-1}$$

$$= 390\ \text{K}$$

Indeed, when the activation energy is lower, a greater temperature change is required to achieve the same increase in the reaction rate.

E10.35 $$k_r(T) = Ae^{-E_a/RT}\quad[10.19]$$

A rate constant responds more strongly to an increase in temperature when the slope dk_r/dT, or equivalently $d(\ln k_r)/dT$, is larger. Taking the natural logarithm of eqn 10.19 and evaluating the temperature derivative gives

$$\ln k_r = \ln A - \frac{E_a}{RT}$$

$$\frac{d\ln k_r}{dT} = \frac{E_a}{RT^2}.$$

Examination of the right side of the above equation shows that at any particular temperature the reaction that has the greater value of E_a has the greater response to an increase in temperature. That is, the reaction for which $\boxed{E_a = 52\ \text{kJ mol}^{-1}}$ responds more strongly than the reaction for which E_a = 25 kJ mol^{-1} even though it may have a smaller rate constant (depending upon relative values of pre-exponential factors).

E10.36 $$\ln\frac{k_r(T')}{k_r(T)} = \frac{E_a}{R}\left(\frac{1}{T} - \frac{1}{T'}\right)\quad[10.20]$$

Solve the above equation for E_a.

$$E_a = \frac{R}{\left(\frac{1}{T} - \frac{1}{T'}\right)}\ln\frac{k_r(T')}{k_r(T)}$$

$$= \left(\frac{8.3145\ \text{J K}^{-1}\ \text{mol}^{-1}}{\frac{1}{293\ \text{K}} - \frac{1}{300\ \text{K}}}\right)\times\ln(1.41)$$

$$= \boxed{35.9\ \text{kJ mol}^{-1}}$$

E10.37 According to eqn 10.18, a plot of ln k_r against $1/T$ will be linear with a slope equal to $-E_a/R$. The plot and linear regression fit of the data is shown in Figure 10.12.

$$E_a = -slope \times R = -(33 \times 10^3 \text{ K}) \times (8.314 \text{ J K}^{-1} \text{ mol}^{-1}) = \boxed{27\bar{4} \text{ kJ mol}^{-1}}$$

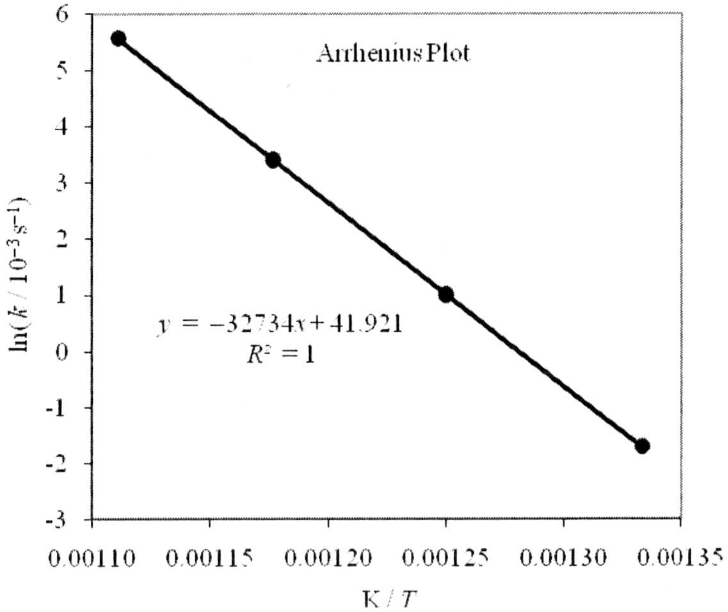

Figure 10.12

E10.38 $$\ln\frac{k_r(T')}{k_r(T)} = \frac{E_a}{R}\left(\frac{1}{T} - \frac{1}{T'}\right) \quad [10.20]$$

Solve the above equation for E_a.

$$E_a = \frac{R}{\left(\frac{1}{T} - \frac{1}{T'}\right)} \ln\frac{k_r(T')}{k_r(T)}$$

$$= \left(\frac{8.3145 \text{ J K}^{-1} \text{ mol}^{-1}}{\frac{1}{277 \text{ K}} - \frac{1}{298 \text{ K}}}\right) \times \ln(40)$$

$$= \boxed{121 \text{ kJ mol}^{-1}}$$

E10.39 $$\ln\frac{k_r(T')}{k_r(T)} = \frac{E_a}{R}\left(\frac{1}{T} - \frac{1}{T'}\right) \quad [10.20]$$

Solve the above equation for E_a.

$$E_a = \frac{R}{\left(\frac{1}{T} - \frac{1}{T'}\right)} \ln\frac{k_r(T')}{k_r(T)}$$

$$= \left(\frac{8.3145 \text{ J K}^{-1} \text{ mol}^{-1}}{\frac{1}{293 \text{ K}} - \frac{1}{300 \text{ K}}}\right) \times \ln\left(\frac{1}{1.23}\right)$$

$$= \boxed{-21.6 \text{ kJ mol}^{-1}}$$

Being negative, this can hardly be interpreted as an activation energy.

E10.40 $$\ln\frac{k_r(T')}{k_r(T)}=\frac{E_a}{R}\left(\frac{1}{T}-\frac{1}{T'}\right)\quad[10.20]$$

Solve the above equation for E_a.

$$E_a=\frac{R}{\left(\frac{1}{T}-\frac{1}{T'}\right)}\ln\frac{k_r(T')}{k_r(T)}$$

Substitute $k_r=\frac{\ln 2}{t_{1/2}}$ [10.16] at each temperature.

$$E_a=\frac{R}{\left(\frac{1}{T}-\frac{1}{T'}\right)}\ln\frac{t_{1/2}(T)}{t_{1/2}(T')}$$

$$=\left(\frac{8.3145\ \text{J K}^{-1}\ \text{mol}^{-1}}{\frac{1}{293\ \text{K}}-\frac{1}{283\ \text{K}}}\right)\times\ln\left(\frac{1}{2}\right)$$

$$=\boxed{47.8\ \text{kJ mol}^{-1}}$$

E10.41 The fraction of collisions f that occur with at least a kinetic energy E_a is $f=e^{-E_a/RT}$ [10.21].

(a) At 293 K, $f=e^{-111000/(8.3145\times293)}=\boxed{1.62\times10^{-20}}$.

(b) At 473 K, $f=e^{-111000/(8.3145\times473)}=\boxed{5.52\times10^{-13}}$.

E10.42 $$A=\sigma\left(\frac{8kT}{\pi\mu}\right)^{1/2}N_A\quad[10.22]$$ where μ is the effective mass

We estimate the collision section of NO_2 as 0.50 nm^2, which is similar to that of other triatomic molecules (see Table 1.3), and the effective mass is about 10^{-26} kg. Using $T = 500$ K,

$$A_{est}=\left(0.5\times10^{-18}\ \text{m}^2\right)\times\left(\frac{8\times\left(1.38\times10^{-23}\ \text{J K}^{-1}\right)\times\left(500\ \text{K}\right)}{\pi\times10^{-26}\ \text{kg}}\right)^{1/2}\times\left(6.022\times10^{23}\ \text{mol}^{-1}\right)$$

$$=4\times10^{8}\ \text{m}^3\ \text{mol}^{-1}\ \text{s}^{-1}=\boxed{4\times10^{11}\ \text{dm}^3\ \text{mol}^{-1}\ \text{s}^{-1}}$$

The estimated value of the pre-exponential factor is much larger than the experimental value of $2\times10^{9}\ \text{dm}^3\ \text{mol}^{-1}\ \text{s}^{-1}$. A possible explanation of the discrepancy is that a steric factor $P\sim5\times10^{-3}$ must be included in the pre-exponential factor to account for the probability that the NO_2 molecules collide in the specific orientation required for reaction.

E10.43 Consider the example discussed in the text.

$$K + Br_2 \rightarrow KBr + Br$$

We will estimate the value of steric factor P based on a harpoon mechanism for this reaction by calculating the distance at which it is energetically favorable for the electron to leap from K to Br_2.

We should begin by identifying all the contributions to the energy of interaction between the colliding species. There are three contributions to the energy of the process $K + Br_2 \rightarrow K^+ + Br_2^-$. The first is the ionization energy I of K. The second is the electron affinity E_{ea} of Br_2. The third is the Coulombic interaction energy between the ions when they have been formed; when their separation is R, this energy is $-e^2/4\pi\varepsilon_0 R$. The electron flips across when the sum of these three contributions changes from positive to negative (that is, when the sum is zero).

The net change in energy when the transfer occurs at a separation R is

$$E = I - E_{ea} - \frac{e^2}{4\pi\varepsilon_0 R}$$

The ionization energy I is larger than E_{ea}, so E becomes negative only when R has decreased to less than some critical value R^* given by

$$\frac{e^2}{4\pi\varepsilon_0 R^*} = I - E_{ea}$$

When the particles are at this separation, the harpoon shoots across from K to Br_2, so we can identify the reactive cross-section as $\sigma^* = \pi R^{*2}$. Then, since $\sigma^* = P\sigma$,

$$P = \frac{\sigma^*}{\sigma} = \frac{R^{*2}}{d^2} = \left\{\frac{e^2}{4\pi\varepsilon_0 d\,(I - E_{ea})}\right\}^2 \qquad \text{where } d = R(\text{K}) + R(\text{Br}_2).$$

With $I = 420$ kJ mol^{-1} (corresponding to 6.97×10^{-19} J), $E_{ea} \approx 250$ kJ mol^{-1} (corresponding to 4.15×10^{-19} J), and $d = 500$ pm, we find $P = 2.7$, in fair agreement with the experimental value of 4.1.

We may use these calculated values to calculate σ^*, the reaction cross-section.

$$\sigma^* = \pi R^{*2} = \pi P d^2$$
$$= \pi \times (2.7) \times (5.0 \times 10^{-10}\ \text{m})^2$$
$$= 2.1 \times 10^{-18}\ \text{m}^2 = \boxed{2.1\ \text{nm}^2}$$

E10.44 We use

$$k_r = \kappa \frac{kT}{h} K^\ddagger\ [10.23] = \frac{\kappa kT}{h} e^{-\Delta^\ddagger G/RT}$$
$$= \frac{kT}{h} e^{-\Delta^\ddagger G/RT}\ \text{[assume that the transmission coefficient equals 1]}$$

Solving for $\Delta^\ddagger G$ gives

$$\Delta^\ddagger G = RT \ln \frac{kT}{hk_r}.$$

At 333 K,

$$\Delta^\ddagger G = (8.3145\ \text{J K}^{-1}\ \text{mol}^{-1}) \times (333\ \text{K}) \ln \frac{(1.381 \times 10^{-23}\ \text{J K}^{-1}) \times (333\ \text{K})}{(6.626 \times 10^{-34}\ \text{J s}) \times (1.2 \times 10^{-7}\ \text{s}^{-1})}$$
$$= 12\overline{6}\ \text{kJ mol}^{-1}.$$

At 343 K,

$$\Delta^{\ddagger}G = (8.3145 \text{ J K}^{-1} \text{ mol}^{-1}) \times (343 \text{ K}) \ln \frac{(1.381 \times 10^{-23} \text{ J K}^{-1}) \times (343 \text{ K})}{(6.626 \times 10^{-34} \text{ J s}) \times (4.6 \times 10^{-7} \text{ s}^{-1})}$$

$$= 12\overline{6} \text{ kJ mol}^{-1}.$$

The rate constant lacks the necessary precision to distinguish between the $\Delta^{\ddagger}G$ values at these two temperatures so we conclude that $\boxed{\Delta^{\ddagger}G = 12\overline{6} \text{ kJ mol}^{-1}}$ in this temperature range.

E10.45

$$k_r = \left(\frac{kT}{h} e^{\Delta^{\ddagger}S/R}\right) e^{-\Delta^{\ddagger}H/RT} \quad [10.25, \text{ assume that the transmission coefficient equals } 1]$$

$$= A e^{-\Delta^{\ddagger}H/RT} \qquad \text{where } A = \frac{kT}{h} e^{\Delta^{\ddagger}S/R}$$

Let us assume that the pre-exponential factor is essentially a constant over this small temperature range (10 K). Then, the equation for the natural logarithm of the ratio of rate constants at two different temperatures can be solved for $\Delta^{\ddagger}H$ to give

$$\Delta^{\ddagger}H = \frac{R}{\left(\dfrac{1}{T} - \dfrac{1}{T'}\right)} \ln \frac{k_r(T')}{k_r(T)}$$

$$= \frac{(8.3145 \text{ J K}^{-1} \text{ mol}^{-1})}{\left(\dfrac{1}{333 \text{ K}} - \dfrac{1}{343 \text{ K}}\right)} \ln \frac{4.6}{1.2} = 12\overline{8} \text{ kJ mol}^{-1}.$$

To evaluate $\Delta^{\ddagger}S$, solve $\Delta^{\ddagger}G = \Delta^{\ddagger}H - T\Delta^{\ddagger}S$ [10.24] for $\Delta^{\ddagger}S$.

$$\Delta^{\ddagger}S = \frac{\Delta^{\ddagger}H - \Delta^{\ddagger}G}{T}$$

In Exercise 10.44 we found that $\Delta^{\ddagger}G = 12\overline{6} \text{ kJ mol}^{-1}$ so the precision of this data does not warrant the claim that the value of $\Delta^{\ddagger}H - \Delta^{\ddagger}G = 2 \text{ kJ mol}^{-1}$. All we really know is that $\Delta^{\ddagger}H - \Delta^{\ddagger}G \approx 0$ so $\boxed{\Delta^{\ddagger}S \approx 0}$. If the rate constants had an additional significant figure so that we could claim that $\Delta^{\ddagger}H - \Delta^{\ddagger}G = 2 \text{ kJ mol}^{-1}$, $\Delta^{\ddagger}S$ in this temperature ranges would have the estimate $\Delta^{\ddagger}S \approx \dfrac{2000 \text{ J mol}^{-1}}{333 \text{ K}} = 6 \text{ J K}^{-1} \text{ mol}^{-1}$.

Answers to projects

P10.46 (a)

$$v = -\frac{d[A]}{dt} = k_r[A]^3$$

$$\frac{d[A]}{[A]^3} = -k_r \, dt$$

$$\int_{[A]_0}^{[A]} \frac{d[A]}{[A]^3} = -k_r \int_0^t dt$$

We now use the standard integral $\displaystyle\int \frac{dx}{x^n} = -\frac{1}{(n-1)x^{n-1}} + \text{constant}$ where $n \geq 2$

which implies that

$$\int_a^b \frac{dx}{x^n} = \left\{-\frac{1}{(n-1)x^{n-1}} + \text{constant}\right\}\Bigg|_b - \left\{-\frac{1}{(n-1)x^{n-1}} + \text{constant}\right\}\Bigg|_a = -\frac{1}{n-1}\left(\frac{1}{b^{n-1}} - \frac{1}{a^{n-1}}\right)$$

$$-\frac{1}{3-1}\left(\frac{1}{[\text{A}]^{3-1}} - \frac{1}{[\text{A}]_0^{3-1}}\right) = -k_r t$$

$$\boxed{\frac{1}{[\text{A}]^2} = \frac{1}{[\text{A}]_0^2} + 2k_r t}.$$

Confirmation of a third-order reaction rate is provided by a linear plot of $\frac{1}{[\text{A}]^2}$ against t.

The third-order rate constant equals the linear regression slope divided by 2.

(b) $\text{A} + \text{B} \rightarrow \text{products}$

$$v = -\frac{d[\text{A}]}{dt} = k_r\,[\text{A}][\text{B}] \qquad \text{(Differential form of rate expression.)}$$

$$\frac{dx}{dt} = k_r\,([\text{A}]_0 - x)([\text{B}]_0 - x) \qquad \text{or} \qquad \frac{dx}{([\text{A}]_0 - x)([\text{B}]_0 - x)} = k_r\,dt$$

$$\int_0^x \frac{dx}{([\text{A}]_0 - x)([\text{B}]_0 - x)} = \int_0^t k_r\,dt$$

$$\int_0^x \frac{dx}{([\text{A}]_0 - x)([\text{B}]_0 - x)} = k_r t$$

(i) $[\text{A}]_0 \neq [\text{B}]_0$. Let $[\text{A}] = [\text{A}]_0 - x$, then $d[\text{A}] = -dx$ and $[\text{B}] = [\text{B}]_0 - x$ because of the 1:1 reaction stoichiometry.

We use the standard integral

$$\int \frac{dx}{(a-x)(b-x)} = \frac{1}{b-a}\left(\ln\frac{1}{a-x} - \ln\frac{1}{b-x}\right) + \text{constant}$$

which implies that

$$\int_0^x \frac{dx}{(a-x)(b-x)} = \left\{\frac{1}{b-a}\left(\ln\frac{1}{a-x} - \ln\frac{1}{b-x}\right) + \text{constant}\right\}\Bigg|_x$$

$$- \left\{\frac{1}{b-a}\left(\ln\frac{1}{a-x} - \ln\frac{1}{b-x}\right) + \text{constant}\right\}\Bigg|_0$$

$$= \frac{1}{b-a}\left(\ln\frac{1}{a-x} - \ln\frac{1}{b-x}\right) - \frac{1}{b-a}\left(\ln\frac{1}{a} - \ln\frac{1}{b}\right)$$

$$= \frac{1}{b-a}\ln\left\{\left(\frac{a}{a-x}\right)\left(\frac{b-x}{b}\right)\right\}$$

and the general form of the integrated rate expression becomes

$$\boxed{\frac{1}{[\text{B}]_0 - [\text{A}]_0}\ln\left\{\left(\frac{[\text{A}]_0}{[\text{A}]_0 - x}\right)\left(\frac{[\text{B}]_0 - x}{[\text{B}]_0}\right)\right\} = k_r t}. \qquad \text{(Integrated form of rate expression.)}$$

(ii) $[A]_0 = [B]_0$. We use the standard integral

$$\int \frac{dx}{(a-x)^2} = \frac{1}{a-x} + \text{constant}$$

which implies that

$$\int_0^x \frac{dx}{(a-x)^2} = \left\{\frac{1}{a-x} + \text{constant}\right\}\Bigg|_x - \left\{\frac{1}{a-x} + \text{constant}\right\}\Bigg|_0$$

$$= \frac{1}{a-x} - \frac{1}{a}$$

and the general form of the integrated rate expression becomes

$$\boxed{\frac{1}{[A]_0 - x} - \frac{1}{[A]_0} = k_r t} \quad \text{or} \quad \boxed{x = \frac{[A]_0^2 k_r t}{1+[A]_0 k_r t}}.$$

P10.47 The reaction is

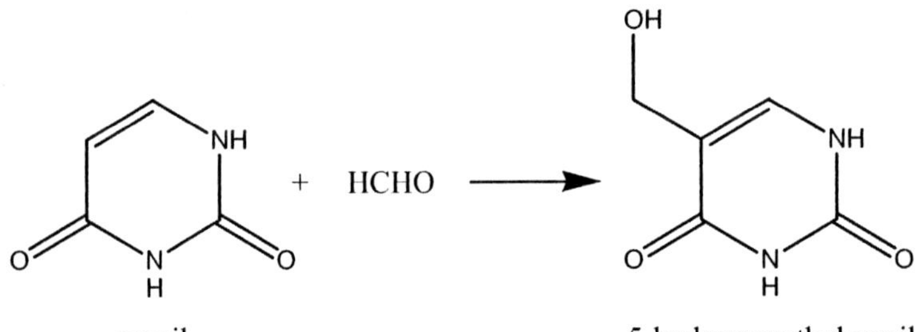

Properties at pH = 7:

$$\log k_r / (\text{dm}^3\ \text{mol}^{-1}\ \text{s}^{-1}) = 11.75 - 5488/(T/\text{K})$$

$$\log K = -1.36 + 1794/(T/\text{K})$$

(a) Plots of k_r and K against T in the range 273 K to 323 K are shown in Figures 10.13 and 10.14. Figure 10.13 shows that, as expected, an increase in T causes an increase in k_r. Figure 10.14 shows a decrease in K with increasing T, a condition that indicates an exothermic reaction. From a kinetic point of view the reaction becomes more favourable at higher temperatures; from a thermodynamic point of view it becomes less favourable.

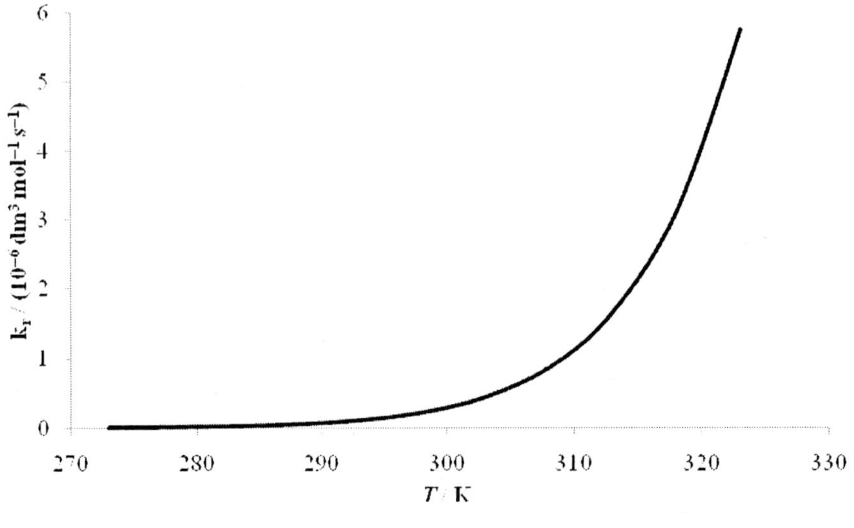

Figure 10.13

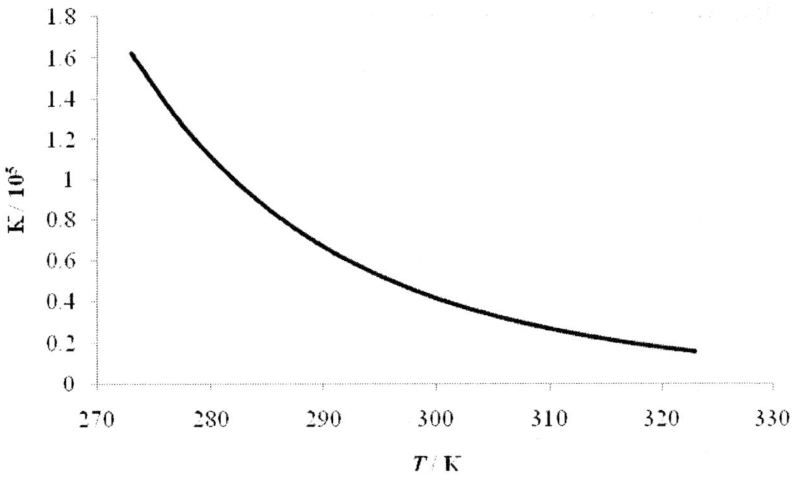

Figure 10.14

(b) Examination of the antilog of the above empirical rate constant relation,

$$k_r/(\mathrm{dm^3\ mol^{-1}\ s^{-1}}) = 10^{11.75-5488/(T/\mathrm{K})} = 10^{11.75}\times 10^{-5488/(T/\mathrm{K})} = (5.62\times 10^{11})\times 10^{-5488/(T/\mathrm{K})},$$

reveals that it has the Arrhenius form

$$k_r = A\,\mathrm{e}^{-E_a/RT} \quad [10.19]$$

with the pre-exponential factor $A = 5.62\times 10^{11}\ \mathrm{dm^3\ mol^{-1}\ s^{-1}}$. The activation energy is deduced by equating the exponentials and solving for E_a.

$$\mathrm{e}^{-E_a/RT} = 10^{-5488/(T/\mathrm{K})}$$

$$-E_a/RT = \ln\left(10^{-5488/(T/\mathrm{K})}\right)$$

$$-E_a/RT = \left(-5488/(T/\mathrm{K})\right)\times \ln 10$$

$$E_a = (\ln 10)\times(5488\ \mathrm{K^{-1}})\times R = (\ln 10)\times(5488\ \mathrm{K})\times(8.3145\ \mathrm{J\ K^{-1}\ mol^{-1}})$$

$$= \boxed{105\ \mathrm{kJ\ mol^{-1}}}$$

The standard biological Gibbs energy (i.e., pH = 7 and other activities equal to 1 in the biological standard state; see Box 7.1) is related to the above empirical expression for K.

$$\Delta_r G^{\oplus} = -RT\ln K\ [7.8] = -RT\ln(10)\log K$$

At 298.15 K,

$$\Delta_r G^{\oplus} = -(8.3145\,\mathrm{J\ K^{-1}\ mol^{-1}})\times(298.15\,\mathrm{K})\times(\ln 10)\times\left(-1.36+\frac{1794}{298.15}\right)$$

$$= \boxed{-26.6\,\mathrm{kJ\ mol^{-1}}}.$$

To find a relation between $\Delta_r H^{\oplus}$ and log$K$, consider the relation

$$\ln K = \frac{-\Delta_r G^{\oplus}}{RT} = -\frac{\Delta_r H^{\oplus}}{RT} + \frac{\Delta_r S^{\oplus}}{R}.$$

We use the approximation that the standard reaction enthalpy and entropy are independent of temperature over the range of interest and take the derivative w/r/t $1/T$. Solving the result for $\Delta_r H^{\oplus}$ yields a very useful relation:

$$\Delta_r H^{\oplus} = -R\frac{\mathrm{d}\ln(K)}{\mathrm{d}(1/T)} = -R\ln(10)\frac{\mathrm{d}\log(K)}{\mathrm{d}(1/T)}.$$

Substitution of the above empirical relation for log$K$ gives the value of $\Delta_r H^{\oplus}$.

$$\Delta_r H^{\oplus} = -R\ln(10)\frac{d}{d(1/T)}\left(-1.36+\frac{1794}{T/K}\right) = -\left(8.3145\,J\,K^{-1}\,mol^{-1}\right)\times\left(\ln 10\right)\times\left(1794\,K\right)$$

$$= \boxed{-34.3\ kJ\ mol^{-1}}$$

(c) The equations for the rate constant k_r and the equilibrium constant K were obtained under conditions corresponding to the biological standard state of pH = 7. Thus the values of $\Delta_r G$ calculated from the equation for K are $\Delta_r G^{\oplus}$ values which can differ significantly from $\Delta_r G^{\ominus}$ for which pH = 1. Prebiotic conditions are more likely to be near pH = 7 than pH = 1 so we expect the part (a) plot of K against T to be relevant to the prebiotic environment. The plot shows that the reaction will be $\boxed{\text{thermodynamically favorable}}$ ($K >> 1$). Because $\Delta_r G = \Delta_r G^{\oplus} + RT\ln Q$ [7.6] and since we might expect $Q < 1$ in a prebiotic environment, $\Delta_r G < \Delta_r G^{\oplus}$. But, as shown in the calculation above, $\Delta_r G^{\oplus}$ is rather large and negative (–26.6 kJ mol^{-1}), so we expect it will still be large and negative under the prebiotic conditions; hence the reaction will be spontaneous for these conditions. We also expect that $\Delta_r H \approx \Delta_r H^{\oplus}$ under prebiotic conditions because enthalpy changes largely reflect bond breakage and bond formation energies.

11 Accounting for the rate laws

Answers to discussion questions

D11.1 Figure 11.1 sketches the concentration variations for second-order reversible steps with an assumed equilibrium constant equal to 1. This figure and Figure 11.2 of the text are qualitatively comparable.

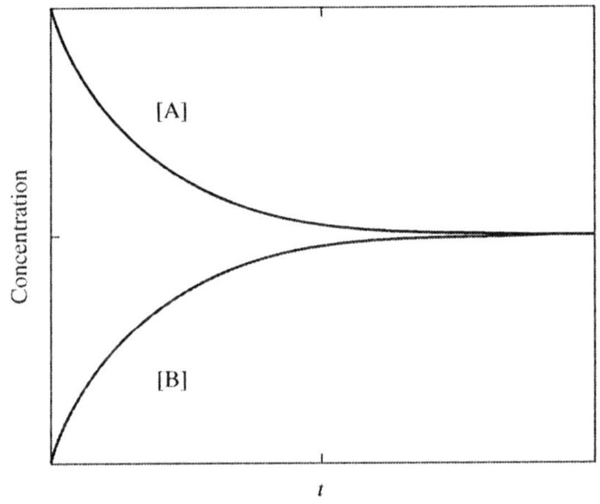

Figure 11.1

D11.2 The rate-determining step is not just the slowest step: it must be slow *and* be a crucial gateway for the formation of products. If a faster reaction can also lead to products, then the slowest step is irrelevant because the slow reaction can then be side-stepped. The rate-determining step is like a slow ferry crossing between two fast highways: the overall rate at which traffic can reach its destination is determined by the rate at which it can make the ferry crossing.

If the first step in a mechanism is the slowest step with the highest activation energy, then it is rate-determining, and the overall reaction rate is equal to the rate of the first step because all subsequent steps are so fast that once the first intermediate is formed it results immediately in the formation of products. Once over the initial barrier, the intermediates cascade into products. However, a rate-determining step may also stem from the low concentration of a crucial reactant or catalyst and need not correspond to the step with highest activation barrier. A rate-determining step arising from the low activity of a crucial enzyme can sometimes be identified by determining whether or not the reactants and products for that step are in equilibrium: if the reaction is not at equilibrium it suggests that the step may be slow enough to be rate-determining.

D11.3 The steady-state approximation is the assumption that the concentrations of all intermediates remain constant and small throughout the reaction (except right at the beginning and right at the end). The mathematical form of the approximation for intermediate I is

$$\text{net rate of formation of I} = 0$$

or with the symbols of calculus

$$(\mathrm{d}[\mathrm{I}]/\mathrm{d}t)_{\mathrm{net}} = 0.$$

A pre-equilibrium approximation is similar in that it is a good approximation when the rate of formation of the intermediate from the reactants and the rate of its reversible decay back to the reactants are both very fast in comparison to the rate of formation of the product from the intermediate. This results in the intermediate being in approximate equilibrium with the reactants over relatively long time periods. Hence the concentration of the intermediate remains approximately constant over the time period that the equilibrium can be considered to be maintained. This allows one to relate the rate constants and concentrations to each other through a constant (the pre-equilibrium constant).

To illustrate the two approximations, consider the symbolic generalization of the gas-phase mechanism for the oxidation of the nitric oxide that is discussed in Section 11.5. The generalization is

$$\mathrm{A}+\mathrm{A} \underset{k_\mathrm{a}'}{\overset{k_\mathrm{a}}{\rightleftharpoons}} \mathrm{I}$$

$$\mathrm{I}+\mathrm{B} \xrightarrow{k_\mathrm{b}} \mathrm{P}$$

Application of the steady-state approximation to intermediate I gives

$$\text{net rate of formation of I} = k_\mathrm{a}[\mathrm{A}]^2 - k_\mathrm{a}'[\mathrm{I}] - k_\mathrm{b}[\mathrm{I}][\mathrm{B}] = 0$$

which implies that

$$[\mathrm{I}] = \frac{k_\mathrm{a}[\mathrm{A}]^2}{k_\mathrm{a}' - k_\mathrm{b}[\mathrm{B}]}$$

It follows that the rate of formation of product is

$$\text{rate of formation of P} = k_\mathrm{b}[\mathrm{I}][\mathrm{B}] = \frac{k_\mathrm{a}k_\mathrm{b}[\mathrm{A}]^2[\mathrm{B}]}{k_\mathrm{a}' - k_\mathrm{b}[\mathrm{B}]} \quad \text{(steady-state approx.)}$$

In contrast, the pre-equilibrium approximation assumes the equilibrium condition that formation rate of I equals the rate at which I decomposes to reactants. Thus, $k_\mathrm{a}[\mathrm{A}]^2 = k_\mathrm{a}'[\mathrm{I}]$ so that

$$\frac{[\mathrm{I}]}{[\mathrm{A}]^2} = \frac{k_\mathrm{a}}{k_\mathrm{a}'} = K$$

which implies that

$$[\mathrm{I}] = K[\mathrm{A}]^2$$

It follows that the rate of formation of product is

$$\text{rate of formation of P} = k_\mathrm{b}[\mathrm{I}][\mathrm{B}] = k_\mathrm{b}K[\mathrm{A}]^2[\mathrm{B}] \quad \text{(pre-equilibrium approx.)}$$

By comparing the rate of formation of P in the steady-state approximation with the rate in the pre-equilibrium approximation, we see that in general they give very different predictions about the concentration dependence of the rate of product formation. They do, however, agree in a special case. When the rate at which I decomposes to reactants is much faster than the rate at which it forms product, $k_\mathrm{a}' \gg k_\mathrm{b}[\mathrm{B}]$ and the second term in the denominator of the steady-state approximation is negligibly small. In this case, the steady-state approximation simplifies to the form provided by the pre-equilibrium approximation. Thus, we see that the steady-state approximation describes a greater range of concentrations and rates than that provided by the pre-equilibrium approximation.

D11.4

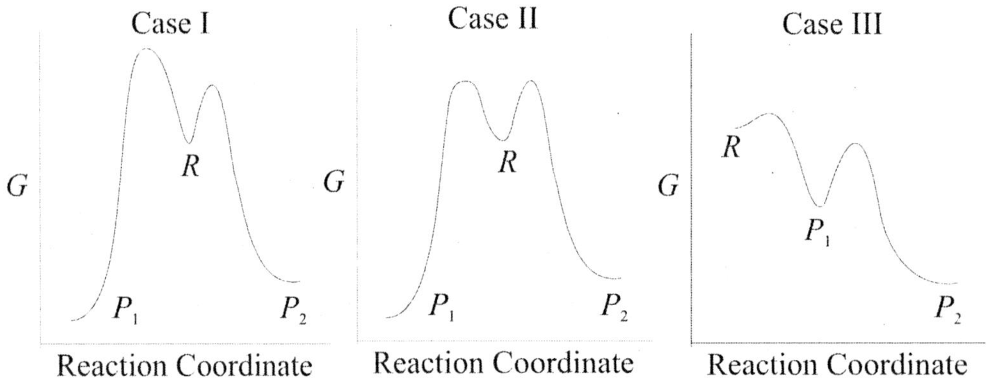

Simple diagrams of Gibbs energy against reaction coordinate are useful for distinguishing between kinetic and thermodynamic control of a reaction. For the simple parallel reactions $R \to P_1$ and $R \to P_2$, shown above as Cases I and II, the product P_1 is thermodynamically favored because the Gibbs energy decreases to a greater extent for its formation. However, the rate at which each product appears does not depend upon thermodynamic favorability. Rate constants depend upon activation energy. In Case I the activation energy for the formation of P_1 is much larger than that for formation of P_2. At low and moderate temperature the large activation energy may not be readily available and P_1 either cannot form or forms at a slow rate. The much smaller activation energy for P_2 formation is available and, consequently, P_2 is produced even though it is not the thermodynamically favor product. This is kinetic control. In this case, $[P_2] / [P_1] = k_{r2}/k_{r1} > 1$ [11.13].

The activation energies for the parallel reactions are equal in Case II and, consequently, the two products appear at identical rates. If the reactions are irreversible, $[P_2] / [P_1] = k_{r2}/k_{r1} = 1$ at all times. The results are very different for reversible reactions. The activation energy for $P_1 \to R$ is much larger than that for $P_2 \to R$ and P_1 accumulates as the more rapid $P_2 \to R \to P_1$ occurs. Eventually the ratio $[P_2] / [P_1]$ approaches the equilibrium value for which

$$\left(\frac{[P_2]}{[P_1]}\right)_{eq} = e^{-(\Delta G_2 - \Delta G_1)/RT} < 1$$

This is thermodynamic control.

Case III above represents an interesting consecutive reaction series $R \to P_1 \to P_2$. The first step has relatively low activation energy and P_1 rapidly appears. However, the relatively large activation energy for the second step is not available at low and moderate temperatures. By using low or moderate temperatures and short reaction times it is possible to produce more of the thermodynamically less favorable P_1. This is kinetic control. High temperatures and long reaction times will yield the thermodynamically favored P_2.

The ratio of reaction products is determined by relative reaction rates in kinetic controlled reactions. Favorable conditions include short reaction times, lower temperatures, and irreversible reactions. Thermodynamic control is favored by long reaction times, higher temperatures, and reversible reactions. The ratio of products depends on the relative stability of products for thermodynamically controlled reactions.

D11.5 The issue of how some gas-phase reactions show first-order kinetics is discussed as an aspect of the **Lindemann mechanism** of unimolecular reactions in text Section 11.9. The reactant molecule acquires the activation energy for reaction in a bimolecular collision. The energized, activated molecule may deactivate by loss of energy in subsequent collisions or it may shake itself apart in a unimolecular elementary step to form product. When the unimolecular step is rate-determining, the overall reaction will have first-order kinetics. Many unimolecular decomposition reactions and isomerization reactions have been observed.

D11.6 The **Michaelis–Menten mechanism** of enzyme activity models the enzyme with one active site, that weakly and reversibly, binds a substrate in homogeneous solution. It is a three-step mechanism. The first and second steps are the reversible formation of the enzyme–substrate complex (ES). The third step is the decay of the complex into the product. The steady-state approximation is applied to the concentration of the intermediate (ES) and its use simplifies the derivation of the final rate expression. However, the justification for the use of the approximation with this mechanism is suspect, in that both rate constants for the reversible step may not be as large, in comparison to the rate constant for the decay to products, as they need to be for the approximation to be valid. The mechanism clearly indicates that the simplest form of the rate law, $v = v_{max} = k_b[E]_0$, occurs when $[S]_0 \gg K_M$ and the general form of the rate law [11.29] does seem to match the principal experimental features of enzyme catalyzed reactions. It provides a mechanistic understanding of both the turnover number and catalytic efficiency. The model may be expanded to include multisubstrate reactions and it must be modified to accommodate the effects of competitive and noncompetitive inhibition.

D11.7 (a) Enzyme-catalyzed reactions that require minutes or hours may be followed with standard spectroscopic methods including uv-visible, infrared, fluorescence, and NMR techniques. It may even be possible to determine the progression of substrate consumption, or product formation, with a chemical titration or radioactivity assay. Electrical conductivity may be used when ions are reaction participants; pH measurements are used to follow the reaction rate when an acid or base is a participant. An inhibitor that binds very strongly to the active site may be used to quench the reaction at any time, thereby, making it possible to separate substrate and product with a chromatographic method; after which spectroscopic techniques, and the application of Beer-Lambert law, may prove useful in the case for which substrate and product have overlapping spectra.

A stopped-flow technique provides for the rapid mixing that is necessary to follow reactions that require milliseconds or minutes. Standard spectroscopic, conductivity, or pH measurements are used to follow the reaction rate.

Very fast reactions may be followed by disturbing an equilibrium system with the excitation energy of flash photolysis or by a very sudden temperature jump initiated with a large current burst through the reaction solution. A pulsed laser beam is subsequently used to generate absorption, emission, or fluorescence spectra and the evolution of such spectra yields reaction rates.

(b) The initial rate of an enzyme-catalyzed reaction is acquired by extrapolation of the time evolution of observed rates to the initial mixing time. When repeated over a range of initial substrate concentrations, it is possible to prepare a double reciprocal, **Lineweaver-Burk plot** of $1/v_0$ against $1/[\mathrm{S}]_0$. It is the intercept and slope of this plot that provides the values of the maximum reaction rate and the Michaelis constant (see eqn 11.30). The manner in which an enzyme inhibitor alters the slope and intercept provides both evidence for the type of inhibition (competitive, uncompetitive, or non-competitive) and values of inhibitor binding constants.

(c) The molecular shape of a strongly enzyme-binding, competitive inhibitor gives clues about the intermediate enzyme-substrate activated complex because, like the inhibitor, the activated ES complex must bind strongly to the active site in order to initiate reaction. Clues include charge distribution, hydrogen bonding, and hydrophobic interactions. The idea of a transition-state intermediate involves a slight modification of the Michaelis–Menten mechanism:

$$\mathrm{E} + \mathrm{S} \rightleftharpoons \mathrm{ES} \rightleftharpoons \mathrm{ES}^* \rightarrow \mathrm{EP} \rightleftharpoons \mathrm{E} + \mathrm{P}.$$

The activated transition-state enzyme-substrate complex, ES*, is a very short lived intermediate because it has the activation energy necessary to react. It has been shown that the enzyme has a higher affinity for the transition-state intermediate than for the substrate and will bind the intermediate more strongly. Good inhibitors are generally transition-state analogs.

Solutions to exercises

E11.1

$$\mathrm{E} + \mathrm{S} \underset{k_\mathrm{r}'}{\overset{k_\mathrm{r}}{\rightleftharpoons}} \mathrm{ES}$$

At equilibrium the forward and reverse rates are equal:

$$k_\mathrm{r}[\mathrm{E}]_\mathrm{eq}[\mathrm{S}]_\mathrm{eq} = k_\mathrm{r}'[\mathrm{ES}]_\mathrm{eq} \quad \text{or} \quad \frac{[\mathrm{ES}]_\mathrm{eq}}{[\mathrm{E}]_\mathrm{eq}[\mathrm{S}]_\mathrm{eq}} = \frac{k_\mathrm{r}}{k_\mathrm{r}'}$$

The latter is closely related to the equilibrium constant.

$$K = \frac{[\mathrm{ES}]_\mathrm{eq}/c^\ominus}{\left([\mathrm{E}]_\mathrm{eq}/c^\ominus\right)\left([\mathrm{S}]_\mathrm{eq}/c^\ominus\right)} = \frac{[\mathrm{ES}]_\mathrm{eq}}{[\mathrm{E}]_\mathrm{eq}[\mathrm{S}]_\mathrm{eq}}c^\ominus = \frac{k_\mathrm{r}}{k_\mathrm{r}'}c^\ominus \quad [11.1]$$

Solve this expression for k_r'.

$$k_\mathrm{r}' = \frac{k_\mathrm{r}c^\ominus}{K} = \frac{\left(1.5\times10^{8}\ \mathrm{dm^3\ mol^{-1}\ s^{-1}}\right)\times\left(1\ \mathrm{mol\ dm^{-3}}\right)}{200} = \boxed{7.5\times10^{5}\ \mathrm{s^{-1}}}$$

E11.2 $NH_3(aq) + H_2O(l) \underset{k_r'}{\overset{k_r}{\rightleftharpoons}} NH_4^+(aq) + OH^-(aq)$

$K_b = [NH_4^+]_{eq}[OH^-]_{eq}/[NH_3]_{eq}\, c^{\ominus} = 1.78 \times 10^{-5}$ at 25°C and $[NH_4^+]_{eq}[OH^-]_{eq} = K_b[NH_3]_{eq}\, c^{\ominus}$

Making the reasonable assumption that both the forward and reverse reactions are second-order kinetics, let $k_{r,eff} = k_r[H_2O]$, where $[H_2O] = 55.6$ mol dm^{-3} for pure water, be the pseudo first-order rate constant for the forward reaction. Then, we write the rate of ammonia formation, $d[NH_3]/dt$, as

$$\frac{d[NH_3]}{dt} = -k_{r,eff}[NH_3] + k_r'[NH_4^+][OH^-]$$

$$= -k_{r,eff}[NH_3] + k_r'[NH_4^+]^2 \text{ (Neglecting the autoprotolysis of water so that } [NH_4^+] = [OH^-])$$

We write $[NH_3] = [NH_3]_{eq} + x$, and $[NH_4^+] = [NH_4^+]_{eq} - x$, and obtain

$$\frac{dx}{dt} = -k_{r,eff}\{[NH_3]_{eq} + x\} + k_r'\{[NH_4^+]_{eq} - x\}^2$$

$$= -k_{r,eff}[NH_3]_{eq} + k_r'[NH_4^+]_{eq}^2 - \{k_{r,eff} + 2k_r'[NH_4^+]_{eq}\}x + k_r'x^2$$

where x is small fraction of $c^{\ominus}$. Consequently, the x^2 term is negligibly smaller than the x term and can be discarded. The first two terms cancel as they represent the fact that the forward and reverse rates are equal at equilibrium. Thus, the working equation becomes

$$\frac{dx}{dt} = -\frac{1}{\tau}x \qquad \text{where } \tau^{-1} = k_{r,eff} + 2k_r'[NH_4^+]_{eq}$$

Solving the equilibrium expression for $[NH_4^+]_{eq}$ gives $[NH_4^+]_{eq} = (K_b[NH_3]_{eq}c^{\ominus})^{1/2}$. Substitution into the expression for the relaxation time τ gives

$$\tau^{-1} = k_{r,eff} + 2k_r'(K_b[NH_3]_{eq}c^{\ominus})^{1/2}$$

The criteria for equilibrium, $k_{r,eff}[NH_3]_{eq} = k_r'[NH_4^+]_{eq}^2$, tells us that

$$k_r' = k_{r,eff}[NH_3]_{eq}/[NH_4^+]_{eq}^2 = k_{r,eff}/(K_b c^{\ominus})$$

We substitute this into the relaxation time expression and solve for $k_{r,eff}$ to find that

$$k_{r,eff} = \frac{\tau^{-1}}{1 + 2\left(\frac{[NH_3]_{eq}}{K_b c^{\ominus}}\right)^{1/2}}$$

$$= \frac{(7.61\times10^{-9}\text{ s})^{-1}}{1 + 2\left(\frac{0.15\text{ mol dm}^{-3}}{1.78\times10^{-5}\text{ mol dm}^{-3}}\right)^{1/2}} = 7.12\times10^5\text{ s}^{-1}$$

$$k_r = \frac{k_{r,eff}}{[H_2O]} = \frac{7.12\times10^5\text{ s}^{-1}}{55.6\text{ mol dm}^{-3}} = \boxed{1.28\times10^4\text{ mol}^{-1}\text{ dm}^3\text{ s}^{-1}}$$

$$k_r' = k_{r,eff}/(K_b c^{\ominus}) = (7.12\times10^5\text{ s}^{-1})/(1.78\times10^{-5}\text{ mol dm}^{-3}) = \boxed{4.00\times10^{10}\text{ mol}^{-1}\text{ dm}^3\text{ s}^{-1}}$$

E11.3 $X \xrightarrow{t_{1/2,a} = 22.5\text{ d}} Y \xrightarrow{t_{1/2,b} = 33.0\text{ d}} Z$

We use equation 11.6 after solving for k_a and k_b from the half-lives.

$$k_a = \frac{\ln 2}{t_{1/2,a}} \,[10.16] = \frac{\ln 2}{22.5\ \text{d}} = 3.08\times10^{-2}\ \text{d}^{-1}$$

$$k_b = \frac{\ln 2}{33.0\ \text{d}} = 2.10\times10^{-2}\ \text{d}^{-1}$$

$$t = \frac{1}{k_a - k_b}\ln\frac{k_a}{k_b}$$

$$= \frac{1}{(3.08-2.10)\times10^{-2}\ \text{d}^{-1}}\ln(3.08/2.10)$$

$$= \boxed{39.1\ \text{d}}$$

E11.4 The first step is rate determining; hence rate = $k_r[H_2O_2][Br^-]$

The reaction is $\boxed{\text{first-order in } H_2O_2 \text{ and in } Br^-}$, and $\boxed{\text{second-order overall}}$.

E11.5

$$A_2 \underset{k_a'}{\overset{k_a}{\rightleftharpoons}} A + A \quad \text{(fast)}$$

$$A + B \xrightarrow{k_b} P \quad \text{(slow)}$$

We assume pre-equilibrium for the fast, reversible step as the slow step is rate-determining, and write

$$K = \frac{[A]^2}{[A_2]c^\ominus} = \frac{k_a}{k_a' c^\ominus}, \quad \text{which implies that } [A] = \left(K[A_2]c^\ominus\right)^{1/2}$$

The rate-determining step then gives

$$\text{rate of product formation} = k_b[A][B] = k_b(Kc^\ominus)^{1/2}[A_2]^{1/2}[B]$$

$$= \boxed{k_{r,\text{eff}}[A_2]^{1/2}[B] \text{ where } k_{r,\text{eff}} = k_b(Kc^\ominus)^{1/2}}$$

E11.6

$$A + B \underset{k_a'}{\overset{k_a}{\rightleftharpoons}} \text{unstable helix} \quad \text{(fast)}$$

$$\text{unstable helix} \xrightarrow{k_b} \text{stable double helix} \quad \text{(slow)}$$

(i) Rate Analysis with Pre-equilibrium Assumption for Formation of Unstable helix

$$K = \frac{[\text{unstable helix}]c^\ominus}{[A][B]} = \frac{k_a c^\ominus}{k_a'}, \text{ indicating that } [\text{unstable helix}] = K[A][B]/c^\ominus$$

The slow rate-determining step then gives

$$\text{rate of double helix formation} = k_b[\text{unstable helix}] = k_b K[A][B]/c^\ominus$$

$$= k_r[A][B] \quad \text{where } k_r = k_b K/c^\ominus \text{ (pre-equilibrium approx.)}$$

(ii) Rate Analysis with the Steady-state Assumption for the Unstable Helix Intermediate

$$\text{net rate of unstable helix formation} = k_a[A][B] - (k_a' + k_b)[\text{unstable helix}] = 0$$

So,

$$[\text{unstable helix}] = k_a[A][B]/(k_a' + k_b)$$

and

$$\text{rate of double helix formation} = k_b[\text{unstable helix}] = k_a k_b[A][B]/(k_a' + k_b)$$

$$= k_r[A][B] \quad \text{where } k_r = k_a k_b/(k_a' + k_b) \text{ (steady-state approx.)}$$

$\boxed{\text{Both the pre-equilibrium approx. and the steady-state approx. predict that the reaction is first-order in A, first-order in B, and second-order overall.}}$ However, the predictions for k_r are generally different functions of the mechanistic rate constants. In the case for which $k_a' \gg k_b$, the predictions for k_r are identical.

E11.7

$$O_3 \underset{k_1'}{\overset{k_1}{\rightleftharpoons}} O_2 + O$$

$$O + O_3 \xrightarrow{k_2} O_2 + O_2$$

The net reaction for the decomposition of atmospheric ozone is $2\,O_3 \rightarrow 3\,O_2$ and the reaction rate, v, is proportional to the net rate of ozone decomposition:

$$v = -\frac{1}{2}\frac{d[O_3]}{dt}$$

In order to simplify the kinetic analysis, we assume that the steady-state approximation applies to [O] (but see the question below). Then,

$$\text{rate of atomic oxygen formation, } \frac{d[O]}{dt} = k_1[O_3] - k_1'[O][O_2] - k_2[O][O_3] = 0$$

Solving for [O],

$$[O] = \frac{k_1[O_3]}{k_1'[O_2] + k_2[O_3]}$$

and

$$\text{net rate of ozone decomposition, } \frac{d[O_3]}{dt} = -k_1[O_3] + k_1'[O][O_2] - k_2[O][O_3]$$

Substituting for [O] from above

$$\frac{d[O_3]}{dt} = -k_1[O_3] + \frac{k_1[O_3](k_1'[O_2] - k_2[O_3])}{k_1'[O_2] + k_2[O_3]}$$

$$= \frac{-k_1[O_3](k_1'[O_2] + k_2[O_3]) + k_1[O_3](k_1'[O_2] - k_2[O_3])}{k_1'[O_2] + k_2[O_3]} = \frac{-2k_1k_2[O_3]^2}{k_1'[O_2] + k_2[O_3]}$$

giving

$$\boxed{v = \frac{k_1k_2[O_3]^2}{k_1'[O_2] + k_2[O_3]}}$$

If the second step is slow, then $k_2[O_3] << k_1'[O_2]$ and the rate reduces to

$$v = \frac{k_1k_2[O_3]^2}{k_1'[O_2]}$$

which is second-order in $[O_3]$ and –1 order in $[O_2]$.

Question. Can you determine the rate law expression if the first step of the proposed mechanism is a rapid pre-equilibrium? Under what conditions does the rate expression above reduce to the case of rapid pre-equilibrium?

E11.8

$$A + M \underset{k_a'}{\overset{k_a}{\rightleftharpoons}} A^* + M$$

$$A^* \xrightarrow{k_b} P$$

This is the Lindemann mechanism for the unimolecular reaction (Section 11.9) in which reactant molecule A becomes energetically excited by collision with any M molecule, which may also be another A molecule provided that M and A have identical rate constants for activation and deactivation. To find a useful analytical form for the rate law, we apply the steady-state approximation to the net rate of formation of the energized species A*.

$$\text{Net rate of A* formation} = k_a[A][M] - (k_a'[M] + k_b)[A^*] = 0$$

This equation solves to

$$[\mathrm{A}^*]=\frac{k_\mathrm{a}[\mathrm{A}][\mathrm{M}]}{k_\mathrm{a}'[\mathrm{M}]+k_\mathrm{b}}$$

so

$$\text{Rate of P formation}=k_\mathrm{b}[\mathrm{A}^*]=\frac{k_\mathrm{a}k_\mathrm{b}[\mathrm{A}][\mathrm{M}]}{k_\mathrm{a}'[\mathrm{M}]+k_\mathrm{b}}$$

$$=\boxed{k_\mathrm{r,eff}[\mathrm{A}]\text{ where }k_\mathrm{r,eff}=\frac{k_\mathrm{a}k_\mathrm{b}[\mathrm{M}]}{k_\mathrm{a}'[\mathrm{M}]+k_\mathrm{b}}}$$

The mathematical form of the effective, unimolecular rate constant $k_\mathrm{r,eff}$ provides insight into an experimental procedure that may either support or refute the mechanism. By taking the inverse of the $k_\mathrm{r,eff}$ expression, we find that

$$\frac{1}{k_\mathrm{r,eff}}=\frac{k_\mathrm{a}'}{k_\mathrm{a}k_\mathrm{b}}+\frac{1}{k_\mathrm{a}[\mathrm{M}]}$$

This suggests that the observed unimolecular rate constant, k_obs, be measured over a range of M concentrations, or pressures. The observed data supports the Lindemann mechanism if a plot of $1/k_\mathrm{obs}$ against 1/[M] is linear. The mechanism is refuted if the plot is non-linear. Rate constant information is extracted from the intercept ($=k_\mathrm{a}'/k_\mathrm{a}k_\mathrm{b}$) and slope ($=1/k_\mathrm{a}$) of a linear plot with the relations $k_\mathrm{a}=1/slope$ and $k_\mathrm{a}'/k_\mathrm{b}=intercept\times k_\mathrm{a}$.

The rate of P formation at the extreme ends of the concentration range are worthy of examination. When [M] is very low, the denominator term $k_\mathrm{a}'[\mathrm{M}]$ is negligibly small compared to k_b and the rate reduces to $k_\mathrm{a}[\mathrm{A}][\mathrm{M}]$, a rate law that is first-order in A, first-order in B, and second-order overall. When [M] is very high, the denominator term k_b is negligibly small compared to $k_\mathrm{a}'[\mathrm{M}]$ and the rate reduces to $\dfrac{k_\mathrm{a}k_\mathrm{b}[\mathrm{A}]}{k_\mathrm{a}'}$, a first-order rate law. The switch from second-order to first-order is evidence for the Lindemann mechanism.

E11.9 Molecules M and molecules A activate and deactivate the activation of A to A* at different rates so we write the Lindemann mechanism for a unimolecular reaction as

$$\mathrm{A}+\mathrm{A}\underset{k_\mathrm{A}'}{\overset{k_\mathrm{A}}{\rightleftharpoons}}\mathrm{A}^*+\mathrm{A}$$

$$\mathrm{A}+\mathrm{M}\underset{k_\mathrm{M}'}{\overset{k_\mathrm{M}}{\rightleftharpoons}}\mathrm{A}^*+\mathrm{M}$$

$$\mathrm{A}^*\xrightarrow{k_\mathrm{b}}\mathrm{P}$$

To find a useful analytical form for the rate law, we apply the steady-state approximation to the net rate of formation of the energized species A*.

$$\text{Net rate of A* formation}=k_\mathrm{A}[\mathrm{A}]^2-\left(k_\mathrm{A}'[\mathrm{A}]+k_\mathrm{b}\right)[\mathrm{A}^*]+k_\mathrm{M}[\mathrm{A}][\mathrm{M}]-k_\mathrm{M}'[\mathrm{M}][\mathrm{A}^*]=0$$

This equation solves to

$$[\mathrm{A}^*]=\frac{(k_\mathrm{A}[\mathrm{A}]+k_\mathrm{M}[\mathrm{M}])[\mathrm{A}]}{k_\mathrm{A}'[\mathrm{A}]+k_\mathrm{M}'[\mathrm{M}]+k_\mathrm{b}}$$

so

$$\text{Rate of P formation}=k_\mathrm{b}[\mathrm{A}^*]=\frac{(k_\mathrm{A}[\mathrm{A}]+k_\mathrm{M}[\mathrm{M}])k_\mathrm{b}[\mathrm{A}]}{k_\mathrm{A}'[\mathrm{A}]+k_\mathrm{M}'[\mathrm{M}]+k_\mathrm{b}}$$

$$=\boxed{k_\mathrm{r,eff}[\mathrm{A}]\text{ where }k_\mathrm{r,eff}=\frac{(k_\mathrm{A}[\mathrm{A}]+k_\mathrm{M}[\mathrm{M}])k_\mathrm{b}}{k_\mathrm{A}'[\mathrm{A}]+k_\mathrm{M}'[\mathrm{M}]+k_\mathrm{b}}}$$

The mathematical form of the effective, unimolecular rate constant $k_{r,eff}$ provides insight into an experimental procedure that may either support or refute the mechanism but the complexity of the above expression forces the examination of the extreme cases for which [M] = 0 and [M] >> [A].

Case for which [M] = 0. This is the case discussed in text Section 11.9. The effective unimolecular rate constant reduces to

$$k_{r,eff} = \frac{k_A k_b [A]}{k_A'[A] + k_b}$$

By taking the inverse of the $k_{r,eff}$ expression, we find that

$$\frac{1}{k_{r,eff}} = \frac{k_A'}{k_A k_b} + \frac{1}{k_A [A]}$$

This suggests that the observed unimolecular rate constant, k_{obs}, be measured over a range of A concentrations, or pressures. The observed data supports the Lindemann mechanism if a plot of $1/k_{obs}$ against 1/[A] is linear. The mechanism is refuted if the plot is non-linear. Rate constant information is extracted from the intercept ($= k_A'/k_A k_b$) and slope ($= 1/k_A$) of a linear plot with the relations $k_A = 1/slope$ and $k_A'/k_b = intercept \times k_A$.

The rate of P formation at the extreme ends of the [A] range are worthy of examination. When [A] is very low, the denominator term $k_A'[A]$ is negligibly small compared to k_b and the rate reduces to $k_A [A]^2$, a second-order rate law. When [A] is very high, the denominator term k_b is negligibly small compared to $k_A'[A]$ and the rate reduces to $\frac{k_A k_b [A]}{k_A'}$, a first-order rate law. The switch from second-order to first-order is evidence for the Lindemann mechanism in A.

Case for which [M] >> [A]. The effective unimolecular rate constant reduces to

$$k_{r,eff} = \frac{k_M k_b [M]}{k_M'[M] + k_b}$$

By taking the inverse of the $k_{r,eff}$ expression, we find that

$$\frac{1}{k_{r,eff}} = \frac{k_M'}{k_M k_b} + \frac{1}{k_M [M]}$$

This suggests that the observed unimolecular rate constant, k_{obs}, be measured over a range of M concentrations, or partial pressures. The observed data supports the Lindemann mechanism in M if a plot of $1/k_{obs}$ against 1/[M] is linear. The mechanism is refuted if the plot is non-linear. Rate constant information is extracted from the intercept ($= k_M'/k_M k_b$) and slope ($= 1/k_M$) of a linear plot with the relations $k_M = 1/slope$ and $k_M'/k_b = intercept \times k_M$. When [M] is so very high that the denominator term k_b is negligibly small compared to $k_M'[M]$, the unimolecular rate reduces to $\frac{k_M k_b [A]}{k_M'}$, which is zero-order in M and first-order in A.

E11.10 For the simple irreversible, parallel reactions $R \rightarrow P_1$ and $R \rightarrow P_2$, shown in Figure 11.2, the product P_1 is thermodynamically favored. However, the rate at which each product appears does not depend upon thermodynamic favorability. Rate constants depend upon activation energy according to the expression $k_r = Ae^{-E_a/RT}$ [10.19]. The activation energy for the formation of P_1 is much larger than that for formation of P_2. At low and moderate temperature, the large activation energy may not be readily available in the environment of the reactant R and P_1 either cannot form or forms at a slow rate. The much smaller activation energy for P_2 formation is readily available and, consequently, P_2 is produced even though it is not the thermodynamically favored product. This kinetic control yields the ratio $[P_1]/[P_2] = k_{r1}/k_{r2} < 1$. Details of the temperature dependence can be mathematically examined.

$$\frac{[P_1]}{[P_2]} = \frac{k_{r1}}{k_{r2}} = \frac{A_1 e^{-E_{a1}/RT}}{A_2 e^{-E_{a2}/RT}} [10.19] = \left(\frac{A_1}{A_2}\right) e^{-(E_{a1}-E_{a2})/RT} = \left(\frac{A_1}{A_2}\right) e^{-\Delta E_a/RT} \quad \text{where } \Delta E_a = E_{a1} - E_{a2} > 0$$

Since as $T \rightarrow$ larger, $\Delta E_a/RT \rightarrow$ smaller and $e^{-\Delta E_a/RT} \rightarrow$ larger (but is always less than 1). Consequently, the above equation indicates that as $T \rightarrow$ larger, the concentration ratio $[P_1]/[P_2] \rightarrow$ larger. This is expected because the high requisite activation energy for the formation of P_1 is available in the high temperature environment of R.

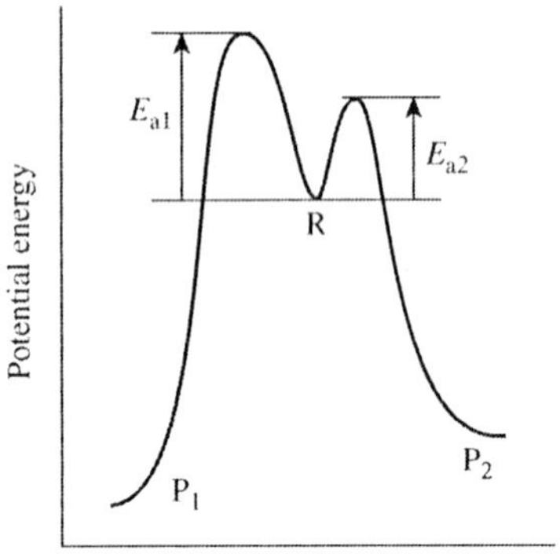

Figure 11.2

E11.11 Since molar concentration is proportional to partial pressure, the Lindemann unimolecular mechanism and rate constants may be written in terms of partial pressures to give (see Derivation 11.4):

$$\text{Rate of product formation} = k_b p_{A*} = k_{r,\text{eff}}\, p_A \qquad \text{where } k_{r,\text{eff}} = \frac{k_a k_b p_A}{k_a' p_A + k_b}$$

Taking the inverse of the effective rate constant gives the relation

$$\frac{1}{k_{r,\text{eff}}} = \frac{k_a'}{k_a k_b} + \frac{1}{k_a p_A}$$

which indicates that a plot of $1/k_{r,\text{eff}}$ against $1/p_A$ is linear with a slope (i.e., $\Delta(1/k_{r,\text{eff}})/\Delta(1/p_A)$) that equals $1/k_a$. Thus,

$$k_a = 1/slope = \frac{\Delta(1/p_A)}{\Delta(1/k_{r,\text{eff}})}$$

$$= \frac{\frac{1}{1.30\times10^3\ \text{Pa}} - \frac{1}{12\ \text{Pa}}}{\frac{1}{2.50\times10^{-4}\ \text{s}^{-1}} - \frac{1}{2.10\times10^{-5}\ \text{s}^{-1}}}$$

$$= \boxed{1.89\times10^{-6}\ \text{Pa}^{-1}\ \text{s}^{-1}}$$

E11.12

$$k_{r,d} = \frac{8RT}{3\eta} [11.17] = \frac{8\times(8.3145\ \text{J K}^{-1}\text{mol}^{-1})\times(298\,\text{K})}{3\eta}$$

$$= \frac{6.61\times10^3\ \text{J mol}^{-1}}{\eta} = \frac{6.61\times10^3\ \text{kg m}^2\text{s}^{-2}\ \text{mol}^{-1}}{(\eta/\text{kg m}^{-1}\text{s}^{-1})\times\text{kg m}^{-1}\text{s}^{-1}} = \frac{6.61\times10^3\ \text{m}^3\ \text{mol}^{-1}\ \text{s}^{-1}}{(\eta/\text{kg m}^{-1}\text{s}^{-1})}$$

(a) Water, $\eta = 1.00\times10^{-3}\ \text{kg m}^{-1}\text{s}^{-1}$

$$k_{r,d} = \frac{6.61\times10^3}{1.00\times10^{-3}}\ \text{m}^3\ \text{mol}^{-1}\ \text{s}^{-1}$$

$$= \boxed{6.61\times10^6\ \text{m}^3\ \text{mol}^{-1}\ \text{s}^{-1}}$$

(b) Pentane, $\eta = 0.22\times10^{-3}\ \mathrm{kg\,m^{-1}s^{-1}}$

$$k_{\mathrm{r,d}} = \frac{6.61\times10^{3}}{2.2\times10^{-4}}\ \mathrm{m^{3}\ mol^{-1}\ s^{-1}}$$

$$= \boxed{3.0\times10^{7}\ \mathrm{m^{3}\ mol^{-1}\ s^{-1}}}$$

E11.13 (a) $J = -D\times(\text{concentration gradient})$ Fick's first law [11.18b]

$$= -\left(5.22\times10^{-10}\ \mathrm{m^{2}\ s^{-1}}\right)\times\left(-0.10\ \mathrm{mol\ dm^{-3}\ m^{-1}}\right)\times\left(\frac{1\ \mathrm{dm^{3}}}{10^{-3}\ \mathrm{m^{3}}}\right)$$

$$= \boxed{-5.2\times10^{-8}\ \mathrm{mol\ m^{-2}\ s^{-1}}}$$

The negative value indicates the flow of mass from high to low concentration.

(b) Let n be the number of moles of molecules that flow down the concentration gradient through area A, which is perpendicular to the concentration gradient, in time Δt. Then, with the gradient express in molar concentration per meter:

$$n = JA\Delta t \quad [11.18\mathrm{a}]$$

$$= \left(5.2\times10^{-8}\ \mathrm{mol\ m^{-2}\ s^{-1}}\right)\times\left(5.0\ \mathrm{mm^{2}}\right)\times\left(60\ \mathrm{s}\right)\times\left(\frac{10^{-6}\ \mathrm{m^{2}}}{1\ \mathrm{mm^{2}}}\right)$$

$$= \boxed{1.6\times10^{-11}\ \mathrm{mol}}$$

E11.14 $t = d^{2}/2D$ [11.20]

$$= \frac{d^{2}}{2\left(5.22\times10^{-10}\ \mathrm{m^{2}\ s^{-1}}\right)} \quad \text{[sucrose, Table 11.1]}$$

(a) $$t = \frac{\left(10\times10^{-3}\ \mathrm{m}\right)^{2}}{2\left(5.22\times10^{-10}\ \mathrm{m^{2}\ s^{-1}}\right)} = 9.6\times10^{4}\ \mathrm{s} = \boxed{27\ \mathrm{h}}$$

(b) $$t = \frac{\left(10\times10^{-2}\ \mathrm{m}\right)^{2}}{2\left(5.22\times10^{-10}\ \mathrm{m^{2}\ s^{-1}}\right)} = \boxed{2.7\times10^{3}\ \mathrm{h}}$$

(c) $$t = \frac{\left(10\ \mathrm{m}\right)^{2}}{2\left(5.22\times10^{-10}\ \mathrm{m^{2}\ s^{-1}}\right)} = \boxed{3.0\times10^{3}\ \mathrm{a}}$$

These times are so long that we obviously do not wait for diffusion to move mass across macroscopic distances. Stirring and convective motion have much higher mixing rates on the macroscopic scale.

E11.15 (a) $$D = \frac{\lambda^{2}}{2\tau} \quad [11.21]$$

$$= \frac{\left(150\times10^{-12}\ \mathrm{m}\right)^{2}}{2\left(1.8\times10^{-12}\ \mathrm{s}\right)} = \boxed{6.3\times10^{-9}\ \mathrm{m^{2}\ s^{-1}}}$$

(b) $$D = \frac{\lambda^{2}}{2\tau} \quad [11.21]$$

$$= \frac{\left(75\times10^{-12}\ \mathrm{m}\right)^{2}}{2\left(1.8\times10^{-12}\ \mathrm{s}\right)} = \boxed{1.6\times10^{-9}\ \mathrm{m^{2}\ s^{-1}}}$$

E11.16 Diffusion is only important in the absence of macroscopic fluid flow or convection or turbulence as may be the case when microscopic spatial constraints exist such as passage of gases through lung alveoli or the passage of neurotransmitter molecules across a synaptic gap. An extraordinarily large time is required for a molecule to move across a lake by diffusion alone. A pollutant with a radius of 100 pm (the radius of a water molecule) requires about 90 millennia to travel across of 100 m lake at the mean surface temperature (15°C).

$$D_{15°C} = \frac{kT}{6\pi\eta a} \quad [11.23]$$

$$= \frac{(1.381\times10^{-23}\ \text{J K}^{-1})\times(288\ \text{K})}{6\pi(1.139\times10^{-3}\ \text{kg m}^{-1}\ \text{s}^{-1})\times(100\times10^{-12}\ \text{m})} \quad \text{[text Figure 11.11]}$$

$$= 1.85\times10^{-9}\ \text{m}^2\ \text{s}^{-1}$$

$$t_{H_2O} \simeq \frac{d^2}{2D}[11.20] = \frac{(100\ \text{m})^2}{2(1.86\times10^{-9}\ \text{m}^2\ \text{s}^{-1})} = 2.70\times10^{12}\ \text{s} = \boxed{8.55\times10^4\ \text{a}}$$

E11.17 Let N be the number of one-dimensional steps of length λ that a molecule takes in time t and let τ be the time each step takes; $\tau = t/N$.

$$D = \frac{\lambda^2}{2\tau}[11.21] = \frac{\lambda^2 N}{2t} \quad \text{or} \quad t = \frac{\lambda^2 N}{2D}$$

Substitution into [11.20] gives:

$$d = (2Dt)^{1/2} = \left(\frac{2D\lambda^2 N}{2D}\right)^{1/2} = (\lambda^2 N)^{1/2} \quad \text{or} \quad N = \left(\frac{d}{\lambda}\right)^2$$

$$\text{For } d = 1000\ \lambda, \qquad N = \left(\frac{1000\ \lambda}{\lambda}\right)^2 = \boxed{1\times10^6\ \text{steps}}$$

E11.18

$$\eta = \eta_0 e^{E_a/RT} \quad [11.24]$$

$$\frac{\eta_{T_1}}{\eta_{T_2}} = \frac{e^{E_a/RT_1}}{e^{E_a/RT_2}} = e^{\frac{E_a}{R}\left(\frac{1}{T_1}-\frac{1}{T_2}\right)}$$

$$\ln\left(\frac{\eta_{T_1}}{\eta_{T_2}}\right) = \frac{E_a}{R}\left(\frac{1}{T_1}-\frac{1}{T_2}\right) \quad \text{or} \quad E_a = R\left(\frac{1}{T_1}-\frac{1}{T_2}\right)^{-1}\ln\left(\frac{\eta_{T_1}}{\eta_{T_2}}\right)$$

$$E_a = (8.3145\ \text{J mol}^{-1}\ \text{K}^{-1})\times\left(\frac{1}{293\ \text{K}}-\frac{1}{303\ \text{K}}\right)^{-1}\ln\left(\frac{1.0019}{0.7982}\right) = \boxed{16.8\ \text{kJ mol}^{-1}}$$

E11.19

$$k_{\text{rate}} = \frac{kT}{h}e^{-\Delta^{\ddagger}G/RT} \quad [10.23 \text{ with } \kappa = 1]$$

The ratio of the rates is then given by

$$\frac{k(\text{cat})}{k(\text{uncat})} = e^{[-(15-150)\ \text{kJ mol}^{-1}/(8.3145\times10^{-3}\ \text{kJ mol}^{-1}\ \text{K}^{-1}\times310\ \text{K})]}$$

$$= \boxed{5.6\times10^{22}}$$

E11.20

$$AH + B \xrightarrow{k_{r1}} BH^+ + A^-$$

$$A^- + BH^+ \xrightarrow{k_{r2}} AH + B$$

$$A^- + HA \xrightarrow{k_{r3}} P$$

Apply the steady-state approximation to the carbanion A^-.

$$\text{Net rate of A}^- \text{ formation} = k_{r1}[HA][B] - k_{r2}[A^-][BH^+] - k_{r3}[A^-][HA] = 0$$

Therefore $$\boxed{[A^-] = \frac{k_{r1}[HA][B]}{k_{r2}[BH^+] + k_{r3}[HA]}}$$

and the rate of formation of product is

$$\text{Net rate of P formation} = k_{r3}[HA][A^-] = \boxed{\frac{k_{r1}k_{r3}[HA]^2[B]}{k_{r2}[BH^+] + k_{r3}[HA]}}$$

E11.21 $$HA + H^+ \underset{k_a'}{\overset{k_a}{\rightleftharpoons}} HAH^+ \quad \text{(fast)}$$

$$HAH^+ + B \xrightarrow{k_b} BH^+ + HA \quad \text{(slow)}$$

The rate of production of the product is

$$\frac{d[BH^+]}{dt} = k_b[HAH^+][B]$$

HAH^+ is an intermediate involved in a rapid pre-equilibrium

$$\frac{[HAH^+]}{[HA][H^+]} = \frac{k_a}{k_a'} \quad \text{so } [HAH^+] = \frac{k_a[HA][H^+]}{k_a'}$$

and $$\frac{d[BH^+]}{dt} = \boxed{\frac{k_a k_b}{k_a'}[HA][H^+][B]}$$

This rate law can be made independent of $[H^+]$, if the source of H^+ is the weak acid HA, for then H^+ is given by another equilibrium.

$$\frac{[H^+][A^-]}{[HA]} = K_a = \frac{[H^+]^2}{[HA]} \quad \text{so } [H^+] = (K_a[HA])^{1/2}$$

and $$\frac{d[BH^+]}{dt} = \boxed{\frac{k_a k_b K_a^{1/2}}{k_a'}[HA]^{3/2}[B]}$$

E11.22 $$E + S \underset{k_a'}{\overset{k_a}{\rightleftharpoons}} ES \xrightarrow{k_b} E + P$$

In the pre-equilibrium approximation, the intermediate ES is in equilibrium with the reactants E and S. This requires equality between the rate for formation of ES and the rate of ES dissociation to reactants.

rate of formation of ES = rate of ES dissociation to reactants

$$k_a[E][S] = k_a'[ES]$$

$$\frac{[ES]}{[E][S]} = \frac{k_a}{k_a'} = K \quad \text{or} \quad [ES] = K[E][S]$$

If $[E]_0$ is the total enzyme concentration, then $[E] = [E]_0 - [ES]$ by conservation of mass.

$$[ES] = K[E][S] = K([E]_0 - [ES])[S]$$

$$[ES] + \frac{[ES]}{K[S]} = [E]_0$$

$$[ES] = \frac{[E]_0}{1 + \frac{1}{K[S]}} = \frac{[E]_0[S]}{[S] + \frac{1}{K}}$$

Substitution in the expression for the rate of formation of P gives:

$$\boxed{\text{rate of formation of P} = k_b[\text{ES}] = \frac{k_b[\text{E}]_0[\text{S}]}{[\text{S}]+\dfrac{1}{K}}} \qquad \text{pre-equilibrium approximation, } K = \frac{k_a}{k_a'}$$

$$\text{rate of formation of P} = k_b[\text{ES}] = \frac{k_b[\text{E}]_0[\text{S}]}{[\text{S}]+K_M} \qquad \text{steady-state approximation [11.25]}$$

Comparison of the pre-equilibrium approximation with the steady-state approximation shows that the two approximations are the same when $\dfrac{1}{K} = K_M$.

Since $\dfrac{1}{K} = \dfrac{k_a'}{k_a}$ and $K_M = \dfrac{k_a'+k_b}{k_a}$ [11.26], the two approximations are identical when $\boxed{k_a' \gg k_b}$.

E11.23 Rate for formation of P, $v = k_r[\text{E}]_0 = \dfrac{k_b[\text{S}][\text{E}]_0}{K_M+[\text{S}]} = \dfrac{[\text{S}]v_{max}}{K_M+[\text{S}]}$ [11.25, 11.31]

Solving the reaction rate expression for v_{max} gives

$$\begin{aligned} v_{max} &= \frac{(K_M+[\text{S}])v}{[\text{S}]} \\ &= \frac{(0.045\ \text{mol dm}^{-3}+0.110\ \text{mol dm}^{-3})\times(1.15\ \text{mmol dm}^{-3}\ \text{s}^{-1})}{0.110\ \text{mol dm}^{-3}} \\ &= \boxed{1.62\ \text{mmol dm}^{-3}\ \text{s}^{-1}} \end{aligned}$$

E11.24 $v_{max} = k_b[\text{E}]_0 = k_{cat}[\text{E}]_0$ [11.31]

Thus,

$$\begin{aligned} k_{cat} &= \frac{v_{max}}{[\text{E}]_0} \\ &= \frac{4.25\times10^{-4}\ \text{mol dm}^{-3}\ \text{s}^{-1}}{3.60\times10^{-9}\ \text{mol dm}^{-3}} = \boxed{1.18\times10^{5}\ \text{s}^{-1}} \end{aligned}$$

$$\begin{aligned} \eta &= \frac{k_{cat}}{K_M} \quad [11.32] \\ &= \frac{1.18\times10^{5}\ \text{s}^{-1}}{0.015\ \text{mol dm}^{-3}} = \boxed{7.9\times10^{6}\ \text{dm}^3\ \text{mol}^{-1}\ \text{s}^{-1}} \end{aligned}$$

This enzyme is not 'catalytically perfect' as the catalytic efficiency is much less than the diffusion controlled rate constant of eqn 11.17, $k_{r,d} = 7.4\times10^{9}\ \text{dm}^3\ \text{mol}^{-1}\ \text{s}^{-1}$.

E11.25 We fit the data to the Lineweaver-Burk equation [11.30].

$$\frac{1}{v} = \frac{1}{v_{max}} + \left(\frac{K_M}{v_{max}}\right)\frac{1}{[\text{S}]} \qquad [11.30]$$

Hence, draw up the following table.

[ATP] / μmol dm^{-3}	0.60	0.80	1.4	2.0	3.0
v / μmol dm^{-3} s^{-1}	0.81	0.97	1.30	1.47	1.69
1/[ATP] / μmol^{-1} dm^3	1.67	1.25	0.71	0.50	0.33
$1/v$ / μmol^{-1} dm^3 s	1.23	1.03	0.769	0.680	0.592

A plot of $1/v$ against $1/[S]$ is shown in the Figure 11.3. The plot is linear and the linear least-squares regression fit is shown as a box insert.

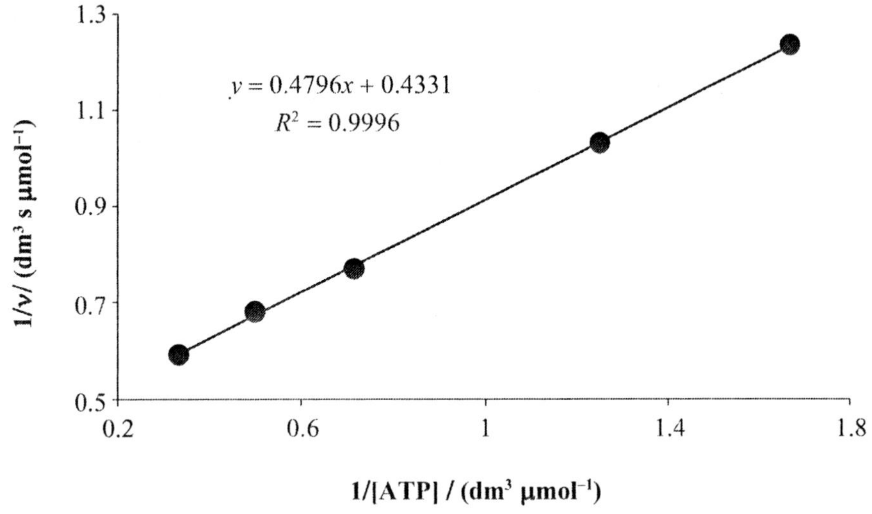

Figure 11.3

The intercept is $1/v_{max} = 0.433\ dm^3\ s\ \mu mol^{-1}$. Therefore, $v_{max} = \boxed{2.31\ \mu mol\ dm^{-3}\ s^{-1}}$

The slope is $K_M / v_{max} = 0.480$ s. Therefore,

$$K_M = (0.480\ s) \times (2.31\ \mu mol\ dm^{-3} s^{-1}) = \boxed{1.11\ \mu mol\ dm^{-3}}$$

The maximum turnover number k_{cat} is

$$k_{cat} = \frac{v_{max}}{[E]_0}\ [11.31] = \frac{2.31\ \mu mol\ dm^{-3}\ s^{-1}}{0.020\ \mu mol\ dm^{-3}} = \boxed{1.2 \times 10^2\ s^{-1}}$$

E11.26 (a) We start with the Lineweaver-Burk expression:

$$\frac{1}{v} = \frac{1}{v_{max}} + \left(\frac{K_M}{v_{max}}\right)\frac{1}{[S]} \quad [11.30]$$

Multiply both sides of this equation by $v \times v_{max}$.

$$v_{max} = v + K_M\left(\frac{v}{[S]}\right) \quad \text{or} \quad \boxed{\frac{v}{[S]} = \frac{v_{max}}{K_M} - \frac{v}{K_M}} \quad \text{Eadie–Hofstee relation}$$

Multiplication of the Lineweaver–Burk expression by [S] gives

$$\boxed{\frac{[S]}{v} = \frac{[S]}{v_{max}} + \frac{K_M}{v_{max}}} \quad \text{[Hane's relation]}$$

(b) Examination of the above Eadie–Hofstee relation reveals that, if the Michaelis–Menten mechanism is valid, a plot of $v/[S]$ against v should be linear with a slope equal to $-1/K_M$ and an intercept equal to v_{max}/K_M. Thus, $K_M = -1/slope$ and $v_{max} = intercept \times K_M$.

The above Hane's relation reveals that a plot of $[S]/v$ against [S] should be linear with a slope equal to $1/v_{max}$ and an intercept equal to $K_M/\ v_{max}$. Thus, $v_{max} = 1/slope$ and $K_M = intercept \times v_{max}$.

(c) An Eadie–Hofstee plot of the appropriately transformed data of Exercise 11.25 is shown in Figure 11.4 while the Hane's plot is shown in Figure 11.5. The plots are linear and the linear least-squares regression fits are shown as box inserts.

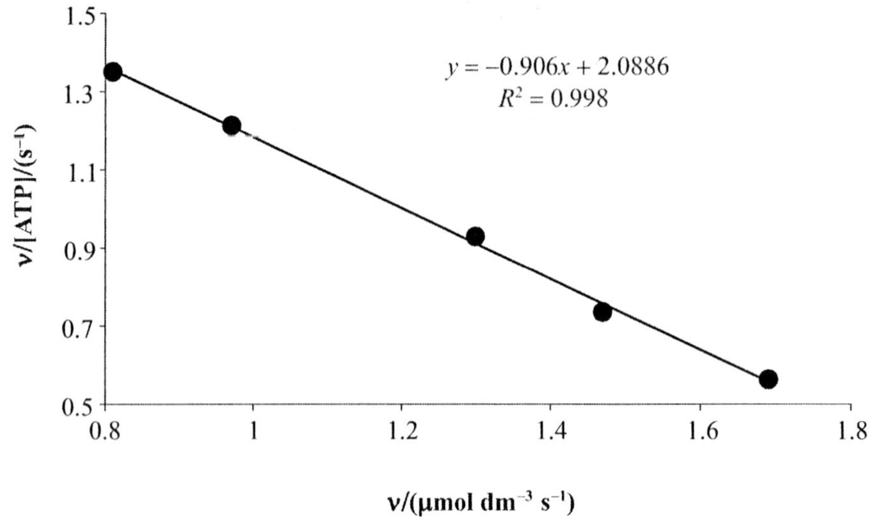

Figure 11.4

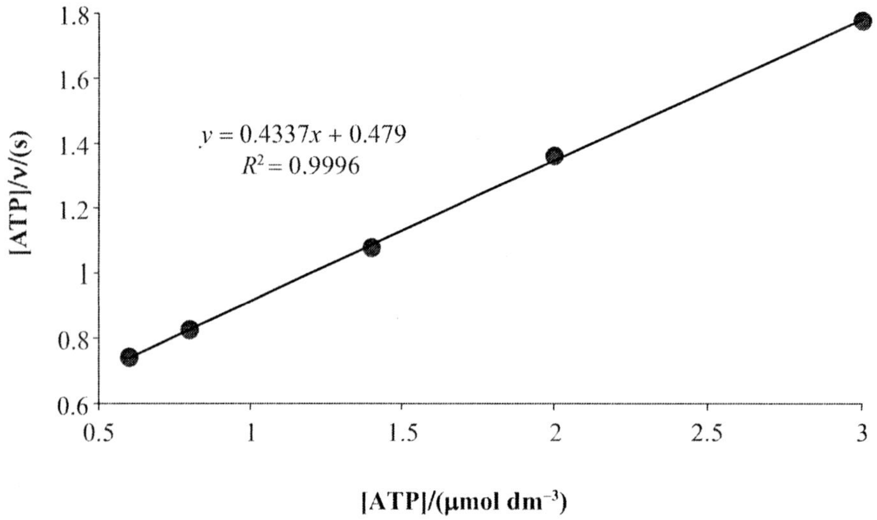

Figure 11.5

Analysis of Eadie–Hofstee plot

$$K_M = -1/slope = -1/(-0.906\ \text{dm}^3\ \mu\text{mol}^{-1}) = \boxed{1.10\ \mu\text{mol dm}^{-3}}$$

$$v_{max} = intercept \times K_M = (2.089\ \text{s}^{-1}) \times (1.10\ \mu\text{mol dm}^{-3}) = \boxed{2.30\ \mu\text{mol dm}^{-3}\ \text{s}^{-1}}$$

Analysis of Hane's plot

$$v_{max} = 1/slope = 1/(0.4337\ \text{s dm}^3\ \mu\text{mol}^{-1}) = \boxed{2.31\ \mu\text{mol dm}^{-3}\ \text{s}^{-1}}$$

$$K_M = intercept \times v_{max} = (0.479\ \text{s}) \times (2.31\ \mu\text{mol dm}^{-3}\ \text{s}^{-1}) = \boxed{1.11\ \mu\text{mol dm}^{-3}}$$

The values for K_M and v_{max} as determined by a Lineweaver-Burk plot (Exercise 11.25), a Eadie–Hofstee plot, and a Hane's plot are in agreement for the data of these exercises. However, if the data were clumped at either low or high substrate concentrations one-or-the-other plot would give better results.

E11.27

$$AH \xrightarrow{k_a} A\cdot + H\cdot \qquad \boxed{\text{initiation}}$$

$$A\cdot \xrightarrow{k_b} B\cdot + C \qquad \boxed{\text{propagation}}$$

$$AH + B\cdot \xrightarrow{k_c} A\cdot + D \qquad \boxed{\text{propagation}}$$

$$A\cdot + B\cdot \xrightarrow{k_d} P \qquad \boxed{\text{termination}}$$

Net rate of AH reaction $= -k_a[AH] - k_c[AH][B\cdot] = \boxed{-k_{r,eff}[AH] \text{ where } k_{r,eff} = k_a + k_c[B\cdot]}$

In order to simplify the analysis of the kinetics, apply the steady-state approximation to radical intermediates $A\cdot$ and $B\cdot$ (except at the beginning and end of the reaction). Thus, since $[B\cdot]$ is approximately constant, we initially expect that the above expression for $k_{r,eff}$ is approximately a constant and the net rate of the AH reaction is $\boxed{\text{first-order in AH}}$. To confirm this, we must carefully check that $[B\cdot]$ is a function of rate constants only and does not depend upon [AH]. It turns out that $[B\cdot]$ is a complicated function of the rate constants only. To find this function, apply the steady-state approximation to $[A\cdot]$ and $[B\cdot]$ and solve the equations for $[B\cdot]$.

(i) Net rate of reaction of $A\cdot = k_a[AH] - k_b[A\cdot] + k_c[AH][B\cdot] - k_d[A\cdot][B\cdot] \approx 0$

(ii) Net rate of $B\cdot$ reaction $= k_b[A\cdot] - k_c[AH][B\cdot] - k_d[A\cdot][B\cdot] \approx 0$

Adding expression (ii) to expression (i) gives

(iii) $$[A\cdot] = \left(\frac{k_a}{2k_d}\right)\frac{[AH]}{[B\cdot]}$$

Subtracting expression (ii) from expression (i) gives

(iv) $$[A\cdot] = \left(\frac{k_a + 2k_c[B\cdot]}{2k_b}\right)[AH]$$

Equating expressions (iii) and (iv) and solving for $[B\cdot]$ gives

$$\boxed{[B\cdot] = \frac{-k_a k_d + \left(k_a^2 k_d^2 + 8k_a k_b k_c k_d\right)^{1/2}}{4k_c k_d}}$$

E11.28

$$R_2 \xrightarrow{k_a} R + R \qquad \text{initiation}$$

$$R + R_2 \xrightarrow{k_b} P_B + R' \qquad \text{propagation}$$

$$R' \xrightarrow{k_c} P_A + R \qquad \text{propagation}$$

$$R + R \xrightarrow{k_d} P_A + P_B \qquad \text{termination}$$

Net rate of R_2 reaction $= -k_a[R_2] - k_b[R][R_2] = -k_{r,eff}[R_2]$ where $k_{r,eff} = k_a + k_b[R]$

In order to simplify the analysis of the kinetics, apply the steady-state approximation to radical intermediates R and R′ (except at the beginning and end of the reaction). Thus, since [R] is approximately constant, the above expression for $k_{r,eff}$ is expected to approximate a constant and the rate expression may approximate first-order kinetics. However, we must carefully check that $k_{r,eff}$ depends upon rate constants only because, if the steady-state approximation predicts that $k_{r,eff}$ has a concentration dependence, the rate may prove not to be first-order. To find a radical-free expression for [R], apply the steady-state approximation to R and R′ and solve the equations for [R].

(i) Net rate of R′ reaction $= k_b[R][R_2] - k_c[R'] = 0$

(ii) Net rate of R reaction $= 2k_a[R_2] - k_b[R][R_2] + k_c[R'] - 2k_d[R]^2 = 0$

Adding relations (i) and (ii) and solving for [R] gives

$$[\mathrm{R}]=\left(\frac{k_\mathrm{a}}{k_\mathrm{d}}[\mathrm{R}_2]\right)^{1/2}$$

Thus,

$$k_{\mathrm{r,eff}}=k_\mathrm{a}+k_\mathrm{b}[\mathrm{R}]=k_\mathrm{a}+k_\mathrm{b}\left(\frac{k_\mathrm{a}}{k_\mathrm{d}}[\mathrm{R}_2]\right)^{1/2}$$

and

$$\text{Net rate of R}_2\text{ reaction }=-k_{\mathrm{r,eff}}[\mathrm{R}_2]=\boxed{-k_\mathrm{a}[\mathrm{R}_2]-k_\mathrm{b}\left(\frac{k_\mathrm{a}}{k_\mathrm{d}}\right)^{1/2}[\mathrm{R}_2]^{3/2}}$$

$k_{\mathrm{r,eff}}$ has proven to be concentration dependent. Thus, the rate law is not first-order. As shown in the above relation for the net rate of R_2 reaction, the rate law is complicated and has an indefinite order.

E11.29 (a) Observed rate of formation of HBr $=\dfrac{k_{\mathrm{r1}}[\mathrm{H_2}][\mathrm{Br_2}]^{3/2}}{[\mathrm{Br_2}]+k_{\mathrm{r2}}[\mathrm{HBr}]}$ [11.33]

The radical chain mechanism for the formation of HBr(g) from the elements is discussed in detail in Section 11.15. Using the steady-state approximation for the intermediate of atomic hydrogen and atomic bromine, the mechanism indicates that

$$\text{rate of formation of HBr}=\frac{2k_\mathrm{b}(k_\mathrm{a}/k_\mathrm{e})^{1/2}[\mathrm{H_2}][\mathrm{Br_2}]^{3/2}}{[\mathrm{Br_2}]+(k_\mathrm{d}/k_\mathrm{c})[\mathrm{HBr}]}\quad[11.35]$$

This rate has the same mathematical form as the experimental rate law and, although the form does not prove that the proposed mechanism is correct, the match does give supporting evidence for the proposed mechanism. Term-by-term comparison of eqns 11.33 and 11.35 indicates that

$$k_{\mathrm{r1}}=2k_\mathrm{b}(k_\mathrm{a}/k_\mathrm{e})^{1/2}\quad\text{and}\quad k_{\mathrm{r2}}=k_\mathrm{d}/k_\mathrm{c}$$

(b) (i) When the concentration of HBr is very low so that k_{r2} [HBr] << [Br_2], the formation rate simplifies.

$$\text{rate of formation of HBr}=\frac{k_{\mathrm{r1}}[\mathrm{H_2}][\mathrm{Br_2}]^{3/2}}{[\mathrm{Br_2}]+k_{\mathrm{r2}}[\mathrm{HBr}]}=\frac{k_{\mathrm{r1}}[\mathrm{H_2}][\mathrm{Br_2}]^{3/2}}{[\mathrm{Br_2}]}=k_{\mathrm{r1}}[\mathrm{H_2}][\mathrm{Br_2}]^{1/2}$$

Under this condition, the rate is first-order in [H_2] and half-order in [Br_2]. The proposed mechanism suggests that the condition is expected when the rate of the propagation step for atomic bromine formation is much greater than the rate of the retardation step for atomic bromine formation, a step that consumes HBr (i.e., k_d [H] [HBr] << k_c [H] [Br_2]).

(ii) When the concentration of HBr is very high so that k_{r2} [HBr] >> [Br_2], the formation rate simplifies.

$$\text{rate of formation of HBr}=\frac{k_{\mathrm{r1}}[\mathrm{H_2}][\mathrm{Br_2}]^{3/2}}{[\mathrm{Br_2}]+k_{\mathrm{r2}}[\mathrm{HBr}]}=\frac{k_{\mathrm{r1}}[\mathrm{H_2}][\mathrm{Br_2}]^{3/2}}{k_{\mathrm{r2}}[\mathrm{HBr}]}$$

$$=k_{\mathrm{r3}}\frac{[\mathrm{H_2}][\mathrm{Br_2}]^{3/2}}{[\mathrm{HBr}]}\quad\text{where }k_{\mathrm{r3}}=k_{\mathrm{r1}}/k_{\mathrm{r2}}$$

Under this condition, the rate is first-order in [H_2], three-halves-order in [Br_2], and negative-first-order in [HBr]. The proposed mechanism suggests that the condition is expected when the rate of the propagation step for atomic bromine formation is much less than the rate of the retardation step for atomic bromine formation, a step that consumes HBr (i.e., k_d [H] [HBr] >> k_c [H] [Br_2]).

Answers to projects

P11.30 (a) $A \underset{k_r'}{\overset{k_r}{\rightleftharpoons}} B$

The rate expressions for [B] and [A] are:

$$\frac{d[B]}{dt} = k_r[A] - k_r'[B] \qquad \text{and} \qquad \frac{d[A]}{dt} = k_r'[B] - k_r[A]$$

and the solutions to be tested are giving in eqn. 11.2

$$[B] = \frac{k_r\left(1 - e^{-(k_r+k_r')t}[A]_0\right)}{k_r + k_r'} \qquad \textit{Note: } \frac{d[B]}{dt} = k_r e^{-(k_r+k_r')t}[A]_0$$

$$[A] = \frac{\left(k_r' + k_r e^{-(k_r+k_r')t}\right)[A]_0}{k_r + k_r'}$$

To test the solutions, we substitute the expressions for [B] and [A] into the rate expression for [B]. The left-side substitution requires that we differentiate

$$k_r[A]_0 e^{-(k_r+k_r')t} = \frac{k_r\left(k_r' + k_r e^{-(k_r+k_r')t}\right)[A]_0}{k_r + k_r'} - \frac{k_r' k_r\left(1 - e^{-(k_r+k_r')t}\right)[A]_0}{k_r + k_r'}$$

which simplifies to

$$k_r[A]_0 e^{-(k_r+k_r')t} = k_r[A]_0 e^{-(k_r+k_r')t}$$

because both sides are equal, the expressions for [A] and [B] prove to satisfy the rate expression for [B]. Similar substitution of the expressions for [A] and [B] into the $\frac{d[A]}{dt}$ expression, demonstrates that [A] and [B] satisfy the rate expression. Thus, the above expressions for [A] and [B] are solutions to this set of rate expressions.

(b) $$\frac{d[A]}{dt} = -k_r[A] + k_r'[B] \qquad \text{and} \qquad \frac{d[B]}{dt} = -k_r'[B] + k_r[A]$$

$[A] + [B] = [A]_0 + [B]_0$ at all times.

Therefore, $\boxed{[B] = [A]_0 + [B]_0 - [A]}$

$$\frac{d[A]}{dt} = -k_r[A] + k_r'\{[A]_0 + [B]_0 - [A]\} = -(k_r + k_r')[A] + k_r'([A]_0 + [B]_0)$$

and $$\int_{[A]_0}^{[A]_0} \frac{d[A]}{(k_r + k_r')[A] + k_r'([A]_0 + [B]_0)} = -\int_0^t dt$$

The solution is $$\boxed{[A] = \frac{k_r'([A]_0 + [B]_0) + (k_r[A]_0 - k_r'[B]_0)e^{-(k_r+k_r')t}}{k_r + k_r'}}$$

Setting $[B]_0 = 0$

$$[A] = \frac{k_r'[A]_0 + k_r[A]_0 e^{-(k+k_r')t}}{k_r + k_r'} = \frac{(k_r' + k_r e^{-(k+k_r')t})[A]_0}{k_r + k_r'}$$

which is the expression for [A] in eqn. 11.2. Then

$$[B] = [A]_0 - [A] = \frac{k_r(1 - e^{-(k_r+k_r')t})[A]_0}{k_r + k_r'}$$

which agrees with [B] in eqn 11.2.

P11.31 $\mathrm{A} \xrightarrow{k_a} \mathrm{I} \xrightarrow{k_b} \mathrm{P}$

The rate laws are

(i) $$\frac{d[\mathrm{A}]}{dt} = -k_a[\mathrm{A}]$$

(ii) $$\frac{d[\mathrm{I}]}{dt} = k_a[\mathrm{A}] - k_b[\mathrm{I}]$$

(iii) $$\frac{d[\mathrm{P}]}{dt} = k_b[\mathrm{I}]$$

Differentiating eqn 11.5a yields (i) directly. Differentiating 11.5c yields (iii) almost as directly.

$$\frac{d[\mathrm{P}]}{dt} = \frac{1}{k_b - k_a}\left(-k_a k_b e^{-k_b t} + k_a k_b e^{-k_a t}\right)[\mathrm{A}]_0$$
$$= \frac{k_a k_b}{k_b - k_a}\left(e^{-k_a t} - e^{-k_b t}\right)[\mathrm{A}]_0$$
$$= k_b[\mathrm{I}], \text{ which is (iii)}$$

From eqn 11.5b, after differentiating, we have

(iv) $$\frac{d[\mathrm{I}]}{dt} = \frac{k_a}{k_b - k_a}\left(-k_a e^{-k_a t} + k_b e^{-k_b t}\right)[\mathrm{A}]_0$$

The reduction of (iv) to (ii) requires a little more algebraic manipulation. In this case, it seems easier to work backwards. Thus, substitute eqn 11.5a and 11.5b into (ii). We obtain

$$\frac{d[\mathrm{I}]}{dt} = k_a e^{-k_a t}[\mathrm{A}]_0 - k_b\left(\frac{k_a}{k_b - k_a}\right)(e^{-k_a t} - e^{-k_b t})[\mathrm{A}]_0$$
$$= \left(\frac{k_a}{k_b - k_a}\right)(k_b - k_a)e^{-k_a t}[\mathrm{A}]_0 - \left(\frac{k_a}{k_b - k_a}\right)k_b(e^{-k_a t} - e^{-k_b t})[\mathrm{A}]_0$$
$$= \left(\frac{k_a}{k_b - k_a}\right)\left[(k_b - k_a)e^{-k_a t}[\mathrm{A}]_0 - k_b(e^{-k_a t} - e^{-k_b t})[\mathrm{A}]_0\right]$$
$$= \left(\frac{k_a}{k_b - k_a}\right)\left[k_b e^{-k_a t}[\mathrm{A}]_0 - k_a e^{-k_a t}[\mathrm{A}]_0 - k_b e^{-k_a t}[\mathrm{A}]_0 + k_b e^{-k_b t}[\mathrm{A}]_0\right]$$

The first and third terms in the brackets cancel leaving

$$\frac{d[\mathrm{I}]}{dt} = \left(\frac{k_a}{k_b - k_a}\right)(-k_a e^{-k_a t} + k_b e^{-k_b t})[\mathrm{A}]_0$$

which is (iv) above.

P11.32 (a) For the mechanism

$$hhhh\ldots \underset{k_a'}{\overset{k_a}{\rightleftharpoons}} hchh\ldots$$

$$hchh\ldots \underset{k_b'}{\overset{k_b}{\rightleftharpoons}} cccc\ldots$$

the rate equations are

$$\frac{d[hhhh\ldots]}{dt} = -k_a[hhhh\ldots] + k_a'[hchh\ldots]$$

$$\frac{\mathrm{d}[hchh\ldots]}{\mathrm{d}t} = k_{\mathrm{a}}[hhhh\ldots] - k'_{\mathrm{a}}[hchh\ldots] - k_{\mathrm{b}}[hchh\ldots] + k'_{\mathrm{b}}[cccc\ldots]$$

$$\frac{\mathrm{d}[cccc\ldots]}{\mathrm{d}t} = k_{\mathrm{b}}[hchh\ldots] - k'_{\mathrm{b}}[cccc\ldots]$$

(b) Apply the steady-state approximation to the intermediate in the above mechanism:

$$\frac{\mathrm{d}[hchh\ldots]}{\mathrm{d}t} = k_{\mathrm{a}}[hhhh\ldots] - k'_{\mathrm{a}}[hchh\ldots] - k_{\mathrm{b}}[hchh\ldots] + k'_{\mathrm{b}}[cccc\ldots] = 0$$

so $$[hchh\ldots] = \frac{k_{\mathrm{a}}[hhhh\ldots] + k'_{\mathrm{b}}[cccc\ldots]}{k'_{\mathrm{a}} + k_{\mathrm{b}}}$$

Therefore, $$\frac{\mathrm{d}[hhhh\ldots]}{\mathrm{d}t} = -\frac{k_{\mathrm{a}}k_{\mathrm{b}}}{k'_{\mathrm{a}} + k_{\mathrm{b}}}[hhhh\ldots] + \frac{k'_{\mathrm{a}}k'_{\mathrm{b}}}{k'_{\mathrm{a}} + k_{\mathrm{b}}}[cccc\ldots]$$

or

$$\boxed{\frac{\mathrm{d}[hhhh...]}{\mathrm{d}t} = -k_{\mathrm{r,eff}}[hhhh...] + k'_{\mathrm{r,eff}}[cccc...] \text{ where } k_{\mathrm{r,eff}} = \frac{k_{\mathrm{a}}k_{\mathrm{b}}}{k'_{\mathrm{a}} + k_{\mathrm{b}}} \text{ and } k'_{\mathrm{r,eff}} = \frac{k'_{\mathrm{a}}k'_{\mathrm{b}}}{k'_{\mathrm{a}} + k_{\mathrm{b}}}}$$

The simple mechanism

$$hhhh\ldots \underset{k'_{\mathrm{r}}}{\overset{k_{\mathrm{r}}}{\rightleftharpoons}} cccc\ldots$$

has the rate law $\frac{\mathrm{d}[hhhh\ldots]}{\mathrm{d}t} = -k_{\mathrm{r}}[hhhh\ldots] + k'_{\mathrm{r}}[cccc\ldots]$, which is equivalent to the above rate law.

(c) It is difficult to make conclusive inferences about intermediates from kinetic data alone. For example, if rate measurements show formation of coils from helices with a single rate constant, they tell us nearly nothing about the mechanism. The rate law

$$\frac{\mathrm{d}[cccc...]}{\mathrm{d}t} = k_{\mathrm{r}}[hhhh...]$$

is consistent with a single-step mechanism, with a two-step mechanism with a rate-determining second step, and with a two-step mechanism with a steady-state intermediate. Even if kinetic monitoring of the product shows production with two rate constants, the rate constants could belong to competing paths or to steps of a single reaction path. The best evidence for an intermediate's participation in a reaction is detection of the intermediate, or at least detection of structural features that can belong to a proposed intermediate but not reactant or product.

P11.33 (a) (i) The illustration in *Box* 11.2 suggests that a chain-branching explosion $\boxed{\text{does not occur}}$ at temperatures as low as 700 K. There may, however, be a thermal explosion regime at pressures in excess of 10^6 Pa.

(ii) At 900 K there is a lower chain-branching explosion limit when

$$\log(p/\mathrm{Pa}) = 2.1 \qquad \text{so } p = 10^{2.1}\ \mathrm{Pa} = \boxed{1.3\times10^2\ \mathrm{Pa}}$$

There does not seem to be a pressure above which a steady reaction occurs. Rather the chain-branching explosion range seems to run into the thermal explosion range around

$$\log(p/\mathrm{Pa}) = 4.5 \qquad \text{so } p = 10^{4.5}\ \mathrm{Pa} = \boxed{3\times10^4\ \mathrm{Pa}}$$

(b) The two equations that describe the time evolution of hydrogen radial concentration and the conditions under which they are applicable (i.e., make [H] a positive value) are

$$[\mathrm{H}] = \frac{v_{\text{initiation}}}{k_{\text{termination}} - k_{\text{branching}}}\left\{1 - \mathrm{e}^{-(k_{\text{termination}} - k_{\text{branching}})t}\right\} \quad \text{when } k_{\text{termination}} > k_{\text{branching}} \text{ and } [\mathrm{O_2}] \text{ is low}$$

$$= \frac{v_{\text{initiation}}}{|\Delta k|}\left\{1 - \mathrm{e}^{-|\Delta k|t}\right\} \quad \text{where } |\Delta k| = |k_{\text{termination}} - k_{\text{branching}}|$$

and

$$[\mathrm{H}] = \frac{v_{\text{initiation}}}{k_{\text{termination}} - k_{\text{branching}}}\left\{1 - \mathrm{e}^{-(k_{\text{termination}} - k_{\text{branching}})t}\right\} \quad \text{when } k_{\text{branching}} > k_{\text{termination}} \text{ and } [\mathrm{O_2}] \text{ is high}$$

$$= \frac{-v_{\text{initiation}}}{k_{\text{branching}} - k_{\text{termination}}}\left\{1 - \mathrm{e}^{(k_{\text{branching}} - k_{\text{termination}})t}\right\}$$

$$= \frac{v_{\text{initiation}}}{k_{\text{branching}} - k_{\text{termination}}}\left\{\mathrm{e}^{(k_{\text{branching}} - k_{\text{termination}})t} - 1\right\}$$

$$= \frac{v_{\text{initiation}}}{|\Delta k|}\left\{\mathrm{e}^{|\Delta k|t} - 1\right\}$$

$[\mathrm{H}]\times|\Delta k|/v_{\text{initiation}}$ is plotted against $|\Delta k|t$ under both conditions in Figure 11.6. The conditions for the rapid growth of [H], and explosion, are $k_{\text{branching}} > k_{\text{termination}}$ and high $[\mathrm{O_2}]$. See P. Atkins' and J. de Paula's, *Physical Chemistry*, 8th ed., W.H. Freeman, New York, 2006, Section 23.2 for a detailed discussion of explosions.

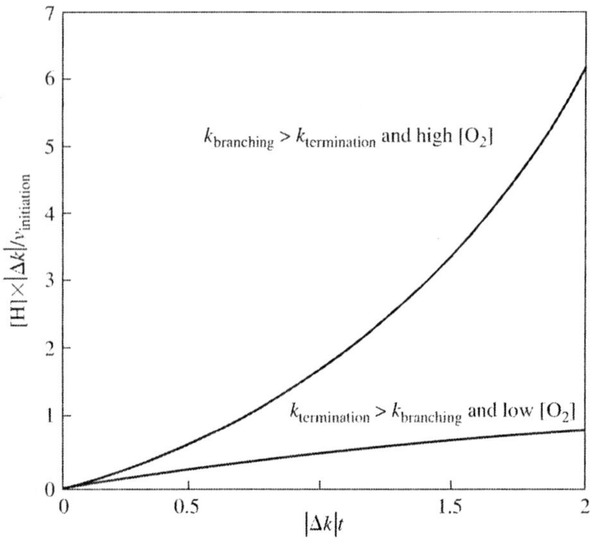

Figure 11.6

12 Quantum theory

Answers to discussion questions

D12.1 At the end of the nineteenth century and the beginning of the twentieth, there were many experimental results on the properties of matter and radiation that could not be explained on the basis of established physical principles and theories. Here we list only some of the most significant.

(1) The photoelectric effect revealed that electromagnetic radiation, classically considered to be a wave, also exhibits the particle-like behavior of photons. Each photon is a discrete unit, or quantum, of energy that is absorbed during collisions with electrons. Photons are never partially absorbed. They either completely give up their energy or they are not absorbed. The energy of a photon can be calculated if either the radiation frequency or wavelength is known: $\boldsymbol{E_{\text{photon}} = h\nu = hc/\lambda}$.

(2) Absorption and emission spectra indicated that atoms and molecules can only absorb or emit discrete packets of energy (i.e., photons). This means that an atom or molecule has specific, allowed energy levels and we say that their energies are quantized. During a spectroscopic transition the atom or molecule gains or loses the energy ΔE by either absorption of a photon or emission of a photon, respectively. Thus, spectral lines must satisfy the **Bohr frequency condition:** $\boldsymbol{\Delta E = h\nu}$.

(3) Neutron and electron diffraction studies indicated that these particles also possess wave-like properties of constructive and destructive interference. The joint particle and wave character of matter and radiation is called **wave-particle duality**. The **de Broglie relation,** $\lambda_{\text{de Broglie}} = h/p$, connects the wave character of a particle ($\lambda_{\text{de Broglie}}$) with its particulate momentum (p).

Evidence that resulted in the development of quantum theory also included:

(4) The energy density distribution of blackbody radiation as a function of wavelength.

(5) The heat capacities of monatomic solids such as copper metal.

D12.2 The wave-particle duality of quantum theory requires a particle wavefunction that does not experience destructive interference upon reflection by a barrier or in motion around a closed loop. These are **boundary conditions** of the wavefunction and they are the cause of energy quantization. The criteria of particle existence and the restrictions of boundary conditions result in quantum conditions that must be satisfied for a wavefunction to be acceptable. The conditions on the wavefunction, ψ, are:

(1) ψ must be single valued at each point.

(2) The probability of finding a particle in a very small subatomic region, $\psi^2\delta V$, cannot exceed 1.

(3) ψ is continuous everywhere.

(4) ψ has a continuous slope everywhere.

When applied to a particle of mass m confined to move in a one-dimensional box of length L, these requirements restrict the de Broglie wavelength to $\lambda = 2L/n$, where the **quantum number** n is a non-zero, positive integer. Then using the relation $E = E_k = p^2/2m$ and the de Broglie relation $\lambda = h/p$, the energy is quantized at $E_n = n^2h^2/8mL^2$ [12.8]. This derivation applies specifically to the particle in a box but the derivation is similar for the particle on a ring; see Section 12.8(a) of the text.

D12.3 The lowest energy level possible for a confined quantum mechanical system is the **zero-point energy**, and zero-point energy is not necessarily zero energy. This lowest, irremovable energy is consistent with the uncertainty principle, $\Delta x \Delta p_x \geq \frac{1}{2}\hbar$ [12.6]. In this relation the uncertainty in position and the uncertainty of the momentum are along the same line of motion and the uncertainties are defined with the relations $(\Delta x)^2 = (x^2)_{mean} - (x)_{mean}^2$ and $(\Delta p_x)^2 = (p_x^2)_{mean} - (p_x)_{mean}^2$. Should a hypothetical quantum state exhibit both $(p_x^2)_{mean} = 0$ and $(p_x)_{mean}^2 = 0$, so that the momentum uncertainty is zero, a finite value of Δx gives $\Delta x \Delta p_x = 0$ and quantum theory declares that, because the uncertainty principle is violated, the state is invalid and will not be observed in nature. However, should Δx be infinitely large the uncertainty principle may be satisfied even though there is no momentum uncertainty.

The zero-point energy of the particle in a box is the energy of the $n = 1$ quantum state of eqn 12.8: $E_1 = h^2/8mL^2$. For an electron in a 1 nm box, the zero-point energy is calculated to be 6.0×10^{-20} J and 36 kJ mol^{-1}, an energy that remains even after cooling to the absolute zero of temperature. This is consistent with the uncertainty principle as quantum theory does not assign a precise location to the particle, it is in the box but the uncertainty of knowing its position equals the length L of the box. Thus, a hypothetical zero-point energy of zero implies zero kinetic energy so that both $(p_x^2)_{mean} = 0$ and $(p_x)_{mean}^2 = 0$. Consequently, there is zero uncertainty in knowledge of the momentum giving $\Delta x \Delta p_x = L \times 0 = 0$ in violation to the uncertainty principle and quantum theory declares that the particle in a box cannot have zero energy.

The zero-point energy of the harmonic oscillator is the energy of the $v = 0$ quantum state of eqn 12.21: $E_0 = \frac{1}{2}h\nu$ where v (italic vee) is the quantum number and ν (Greek nu) is the frequency of the oscillator. A typical chemical bond has a vibrational frequency of 3.0×10^{13} Hz (corresponding to a wavenumber of 1000 cm^{-1}) so the zero-point energy of molecular vibration is typically about 1×10^{-20} J. In this case, a hypothetical zero-point energy of zero implies precise knowledge of the oscillator position; it is at the bottom of the harmonic potential where the potential is zero and the displacement is exactly zero. Zero energy implies zero uncertainty in knowledge of position. It also implies a precise momentum equal to zero giving $\Delta x \Delta p_x = 0 \times 0 = 0$ in violation to the uncertainty principle. Consequently, quantum theory declares that the harmonic oscillator cannot have zero energy.

The energy levels for a particle confined on a ring of constant potential, see text Section 12.8(a) and Figure 12.1, are given by:

$$E_{m_l} = m_l^2\hbar^2/2I \ [12.16] \qquad \text{where } m_l = 0,\ \pm 1,\ \pm 2, \ldots$$

so the zero-point energy equals zero because it is the state for which $m_l = 0$. This does not violate the uncertainty principle. To see this, we recognize that the complementary variables of this two-dimensional quantum system are the position angle ϕ and the angular momentum J_z and we write the uncertainty principle as $\Delta\phi\Delta J_z \geq \frac{1}{2}\hbar$. As the classical particle travels through one cycle of rotation, the angle sweeps from zero through 2π radians. The second cycle takes the particle through 4π radian, the third cycle through 6π radians, etc. In quantum theory, however, we cannot precisely know either the angle within the first cycle or the cycle of rotation. The angular uncertainty is not 2π. It is infinitely large. Thus, even though J_z is precisely known as zero at the zero-point, the uncertainty principle is not violated.

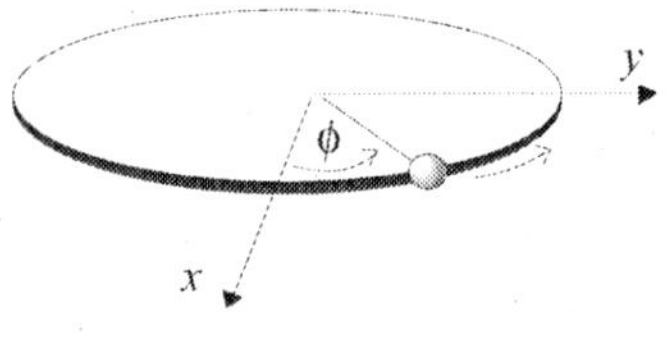

Figure 12.1

D12.4 With the scientifically observed failures of classical physics it became apparent during about the first-quarter of the twentieth century that the Newtonian concept of a deterministic particle path over which the precise position and momentum could be specified at each instant is wrong at the atomic/molecular level. The Schrödinger equation, discovered in 1926, resolved the theoretical difficulties of the particulate-wave duality of nature at the microscopic scale by providing a quantum method for calculating atomic and molecular energy levels along with other observable properties. The method involved the calculation and use of the wavefunction ψ, a property that is often imaginary and, consequently, non-observable, and an intellectual question arose 'Is the wavefunction an artifact that provides a method for calculating observables or is there a useful physical interpretation for it?' Max Born suggested the interpretation that came to be accepted, and found to be very useful, by the scientific community. It is: The probability of finding a particle in a small region of space of volume δV is proportional to $|\psi|^2\,\delta V$, where ψ is the value of the wavefunction in the region. (The modulus square of the wavefunction, $|\psi|^2 = \psi^* \times \psi$, is always real.) This is the probabilistic **Born interpretation** of the wavefunction and $|\psi|^2$ is called the **probability density** as its SI unit is m^{-3} for one particle quantum systems in three dimensions (like the probability density for an electron in a hydrogen atom). In addition to being an integral part of the quantum methodology, the probability density and Born interpretation are regularly used to provide a rationale for quick understanding of molecular bond angles, bond lengths, bonding types, sites of electrophilic and nucleophilic attack, and more.

The analogy between the Born interpretation of the probability density and the square of the amplitude of an electromagnetic wave is very supportive of the Born interpretation. In classical electromagnetic theory the square of the amplitude is the radiation intensity and therefore proportional to the number of photons present.

D12.5 The uncertainty principle, $\Delta x \Delta p_x \geq \frac{1}{2}\hbar$ [12.6] implies that complementary variables like x and p_x cannot be simultaneously measured with exact precision. An attempt to increase the measurement precision in one variable (i.e., lower the measurement uncertainty) results in a simultaneous decrease of the precise knowledge of the other variable (i.e., an increase in uncertainty). This is a natural consequence of the wave nature of matter and cannot be circumvented by the invention of perfect measurement tools.

D12.6 The physical origin of tunnelling is related to the probability density of the particle, which according to the Born interpretation is the square of the wavefunction that represents the particle. This interpretation requires that the wavefunction of the system be everywhere continuous, even at barriers. Therefore, if the wavefunction is non-zero on one side of a barrier it must be non-zero on the other side of the barrier and this implies that the particle has tunnelled through the barrier. The transmission probability depends upon the mass of the particle (specifically $m^{1/2}$): the greater the mass the smaller the probability of tunnelling. Electrons and protons have small masses, molecular groups large masses; therefore, tunnelling effects are more observable in process involving electrons and protons. An electron tunnels more readily than a proton and a proton tunnels more readily than a deuteron.

The very rapid equilibration of proton transfer reactions is a manifestation of the ability of protons to tunnel through barriers and transfer quickly from an acid to a base. Tunnelling of protons between acidic and basic groups is also an important feature of the mechanism of some enzyme-catalysed reactions. Electron tunnelling is one of the factors that determine the rates of electron transfer reactions at electrodes and in biological systems.

D12.7 The time-independent wavefunction in three-dimensional space is a function of x, y, and z so we write $\psi(x, y, z)$ with each variable ranging from $-\infty$ to $+\infty$ in the general case. The time-independent wavefunction is said to be a **stationary state**. It is reasonable to expect that for special quantum systems the probability densities in each of the three independent directions should be mutually independent. This implies that the probability density for the time-independent wavefunction $\psi(x, y, z)$ should be the product of three probability densities, one for each coordinate: $|\psi(x, y, z)|^2 \propto |X(x)|^2 \times |Y(y)|^2 \times |Z(z)|^2$. Subsequently, we see that the wavefunction is the product of three independent wavefunctions in and we write $\psi(x, y, z) \propto X(x) \times Y(y) \times Z(z)$. Such a wavefunction is said to exhibit the **separation of variables**. When a wavefunction can be separated in this manner, the solution to the Schrödinger equation is greatly simplified as demonstrated by the particle in the three-dimensional box.

For a particle free to move within a cube of volume L^3 under the influence of a constant potential, which is assigned the value zero, we may generalize the simple wavefunction provided by the one-dimensional solution discussed in Section 12.7 of the text. Using the method of separation of variables, we find that each dimension has a solution analogous to that of the one-dimensional problem: $X(x) \propto \sin(n_x\pi x/L)$, $Y(y) \propto \sin(n_y\pi y/L)$ and $Z(z) \propto \sin(n_z\pi z/L)$. The wavefunction is then given by $\psi(x, y, z) \propto \sin(n_x\pi x/L) \times \sin(n_y\pi y/L) \times \sin(n_z\pi z/L)$ with three independent quantum numbers each of which is an integer that can range between 1 and ∞. Furthermore, the total energy is the sum of the energies that originate from motion along each of the independent variables. The energy along each coordinate is analogous to that of the one-dimensional problem giving:

$$E_{n_x,n_y,n_z} = E_{n_x} + E_{n_y} + E_{n_z} = \frac{n_x^2h^2}{8mL^2} + \frac{n_y^2h^2}{8mL^2} + \frac{n_z^2h^2}{8mL^2} = \left(n_x^2 + n_y^2 + n_z^2\right)\frac{h^2}{8mL^2}$$

(Remarkably, when the potential energy term of the hamiltonian is either zero or a constant value throughout space, the time-independent wavefunction does not depend upon the particle mass! Similarly, electrical charge does not appear in the time-independent wavefunction in this particular quantum system.)

The method of separation of variables is not generally valid and a simple analytical solution to the Schrödinger equation does not usually exist. This means that an attempt must generally be made to find solutions with rather difficult numerical methods. The question then becomes 'When is a solution by separation of variables possible?' The answer is: When motion along a particular variable does not change the potential energy of the particle, that variable may be separated from the others.

For example, consider an isolated hydrogen atom. This quantum system has a proton nucleus and one electron. The Cartesian coordinate system is centered on the nucleus and we want to know whether the wavefunction for the electron, $\psi(x, y, z)$, is amenable to the method of separation of variables in Cartesian coordinates. To address this question, we remember that the electron and proton exhibit an electrostatic attraction that has a potential energy that is inversely proportion to the distance r between them: $V \propto 1/r$ where $r = (x^2 + y^2 + z^2)^{1/2}$. Examination of the potential reveals that when the electron's x position, or y position, or z position change, the electron potential changes and we conclude that none of these variables may be separated. However, we remember that r is one of the three independent variables of the spherical coordinate system and we write the wavefunction as $\psi(r, \theta, \phi)$. The electron potential does not change when either of the angles changes so functions for each of the angles may be separated giving a wavefunction that is the product of functions for each variable. It has the form: $\psi(r, \theta, \phi) \propto R(r) \times \Theta(\theta) \times \Phi(\phi)$ where the functions R, Θ, and Φ are each functions of a single variable only. This provides the complete separation of variables. It greatly simplifies the solution to the Schrödinger equation and an analytical, non-numeric, solution becomes possible for the hydrogen atom.

Solutions to exercises

E12.1 The key relations are (i) $E_{\text{photon}} = h\nu$ and (ii) $\lambda\nu = c$. Solving (ii) for frequency and substitution into (i) gives

$$E_{\text{photon}} = \frac{hc}{\lambda}$$

$$= \frac{\left(6.626\times10^{-34}\ \text{J s}\right)\times\left(2.998\times10^{8}\ \text{m s}^{-1}\right)}{652\times10^{-9}\ \text{m}} = \boxed{3.05\times10^{-19}\ \text{J}}$$

E12.2 The key relations are (i) $E_{\text{photon}} = h\nu$, (ii) $\lambda\nu = c$, and the definition of wavenumber (iii) $\tilde{\nu} = 1/\lambda$. Solving (ii) for frequency and substituting (iii) for $1/\lambda$ gives $\nu = c\tilde{\nu}$, which, upon substitution into (i) gives

(iv) $E_{\text{photon}} = hc\tilde{\nu}$

Solving eqn (iv) for wavenumber gives

$$\tilde{\nu} = \frac{E_{\text{photon}}}{hc}$$

$$= \frac{1.634\times10^{-18}\ \text{J}}{(6.626\times10^{-34}\ \text{J s})\times(2.998\times10^{8}\ \text{m s}^{-1})} = 8.226\times10^{6}\ \text{m}^{-1} = \boxed{8.226\times10^{4}\ \text{cm}^{-1}}$$

COMMENT. When there is a need to regularly perform quantum and spectroscopic calculations, it becomes very helpful to remember eqns (i)–(iv) and to be able to perform quick, effortless substitutions.

E12.3 The definition of power P is $P = E/t$ and the energy transported by N photons in time t is $E = Nh\nu = Nhc/\lambda$. Thus, $P = Nhc/\lambda t$ and solving for the rate N/t gives

$$\frac{N}{t} = \frac{P\lambda}{hc}$$

$$= \frac{(0.68\times10^{-6}\ \text{J s}^{-1})\times(245\times10^{-9}\ \text{m})}{(6.626\times10^{-34}\ \text{J s})\times(2.998\times10^{8}\ \text{m s}^{-1})} = \boxed{8.4\times10^{11}\ \text{s}^{-1}}$$

E12.4 (a) For an electronic transition of known frequency the transition quanta is the corresponding photon energy. Thus,

$$E_{\text{photon}} = h\nu$$

$$= (6.626\times10^{-34}\ \text{J s})\times(1.0\times10^{15}\ \text{s}^{-1}) = \boxed{6.6\times10^{-19}\ \text{J}}$$

and

$$E_{\text{m}} = N_{\text{A}}h\nu$$

$$= (6.022\times10^{23}\ \text{mol}^{-1})\times(6.626\times10^{-34}\ \text{J s})\times(1.0\times10^{15}\ \text{s}^{-1}) = \boxed{4.0\times10^{2}\ \text{kJ mol}^{-1}}$$

(b) The harmonic oscillator is used as the model for the quantum motion of molecular vibration and eqn 12.21 indicates that quantum states are separated by the energy quantum $\Delta E = h\nu = h/T$ where the period T is defined to be the inverse of frequency ($T = 1/\nu$).

$$\Delta E = h/T$$

$$= (6.626\times10^{-34}\ \text{J s})/(20\times10^{-15}\ \text{s}) = \boxed{3.3\times10^{-20}\ \text{J}}$$

$$\Delta E_{\text{m}} = N_{\text{A}}E$$

$$= (6.022\times10^{23}\ \text{mol}^{-1})\times(3.3\times10^{-20}\ \text{J}) = \boxed{20.\ \text{kJ mol}^{-1}}$$

(c) The harmonic oscillator is also used as the model for the quantum states of pendulum motion. So, like part (b) eqn 12.20 indicates that quantum states are separated by the energy quantum $\Delta E = h\nu = h/T$.

$$\Delta E = h/T$$

$$= (6.626\times10^{-34}\ \text{J s})/(0.50\ \text{s}) = \boxed{1.3\times10^{-33}\ \text{J}}$$

$$\Delta E_{\text{m}} = N_{\text{A}}E$$

$$= (6.022\times10^{23}\ \text{mol}^{-1})\times(1.3\times10^{-33}\ \text{J}) = \boxed{7.8\times10^{-13}\ \text{kJ mol}^{-1}}$$

This extraordinarily small separation is caused by the macroscopic, large mass characteristics of a pendulum. The energy levels are so close together that the pendulum energies appear as a continuum of values that are successfully described by the classical laws of physics.

E12.5 The definition of power P is $P = E/t$ and the energy transported by N photons in time t is $E = Nh\nu = Nhc/\lambda$. Thus, $P = Nhc/\lambda t$ and solving for the rate N/t gives

$$\frac{N}{t} = \frac{P\lambda}{hc}$$

(a) $P = 1.00$ W and $\lambda = 380$ nm

$$\frac{N}{t} = \frac{(1.00\ \text{J s}^{-1})\times(380\times10^{-9}\ \text{m})}{(6.626\times10^{-34}\ \text{J s})\times(2.998\times10^{8}\ \text{m s}^{-1})} = \boxed{1.91\times10^{18}\ \text{s}^{-1}}$$

(b) $P = 100$ W and $\lambda = 380$ nm

$$\frac{N}{t} = \frac{(100\ \text{J s}^{-1})\times(380\times10^{-9}\ \text{m})}{(6.626\times10^{-34}\ \text{J s})\times(2.998\times10^{8}\ \text{m s}^{-1})} = \boxed{1.91\times10^{20}\ \text{s}^{-1}}$$

E12.6 The definition of power P is $P = E/t$ and the energy transported by N photons in time t is $E = Nh\nu = Nhc/\lambda$. Thus, $P = Nhc/\lambda t$ and solving for the rate t gives

$$t = \frac{Nhc}{P\lambda}$$
$$= \frac{(6.022\times10^{23})\times(6.626\times10^{-34}\ \text{J s})\times(2.998\times10^{8}\ \text{m s}^{-1})}{(100\ \text{J s}^{-1})\times(590\times10^{-9}\ \text{m})} = \boxed{2.03\times10^{3}\ \text{s}}$$

E12.7 $$\frac{N}{t} = \frac{P}{h\nu} = \frac{45\times10^{3}\ \text{J s}^{-1}}{(6.626\times10^{-34}\ \text{J s})\times(98.4\times10^{6}\ \text{s}^{-1})} = \boxed{6.90\times10^{29}\ \text{s}^{-1}}$$

E12.8 $$\Phi = 2.14\ \text{eV} = (2.14\ \text{eV})\times(1.602\times10^{-19}\ \text{J eV}^{-1}) = 3.43\times10^{-19}\ \text{J}$$

$$E_k = h\nu - \Phi = \frac{hc}{\lambda} - \Phi \quad [12.2]$$

and, since $E_k = \tfrac{1}{2}m_e v^2$,

$$v = (2E_k / m_e)^{1/2}$$

(a) $$\frac{hc}{\lambda} = \frac{(6.626\times10^{-34}\ \text{J s})\times(2.998\times10^{8}\ \text{m s}^{-1})}{(750\times10^{-9}\ \text{m})} = 2.65\times10^{-19}\ \text{J} < \Phi$$

Therefore, no ejection occurs.

(b) $$\frac{hc}{\lambda} = \frac{(6.626\times10^{-34}\ \text{J s})\times(2.998\times10^{8}\ \text{m s}^{-1})}{(250\times10^{-9}\ \text{m})} = 7.95\times10^{-19}\ \text{J} > \Phi$$

Therefore, ejection occurs.

Hence, $E_k = (7.95 - 3.43)\times10^{-19}\ \text{J} = \boxed{4.52\times10^{-19}\ \text{J}}$

$$v = \left(\frac{2\times(4.52\times10^{-19}\ \text{J})}{9.109\times10^{-31}\ \text{kg}}\right)^{1/2} = \boxed{996\ \text{km s}^{-1}}$$

E12.9 $$E_k = h\nu - \Phi = \frac{hc}{\lambda} - \Phi \quad [12.2]$$

The data provided in the problem is used to prepare the plot of E_k against $1/\lambda$ shown in Figure 12.2. The plot is linear with a linear least squares regression fit that is shown in the box insert of the figure. According to eqn 12.2, the slope equals hc and the intercept is the negative of the work function Φ.

$$h = \frac{slope}{c}$$

$$= \frac{(1.2403\times10^3\ \text{eV nm})\times(1.6022\times10^{-19}\ \text{J eV}^{-1})\times(10^{-9}\ \text{m nm}^{-1})}{2.998\times10^8\ \text{m s}^{-1}} = \boxed{6.63\times10^{-34}\ \text{J s}}$$

$$\Phi = -intercept = \boxed{2.52\ \text{eV}}$$

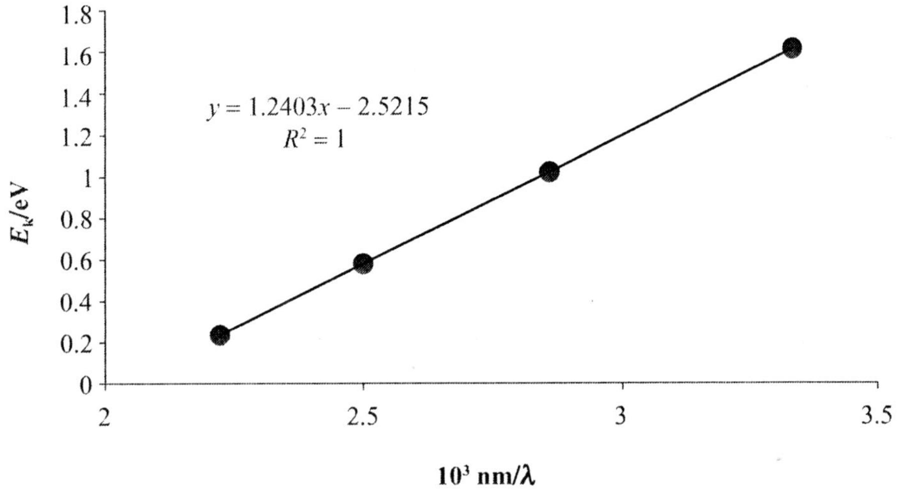

Figure 12.2

E12.10 $p = m_e v$ and $p = \dfrac{h}{\lambda}$

Therefore,

$$v = \frac{h}{m_e\lambda} = \frac{(6.626\times10^{-34}\ \text{J s})}{(9.109\times10^{-31}\ \text{kg})\times(550\times10^{-12}\ \text{m})} = \boxed{1.32\times10^6\ \text{m s}^{-1}}$$

E12.11 $\lambda = \dfrac{h}{p} = \dfrac{h}{mv}$

(a) $$\lambda = \frac{(6.626\times10^{-34}\ \text{J s})}{(1.00\ \text{m s}^{-1})\times(1.0\times10^{-3}\ \text{kg})} = \boxed{6.6\times10^{-31}\ \text{m}}$$

(b) $$\lambda = \frac{(6.626\times10^{-34}\ \text{J s})}{(1.0\times10^8\ \text{m s}^{-1})\times(1.0\times10^{-3}\ \text{kg})} = \boxed{6.6\times10^{-39}\ \text{m}}$$

(c) $$\lambda = \frac{(6.626\times10^{-34}\ \text{J s})}{4.003\times(1.6605\times10^{-27}\ \text{kg})\times(1.0\times10^3\ \text{m s}^{-1})} = \boxed{99.7\ \text{pm}}$$

E12.12 In order to avoid confusion between the potential difference V and the unit volt, we represent the potential difference below as $\Delta\phi$. Then

$$\tfrac{1}{2}m_e v^2 = e\Delta\phi,\ \text{implying that}\ v = \left(\frac{2e\Delta\phi}{m_e}\right)^{1/2} \quad \text{and} \quad p = m_e v = (2m_e e\Delta\phi)^{1/2}$$

Therefore,

$$\lambda = \frac{h}{p} = \frac{h}{(2m_e e\Delta\phi)^{1/2}}$$

$$= \frac{(6.626\times10^{-34}\text{J s})}{\{2\times(9.109\times10^{-31}\ \text{kg})\times(1.602\times10^{-19}\ \text{C}\times\Delta\phi)\}^{1/2}} = \frac{1.226\ \text{nm}}{(\Delta\phi/\text{V})^{1/2}}\quad [1\ \text{J} = 1\ \text{C V}]$$

(a) $\Delta\phi = 1.00\,\text{V}, \quad \lambda = \boxed{1.23\ \text{nm}}$

(b) $\Delta\phi = 1.0\,\text{kV}, \quad \lambda = \dfrac{1.226\ \text{nm}}{31.6} = \boxed{39\ \text{pm}}$

(c) $\Delta\phi = 100\ \text{kV}, \quad \lambda = \dfrac{1.226\ \text{nm}}{316.2} = \boxed{3.88\ \text{pm}}$

E12.13 $m = 85\ \text{kg} \qquad v = 8.0\ \text{km h}^{-1}$

$$\lambda = \frac{h}{p} = \frac{h}{mv} \quad \text{de Broglie relation [12.3]}$$

$$= \frac{6.626\times10^{-34}\ \text{J s}}{(85\ \text{kg})\times(8.0\times10^{3}\ \text{m h}^{-1})}\left(\frac{3600\ \text{s}}{1\ \text{h}}\right) = \boxed{3.5\times10^{-36}\ \text{m}}$$

This extraordinarily small wavelength is much, much smaller than the diameter of a hydrogen nucleus and the calculation illustrates the hopelessness of measuring the de Broglie wavelength of a macroscopic object. The de Broglie wavelength does increase as the speed of an object decreases and, according to the quantum behavior of a particle in a one-dimensional box of length L, the de Broglie wavelength may be as long as $2L$.

E12.14 $p = \dfrac{h}{\lambda} \quad [12.3]$

(a) $p = \dfrac{(6.626\times10^{-34}\ \text{J s})}{(600\times10^{-9}\ \text{m})} = \boxed{1.10\times10^{-27}\ \text{kg m s}^{-1}}$

(b) $p = \dfrac{(6.626\times10^{-34}\ \text{J s})}{(70\times10^{-12}\ \text{m})} = \boxed{9.5\times10^{-24}\ \text{kg m s}^{-1}}$

(c) $p = \dfrac{(6.626\times10^{-34}\ \text{J s})}{(200\ \text{m})} = \boxed{3.31\times10^{-36}\ \text{kg m s}^{-1}}$

E12.15 $E = h\nu = \dfrac{hc}{\lambda}$

$$hc = (6.6261\times10^{-34}\ \text{J s})\times(2.99792\times10^{8}\ \text{m s}^{-1}) = 1.986\times10^{-25}\,\text{J m}$$

$$N_A hc = (6.02214\times10^{23}\ \text{mol}^{-1})\times(1.986\times10^{-25}\ \text{J m})$$

$$= 0.1196\ \text{J m mol}^{-1}$$

We can therefore draw up the following table:

λ	E/J	E/(kJ mol^{-1})
(a) 600 nm	3.31×10^{-19}	199
(b) 550 nm	3.61×10^{-19}	218
(c) 400 nm	4.97×10^{-19}	299
(d) 200 nm	9.93×10^{-19}	598
(e) 150 pm	1.32×10^{-15}	7.98×10^{5}
(f) 1.00 cm	1.99×10^{-23}	0.012

E12.16

$$p = mv = \frac{h}{\lambda} \quad [12.3]$$

$$v = \frac{h}{\lambda m}$$

$$v = \frac{(6.626\times10^{-34}\ \text{J s})}{(300\times10^{-9}\ \text{m})\times1.0\times10^{-3}\ \text{kg}}$$

$$= \boxed{2.2\times10^{-24}\,\text{m s}^{-1}}$$

E12.17 The momentum per photon of wavelength 650 nm is

$$p_{\text{photon}} = \frac{h}{\lambda}\ [12.3] = \frac{6.626\times10^{-34}\ \text{J s}}{650\times10^{-9}\ \text{m}} = 1.02\times10^{-27}\,\text{kg m s}^{-1}$$

and this is also the change of momentum per photon absorbed by the fabric. The laser power P is not given in the exercise so we will assume that the laser produces a hefty N_A photons per second and that all photons are absorbed by the spacecraft sail. The power P of this 650 nm laser is

$$P = (N_A\ \text{s}^{-1})\times E_{\text{photon}} = (N_A\ \text{s}^{-1})\times hc/\lambda$$
$$= (6.022\times10^{23}\ \text{s}^{-1})\times(6.626\times10^{-34}\ \text{J s})\times(2.998\times10^{8}\ \text{m s}^{-1})/(650\times10^{-9}\ \text{m}) = 184\ \text{kW}$$

(a) The force F in SI units on the sail is the change in momentum experienced by the sail per second. This is equal to the photon flux, $N_A\ \text{s}^{-1}$, multiplied by the momentum lost by a photon.

$$F = (N_A\ \text{s}^{-1})\times p_{\text{photon}}$$
$$= (6.022\times10^{23}\ \text{s}^{-1})\times(1.02\times10^{-27}\,\text{kg m s}^{-1}) = \boxed{6.14\times10^{-4}\ \text{N}}$$

(b) The pressure exerted by the radiation equals the force F divided by the sail area A.

$$F/A = (6.14\times10^{-4}\ \text{N})/(1.0\times10^{6}\ \text{m}^2) = \boxed{614\ \text{pPa}}$$

(c) $$t = \left(\frac{mv}{F}\right)_{\text{spacecraft}} = \frac{(1.0\ \text{kg})\times(1.0\ \text{m s}^{-1})}{6.14\times10^{-4}\ \text{N}} = 1.63\times10^{3}\ \text{s} = \boxed{0.452\ \text{h}}$$

E12.18 This is essentially the photoelectric effect with the work function Φ being the ionization energy I. Hence,

$$\tfrac{1}{2}m_e v^2 = h\nu - I = \frac{hc}{\lambda} - I$$

Solving for λ

$$\lambda = \frac{hc}{I + \tfrac{1}{2}m_e v^2}$$

$$= \frac{(6.626\times10^{-34}\ \text{J s})\times(2.998\times10^{8}\,\text{m s}^{-1})}{(3.44\times10^{-18}\ \text{J}) + \tfrac{1}{2}(9.109\times10^{-31}\ \text{kg})\times(1.03\times10^{6}\ \text{m s}^{-1})^2} = 5.06\times10^{-8}\ \text{m} = \boxed{50.6\ \text{nm}}$$

Question. What is the energy of the photon?

E12.19

$$\tfrac{1}{2}m_e v^2 = h\nu - I = \frac{hc}{\lambda} - I$$

$$I = \frac{hc}{\lambda} - \tfrac{1}{2}m_e v^2$$

$$= \frac{(6.626\times10^{-34}\ \text{J s})\times(2.998\times10^{8}\ \text{m s}^{-1})}{100\times10^{-12}\ \text{m}} - \tfrac{1}{2}(9.109\times10^{-31}\,\text{kg})\times(2.34\times10^{7}\ \text{m s}^{-1})^2$$

$$= \boxed{1.74\times10^{-15}\ \text{J}}$$

E12.20 For a single particle of mass m moving with energy E in one-dimension and under the influence of a potential energy that varies as ax^4 the Schrödinger equation [12.4a] is

$$-\frac{\hbar^2}{2m}\frac{\mathrm{d}^2\psi}{\mathrm{d}x^2} + ax^4\psi = E\psi$$

E12.21 Figure 12.3 shows a sketch of the wavefunction $\psi(x) = N\mathrm{e}^{-ax^2}$ where N is the normalization constant and a is also a constant. This type of function is said to be a **bell-shaped function** or a **Gaussian function**.

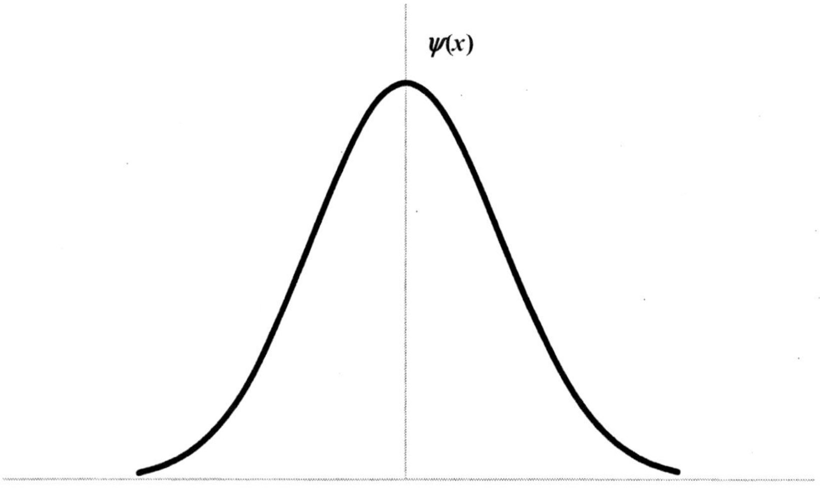

Figure 12.3

The probability distribution for finding the particle is $|\psi(x)|^2 = N^2\mathrm{e}^{-2ax^2}$. This is also a Gaussian function and, like analogous probability distributions in non-science fields, is often said to be a **normal distribution**. It has a peak at $\boxed{x = 0}$ so that is where the particle is most likely to be found. The value of x for which the probability of finding the particle reduced by 50% from its maximum value is determined by the relation:

$$\frac{|\psi(x)|^2}{|\psi(0)|^2} = \frac{N^2\mathrm{e}^{-2ax^2}}{N^2} = \mathrm{e}^{-2ax^2} = 0.500$$

$$-2ax^2 = \ln(0.500)$$

$$ax^2 = 0.3466$$

$$x = \boxed{\pm 0.5887 / a^{1/2}}$$

E12.22 The Born interpretation (Section 12.5) of a normalized wavefunction states that the probability, P, of finding a particle in a very small region equals $\psi^2\delta V$. When consideration is given to a particle in a one-dimensional box, this becomes $P = \psi^2\delta x$ where ψ [12.7] is evaluated at the mid-point of the δx range. This method is an estimate for which improvements require the use of calculus (see Exercise 12.42).

$$P=\psi^2\delta x=\left\{\left(\frac{2}{L}\right)^{1/2}\sin\left(\frac{2\pi x}{L}\right)\right\}^2\delta x=\left(\frac{2}{L}\right)\sin^2\left(\frac{2\pi x}{L}\right)\delta x$$

(a) $$P=\left(\frac{2}{10\text{ nm}}\right)\times\sin^2\left(\frac{2\pi\times0.15\text{ nm}}{10\text{ nm}}\right)\times(0.2\text{ nm}-0.1\text{ nm})=\boxed{1.77\times10^{-4}}$$

(b) $$P=\left(\frac{2}{10\text{ nm}}\right)\sin^2\left(\frac{2\pi\times5.05\text{ nm}}{10\text{ nm}}\right)(5.2\text{ nm}-4.9\text{ nm})=\boxed{5.92\times10^{-5}}$$

E12.23 $\Delta p=1.00\times10^{-4}p$ [i.e., 0.0100% of p] $=1.00\times10^{-4}m_\text{p}v$

$$\Delta x=\frac{\hbar}{2\Delta p}[12.6]=\frac{\hbar}{2\times(1.00\times10^{-4})\times m_\text{p}v}$$

$$=\frac{(1.055\times10^{-34}\text{ J s})}{2\times(1.00\times10^{-4})\times(1.673\times10^{-27}\text{ kg})\times(3.5\times10^{5}\text{ m s}^{-1})}$$

$$=9.0\times10^{-10}\text{ m, or }\boxed{0.90\text{ nm}}$$

E12.24 $\Delta p\Delta x\ge\frac{1}{2}\hbar$ [12.6] $\quad\Delta p=m\Delta v$

$$\Delta v_\text{min}=\frac{\hbar}{2m\Delta x}=\frac{1.055\times10^{-34}\text{ J s}}{2\times(0.500\text{ kg})\times(5.0\times10^{-6}\text{ m})}=\boxed{2.1\times10^{-29}\text{ m s}^{-1}}$$

E12.25 $\Delta p\Delta x\ge\frac{1}{2}\hbar$ [12.6] $\quad\Delta p=m\Delta v$

$$\Delta x_\text{min}=\frac{\hbar}{2m\Delta v}=\frac{1.055\times10^{-34}\text{ J s}}{2\times(0.0050\text{ kg})\times(1.0\times10^{-6}\text{ m s}^{-1})}=\boxed{1.0\times10^{-26}\text{ m}}$$

E12.26 The minimum uncertainty in position is $\boxed{100\text{ pm}}$. Therefore, because $\Delta x\Delta p\ge\frac{1}{2}\hbar$ [12.6]

$$\Delta p\ge\frac{\hbar}{2\Delta x}=\frac{1.0546\times10^{-34}\text{ J s}}{2(100\times10^{-12}\text{ m})}=5.3\times10^{-25}\text{ kg m s}^{-1}$$

$$\Delta v=\frac{\Delta p}{m_\text{e}}=\frac{5.3\times10^{-25}\text{ kg m s}^{-1}}{9.11\times10^{-31}\text{ kg}}=\boxed{5.8\times10^{5}\text{ m s}^{-1}}$$

E12.27 $$\psi_n(x)=\left(\frac{2}{L}\right)^{1/2}\sin\left(\frac{n\pi x}{L}\right)\text{ [12.7]}\quad\text{and}\quad|\psi_n(x)|^2=\left(\frac{2}{L}\right)\sin^2\left(\frac{n\pi x}{L}\right)$$

For $n=1$ and $L=100$ pm, the wavefunction and probability density are

$$\psi_1(x)=(1.414\times10^{5}\text{ m}^{-1/2})\sin\left(\frac{\pi x/\text{pm}}{100}\right)\quad\text{and}\quad|\psi_1(x)|^2=(2.00\times10^{10}\text{ m}^{-1})\sin^2\left(\frac{\pi x/\text{pm}}{100}\right)$$

(a) $x=10$ pm

$\psi_1(10\text{ pm})=4.37\times10^{4}\text{ m}^{-1/2}$ and $|\psi_1(10\text{ pm})|^2=1.91\times10^{9}\text{ m}^{-1}$

(b) $x=50$ pm

$\psi_1(50\text{ pm})=1.41\times10^{5}\text{ m}^{-1/2}$ and $|\psi_1(50\text{ pm})|^2=2.00\times10^{10}\text{ m}^{-1}$

(c) $x=100$ pm

$\psi_1(100\text{ pm})=0.00$ and $|\psi_1(50\text{ pm})|^2=0.00$

There is a node when $x=L$.

E12.28
$$E_n = \frac{n^2h^2}{8\,mL^2} \quad [12.8] \qquad \text{where } n = 1,2,3,\ldots$$

$$E_2 - E_1 = \frac{(2)^2 h^2}{8\,mL^2} - \frac{(1)^2 h^2}{8\,mL^2} = \frac{3h^2}{8\,mL^2}$$

$$= \frac{3\times(6.626\times10^{-34}\,\text{J s})^2}{8(1.674\times10^{-27}\,\text{kg})(1.0\times10^{-9}\,\text{m})^2}$$

$$= \boxed{9.84\times10^{-23}\,\text{J}}$$

E12.29
$$\psi_n(x) = \left(\frac{2}{L}\right)^{1/2} \sin\left(\frac{n\pi x}{L}\right) \quad [12.7]$$

Examination of the wavefunction for the n = 1 state, shown in text Figure 12.19, reveals that the maximum probability occurs when $x = ½\,L$.

At that location the probability density is given by

$$|\psi(½L)|^2 = \left(\frac{2}{L}\right)\sin^2\frac{\pi}{2} = \frac{2}{L}$$

We need to find the location(s) at which the probability density equals half of the maximum probability density (i.e., $1/L$).

$$|\psi(x)|^2 = \left(\frac{2}{L}\right)\sin^2\frac{\pi x}{L} = \frac{1}{L} \qquad \text{or} \qquad \sin\frac{\pi x}{L} = \frac{\sqrt{2}}{2}$$

The above relation is satisfied at two values of $\pi x/L$: $\pi/4$ and $3\pi/4$. Thus, $x = \boxed{L/4 \text{ and } 3L/4}$.

E12.30
$$E_n = \frac{n^2h^2}{8\,mL^2} \quad [12.8] \quad \text{where } n = 1,2,3,\ldots$$

(a)
$$\Delta E = E_5 - E_4$$

$$= (5^2 - 4^2)\,\frac{h^2}{8\,mL^2} = \frac{9\,h^2}{8\,mL^2}$$

$$= \frac{9\times(6.626\times10^{-34}\ \text{J s})^2}{8\times(9.109\times10^{-31}\ \text{kg})(5.0\times10^{-9}\ \text{m})^2}$$

$$= \boxed{2.1\bar{7}\times10^{-20}\,\text{J}}$$

(b)
$$\Delta E = h\nu = \frac{hc}{\lambda}$$

$$\lambda = \frac{hc}{\Delta E}$$

$$= (6.62608\times10^{-34}\ \text{J s})\times(2.99792\times10^{8}\ \text{m s}^{-1})/(2.1\bar{7}\times10^{-20}\ \text{J})$$

$$= 9.1\bar{6}\times10^{-6}\ \text{m} = \boxed{9.2\ \mu\text{m}}$$

E12.31
$$\int_{-\infty}^{\infty}\psi^2 dx = \int_0^L \psi^2 dx = \int_0^L A^2 dx = A^2\int_0^L dx = A^2 x\Big|_0^L = A^2L = 1 \text{ [the normalization condition]}$$

Therefore, $A = \left(\frac{1}{L}\right)^{1/2}$ and the normalized wave function is $\boxed{\psi = \left(\frac{1}{L}\right)^{1/2}}$.

E12.32 Carotene is a chain of 22 carbon atoms; therefore, there are 21 alternating single and double bonds. There are 11 double bonds, hence 11 π-bonds, each of which has two electrons.

$$L = 21\times(140\times10^{-12}\ \text{m}) = 2.94\times10^{-9}\ \text{m}$$

$$E_n = \frac{n^2h^2}{8\ mL^2}\quad [12.8] \qquad \text{where } n = 1, 2, 3, ...$$

$$\Delta E = E_{12} - E_{11} = (12^2 - 11^2) \times \frac{h^2}{8m_eL^2} = \frac{23h^2}{8m_eL^2}$$

$$= \frac{23 \times (6.626 \times 10^{-34}\ \text{J s})^2}{8 \times (9.109 \times 10^{-31}\ \text{kg}) \times (2.94 \times 10^{-9})^2}$$

$$= 1.60 \times 10^{-19}\ \text{J}$$

$$\nu = \frac{\Delta E}{h} = \frac{1.60 \times 10^{-19}\ \text{J}}{6.626 \times 10^{-34}\ \text{J s}} = 2.4 \times 10^{14}\ \text{s}^{-1} = 2.41 \times 10^{14}\ \text{Hz}$$

$$\lambda = \frac{c}{\nu} = \frac{2.998 \times 10^8\ \text{m s}^{-1}}{2.41 \times 10^{14}\ \text{s}^{-1}} = 1.24 \times 10^{-6}\ \text{m} = \boxed{1.24\ \mu\text{m}}$$

COMMENT. The observed wavelength for this transition is 450 nm (0.450 μm), so our crude model is off by more than a factor of 2. But it is the right order of magnitude.

E12.33 Your sketch should look like text Figure 12.21 with the potential energy barrier beginning at $x = 0$ and ending at $x = L$.

E12.34

$$E_{n_X n_Y} = \left(\frac{n_X^{\ 2}}{L_X^{\ 2}} + \frac{n_Y^{\ 2}}{L_Y^{\ 2}}\right)\frac{h^2}{8m}\quad [12.12b] \qquad \text{where } n_X, n_Y = 1, 2, 3, ...$$

When $L_X = L$ and $L_Y = 2L$, the energy expression becomes

$$E_{n_X n_Y} = \left(n_X^{\ 2} + \frac{n_Y^{\ 2}}{4}\right)\frac{h^2}{8mL^2}$$

Degeneracy occurs whenever there are multiple (n_X, n_Y) quantum number sets that give identical values of $n_X^{\ 2} + \frac{n_Y^{\ 2}}{4}$. We draw the following table to calculate and compare $n_X^{\ 2} + \frac{n_Y^{\ 2}}{4}$ values for different sets of quantum numbers.

n_X	n_Y	$n_X^{\ 2} + \frac{n_Y^{\ 2}}{4}$
1	1	5/4
	2	8/4
	3	13/4
	4	20/4
	5	29/4
	6	40/4
2	1	17/4
	2	20/4
	3	25/4
	4	32/4
	5	41/4
3	1	37/4
	2	40/4
	3	45/4
	4	52/4

Inspection of the table reveals that the states $\boxed{(n_X,n_Y) = (1,4) \text{ and } (2,2)}$ have the same energy so these states are degenerate. Also, the states $\boxed{(n_X,n_Y) = (1,6) \text{ and } (3,2)}$ have the same energy so these states are degenerate.

E12.35 (a) $I = m_H r^2$ [text Section 12.8a]

$$= (1.008\text{ u})\times(1.6605\times10^{-27}\text{ kg/u})\times(161\times10^{-12}\text{ m})^2$$

$$= \boxed{4.34\times10^{-47}\text{ kg m}^2}$$

(b) $E_{m_l} = \dfrac{m_l^2\hbar^2}{2I}$ [12.16] where $m_l = 0, \pm1, \pm2, \ldots$

$$\Delta E = E_1 - E_0 = \frac{h^2}{8\pi^2 I} \quad \text{and} \quad \Delta E = h\nu = \frac{hc}{\lambda}$$

$$\lambda = \frac{hc}{\Delta E} = \frac{8\pi^2 cI}{h}$$

$$= \frac{8\pi^2\times(2.998\times10^8\text{ m s}^{-1})\times(4.34\times10^{-47}\text{ kg m}^2)}{6.626\times10^{-34}\text{ J s}}$$

$$= 1.55\times10^{-3}\text{ m} = \boxed{1.55\text{ mm}}$$

This wavelength is in the microwave region of the electromagnetic spectrum.

E12.36 The angular speed ω in radians per second is related to the angular momentum (J_z) and moment of inertia (I) through the expression $\omega = J_z/I$, which comes from the classical, Newtonian physics. Since there are 2π radians per revolution, the number of revolutions per second around the axis that bisects the HOH angles equals $\omega/2\pi$. Furthermore, the minimum, non-zero value of J_z equals $\hbar$ according to quantum theory. Thus, the minimum number of revolutions per second is given by

$$\text{revolutions per second} = \omega/2\pi = J_z/2\pi I = \hbar/2\pi I = \frac{h}{(2\pi)^2 I}$$

$$= \frac{6.626\times10^{-34}\text{ J s}}{(2\pi)^2\times(1.91\times10^{-47}\text{ kg m}^2)} = \boxed{8.79\times10^{11}\text{ s}^{-1}}$$

E12.37 The energy needed to excite the molecule from a state that has no angular momentum to the state for which the angular momentum is given by $J_z = \hbar$ is

$$E = \frac{J_z^2}{2I}\text{ [12.15]} = \frac{\hbar^2}{2I} = \frac{h^2}{8\pi^2 I}$$

$$= \frac{(6.626\times10^{-34}\text{ J s})^2}{8\pi^2\times(1.91\times10^{-47}\text{ kg m}^2)} = \boxed{2.91\times10^{-22}\text{ J}}$$

E12.38 The methane molecule is an example of a spherical rotor, a molecule that has three equal moments of inertia because of its molecular symmetry. The quantized rotational energies of this molecular class is given by eqn 12.18 with a rotational quantum number that is denoted by J and a quantum number m_J that tells us the value, as $m_J\hbar$, of the rotational momentum around the z-axis (the polar axis of a sphere).

$J = 0, 1, 2,\ldots \quad m_J = J, J-1,\ldots, -J$

$$E_J = J(J+1)\frac{\hbar^2}{2I}\text{ [12.18]}$$

The minimum rotational energy (other than zero) is the state $J = 1$ and $m_{J=1} = 1, 0,$ and -1 giving a $\boxed{\text{degeneracy of 3}}$.

$$E_{J=1} = \frac{\hbar^2}{2I} = \frac{h^2}{8\pi^2 I}$$

With $I = \tfrac{8}{3} m_{\mathrm{H}} R^2$ and $m_{\mathrm{H}} = M_{\mathrm{H}}/N_{\mathrm{A}}$,

$$\begin{aligned} E_{J=1} &= \frac{3N_{\mathrm{A}}h^2}{64\pi^2 M_{\mathrm{H}} R^2} \\ &= \frac{3\times(6.022\times10^{23}\ \mathrm{mol^{-1}})\times(6.626\times10^{-34}\ \mathrm{J\ s})^2}{64\pi^2\times(0.0010079\ \mathrm{kg\ mol^{-1}})\times(109\times10^{-12}\ \mathrm{m})^2} \\ &= \boxed{1.05\times10^{-22}\ \mathrm{J}} \end{aligned}$$

E12.39 A period T is the reciprocal of the frequency ν. Therefore, since $T = 1$ s, $\nu = 1\ \mathrm{s^{-1}}$

$$\nu = \frac{1}{2\pi}\left(\frac{k}{m}\right)^{1/2} \quad [12.21]$$

Solve for the force constant, k.

$$\begin{aligned} k &= (2\pi\nu)^2 m \\ &= (2\pi\times1\ \mathrm{s^{-1}})^2\times(1\times10^{-3}\ \mathrm{kg}) \\ &= 0.04\ \mathrm{kg\ s^{-2}} = \boxed{0.04\ \mathrm{N\ m^{-1}}} \end{aligned}$$

E12.40 (a) $$\nu = \frac{1}{2\pi}\left(\frac{k}{m}\right)^{1/2} \quad [12.21]$$

$$= \frac{1}{2\pi}\left(\frac{314\ \mathrm{N\ m^{-1}}}{(1.0079\ \mathrm{u})\times(1.6605\times10^{-27}\ \mathrm{kg\ u^{-1}})}\right)^{1/2} = \boxed{6.89\times10^{13}\,\mathrm{s^{-1}}}$$

(b) $$\lambda = \frac{c}{\nu} = \frac{2.998\times10^{8}\ \mathrm{m\ s^{-1}}}{6.89\times10^{13}\ \mathrm{s^{-1}}} = 4.35\times10^{-6}\ \mathrm{m} = \boxed{4.35\ \mu\mathrm{m}}$$

E12.41 The D—I bond and the H—I bond are expected to have almost identical bond strengths and identical bonding force constants, k, because bonding is an electronic, not a mass/isotopic, property. However, the vibrational frequency does have a mass dependence.

$$\nu = \frac{1}{2\pi}\left(\frac{k}{m}\right)^{1/2} \quad [12.21]$$

$$\frac{\nu_{\mathrm{DI}}}{\nu_{\mathrm{HI}}} = \frac{\dfrac{1}{2\pi}\left(\dfrac{k}{m_{\mathrm{D}}}\right)^{1/2}}{\dfrac{1}{2\pi}\left(\dfrac{k}{m_{\mathrm{H}}}\right)^{1/2}} = \left(\frac{m_{\mathrm{H}}}{m_{\mathrm{D}}}\right)^{1/2} = \left(\frac{1}{2}\right)^{1/2} = 0.707$$

When hydrogen-1 is replaced by hydrogen-2 (deuterium) in H—I, the vibrational frequency decreases by a factor or 0.707.

Answers to projects

P12.42 (a) When the Born interpretation (Section 12.5) describes the infinitesimally small probability, dP, of finding a particle in an infinitely small region dV, it is written as the differential equation $dP = \psi^2 dV$ where the coordinates of ψ are those of the infinitesimal volume dV. To find the probability that the particle will be found in the region of V, integration must be performed over the region.

$$P = \int_{\text{region}} dV = \int_{\text{region}} \psi^2\, dV$$

When consideration is given to a particle in a one-dimensional box, where the region is between x_1 and x_2, this becomes

$$P = \int_{x_1}^{x_2} \psi^2\, dx \qquad \text{where } \psi = \left(\frac{2}{L}\right)^{1/2} \sin\left(\frac{2\pi x}{L}\right)$$

$$P = \int_{x_1}^{x_2} \left\{\left(\frac{2}{L}\right)^{1/2} \sin\left(\frac{2\pi x}{L}\right)\right\}^2 dx = \left(\frac{2}{L}\right)\int_{x_1}^{x_2} \sin^2\left(\frac{2\pi x}{L}\right) dx$$

Using the standard integral $\int \sin^2(ax)\, dx = \frac{x}{2} - \frac{\sin(2ax)}{4a}$, the working equation becomes

$$P = \left(\frac{2}{L}\right)\left[\frac{x}{2} - \frac{\sin\left(2\left(\frac{2\pi}{L}\right)x\right)}{4\left(\frac{2\pi}{L}\right)}\right]_{x_1}^{x_2} = \left[\frac{x}{L} - \frac{1}{4\pi}\sin\left(\frac{4\pi x}{L}\right)\right]_{x_1}^{x_2}$$

(i) $0.1\text{ nm} \le x \le 0.2\text{ nm}$

$$P = \left[\frac{x}{10\text{ nm}} - \frac{1}{4\pi}\sin\left(\frac{4\pi x}{10\text{ nm}}\right)\right]_{0.1\text{ nm}}^{0.2\text{ nm}} = 1.\overline{8}4\times10^{-4}$$

Error of the Exercise 12.22 approximation: $\dfrac{1.\overline{8}4\times10^{-4} - 1.\overline{7}7\times10^{-4}}{1.\overline{8}4\times10^{-4}}\times100 = 12.9\%$

(ii) $4.9\text{ nm} \le x \le 5.2\text{ nm}$

$$P = \left[\frac{x}{10\text{ nm}} - \frac{1}{4\pi}\sin\left(\frac{4\pi x}{10\text{ nm}}\right)\right]_{4.9\text{ nm}}^{5.2\text{ nm}} = 2.\overline{3}6\times10^{-4}$$

Error of the Exercise 12.24 approximation: $\left|\dfrac{2.\overline{3}6\times10^{-4} - 5.\overline{9}2\times10^{-5}}{2.\overline{3}6\times10^{-4}}\right|\times100 = 74.9\%$

(b) $\psi = \left(\frac{2}{L}\right)^{1/2} \sin\left(\frac{\pi x}{L}\right)$ for the $n = 1$ state

$$P = \int_{x_1}^{x_2} \psi^2\, dx$$

$$P = \int_{x_1}^{x_2} \left\{\left(\frac{2}{L}\right)^{1/2} \sin\left(\frac{\pi x}{L}\right)\right\}^2 dx = \left(\frac{2}{L}\right)\int_{x_1}^{x_2} \sin^2\left(\frac{\pi x}{L}\right) dx$$

Using the standard integral $\int \sin^2(ax)\, dx = \frac{x}{2} - \frac{\sin(2ax)}{4a}$, the working equation becomes

$$P = \left(\frac{2}{L}\right)\left[\frac{x}{2} - \frac{\sin\left(2\left(\frac{\pi}{L}\right)x\right)}{4\left(\frac{\pi}{L}\right)}\right]_{x_1}^{x_2} = \left[\frac{x}{L} - \frac{1}{2\pi}\sin\left(\frac{2\pi x}{L}\right)\right]_{x_1}^{x_2}$$

(i) Probability that the particle is in the left one-third of the box; $0 \le x \le L/3$

$$P = \left[\frac{x}{L} - \frac{1}{2\pi}\sin\left(\frac{2\pi x}{L}\right)\right]_{0*L}^{L/3} = \left[x - \frac{1}{2\pi}\sin(2\pi x)\right]_{0}^{1/3} = \boxed{0.196}$$

(ii) Probability that the particle is in the central one-third of the box; $L/3 \le x \le 2L/3$

$$P = \left[\frac{x}{L} - \frac{1}{2\pi}\sin\left(\frac{2\pi x}{L}\right)\right]_{L/3}^{2L/3} = \left[x - \frac{1}{2\pi}\sin(2\pi x)\right]_{1/3}^{2/3} = \boxed{0.609}$$

(iii) Probability that the particle is in the right one-third of the box; $2L/3 \le x \le L$

$$P = \left[\frac{x}{L} - \frac{1}{2\pi}\sin\left(\frac{2\pi x}{L}\right)\right]_{2L/3}^{L} = \left[x - \frac{1}{2\pi}\sin(2\pi x)\right]_{2/3}^{1} = \boxed{0.196}$$

Note that the probabilities sum to 1.

P12.43 (a) $\psi = N\mathrm{e}^{-ax^2/2} \qquad -\infty < x < \infty$

$$N^2\int_{-\infty}^{\infty}\left(\mathrm{e}^{-ax^2/2}\right)^2\mathrm{d}x = 1 \quad \text{(See Derivation 12.2)}$$

$$N^2\int_{-\infty}^{\infty}\mathrm{e}^{-ax^2}\mathrm{d}x = 1$$

(i) Using the standard, definite integral $\int_{-\infty}^{\infty}\mathrm{e}^{-ax^2}\mathrm{d}x = \left(\frac{\pi}{a}\right)^{1/2}$, we find that

$$N^2\left(\frac{\pi}{a}\right)^{1/2} = 1 \quad \text{or} \quad \boxed{N = \left(\frac{a}{\pi}\right)^{1/4}}$$

The normalized wavefunction is $\psi = \left(\frac{a}{\pi}\right)^{1/4}\mathrm{e}^{-ax^2/2}$.

(ii) The function $\psi = \left(\frac{a}{\pi}\right)^{1/4}\mathrm{e}^{-ax^2/2}$ is a "bell" or "Gaussian" function with a maximum $x = 0$.

Consequently, ψ^2 has a maximum value at the displacement $x = 0$ and the Born interpretation of ψ^2 (see Section 12.5) indicates that the displacement $x = 0$ is the most probable displacement. It is instructive to use analytic geometry to find the maximum. This requires identification of the displacement for which $\mathrm{d}\psi/\mathrm{d}x = 0$ and showing that $\mathrm{d}\psi/\mathrm{d}x > 0$ before the maximum and $\mathrm{d}\psi/\mathrm{d}x < 0$ after the maximum

$$\frac{\mathrm{d}\psi}{\mathrm{d}x} = \frac{\mathrm{d}}{\mathrm{d}x}\left\{\left(\frac{a}{\pi}\right)^{1/4}\mathrm{e}^{-ax^2/2}\right\} = \left(\frac{a}{\pi}\right)^{1/4}\frac{\mathrm{d}}{\mathrm{d}x}\mathrm{e}^{-ax^2/2} = -\left(\frac{a}{\pi}\right)^{1/4}ax\,\mathrm{e}^{-ax^2/2}$$

The factor of x in the first derivative indicates that the derivative equals zero when $x = 0$. Furthermore, the formula for the derivative clearly shows that the derivative is positive when $x < 0$ and the derivative is negative when $x > 0$. The function is a maximum at $x = 0$.

(b) $\psi = Nx\mathrm{e}^{-ax^2/2} \qquad -\infty < x < \infty$

$$N^2\int_{-\infty}^{\infty}\left(x\mathrm{e}^{-ax^2/2}\right)^2\mathrm{d}x = 1 \quad \text{(See Derivation 12.2)}$$

$$N^2\int_{-\infty}^{\infty}x^2\mathrm{e}^{-ax^2}\mathrm{d}x = 1$$

(i) Using the standard, definite integral $\int_{-\infty}^{\infty}x^2\mathrm{e}^{-ax^2}\mathrm{d}x = \frac{1}{2}\left(\frac{\pi}{a^3}\right)^{1/2}$, we find that

$$\frac{N^2}{2}\left(\frac{\pi}{a^3}\right)^{1/2}=1 \quad \text{or} \quad \boxed{N=\left(\frac{4a^3}{\pi}\right)^{1/4}}$$

The normalized wavefunction is $\psi=\left(\frac{4a^3}{\pi}\right)^{1/4}xe^{-ax^2/2}$.

(ii) Unlike the function of part (a), the function $\psi=\left(\frac{4a^3}{\pi}\right)^{1/4}xe^{-ax^2/2}$ has a node at $x=0$ so we find displacements at which the oscillator is most likely to be found by finding the displacements for which $\mathrm{d}|\psi^2|/\mathrm{d}x=0$. These are the displacements at which the probability density is a maximum.

$$\frac{\mathrm{d}\psi^2}{\mathrm{d}x}=2\psi\frac{\mathrm{d}\psi}{\mathrm{d}x}=2N^2xe^{-ax^2/2}\frac{\mathrm{d}}{\mathrm{d}x}\left\{xe^{-ax^2/2}\right\}=2N^2xe^{-ax^2/2}\left\{1-ax^2\right\}e^{-ax^2/2}=0$$

This implies that the maxima occur when

$$1-ax^2=0 \quad \text{or} \quad \boxed{x=\pm(1/a)^{1/2}}$$

COMMENT. Can you sketch a graph of the displacement dependence of both the wavefunction and probability density for parts (a) and (b)?

P12.44 $\nu=\frac{1}{2\pi}\left(\frac{k}{\mu}\right)^{1/2}$ [12.21] where $\mu=m_{\mathrm{A}}m_{\mathrm{B}}/(m_{\mathrm{A}}+m_{\mathrm{B}})=M_{\mathrm{A}}M_{\mathrm{B}}/\{N_{\mathrm{A}}(M_{\mathrm{A}}+M_{\mathrm{B}})\}$

(a) $$\mu_{^{12}\mathrm{C}^{16}\mathrm{O}}=\frac{(12.00)\times(16.00)\times10^{-3}\ \mathrm{kg}}{(6.0221\times10^{23})\times(28.00)}=1.139\times10^{-26}\ \mathrm{kg}$$

$$\nu_{^{12}\mathrm{C}^{16}\mathrm{O}}=\frac{1}{2\pi}\left(\frac{1860\ \mathrm{N\ m^{-1}}}{1.139\times10^{-26}\ \mathrm{kg}}\right)^{1/2}=\boxed{6.432\times10^{13}\ \mathrm{s^{-1}}}$$

(b) $$\tilde{\nu}_{^{12}\mathrm{C}^{16}\mathrm{O}}=1/\lambda_{^{12}\mathrm{C}^{16}\mathrm{O}}=\frac{\nu_{^{12}\mathrm{C}^{16}\mathrm{O}}}{c}$$

$$=\frac{6.432\times10^{13}\ \mathrm{s^{-1}}}{2.9979\times10^{8}\ \mathrm{m\ s^{-1}}}=2.146\times10^{5}\ \mathrm{m^{-1}}=\boxed{2146\ \mathrm{cm^{-1}}}$$

(c) Computations like those of part (a) and part (b) can be repeated for additional isotopes of carbon monoxide. We will take a short-cut by recognizing that these isotopic CO molecules have identical force constants and differ only in their reduced mass. Consequently, since both the frequency and wavenumber are inversely proportional to $\mu^{1/2}$, we write

$$\mu_{^{13}\mathrm{C}^{16}\mathrm{O}}=\frac{(13.00)\times(16.00)\times10^{-3}\ \mathrm{kg}}{(6.0221\times10^{23})\times(29.00)}=1.191\times10^{-26}\ \mathrm{kg}$$

$$\tilde{\nu}_{^{13}\mathrm{C}^{16}\mathrm{O}}=\left(\frac{\mu_{^{12}\mathrm{C}^{16}\mathrm{O}}}{\mu_{^{13}\mathrm{C}^{16}\mathrm{O}}}\right)^{1/2}\tilde{\nu}_{^{12}\mathrm{C}^{16}\mathrm{O}}$$

$$=\left(\frac{1.139}{1.191}\right)^{1/2}\times(2146\ \mathrm{cm^{-1}})=\boxed{2099\ \mathrm{cm^{-1}}}$$

Similarly,

$$\mu_{^{12}\mathrm{C}^{18}\mathrm{O}}=1.196\times10^{-26}\ \mathrm{kg} \quad \text{and} \quad \boxed{\tilde{\nu}_{^{12}\mathrm{C}^{18}\mathrm{O}}=2094\ \mathrm{cm^{-1}}}$$

$$\mu_{^{13}\mathrm{C}^{18}\mathrm{O}}=1.253\times10^{-26}\ \mathrm{kg} \quad \text{and} \quad \boxed{\tilde{\nu}_{^{13}\mathrm{C}^{18}\mathrm{O}}=2046\ \mathrm{cm^{-1}}}$$

13 Atomic structure

Answers to discussion questions

D13.1 (1) The **principal quantum number** n determines the energy of the atomic orbitals in a hydrogenic shell through eqn 13.4a. The shells K, L, M, and N correspond to the principal quantum numbers n = 1, 2, 3, and 4. Successive shells are further away from the nucleus on average and successively higher in energy. The permitted orbital energies approach zero as n becomes very large because this is defined to be the minimum energy at which the electron and nucleus are infinitely separated. The principal quantum numbers also describe hydrogenic transitions through eqn 13.1.

(2) The **orbital angular momentum quantum number** l, also called the azimuthal quantum number, determines the magnitude of the orbital angular momentum of a hydrogenic atomic orbital through the formula $\{l(l+1)\}^{1/2}\hbar$. The permitted values of l are 0, 1, 2, 3,..., $n - 1$ for the nth shell and these correspond to the s, p, d, f,... subshells. The degeneracy of a subshell is $2l + 1$ because this is the number of orbitals in each subshell. Thus, the s, p, d, and f subshells consist of 1, 3, 5, and 7 orbitals, respectively. In many-electron atoms a maximum of two electrons can be in an orbital; this is the **Pauli exclusion principle**.

(3) The **magnetic quantum number** m_l determines the z-component of the angular momentum of a hydrogenic orbital through the formula $m_l\hbar$. The permitted values for subshell l are l, $l - 1$, $l - 2$, ..., $-l$, which accounts for the orbital degeneracy of the subshell.

(4) The **spin quantum number** s determines the magnitude of the electron spin angular momentum through the formula $\{s(s+1)\}^{1/2}\hbar$. For hydrogenic atomic orbitals, s can only be $1/2$.

(5) The **spin quantum number** m_s determines the z-component of the spin angular momentum through the formula $m_s\hbar$. m_s can only be ±½. m_s = +½ corresponds to the α or ↑ spin; m_s = −½ corresponds to the β or ↓ spin.

D13.2 (a) A **boundary surface** for a hydrogenic orbital is drawn to contain most (say 90%) of the probability density of an electron in that orbital. Its shape varies from orbital to orbital because the electron density distribution is different for different orbitals. Example boundary surfaces are shown in text figures 13.7 (s orbital), 13.10 (p orbitals), and 13.11 (d orbitals).

(b) The **radial distribution function** gives the probability that the electron will be found anywhere within a shell of radius r around the nucleus (see text Figure 13.8). It gives a better picture of where the electron is likely to be found with respect to the nucleus than the probability density, which is the square of the wavefunction. The radial distribution function for an s orbital is

$$P(r) = 4\pi r^2 \psi_{ns}^2 \quad [13.7a]$$

The more general form, which also applies to orbitals that depend on angle, is

$$P(r) = r^2 R_{n,l}(r)^2 \quad [13.7b] \qquad \text{where } R_{n,l}(r) \text{ is the radial wavefunction}$$

The usefulness of the radial distribution function is illustrated in text Figure 13.16 which shows the relative penetration of the inner core by subshell orbitals of the M shell (n = 3). The order of penetration is 3s > 3p > 3d and, consequently, electrons in these subshells have the same order of relative effective nuclear charge attracting them to the nucleus. High effective nuclear charge means lower orbital energy, so the order of subshell energy is 3s < 3p < 3d, a fact that is very important when use the building-up principle to determine the ground electron configuration of many-electron atoms.

D13.3 In the crudest form of the **orbital approximation**, the many-electron wavefunction for an atom is represented as a simple product of one-electron wavefunctions (see eqn 13.8), each of which has the form of a hydrogenic atomic orbital. This is said to be the independent-electron model. For example, the orbital approximation for the lithium atom ground state wavefunction is the product of the orbitals for each of the three lithium electrons:

$$\psi_{\text{Li}} = \psi_{1s}(1)\alpha(1)\times\psi_{1s}(2)\beta(2)\times\psi_{2s}(3)\alpha(3)$$

This is synonymous with the ground state electronic configuration given by the building-up principle. For the lithium atom it is $1s^2 2s$. The simplest form of this approximation neglects electron repulsions in many-electron atoms so it does not give a very good estimate of the atomic energy. It does, however, provide concepts for the quick analysis of a great many atomic and molecular problems in chemistry and biochemistry.

At a somewhat more sophisticated level, the many electron wavefunctions are written as linear combinations of such simple product functions that explicitly satisfy the Pauli exclusion principle. Relatively good one-electron functions are generated by the Hartree–Fock self-consistent field method described in Section 13.14 in which an electron moves in the average electron repulsion potential field of all other electrons. We can in principle obtain exact energies and wavefunctions with such numerical methods; however, there are significant numerical challenges.

D13.4 The relationship between the location of a many-electron atom in the periodic table and its electron configuration is fundamentally useful, and even a guiding principle, within the sciences. It provides quick information about valence electrons and bonding characteristics of the elements. Relevant discussion is found in text sections 13.11, 13.15, and 13.16. If you feel that your understanding is somewhat weak, do not hesitate to consult a general chemistry textbook at the library. We recommend Sections 1.11–1.19 of P. Atkins and L. Jones, *Chemical Principles: The Quest for Insight*, 3rd edn, W.H. Freeman, and Co., New York (2005).

Examination of the placement of oxygen in the period table of elements illustrates the relation between position and ground electron configuration. Oxygen is in the second period so its inner core electronic structure is [He], the electron configuration of the inert gas of the prior period, and its valence electrons are those of the L shell for which $n = 2$. Finally, being in group VIA it has six valence electron, two of which fill the 2s subshell leaving four in the unfilled 2p subshell. Oxygen has a strong tendency to gain two electrons and, thereby, fill the 2p subshell.

$$\text{O} + 2\,\text{e}^- \rightarrow \text{O}^{2-}$$

$$[\text{He}]2s^2 2p^4 \qquad [\text{He}]2s^2 2p^6 = [\text{Ne}]$$

group→ period↓	1 IA	2 IIA	3 IIIB		12 IIB	13 IIIA	14 IVA	15 VA	16 VIA	17 VIIA	18 VIIIA
1	$_1$H $1s^1$										$_2$He $1s^2$
2	$_3$Li $2s^1$	$_4$Be $2s^2$				$_5$B $2s^2 2p^1$	$_6$C $2s^2 2p^2$	$_7$N $2s^2 2p^3$	$_8$O $2s^2 2p^4$	$_9$F $2s^2 2p^5$	$_{10}$Ne $2s^2 2p^6$
3	$_{11}$Na $3s^1$	$_{12}$Mg $3s^2$				$_{13}$Al $3s^2 3p^1$	$_{14}$Si $3s^2 3p^2$	$_{15}$P $3s^2 3p^3$	$_{16}$S $3s^2 3p^4$	$_{17}$Cl $3s^2 3p^5$	$_{18}$Ar $3s^2 3p^6$
4	$_{19}$K $4s^1$	$_{20}$Ca $4s^2$	$_{21}$Sc $3d^1 4s^2$		$_{30}$Zn $3d^{10} 4s^2$	$_{31}$Ga $4s^2 4p^1$	$_{32}$Ge $4s^2 4p^2$	$_{33}$As $4s^2 4p^3$	$_{34}$Se $4s^2 4p^4$	$_{35}$Br $4s^2 4p^5$	$_{36}$Kr $4s^2 4p^6$

Inspection of the first periods of the period table shows that elements of a group have basically the same valence electron structure, the difference being that in going down a group the valence electrons are in successive shells. When going across a period successive subshells are filled. Going left to right along the second period, we see that the 2s subshell fills; then, the 2p subshell fills. This is exactly the filling order of the building-up principle when placing electrons in the orbital energy level diagram. When going left to right across period 4, the 4s subshell fills, followed by the 3d subshell, then the 4p subshell fills. An f subshell becomes available in the building-up scheme when filling the $n = 6$ shell so the order of filling subshells is 6s, 4f, 5d, 6p.

D13.5 The first ionization energies, shown in text Figure 13.22, increase markedly from Li to Be, decrease slightly from Be to B, again increase markedly from B to N, again decrease slightly from N to O, and finally increase markedly from O to Ne. The general trend is an overall increase of I_1 with atomic number across the period. That is to be expected since the principal quantum number (electron shell) of the outer electron remains the same, while its attraction to the nucleus increases. The slight decrease from Be to B is a reflection of the outer electron being in a higher energy subshell (larger l value) in B than in Be. The slight decrease from N to O is due to the half-filled subshell effect; half-filled subshells have increased stability. O has one electron outside of the half-filled p subshell and that electron must pair with another resulting in strong electron–electron repulsions between them. Period 3 elements mirror this pattern in going from left to right across the periodic row for the same reasons.

D13.6 An electron in a p or d or f orbital has a magnetic moment and magnetic field due to its orbital angular momentum (s orbital electrons have no orbital angular momentum because $l = 0$) and every electron has a magnetic moment and magnetic field due to its spin angular momentum. The magnetic fields of the orbital motion and the spin motion are represented in Figure 13.1 as small bar magnets near the classical angular momentum vector for each motion. In Figure 13.1(a) the magnet ends are oriented north-to-north and south-to-south – a repulsive orientation that increases energy w/r/t the summed energies of the separate motions. In Figure 13.1(b) the magnet ends are oriented north-to-south and south-to-north – a energy lowering, attractive orientation. Thus, this interaction, called **spin-orbit coupling**, separates the energies of states that have different orientations of orbital angular momentum and spin angular momentum giving rise to multiple spectral lines where only one would be expected in the absence of the coupling. An example is provided by the two intense yellow emission lines of sodium. An electric discharge or the high temperatures of a flame can excited the $^2S_{1/2}$ ground level of sodium to either a $^2P_{1/2}$ antiparallel angular momentum level or a $^2P_{3/2}$ parallel angular momentum level. Spontaneous emission of radiation occurs from these excited levels at the wavelengths 589.76 nm and 589.16 nm as the excited electron returns to the ground level, a spectral line difference of 17 cm^{-1}.

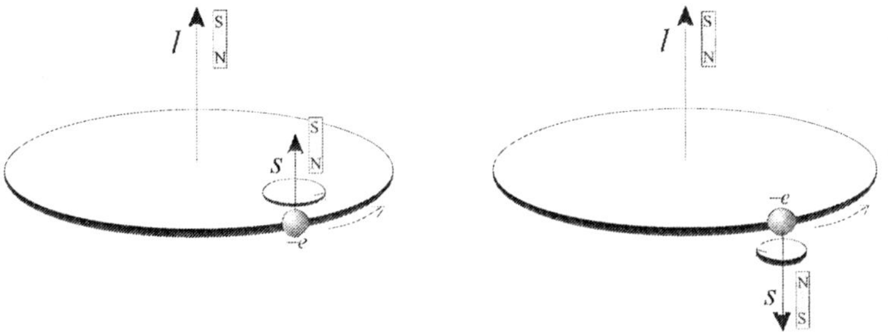

(a) Parallel angular momentum magnetic fields repel high energy

(b) Antiparallel angular momentum magnetic fields attract low energy

Figure 13.1

D13.7 (a) The selection rules for hydrogenic atoms are:

$$\Delta n = \pm 1, \pm 2, \ldots \qquad \Delta l = \pm 1 \qquad \Delta m_l = 0, \pm 1$$

In a spectroscopic transition, the atom emits or absorbs a photon. Photons have a spin angular momentum of 1. Therefore, because of the transition, the angular momentum of the electromagnetic field has changed by $\pm 1\hbar$. The principle of the conservation of angular momentum then requires that the angular momentum of the atom has undergone an equal and opposite change in angular momentum. Hence, the selection rule is $\Delta l = \pm 1$. The principle quantum number n can change by any amount since n does not directly relate to angular momentum. The selection rule on Δm_l is harder to account for on basis of these simple considerations alone. One has to evaluate the transition dipole moment between the wavefunctions representing the initial and final states involved in the transition.

(b) The selection rules for relatively light many-electron atoms are:

$$\Delta S = 0 \qquad \Delta L = 0,\ \pm 1 \qquad \Delta l = \pm 1$$

$$\Delta J = 0,\ \pm 1, \text{ but } J = 0 \leftrightarrow J = 0 \text{ is forbidden}$$

A change in the total spin angular momentum is forbidden in an electronic transition because light does not directly affect spin. This important selection rule applies to both atoms and molecules. Additional rules arise from the conservation of total angular momentum in the atom-radiation system. As discussed in part (a) the orbital angular momentum of an individual electron must change. This may, or may not, affect the total orbital angular momentum.

Solutions to exercises

E13.1 The lower level of the Balmer series of the spectrum of atomic hydrogen has $n = 2$. The wavenumber of the transition from the $n = 6$ level is

$$\tilde{\nu} = R_{\mathrm{H}}\left(\frac{1}{n_1^2}-\frac{1}{n_2^2}\right)\ [13.1]$$

$$= \left(109\ 677\ \mathrm{cm^{-1}}\right)\times\left(\frac{1}{2^2}-\frac{1}{6^2}\right) = 2.43727\times10^4\,\mathrm{cm^{-1}}$$

$$\lambda = \frac{1}{\tilde{\nu}}$$

$$= \frac{1}{2.43727\times10^4\,\mathrm{cm^{-1}}} = \boxed{410.296\ \mathrm{nm}}$$

E13.2 The lower level of the Paschen series of the spectrum of atomic hydrogen has $n = 3$. Thus, the upper level, n_2, is given by

$$\tilde{\nu} = \frac{\nu}{c} = R_{\mathrm{H}}\left(\frac{1}{n_1^2}-\frac{1}{n_2^2}\right)\ [13.1]$$

$$\frac{1}{n_2^2} = \frac{1}{n_1^2} - \frac{\nu}{cR_{\mathrm{H}}}$$

$$n_2 = \left(\frac{1}{3^2} - \frac{2.7415\times10^{14}\ \mathrm{s^{-1}}}{\left(2.9979\times10^8\ \mathrm{m\ s^{-1}}\right)\times\left(1.09677\times10^7\ \mathrm{m^{-1}}\right)}\right)^{-1/2}$$

$$= 6.005 = \boxed{6}$$

E13.3 The conservation of energy, not wavelength, means that spectral computations must involve either the additions or subtractions of energy directly or properties like frequency or wavenumber that are proportional to energy. The wavenumber of the emission is

$$\tilde{\nu} = \frac{1}{\lambda} = \frac{1}{486.1\times10^{-7}\ \mathrm{cm}} = 2057\bar{2}\ \mathrm{cm^{-1}}$$

(a) Since the upper term is at 27414 $\mathrm{cm^{-1}}$, the lower term lies at

$$27414\ \mathrm{cm^{-1}} - 2057\bar{2}\ \mathrm{cm^{-1}} = 684\bar{2}\ \mathrm{cm^{-1}}$$

(b) The energy (in the SI unit) of the lower term is

$$E = hc\tilde{\nu}$$

$$= \left(6.626\times10^{-34}\ \mathrm{J\ s}\right)\times\left(2.998\times10^8\ \mathrm{m\ s^{-1}}\right)\times\left(684\bar{2}\ \mathrm{cm^{-1}}\right)\times\left(100\ \mathrm{cm/m}\right)$$

$$= \boxed{1.36\times10^{-19}\ \mathrm{J}}$$

E13.4 According to eqn 13.4a the emission wavenumber for the hydrogenic transition from level n_2 to level n_1 is

$$\tilde{\nu} = \tilde{E}_2 - \tilde{E}_1 = RZ^2\left(\frac{1}{n_1^2}-\frac{1}{n_2^2}\right) \quad \text{where } R = \frac{\mu e^4}{8\varepsilon_0^2 h^3 c} \quad \text{and} \quad \mu = \frac{m_{\mathrm{e}}m_{\mathrm{p}}}{m_{\mathrm{e}}+m_{\mathrm{p}}}\ [13.4\mathrm{a\ and\ b}]$$

Thus, the wavenumber difference between identical transitions of different hydrogenic atoms A and B is

$$\Delta\tilde{\nu} = \tilde{\nu}_B - \tilde{\nu}_A = \left\{\left(RZ^2\right)_B - \left(RZ^2\right)_A\right\}\times\left(\frac{1}{n_1^2} - \frac{1}{n_2^2}\right)$$

The Rydberg constant for the hydrogen-1 atom is $R_H = 109677\ \text{cm}^{-1}$ but other hydrogenic atoms have slightly different Rydberg constants because, as shown by eqn 13.4b, R is proportional to the reduced mass μ. In fact, the proportionality means that R_B, the Rydberg constant for hydrogenic atom B, and R_H are related by

$$R_B = \left(\frac{\mu_B}{\mu_H}\right)R_H$$

To calculate the Rydberg constant for deuterium (^{2}H), we first calculate the reduced masses of ^{1}H and ^{2}H. These are then substituted into the above expression.

$$\mu_{^1\text{H}} = \frac{m_e m_p}{m_e + m_p} \quad [13.4b]$$

$$= \frac{(9.109390\times10^{-31}\ \text{kg})\times(1.672623\times10^{-27}\ \text{kg})}{(9.109390\times10^{-31}\ \text{kg})+(1.672623\times10^{-27}\ \text{kg})} = (9.104431\times10^{-31}\ \text{kg})$$

$$\mu_{^2\text{H}} = \frac{m_e(m_p + m_n)}{m_e + m_p + m_n}$$

$$= \frac{(9.109390\times10^{-31}\ \text{kg})\times(1.672623\times10^{-27}\ \text{kg}+1.674929\times10^{-27}\ \text{kg})}{(9.109390\times10^{-31}\ \text{kg})+(1.672623\times10^{-27}\ \text{kg}+1.674929\times10^{-27}\ \text{kg})}$$

$$= (9.106912\times10^{-31}\ \text{kg})$$

$$\frac{\mu_{^2\text{H}}}{\mu_{^1\text{H}}} = \frac{9.106912\times10^{-31}\ \text{kg}}{9.104431\times10^{-31}\ \text{kg}} = 1.000273$$

$$R_{^2\text{H}} = (1.000273)\times(109677\ \text{cm}^{-1}) = 109707\ \text{cm}^{-1}$$

Thus, since $Z = 1$ for isotopes of hydrogen, the wavenumber difference between the 3p → 1s transition of ^{1}H and ^{2}H is

$$\Delta\tilde{\nu} = \tilde{\nu}_{^2\text{H}} - \tilde{\nu}_{^1\text{H}}$$

$$= (109707\ \text{cm}^{-1} - 109677\ \text{cm}^{-1})\times\left(\frac{1}{1^2} - \frac{1}{3^2}\right) = \boxed{27\ \text{cm}^{-1}}$$

Note that the actual hydrogen-1 transition occurs at

$$\tilde{\nu}_{^1\text{H}} = (109677\ \text{cm}^{-1})\times\left(\frac{1}{1^2} - \frac{1}{3^2}\right) = 97491\ \text{cm}^{-1}$$

so the difference between the isotopic hydrogen emissions is only about 0.03% of the emission energy.

E13.5 According to eqn 13.4a the emission wavenumber for the hydrogenic transition from level n_2 to level n_1 is

$$\tilde{\nu} = \tilde{E}_2 - \tilde{E}_1 = RZ^2\left(\frac{1}{n_1^2} - \frac{1}{n_2^2}\right) \quad \text{where } R = \frac{\mu e^4}{8\varepsilon_0^2 h^3 c} \quad \text{and} \quad \mu = \frac{m_e m_p}{m_e + m_p} \quad [13.4\text{a and b}]$$

In exercise 13.4 we see that, because the reduced mass is nearly identical for hydrogenic atoms, the Rydberg constant for all of them may be estimated to equal R_H so in this exercise we use the approximation $R_{He^+} = R_{^1H}$. We want to find the He^+ transition such that

$$\tilde{\nu}_{He^+} = \tilde{\nu}_H$$

$$4R_{He^+} \times \left(\frac{1}{n_1^2} - \frac{1}{n_2^2}\right)_{He^+} = R_H \times \left(\frac{1}{1^2} - \frac{1}{2^2}\right)$$

$$\left(\frac{1}{n_1^2} - \frac{1}{n_2^2}\right)_{He^+} = \frac{3}{16}$$

By inspection we find that $\boxed{n_1 = 2 \text{ and } n_2 = 4}$.

E13.6 When a stellar surface temperature is in the range 3000 K–4000 K (a "red star"), the star surface doesn't have the energetic particles and photons that are required for either the collisional or radiation excitation of a neutral hydrogen atom. In the absence of excitation, the atoms neither affect the absorption nor the emission lines of these stars. In contrast, a star with a surface temperature of 8000 K–10000 K has a temperature low enough to avoid complete hydrogen ionization but high enough for the radiation environment to cause electronic transitions of atomic hydrogen. Hydrogen spectral lines are intense for these stars.

"Blue stars" have surface temperatures of 15000 K–20000 K. Both the collision energies and the radiation energy environment of the hydrogen atoms are great enough to ionize a significant fraction of the hydrogen atoms and lacking an electron, the remaining proton cannot affect absorption and emission lines. Atomic hydrogen spectral lines are less intense for these stars.

When a star has a surface temperature above 25000 K, the collision energies and radiation environment are high enough to ionize almost all hydrogen atoms. The percentage of unionized hydrogen atoms is so small that their effects cannot be seen in the spectrum of the star. This statement can be verified by comparing the ionization energy of the hydrogen atom (13.6 eV) with the radiation energy at the maximum of the radiation energy distribution. The **Wien displacement law** relates wavelength λ_{max} at the maximum of the radiation energy distribution to the temperature T.

$$\lambda_{max} = \tfrac{1}{5} c_2 / T \qquad \text{where } c_2 = 1.44 \text{ cm K}$$

$$\tilde{\nu}_{max} = 1/\lambda_{max} = 5T/c_2$$

At 25000 K,

$$\tilde{\nu}_{max} = 5 \times (25000 \text{ K})/(1.44 \text{ cm K}) = 8.68 \times 10^4 \text{ cm}^{-1}$$

This maximum corresponds to 10.8 eV, which is a little less than the hydrogen ionization energy. It is however enough energy to ionize hydrogen atoms that have been collisionally excited and, since the radiation energy distribution shows a significant intensity of radiation at somewhat higher values, the hydrogen atoms are largely ionized. For a discussion of the Wien displacement law, radiation energy distribution, and black-body radiation see Section 11.1a of P. Atkins and J. de Paula, *Physical Chemistry*, 7th ed., W.H. Freeman and Company, New York (2002).

E13.7 According to eqn 13.4a the emission wavenumber for the hydrogenic transition from level n_2 to level n_1 is

$$\tilde{\nu} = \tilde{E}_2 - \tilde{E}_1 = RZ^2\left(\frac{1}{n_1^2} - \frac{1}{n_2^2}\right) \qquad \text{where } R = \frac{\mu e^4}{8\varepsilon_0^2 h^3 c} \quad \text{and} \quad \mu = \frac{m_e m_p}{m_e + m_p} \quad [13.4\text{a and b}]$$

The Rydberg constant for the hydrogen-1 atom is R_H = 109677 cm^{-1} but other hydrogenic atoms have slightly different Rydberg constants because, as shown by eqn 13.4b, R is proportional to the reduced mass μ. In fact, the proportionality means that R_B, the Rydberg constant for hydrogenic atom B, and R_H are related by

$$R_B = \left(\frac{\mu_B}{\mu_H}\right) R_H$$

To calculate the Rydberg constant for $^4He^+$ and $^3He^+$, we first calculate the reduced masses. These are then substituted into the above expression.

$$\mu_{^1H} = \frac{m_e m_p}{m_e + m_p} \quad [13.4b]$$

$$= \frac{(9.109390\times10^{-31}\ \text{kg})\times(1.672623\times10^{-27}\ \text{kg})}{(9.109390\times10^{-31}\ \text{kg})+(1.672623\times10^{-27}\ \text{kg})} = (9.104431\times10^{-31}\ \text{kg})$$

$$\mu_{^4He^+} = \frac{m_e(2m_p + 2m_n)}{m_e + 2m_p + 2m_n}$$

$$= \frac{(9.109390\times10^{-31}\ \text{kg})\times(2\times1.672623\times10^{-27}\ \text{kg}+2\times1.674929\times10^{-27}\ \text{kg})}{(9.109390\times10^{-31}\ \text{kg})+(2\times1.672623\times10^{-27}\ \text{kg}+2\times1.674929\times10^{-27}\ \text{kg})}$$

$$= (9.108150\times10^{-31}\ \text{kg})$$

$$\frac{\mu_{^4He^+}}{\mu_{^1H}} = \frac{9.108150\times10^{-31}\ \text{kg}}{9.104431\times10^{-31}\ \text{kg}} = 1.000409$$

$$R_{^4He^+} = (1.000409)\times(109677\ \text{cm}^{-1}) = 109722\ \text{cm}^{-1}$$

$$\mu_{^3He^+} = \frac{m_e(2m_p + m_n)}{m_e + 2m_p + 2m_n}$$

$$= \frac{(9.109390\times10^{-31}\ \text{kg})\times(2\times1.672623\times10^{-27}\ \text{kg}+1.674929\times10^{-27}\ \text{kg})}{(9.109390\times10^{-31}\ \text{kg})+(2\times1.672623\times10^{-27}\ \text{kg}+1.674929\times10^{-27}\ \text{kg})}$$

$$= (9.107737\times10^{-31}\ \text{kg})$$

$$\frac{\mu_{^3He^+}}{\mu_{^1H}} = \frac{9.107737\times10^{-31}\ \text{kg}}{9.104431\times10^{-31}\ \text{kg}} = 1.000363$$

$$R_{^3He^+} = (1.000363)\times(109677\ \text{cm}^{-1}) = 109717\ \text{cm}^{-1}$$

Thus, since $Z = 2$ for isotopes of helium, the transitions occur at

$$\tilde{\nu}_{^4He^+} = 4R_{^4He^+}\left(\frac{1}{n_1^2} - \frac{1}{n_2^2}\right) \qquad \text{where } R_{^4He^+} = 109722\ \text{cm}^{-1}$$

$$\tilde{\nu}_{^3He^+} = 4R_{^4He^+}\left(\frac{1}{n_1^2} - \frac{1}{n_2^2}\right) \qquad \text{where } R_{^3He^+} = 109717\ \text{cm}^{-1}$$

For the $n = 3 \rightarrow n = 2$ transition:

$$\tilde{\nu}_{^4He^+} = 4\times(109722\ \text{cm}^{-1})\times\left(\frac{1}{2^2} - \frac{1}{3^2}\right) = \boxed{60956.7\ \text{cm}^{-1}}$$

$$\tilde{\nu}_{^3He^+} = 4\times(109717\ \text{cm}^{-1})\times\left(\frac{1}{2^2} - \frac{1}{3^2}\right) = \boxed{60953.9\ \text{cm}^{-1}}$$

For the $n = 2 \rightarrow n = 1$ transition:

$$\tilde{\nu}_{^4\mathrm{He}^+} = 4\times\left(109722\ \mathrm{cm}^{-1}\right)\times\left(\frac{1}{1^2}-\frac{1}{2^2}\right) = \boxed{329166\ \mathrm{cm}^{-1}}$$

$$\tilde{\nu}_{^3\mathrm{He}^+} = 4\times\left(109717\ \mathrm{cm}^{-1}\right)\times\left(\frac{1}{1^2}-\frac{1}{2^2}\right) = \boxed{329151\ \mathrm{cm}^{-1}}$$

The lines for the $n = 3 \rightarrow n = 2$ transition are separated by only 2.8 cm^{-1} while those for the $n = 2 \rightarrow n = 1$ transition are separated by only 15 cm^{-1}. Very sensitive spectroscopic instruments are able to view lines of such small separations and the ratios of the line intensities gives the abundance ratio of the isotopic species.

E13.8 The reduced mass is nearly identical for hydrogenic atoms so the Rydberg constant for all of them may be estimated to equal R_H. Also, the ionization energy is equivalent to the energy of the transition $n = 1 \rightarrow n = \infty$ so we write

$$I = E_{n=\infty} - E_{n=1} = -hcR_\mathrm{H}Z^2\left(\frac{1}{\infty^2}-\frac{1}{1^2}\right) = hcR_\mathrm{H}Z^2$$

Since $I_{\mathrm{He}^+} = 4hcR_\mathrm{H}$, we may write $I_{\mathrm{Li}^{2+}} = 9hcR_\mathrm{H} = \frac{9}{4}I_{\mathrm{He}^+}$. With the ionization of He^+ given as 54.36 eV, the ionization energy of Li^{2+} is $\frac{9}{4}\times(54.36\ \mathrm{eV}) = \boxed{122.31\ \mathrm{eV}}$.

E13.9 $n = 4$ for the N shell

$n^2 = 4^2 = \boxed{16\ \text{orbitals}}$

E13.10 All lines in the hydrogen spectrum fit the Rydberg formula

$$\frac{1}{\lambda} = R_\mathrm{H}\left(\frac{1}{n_1^2}-\frac{1}{n_2^2}\right)\ [13.1,\ \text{with}\ \tilde{\nu} = 1/\lambda] \qquad \text{where}\ R_\mathrm{H} = 109\ 677\ \mathrm{cm}^{-1}$$

(a) Find n_1 from the value of λ_max, which arises from the transition $n_1 + 1 \rightarrow n_1$

$$\frac{1}{\lambda_\mathrm{max}R_\mathrm{H}} = \frac{1}{n_1^2} - \frac{1}{(n_1+1)^2} = \frac{2n_1+1}{n_1^2\,(n_1+1)^2}$$

$$\lambda_\mathrm{max}R_\mathrm{H} = \frac{n_1^2(n_1+1)^2}{2n_1+1} = (12\ 368\times10^{-9}\ \mathrm{m})\times(109\ 677\times10^2\ \mathrm{m}^{-1}) = 135.65$$

Because $n_1 = 1, 2, 3$, and 4 have already been accounted for, try $n_1 = 5, 6, \ldots$.

With $n_1 = 6$ we get $\dfrac{n_1^2\,(n_1+1)^2}{2n_1+1} = 135.69$. Hence, the Humphreys series is $\boxed{n_2 \rightarrow 6}$ and the transitions are given by

(b) $$\frac{1}{\lambda} = (109\ 677\ \mathrm{cm}^{-1})\times\left(\frac{1}{36}-\frac{1}{n_2^2}\right), \qquad n_2 = 7, 8, \ldots$$

and occur at $\boxed{\text{12372 nm, 7503 nm, 5908 nm, 5129 nm, … 3908 nm (at } n_2 = 15\text{), converging to 3282 nm as } n_2 \rightarrow \infty}$, in agreement with the quoted experimental result.

E13.11 In exercise 13.4 we see that, because the reduced mass is nearly identical for hydrogenic atoms, the Rydberg constant for all of them may be estimated to equal R_H so in this exercise we use the approximation $R_{\mathrm{He}^+} = R_\mathrm{H}$. We want to find the He^+ line for the Humphreys $n = 7 \rightarrow n = 6$ transition as this gives the longest wavelength, lowest energy, transition of the series.

According to eqn 13.4a

$$\tilde{\nu} = \tilde{E}_2 - \tilde{E}_1 = RZ^2\left(\frac{1}{n_1^2} - \frac{1}{n_2^2}\right) = R_{\mathrm{H}}Z^2\left(\frac{1}{n_1^2} - \frac{1}{n_2^2}\right)$$

Therefore,

$$\tilde{\nu}_{\mathrm{He}^+} = R_{\mathrm{H}}Z^2\left(\frac{1}{n_1^2} - \frac{1}{n_2^2}\right) = 4\times\left(109677\ \mathrm{cm}^{-1}\right)\times\left(\frac{1}{6^2} - \frac{1}{7^2}\right) = 3233.11\ \mathrm{cm}^{-1}$$

$$\lambda_{\mathrm{He}^+} = \frac{1}{\tilde{\nu}_{\mathrm{He}^+}}$$

$$= \frac{1}{3233.11\ \mathrm{cm}^{-1}} = \boxed{3093.00\ \mathrm{nm}}$$

E13.12 (a) All lines in the hydrogen spectrum fit the Rydberg formula

$$\frac{1}{\lambda} = R_{\mathrm{H}}\left(\frac{1}{n_1^2} - \frac{1}{n_2^2}\right) \quad [13.1,\ \text{with}\ \tilde{\nu} = 1/\lambda] \qquad \text{where}\ R_{\mathrm{H}} = 109\ 677\ \mathrm{cm}^{-1}$$

Find n_1 from the value of λ_{max}, which arises from transition $n_1 + 1 \rightarrow n_1$

$$\frac{1}{\lambda_{\mathrm{max}} R_{\mathrm{H}}} = \frac{1}{n_1^2} - \frac{1}{(n_1+1)^2} = \frac{2n_1+1}{n_1^2(n_1+1)^2}$$

$$\lambda_{\mathrm{max}} R_{\mathrm{H}} = \frac{n_1^2(n_1+1)^2}{2n_1+1} = (656.46\times10^{-9}\ \mathrm{m})\times(109\ 677\times10^{2}\ \mathrm{m}^{-1}) = 7.20$$

and hence $n_1 = 2$, as determined by trial and error substitution. Therefore, the transitions are given by

$$\tilde{\nu} = \frac{1}{\lambda} = (109\ 677\ \mathrm{cm}^{-1})\times\left(\frac{1}{4} - \frac{1}{n_2^2}\right), \qquad n_2 = 3, 4, 5, 6$$

The next line has $n_2 = 7$, and occurs at

$$\tilde{\nu} = \frac{1}{\lambda} = (109\ 677\ \mathrm{cm}^{-1})\left(\frac{1}{4} - \frac{1}{49}\right) = 2.5181\times10^{4}\ \mathrm{cm}^{-1},\ \text{which corresponds to}\ \boxed{397.13\ \mathrm{nm}}$$

(b) The energy required to ionize the atom is obtained by letting $n_2 \rightarrow \infty$. Then

$$\tilde{\nu}_{\infty} = (109\ 677\ \mathrm{cm}^{-1})\times\left(\frac{1}{4} - 0\right) = \boxed{274\ 19\ \mathrm{cm}^{-1}}\ \text{or}\ \boxed{3.40\ \mathrm{eV}}$$

(The answer, 3.40 eV, is the ionization energy of an H atom that is already in an excited state, with $n = 2$.)

COMMENT. The series with $n_1 = 2$ is the Balmer series.

E13.13 A Lyman series corresponds to $n_1 = 1$ and for lithium $Z = 3$; hence

$$\tilde{\nu} = \tilde{E}_n - \tilde{E}_{n_1} = 9R_{\mathrm{Li}^{2+}}\left(1 - \frac{1}{n^2}\right) \quad [13.4\mathrm{a}] \qquad \text{where}\ n = 2, 3, \ldots$$

(a) Therefore, if the formula is appropriate, we expect to find that $\frac{1}{9}\tilde{\nu}\times\left(1 - \frac{1}{n^2}\right)^{-1}$ is a constant ($R_{\mathrm{Li}^{2+}}$). We therefore draw up the following table.

n	2	3	4
$\tilde{\nu}$ /cm^{-1}	740 747	877 924	925 933
$\frac{1}{9}\tilde{\nu}\times\left(1-\frac{1}{n^2}\right)^{-1}$ / cm^{-1}	109 740	109 741	109 740

Hence, the formula does describe the transitions, and $\boxed{R_{\text{Li}^{2+}} = 109\ 740\ \text{cm}^{-1}}$

(b) The Balmer transitions lie at

$$\tilde{\nu} = \tilde{E}_n - \tilde{E}_{n_1} = 9R_{\text{Li}^{2+}}\left(\frac{1}{2^2}-\frac{1}{n^2}\right)\ [13.4\text{a}] \qquad \text{where } n = 3,\ 4,\ldots$$

$$= 9\times\left(109740\ \text{cm}^{-1}\right)\times\left(\frac{1}{4}-\frac{1}{n^2}\right) = \boxed{137\ 175\ \text{cm}^{-1},\ 185\ 186\ \text{cm}^{-1},\ldots}$$

(c) The ionization energy of the ground-state ion is given by

$$\tilde{\nu} = \tilde{E}_{n=\infty} - \tilde{E}_{n=1} = 0 - \tilde{E}_{n=1} = 9R_{\text{Li}^{2+}}\ [13.4\text{a}]$$

$$= 9\times\left(109740\ \text{cm}^{-1}\right) = \boxed{987660\ \text{cm}^{-1}\ \text{or}\ 122.45\ \text{eV}}$$

E13.14 The probability density for the hydrogen 1s orbital varies as

$$\psi^2 = \left(\frac{1}{(\pi a_0^3)^{1/2}}\,\text{e}^{-r/a_0}\right)^2\ [13.6] = \frac{1}{\pi a_0^3}\,\text{e}^{-2r/a_0}$$

The maximum value, $1/\pi a_0^3$, is at $r = 0$ and ψ^2 is 30% of the maximum when $\text{e}^{-2r/a_0} = 0.30$, so that $r = -\frac{1}{2}a_0 \ln(0.30)$, which is at $\boxed{r = 0.602\ a_0,}$ which corresponds to 31.8 pm.

E13.15 The radial distribution function for the hydrogen 1s orbital varies as

$$P = 4\pi r^2\psi^2\ [13.7\text{a}] = 4\pi r^2\left(\frac{1}{(\pi a_0^3)^{1/2}}\,\text{e}^{-r/a_0}\right)^2\ [13.6] = \frac{4r^2}{a_0^3}\text{e}^{-2r/a_0}$$

This formula is used to plot a_0P against r/a_0 in Figure 13.2.

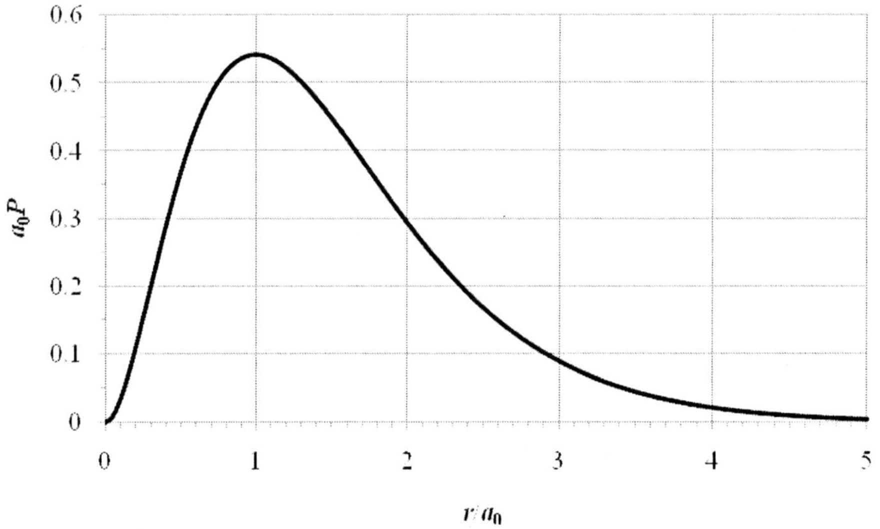

Figure 13.2

We read from the plot of Figure 13.2 that the maximum value of a_0P is $(a_0P)_{\max} = 0.54$

(a) When $a_0P = 0.30 \times (a_0P)_{\max} = 0.16$, the plot reads $r/a_0 \sim 0.25$ and 2.5. Thus, $r \sim \boxed{0.25\ a_0 \text{ and } 2.5\ a_0}$.

(b) When $a_0P = 0.05\times(a_0P)_{\max} = 0.027$, the plot reads $r/a_0 \sim 0.1$ and 3.8. Thus, $r \sim \boxed{0.1\ a_0 \text{ and } 3.8\ a_0}$.

The above effort has solved the exercise with a graphical technique. An alternative method, pursued below, is to attempt an analytical solution.

The maximum of P occurs at the value of r, r_{max}, for which $dP/dr = 0$.

$$\frac{dP}{dr}=\frac{d\left(4\pi r^2\psi^2\right)}{dr}=4\pi\left(2r\psi^2+r^2\frac{d\psi^2}{dr}\right)=4\pi\left(2r\psi^2+2r^2\psi\frac{d\psi}{dx}\right)$$

The above expression equals zero when

$$\psi+r\frac{d\psi}{dr}=0$$

$$\frac{1}{\left(\pi a_0^3\right)^{1/2}}\,e^{-r/a_0}+r\frac{d}{dr}\left(\frac{1}{\left(\pi a_0^3\right)^{1/2}}\,e^{-r/a_0}\right)=0$$

$$\frac{1}{\left(\pi a_0^3\right)^{1/2}}\,e^{-r/a_0}-r\left(\frac{1}{\left(\pi a_0^3\right)^{1/2}a_0}\,e^{-r/a_0}\right)=0$$

$$1-\frac{r}{a_0}=0$$

Thus, $r_{max} = a_0$ and $P_{max} = 4e^{-2}/a_0$.

P falls to a fraction f of its maximum when

$$f=\frac{\dfrac{4r^2}{a_0^3}\,e^{-2r/a_0}}{\dfrac{4}{a_0}\,e^{-2}}=\frac{r^2}{a_0^2}\,e^2e^{-2r/a_0}$$

Let $x = r/a_0$ and rearranging the working equation of f gives a transcendental function for x that does not admit a closed form analytic solution.

$$xe^{-x}=e^{-1}f^{1/2}$$

We must either use a numeric solution to find x for a given f or a graphical method which plots xe^{-x} against x to find x at the desired value of xe^{-x}. We simply use the numeric solve function of the modern scientific calculator to find the x value that when placed in the left side of the above equation yields the value of the right side. Care must be taken because there are two real and positive values of x that satisfy the equation. One value occurs when $r < a_0$ (i.e., $x < 1$) because the maximum occurs at $r = a_0$; the other value occurs when $r > a_0$ (i.e., $x > 1$).

(a) $f = 0.30$; $xe^{-x}=e^{-1}f^{1/2}=e^{-1}(0.30)^{1/2}=0.20150$

This solves to $x = 0.2618$ and 2.5303. Thus, r = $0.2618\,a_0$ and $2.5303\,a_0$.

(b) $f = 0.05$; $xe^{-x}=e^{-1}f^{1/2}=e^{-1}(0.05)^{1/2}=0.08226$

This solves to $x = 0.0900$ and 3.8445. Thus, r = $0.0900\,a_0$ and $3.8445\,a_0$.

Question. What are the advantages to solving this exercise with the graphical method? What are the advantages of the numeric solution?

E13.16 There are 2 lobes to a p-orbital so the probability that a p-orbital electron will be found in one-or-the-other lobe is $\frac{1}{2}$. However, there are 3 degenerate orbitals in a p subshell so the probability of finding a p subshell electron in one-or-another p-orbital lobe of the subshell is $^1/_6$.

E13.17 $V = 6.5 \text{ pm}^3 = \frac{4}{3}\pi r^3$ [assume a spherical volume]

$$r = \left(\frac{3V}{4\pi}\right)^{1/3} = \left(\frac{3\times 6.5\times 10^{-36}\ \text{m}^3}{4\pi}\right)^{1/3} = 1.16\times 10^{-12}\ \text{m} = 1.16\ \text{pm}$$

$$r / a_0 = (1.16\ \text{pm}) / (52.9\ \text{pm}) = 0.0219$$

As this is much smaller than most probable radius of the electron in the ground state of a hydrogen like-atom (a_0, see exercises 13.15 and 13.33), it is probably safe to assume that $\psi^2(r)$ is a constant within the spherical volume with $r = 1.16$ pm, Therefore,

$$\psi^2(r) = \frac{Z^2}{\pi a_0^3} e^{-2Zr/a_0} = \text{constant}$$

$$\text{Probability} \approx \psi^2 \delta V = \frac{Z^2}{\pi a_0^3} e^{-2Zr/a_0} \times \left(\tfrac{4}{3}\pi r^3\right) = \tfrac{4}{3} Z^2 \left(\frac{r}{a_0}\right)^3 e^{-2Zr/a_0}$$

(a) For the hydrogen atom $Z = 1$.

$$\text{Probability} \approx \psi^2 \delta V = \tfrac{4}{3}(1)^2 (0.0219)^3 e^{-2\times 1\times(0.0219)} = \boxed{1.3\times 10^{-5}}$$

(b) For the He^+ ion $Z = 2$.

$$\text{Probability} \approx \psi^2 \delta V = \tfrac{4}{3}(2)^2 (0.0219)^3 e^{-2\times 2\times(0.0219)} = \boxed{5.1\times 10^{-5}}$$

The probability for He^+ is larger than for the hydrogen atom because helium's larger nuclear charge, and stronger Coulomb attraction for the electron, has contracted the 1s orbital.

E13.18 There are two methods that can be used to locate the radial nodes of a hydrogen atom orbital. We could find a textbook plot of the radial function $R_{n,l}$ against r/a_0 and read the node values directly from the plot. Alternatively, we could search an advanced textbook on physical chemistry to find the mathematical form for $R_{n,l}$ and analyze the function for its nodes.

(a) Figure 13.3 is a typical plot for the 3s hydrogen orbital. The nodes occur when the radial function passes through zero. These occur when $r = \boxed{1.9\ a_0 \text{ and } 7.1\ a_0}$.

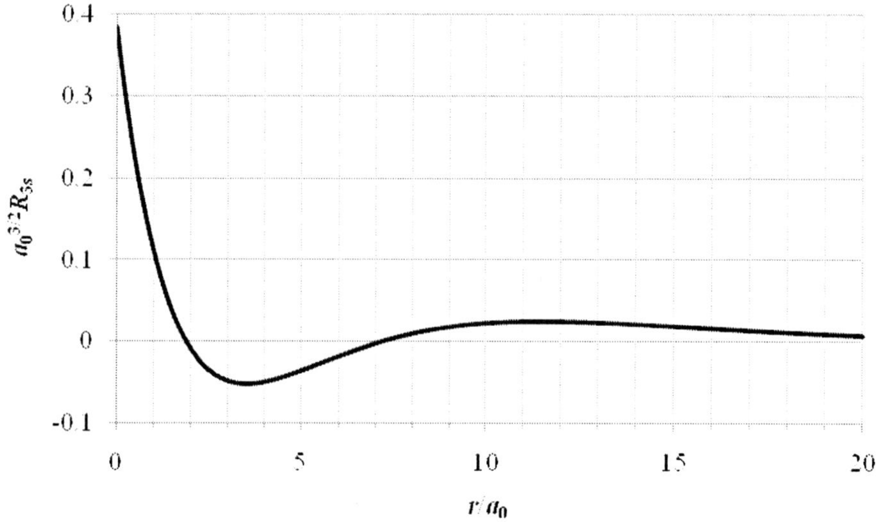

Figure 13.3

(b) Finding the nodes of the 4s orbital of a hydrogen atom is a greater challenge. Few textbooks display a plot of this radial wavefunction so it becomes necessary to find the mathematical form of this function. Eqn 10.14 of P. Atkins and J. de Paula, *Physical Chemistry*, 8th ed., Freeman, 2006 reports that the radial function is

$$R_{n,l} = N_{n,l}\rho^l L_{n+1}^{2l+1}(\rho) e^{-\rho/2} \qquad \text{where } \rho = \frac{2r}{na_0}$$

$L_{n+1}^{2l+1}(\rho)$ is the associated Laguerre polynomial and examination of the above radial function indicates that the radial nodes occur when $L_{n+1}^{2l+1}(\rho)$ equals zero. Thus, we must solve for the zeros of the associated Laguerre polynomial for the 4s orbital. L. Pauling and E. Wilson Jr., *Introduction to Quantum Mechanics with Applications to Chemistry*, McGraw-Hill, New York (1935) report the following form for the associated Laguerre polynomial.

$$\frac{L_{n+1}^{2l+1}(\rho)}{\{(n+l)!\}^2}=\sum_{k=0}^{n-l-1}\frac{(-1)^{k+1}\rho^k}{(n-l-1-k)!(2l+1+k)!k!} \quad \text{where } 0!=1 \text{ and } m!=m\times(m-1)\times(m-2)\times\ldots\times 1$$

For the 4s orbital we use $n = 4$, $l = 0$, and $\rho=\dfrac{r}{2a_0}$.

$$\frac{L_5^1(\rho)}{\{4!\}^2}=\sum_{k=0}^{3}\frac{(-1)^{k+1}\rho^k}{(3-k)!(1+k)!k!}$$

$$=\frac{(-1)^1\rho^0}{(3)!(1)!0!}+\frac{(-1)^2\rho^1}{(2)!(2)!1!}+\frac{(-1)^3\rho^2}{(1)!(3)!2!}+\frac{(-1)^4\rho^3}{(0)!(4)!3!}$$

$$=-\frac{1}{6}+\frac{\rho}{4}-\frac{\rho^2}{12}+\frac{\rho^3}{144}=\frac{1}{144}(\rho^3-12\rho^2+36\rho-24)$$

Thus, the nodes of the 4s orbital occur when

$$\rho^3-12\rho^2+36\rho-24=0$$

The roots of this polynomial can be found with either the root function or the numeric solver of the modern scientific calculator. The roots found are $\rho = 0.936$, 3.305, and 7.759. Since $r = 2a_0\rho$, the nodes are at

$$r = \boxed{1.87\,a_0,\ 6.61\,a_0,\ \text{and}\ 15.5\,a_0}$$

The plot of R_{4s} against r/a_0, shown in Figure 13.4, confirms this result.

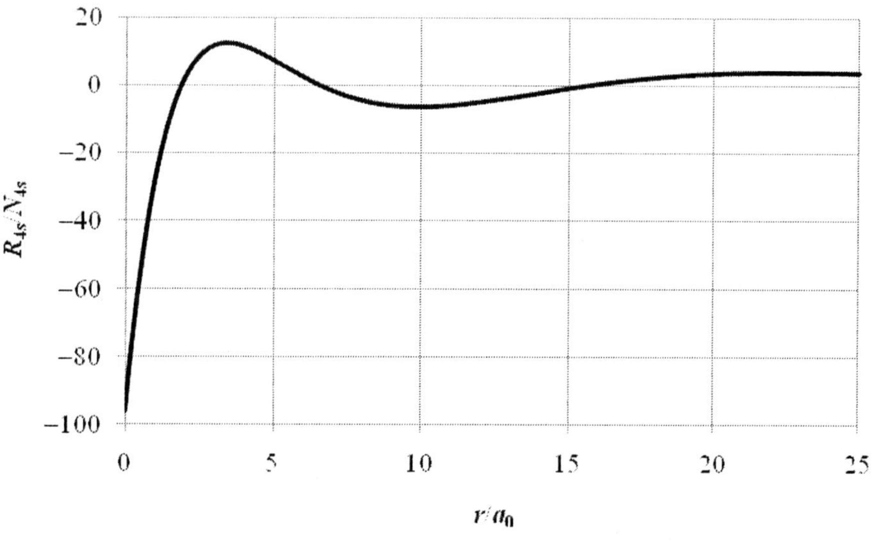

Figure 13.4

E13.19 Look for values of θ for which sin θ or cos θ go to zero. Sin θ goes to zero at $\theta = \boxed{0° \text{ and } 180°;}$ cos θ at $\boxed{90° \text{ and } 270°.}$

E13.20 Identify l and use angular momentum $= \{l\,(l+l)\,\}^{1/2}$

(a) $l = 0$, so $\boxed{\text{ang. mom.} = 0}$

(b) $l = 0$, so $\boxed{\text{ang. mom.} = 0}$

(c) $l = 2$, so $\boxed{\text{ang. mom.} = \sqrt{6}\,\hbar}$

(d) $l = 1$, so $\boxed{\text{ang. mom.} = \sqrt{2}\,\hbar}$

(e) $l = 1$, so $\boxed{\text{ang. mom.} = \sqrt{2}\,\hbar}$

The total number of nodes is equal to $n - 1$, and the number of angular nodes is equal to l; hence the number of radial nodes is equal to $n - 1 - l$. We can draw up the following table:

	1*s*	3*s*	3*d*	2*p*	3*p*
n, l	1, 0	3, 0	3, 2	2, 1	3, 1
Ang. nodes	0	0	2	1	1
Rad. Nodes	0	2	0	0	1

E13.21 The energies are $E_n = -\dfrac{hcR_H}{n^2}$ and the orbital degeneracy g of an energy level of principal quantum number n is $g = n^2$.

(a) $E = -hcR_H$ implies that $n = 1$, so $\boxed{g = 1}$ (the 1s orbital)

(b) $E = -\frac{1}{9}hcR_H$ imples that $n = 3$, so $\boxed{g = 9}$ (the 3s orbital, the three 3p orbitals, and the five 3d orbitals)

(c) $E = -\frac{1}{49}hcR_H$ implies that $n = 7$, so $\boxed{g = 49}$ (the 7s orbital, the three 7p orbitals, the five 7d orbitals, the seven 7f orbitals, the nine 7g orbitals, the eleven 7h orbitals, and the thirteen 7i orbitals).

E13.22 For a given l there are $2l + 1$ values of m_l and hence $2l + 1$ orbitals. Each orbital may be occupied by two electrons. Therefore the maximum occupancy is $2(2l + 1)$.

	l	$2(2l + 1)$
(a)	0	2
(b)	3	14
(c)	5	22

E13.23 By analogy with equation [13.9] the ionization energy of an anion, I_1^-, is for the reaction

$$X^-(g) \rightarrow X(g) + e^-(g) \qquad I_1^- = E(X) - E(X^-)$$

while the electron affinity of equation [13.10a] is for the reaction

$$X(g) + e^-(g) \rightarrow X^-(g) \qquad E_{ea} = -\{E(X^-) - E(X)\} = E(X) - E(X^-)$$

The above expression for the electron affinity E_{ea} is $-\{E(X^-) - E(X)\}$ because the **electron affinity is the energy released** when an electron attaches to a gas-phase atom. Inspection of the expressions for I_1^- and E_{ea} indicates that $\boxed{I_1^- = E_{ea}}$. However, the reaction equation for the electron affinity is the reverse of the ionization reaction of X^-.

E13.24

$$h\nu = \tfrac{1}{2}m_e v^2 + I$$

$$I = h\nu - \tfrac{1}{2}m_e v^2 = \frac{hc}{\lambda} - \tfrac{1}{2}m_e v^2$$

$$= \frac{(6.626\times10^{-34}\ \mathrm{J\ s})\times(2.998\times10^{8}\ \mathrm{m\ s^{-1}})}{58.4\times10^{-9}\ \mathrm{m}} - \tfrac{1}{2}\times(9.109\times10^{-31}\ \mathrm{kg})\times(1.59\times10^{6}\ \mathrm{m\ s^{-1}})^2$$

$$= 2.25\times10^{-18}\ \mathrm{J}, \text{ corresponding to } \boxed{14.0\ \mathrm{eV}}$$

E13.25 The ground configurations of iron and its cations are:

Fe: $[Ar]3d^6 4s^2$

Fe^{2+}: $[Ar]3d^6$

Fe^{3+}: $[Ar]3d^5$

The Fe^{2+} ion has a greater radius than the radius of Fe^{3+} because the repulsions of six 3d subshell electrons is greater than five. Also, two of the iron(II) cation 3d electrons are paired in a single orbital but none of the 3d electrons are paired in iron(III). Paired electrons repel more strong than unpaired especially when in a single orbital.

E13.26 The inner closed orbitals represented by $[Ne]2s^2$ have zero total spin and total orbital angular momentum so they need not be considered when evaluating the terms of the $[Ne]2s^22p^13d^1$ excited state. The two electrons of the $2p^13d^1$ orbitals have either parallel ($S = 1$) or paired spins ($S = 0$). Thus, triplet and singlet spin multiplicities are allowed because the spin multiplicity has the formula $2S + 1$.

Since $l = 1$ for the p orbital and $l = 2$ for the d orbital, the allowed total orbital angular momentum terms are given by $L = l_1 + l_2, \ldots, |l_1 - l_2| = 3, 2, 1$. These are F, D and P terms.

Triplet terms: ^3F, ^3D, ^3P Singlet terms: ^1F, ^1D, ^1P

Possible values of J (using Russell-Saunders coupling) are 3, 2, and 1 ($S = 0$) and 4, 3, 2, 1, and 0 ($S = 1$). The term symbols are

$$\boxed{^1F_3;\ ^3F_4,\ ^3F_3,\ ^3F_2;\ ^1D_2;\ ^3D_3,\ ^3D_2,\ ^3D_1;\ ^1P_1;\ ^3P_2,\ ^3P_1,\ ^3P_0}.$$

Hund's rules state that the lowest energy level has maximum multiplicity. Consideration of spin-orbit coupling says the lowest energy level has the lowest value of $J(J + 1) - L(L + 1) - S(S + 1)$. So the lowest energy level is $\boxed{^3F_2.}$

E13.27 The four electron spins may be aligned (if they are in different orbitals) giving a maximum total spin angular momentum quantum number S equal to $4 \times s = 4 \times ½ = 2$ because each electron has a spin quantum number of $s = ½$. Another possibility is that three of the four electrons are aligned with the remaining electron in an antiparallel alignment. This is equivalent to having two electron spins paired (giving no net spin contribution from the two) with the remaining two aligned (unpaired), thereby, giving $S = 3 \times s - s = 2 \times ½ = 1$. Finally, two electrons may have their spins aligned while the remaining two electrons have apposing, antiparallel, spins. This, being equivalent to two sets of paired electrons, gives no net spin. Thus, the possible values for S are $\boxed{2, 1, \text{and } 0}$.

E13.28

(a) $L = 0, S = 0, J = 0$ $\boxed{1}$ level

(b) $L = 3, S = 1, J = 4, 3, 2$ $\boxed{3}$ levels

(c) $L = 0, S = 2, J = 2$ $\boxed{1}$ level

(d) $L = 1, S = 2, J = 3, 2, 1$ $\boxed{3}$ levels

E13.29 Ti^{2+}: $[Ar]3d^2$

(a) The term of lowest energy will be a term that obeys Hund's rule, which states that the **terms with largest multiplicity ($2S + 1$) arising from terms with the maximum value of total spin, S, lie lowest in energy**. Here

$$S = s_1 + s_2 = (+½) + (+½) = 1$$

is the maximum allowed spin. The corresponding multiplicity is $2S + 1 = 3$.

So our term of lowest energy will be a triplet term.

For terms of the same multiplicity, **the term with the largest total orbital angular momentum, L, lies lowest in energy**. With two electrons in the 3d subshell $L = l_1 + l_2 = 2 + 2 = 4$ is the maximum total orbital angular momentum. For $L = 4$ and $S = 1$, the term symbol would be ^3G.

But for the configuration $3d^2$, this term is not allowed by the Pauli Principle, for in order to achieve $L = 4$, two electrons would have to occupy either the $3d_{m_l=+2}$ orbital or the $3d_{m_l=-2}$ orbital. According to the Pauli Principle, this can only happen if the spins of the two electrons are antiparallel, namely $S = s_1 + s_2 = (+½) + (-½) = 0$.

So a singlet G term ($2S + 1 = 1$) can arise, but not a triplet G term.

This problem does not occur for F terms because here $L = 3$, which can arise from

$$L = l_1 + l_2 = 2 + 1 = 3$$

where the two electrons are in different orbitals, $3d_{m_l=+2}$ and $3d_{m_l=+1}$, for example. Therefore, the term of lowest energy is $\boxed{^3F}$

The possible J values are obtained from the Clebsch-Gordon series and are:

$$J = L + S,...,|L - S| = 3 + 1,...,3 - 1 = 4, 3, \text{ and } 2$$

These values of J give rise to the terms

$$^3F_4, \ ^3F_3, \ ^3F_2,$$

of these the lowest lying is $^3F_2$ because **for atoms with shells that are less than half full, the level with lowest J lies lowest in energy**. Also, notice that, when $L \geq S$, the multiplicity is equal to the number of levels.

(b) Each level with a quantum number J consists of $2J + 1$ individual states distinguished by the quantum number M_J, which accounts for the different quantum projections of total angular momentum onto an axis (z-axis). Since $J = 2$ for the lowest term, $2J + 1 = \boxed{5}$.

E13.30 The spectroscopic selection rules are Δn = any integer and $\Delta l = \pm 1$ where l = 0, 1, 2, 3 corresponds to s, p, d, and f subshells.

(a) $2s \rightarrow 1s$ $\quad \Delta l = 0, \ \boxed{\text{forbidden}}$

(b) $2p \rightarrow 1s$ $\quad \Delta l = -1, \ \boxed{\text{allowed}}$

(c) $3d \rightarrow 2p$ $\quad \Delta l = -1, \ \boxed{\text{allowed}}$

(d) $5d \rightarrow 2s$ $\quad \Delta l = -2, \ \boxed{\text{forbidden}}$

(e) $5p \rightarrow 3s$ $\quad \Delta l = -1, \ \boxed{\text{allowed}}$

(f) $6f \rightarrow 4p$ $\quad \Delta l = -2, \ \boxed{\text{forbidden}}$

E13.31 The spectroscopic selection rules are Δn = any integer and $\Delta l = \pm 1$ so a 5f electron may make a transition to any d and g orbitals.

E13.32 The spectroscopic selection rules are $\Delta S = 0$, $\Delta L = 0, \pm 1$, and $\Delta J = 0, \pm 1$ (but $J = 0 \leftrightarrow J = 0$ is forbidden) for the Russell–Saunders coupling scheme.

(a) $^3D_2 \rightarrow {}^3P_1$ $\quad \boxed{\text{allowed}}$

(b) $^3P_2 \rightarrow {}^1S_0$ $\quad \Delta S = -2, \ \boxed{\text{forbidden}}$

(c) $^3F_4 \rightarrow {}^3D_3$ $\quad \boxed{\text{allowed}}$

Answers to projects

P13.33 (a) The most probable distance of a $1s$ electron from the nucleus occurs when the first derivative of the radial distribution function equals zero.

$$P_{1s} = 4\pi r^2 \psi_{1s}^2 \ [13.7\text{a}] = 4\pi r^2 \left(N e^{-r/a_0}\right)^2 \ [13.6] = 4\pi N^2 \left(r^2 e^{-2r/a_0}\right)$$

$$\frac{dP_{1s}}{dr} = 4\pi N^2 \frac{d\left(r^2 e^{-2r/a_0}\right)}{dr} = 4\pi N^2 \left\{2re^{-2r/a_0} + r^2\left(-\frac{2}{a_0}e^{-2r/a_0}\right)\right\} = 8\pi N^2 \left\{1 - \frac{r}{a_0}\right\} re^{-2r/a_0}$$

The derivative equals zero when the factor $1 - r/a_0$ equals zero. Therefore, $\boxed{r_{\max} = a_0}$.

(b) We will make the assumption that ψ^2 is a constant within this very small volume. Then

$$\text{Probability} = \int \psi^2(r)\delta V \approx \psi^2 \delta V \text{ with } \delta V = 1.0 \text{ pm}^3$$

$$\psi^2 = \frac{1}{32\pi a_0^3}\left(2 - \frac{r}{a_0}\right)^2 e^{-r/a_0} = 6.72 \times 10^{-8} \text{ pm}^{-3}\left(2 - \frac{r}{a_0}\right)^2 e^{-r/a_0}$$

(i) $\psi^2 = 6.72 \times 10^{-8} \text{ pm}^{-3} \times 2^2 \times 1 = 2.7 \times 10^{-7} \text{ pm}^{-3}$

$\psi^2 \delta V = \boxed{2.7 \times 10^{-7}}$

(ii) $\psi^2 = 6.72 \times 10^{-8}\ \text{pm}^{-3} \times 1 \times e^{-1} = 9.9 \times 10^{-8}\ \text{pm}^3$

$\psi^2\delta V = \boxed{9.9 \times 10^{-8}}$

(iii) $\psi^2 = 0$, $\psi^2\delta V = \boxed{0}$ This is a radial node.

(c) The most probable distance of a 2*s* electron from the nucleus may be determined by plotting the radial distribution function against r/a_0 and using the trace function of the plotting software to evaluate the coordinates of the maximum. The following function is plotted in Figure 13.5. The plot reveals that $\boxed{r_{max} = 5.235\ a_0}$.

$$P_{2s} = 4\pi r^2 \psi_{2s}^2\ [13.7a] = 4\pi r^2 \left\{\left(\frac{1}{32\pi a_0^3}\right)^{1/2}\left(2 - \frac{r}{a_0}\right)e^{-r/2a_0}\right\}^2$$

$$= \frac{1}{8a_0^3}\left(2 - \frac{r}{a_0}\right)^2 r^2 e^{-r/a_0} = \boxed{\frac{1}{8a_0}(2-x)^2 x^2 e^{-x} \text{ where } x = r/a_0}$$

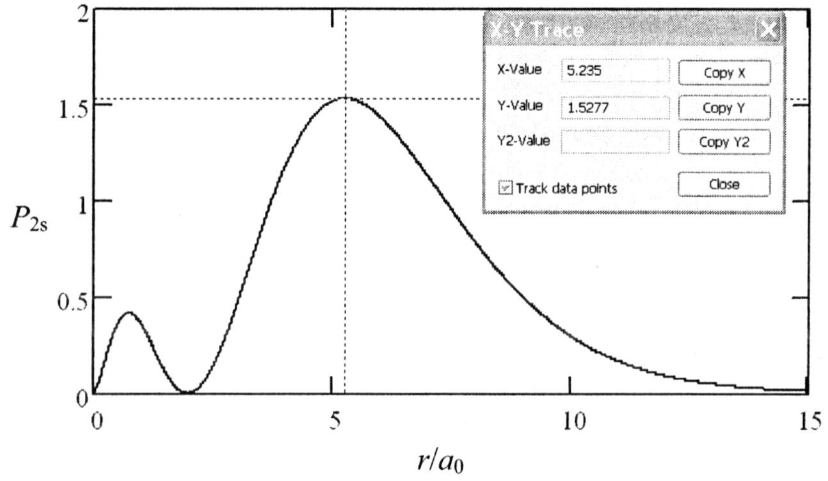

Figure 13.5

(d) Let $x = r/a_0$, then

$$P_{2s} = \frac{1}{8a_0}\left\{x^2(2-x)^2 e^{-x}\right\} = \frac{1}{8a_0}\left(4x^2 - 4x^3 + x^4\right)e^{-x}$$

$$\frac{dP_{2s}}{dr} = \frac{dx}{dr}\frac{dP_{2s}}{dx} = \frac{1}{8a_0^2}\frac{d\left\{\left(4x^2 - 4x^3 + x^4\right)e^{-x}\right\}}{dx}$$

$$= \frac{1}{8a_0^2}\left\{\left(8x - 12x^2 + 4x^3\right)e^{-x} + \left(4x^2 - 4x^3 + x^4\right)\left(-e^{-x}\right)\right\} = \frac{1}{8a_0^2}\left\{8 - 16x + 8x^2 - x^3\right\}xe^{-x}$$

$$= -\frac{1}{8a_0^2}\times(x-2)\times\left(x^2 - 6x + 4\right)xe^{-x}$$

The derivative equals zero when $x = r/a_0 = 0,\ 3 - 5^{1/2},\ 2,\ 3 + 5^{1/2}$, and ∞. These correspond to the radial distribution function being a minimum, a maximum, a minimum, a maximum, and a minimum, respectively. The following ratio identifies the maximum that is most probable.

$$\frac{P_{2s}\left(x = 3 + 5^{1/2}\right)}{P_{2s}\left(x = 3 - 5^{1/2}\right)} = \frac{\left[\left(4x^2 - 4x^3 + x^4\right)e^{-x}\right]_{x=3+5^{1/2}}}{\left[\left(4x^2 - 4x^3 + x^4\right)e^{-x}\right]_{x=3-5^{1/2}}} = \frac{1.528}{0.415} = 3.68$$

The most probable distance is $\boxed{(3 + 5^{1/2})a_0}$.

P13.34 See Figure 13.6 for the ionization trends of Group 13.

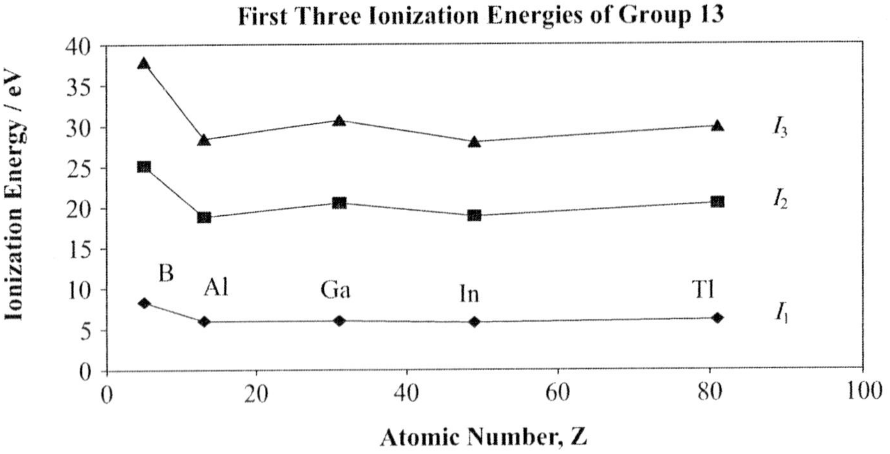

Figure 13.6

Trends:

(i) $I_1 < I_2 < I_3$ because of decreased nuclear shielding as each successive electron is removed.

(ii) The ionization energies of boron are much larger than those of the remaining group elements because the valence shell of boron is very small and compact with little nuclear shielding. The boron atom is much smaller than the aluminum atom.

(iii) The ionization energies of Al, Ga, In, and Tl are comparable even though successive valence shells are further from the nucleus because the ionization energy decrease expected from large atomic radii is balanced by an increase in effective nuclear charge.

P13.35 (a) Let ν_{star} be the emission frequency of star HDE 271 182 while ν is the Earth-bound iron arc. Compute the ratios ν_{star}/ν for all three lines. We are given wavelength data, so we can use:

$$\frac{\nu_{\text{star}}}{\nu} = \frac{\lambda}{\lambda_{\text{star}}}.$$

The ratios are:

$$\frac{438.392\text{ nm}}{438.882\text{ nm}} = 0.998884, \quad \frac{440.510\text{ nm}}{441.000\text{ nm}} = 0.998889, \quad \text{and} \quad \frac{441.510\text{ nm}}{442.020\text{ nm}} = 0.998846.$$

The frequencies of the stellar lines are all less than those of the stationary lines, so we infer that the star is $\boxed{\text{receding}}$ from earth. The Doppler effect follows:

$$\nu_{\text{receding}} = \nu f \quad \text{where } f = \left(\frac{1 - s/c}{1 + s/c}\right)^{1/2}, \quad \text{so}$$

$$f^2(1 + s/c) = (1 - s/c), \quad (f^2 + 1)s/c = 1 - f^2, \quad s = \frac{1 - f^2}{1 + f^2}c$$

Our average value of f is 0.998873. (*Note:* the uncertainty is actually greater than the significant figures here imply, and a more careful analysis would treat uncertainty explicitly.) So the speed of recession with respect to the earth is:

$$s = \left(\frac{1 - 0.997747}{1 + 0.997747}\right)c = \boxed{1.128 \times 10^{-3}\ c} = \boxed{3.381 \times 10^5\ \text{m s}^{-1}}$$

(b) One could compute the star's radial velocity with respect to the sun if one knew the earth's speed with respect to the sun along the sun–star vector at the time of the spectral observation. This could be estimated from quantities available through astronomical observation: the earth's orbital velocity times the cosine of the angle between that velocity vector and the earth–star vector at the time of the spectral observation. (The earth–star direction, which is observable by earth-based astronomers, is practically identical to the sun–star direction, which is technically the direction needed.) Alternatively, repeat the experiment half a year later. At that time, the earth's motion with respect to the sun is approximately equal in magnitude and opposite in direction compared to the original experiment. Averaging f values over the two experiments would yield f values in which the earth's motion is effectively averaged out.

14 The chemical bond

Answers to discussion questions

D14.1 Our comparison of the two theories will focus on the manner of construction of the trial wavefunctions for the hydrogen molecule in the simplest versions of both theories. In the valence bond method, the trial function is a linear combination of two simple product wavefunctions, in which one electron resides totally in an atomic orbital (AO) on atom A, and the other totally in an orbital on atom B. See Derivation 14.1 and text Fig. 14.3. There is no contribution to the wavefunction from products in which both electrons reside on either atom A or B.

$$\psi_{\text{H-H}}(1,2) = \psi_{\text{A}}(1)\psi_{\text{B}}(2) + \psi_{\text{A}}(2)\psi_{\text{B}}(1) \quad [14.1]$$

So the valence bond approach undervalues, by totally neglecting, any ionic contribution to the trial function. It is a totally covalent function.

The modern one-electron molecular orbital (MO) extends throughout the molecule and is written as a linear combination of atomic orbitals (LCAO).

$$\psi_{\text{MO}}(1) = c_{\text{A}}\psi_{\text{A}}(1) + c_{\text{B}}\psi_{\text{B}}(1) \quad [14.9a]$$

The squares of the coefficients give the relative proportions of the AO contributing to the MO.

The two-electron molecular orbital function for the hydrogen molecule is a product of two one-electron MOs. That is

$$\begin{aligned}\psi &= [c_{\text{A}}\psi_{\text{A}}(1) + c_{\text{B}}\psi_{\text{B}}(1)]\times[c_{\text{A}}\psi_{\text{A}}(2) + c_{\text{B}}\psi_{\text{B}}(2)] \\ &= c_{\text{A}}^2\psi_{\text{A}}(1)\psi_{\text{A}}(2) + c_{\text{B}}^2\psi_{\text{B}}(1)\psi_{\text{B}}(2) + c_{\text{A}}c_{\text{B}}\psi_{\text{A}}(1)\psi_{\text{B}}(2) + c_{\text{A}}c_{\text{B}}\psi_{\text{A}}(2)\psi_{\text{B}}(1)\end{aligned}$$

The first two terms are ionic forms for which both electrons on either on atom A or on atom B. The molecular orbital approach greatly overvalues the ionic contributions. At these crude levels of approximation, the valence bond method gives dissociation energies closer to the experimental values. However, more sophisticated versions of the molecular orbital approach are the method of choice for obtaining quantitative results on both diatomic and polyatomic molecules.

D14.2 Consider the case of the carbon atom. Mentally we break the process of hybridization into two major steps. The first is promotion, in which we imagine that one of the electrons in the 2s orbital of carbon ($2s^2 2p^2$) is promoted to the empty 2p orbital giving the configuration $2s2p^3$. In the second step we mathematically mix the four orbitals by way of the specific linear combinations in eqn 14.4 corresponding to the sp^3 hybrid orbitals. There is a principle of conservation of orbitals that enters here. If we mix four unhybridized atomic orbitals we must end up with four hybrid orbitals. In the construction of the sp^2 hybrids we start with the 2s orbital and two of the 2p orbitals, and after mixing we end up with three sp^2 hybrid orbitals. In the sp case we start with the 2s orbital and one of the 2p orbitals. Eqns 14.5 and 14.6 present the LCAO for the sp^2 and sp hybrid orbitals; text Figs 14.6 and 14.7 describe the orientations of sp^3 and sp^2 hybrid orbitals. The justification for all of this is in a sense the first law of thermodynamics. Energy is a state function and therefore its value is determined only by the final state of the system, not by the path taken to achieve that state, and the path can even be imaginary.

D14.3 In valence bond (VB) theory, a bond is regarded as forming when an electron in a valence atomic orbital (AO) on one atom overlaps with a valence AO on an adjacent atom to form a wavefunction orbital that is more diffuse than the individual AO. If constructive interference occurs in the overlap, the wavefunction orbital is a bonding orbital that is lower in total energy than the AO. If the overlap gives destructive interference, the wavefunction is antibonding and higher in total energy than the AO. Sigma bond orbitals (σ) and sigma antibond orbitals (σ*) have cylindrical symmetry around the internuclear axis; pi bond orbitals (π) and pi antibond orbitals (π*) have a nodal plane along the internuclear axis. AO overlap examples that yield these typical bond types are shown in Figure 14.1 and in text Figures 14.2, 14.4, 14.5, and 14.8. The + and – signs in these figures indicate whether an orbital lobe exhibits a positive wavefunction sign or negative wavefunction sign. Two electrons at most may occupy any one of these orbitals and their spins must be paired (Pauli exclusion principle).

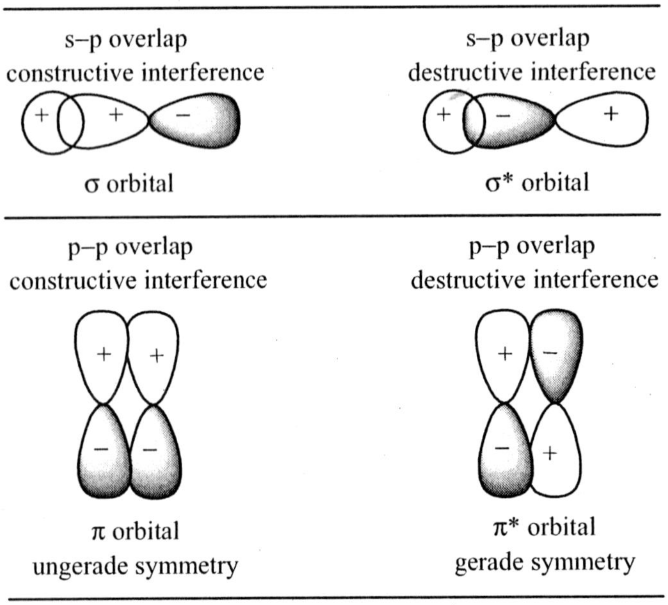

Figure 14.1

In addition to the overlap of combinations of s, p, and d orbitals, VB theory introduces the sp^3, sp^2, and sp hybrid AO and envisions their overlap with adjacent AO. Simple alkanes exhibit sp^3 hybridiation (text Figure 14.6) while alkenes exhibit sp^2 hydridization (text Figure 14.7) and alkynes exhibit sp hybridization (text Figure 14.9). These valence bond concepts are powerful tools that often provide insight into bond strengths, bond lengths, bond angles, and bonding sites of high and low reactivity within molecules.

Molecular orbital theory provides a perspective in which each molecular orbital extends throughout the molecule. This removes the restriction that a bonding electron pair be localized at two adjacent atoms. They are now spread over the whole molecule. For example, Figure 14.2 shows the two highest energy MOs, each occupied by an electron pair, of 1-butene. They are the HOMO (highest occupied molecular orbital) and HOMO{−1}. Both orbitals clearly show a delocalized MO that spreads over the whole molecule. The portion of the HOMO between carbon atom 1 and carbon atom 2 looks very much like a π bond but, because of the delocalized nature of the MO, the probability of finding the electron pair at this location is less than 1.

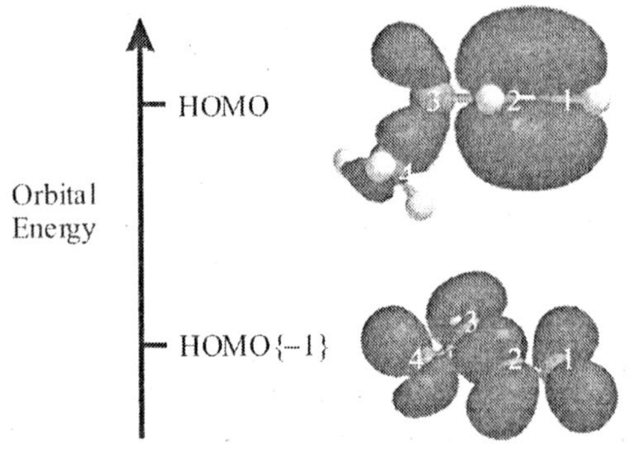

Figure 14.2

Every MO electron contributes to the strength of every bond and the total energy determines whether the bonded chemical species will be stable w/r/t to another bonding pattern, a consideration that gives emphasis to the MO energy level diagram and the electron configuration.

In the discussion of diatomic molecules and ions, the sigma and pi bonding and antibonding descriptors are common to both VB theory and MO theory because the molecule consists of two adjacent atoms only. However, the MOs of larger molecules cannot usually employ these descriptors. For examples of the symmetry-based MO descriptors see text Figures 14.34, 14.35, and 14.37. Even when discussing diatomic molecules, MO theory does not employ VB terms like single bond and double bond. MO theory uses the **bond order**, b, which is defined as

$$b = \tfrac{1}{2}(N - N^*) \quad [14.12]$$

where N is the number of electrons in bonding orbitals and N^* is the number of electrons in antibonding orbitals.

D14.4 For technical reasons related to the Pauli exclusion principle, a bonding orbital can only exist if the two electrons it describes have opposite spins. It follow that the merging of atomic orbitals that gives rise to a bond is accompanied by the pairing of the two electrons that contribute to it. Bonds do not form because electrons tend to pair: bonds are allowed to form by the electrons pairing their spins. Therein lays the reason behind the central concept of the electron pair in bonding theory.

D14.5 Both the Pauling and Mulliken methods for measuring the attracting power of atoms for electrons seem to make good chemical sense.

Pauling electronegativity scale:

$$|\chi_A - \chi_B| = 0.102 \times (\Delta E / \text{kJ mol}^{-1})^{1/2} \quad [14.13]$$

where $\Delta E = E(\text{A—B}) - \tfrac{1}{2}\{E(\text{A—A}) + E(\text{B—B})\}$

Mulliken electronegativity scale:

$$\chi = \tfrac{1}{2}(I + E_{ea}) \quad [14.14]$$

If we look at the Pauling scale, we see that if $E(\text{A—B})$ were equal to $\tfrac{1}{2}[E(\text{A—A}) + E(\text{B—B})]$ the calculated electronegativity difference would be zero, as expected for completely non-polar bonds. Hence, any increased strength of the A—B bond over the average of the A—A and B—B bonds, can reasonably be thought of as being due to the polarity of the A—B bond, which in turn is due to the difference in electronegativity of the atoms involved. Therefore, this difference in bond strengths can be used as a measure of electronegativity difference. To obtain numerical values for individual atoms, a reference state (atom) for electronegativity must be established. The value for fluorine is arbitrarily set at 4.0.

The Mulliken scale may be more intuitive than the Pauling scale because we are used to thinking of ionization energies and electron affinities as measures of the electron attracting powers of atoms. The choice of factor ½, however, is arbitrary, though reasonable, and no more arbitrary than the specific form that defines the Pauling scale.

D14.6 Consider the simple MO wavefunction given by eqn 14.9a for the H_2^+ ion:

$$\psi = \psi_A + \psi_B \quad [14.9a]$$

where ψ_A and ψ_B are atomic orbitals centered on A and B. Consider that the atomic orbitals constructively interfere and, thereby, form a bonding MO that has a lower energy than the separated atoms. The probability density is

$$\psi^2 = \psi_A^{\ 2} + \psi_B^{\ 2} + 2\psi_A\psi_B$$

The term $\psi_A^{\ 2}$ is the probability density if the electron is confined to atomic orbital A and $\psi_B^{\ 2}$ is the probability density if the electron is confined to atomic orbital B. Since these two isolated atom probability density terms cannot explain the electron sharing of a covalent bond, the extra contribution that $2\psi_A\psi_B$ gives to the density must be the origin of bonding. This is called the **overlap density** because it represents an enhancement of the probability of finding the electron in the internuclear region due to orbital overlap. As the two atoms approach from infinite separation the overlap density increases from zero as the orbital energy decreases to the minimum value at the equilibrium nuclear separation of the stable molecular species. Thus, orbital overlap is the guide to assessing the bond strength. However, as the atoms approach more closely, severe nuclear repulsion causes the orbital energy to rapidly increase to a greater value than that of the separate atoms, see text Figure 14.1, and overlap is no longer an indicator of bond strength.

D14.7 The following bullet list identifies, and justifies, the approximations used in the simple Hückel theory of hydrocarbon π-electron systems.

- Only the carbon p_z valence orbitals of sp^2 hybridized carbon atoms contribute to the LCAO of the π system. This is justified to an extent because the hybrization approximation gives reasonable estimates in many instances and p_z orbitals do not overlap with sp^2 hybridized orbitals.
- All overlap integrals are set equal to zero. Overlap integrals have small values and their neglect eases the mathematics so that an indication of the molecular orbital energy level diagram can be obtained.
- All terms of the form H_{AA} equal α (a negative quantity). The electronic environments of each sp^2 hybridized carbon are very similar, thereby, making all p_z valence orbitals equal in size and energy.
- All terms of the form H_{AB} equal β (a negative quantity) if the atoms are neighbours and to zero otherwise. In addition to a justification similar to that for H_{AA}, when A and B are not neighbours the p_{zA} orbital overlap with the p_{zB} orbital is negligibly small.

These approximations are obviously very severe, but they let us calculate at least a general picture of the molecular orbital energy levels with very little work.

D14.8 In *ab initio* methods an attempt is made to evaluate the Schrödinger equation numerically without employing empirical information. Approximations are employed, but these are mainly associated with the construction of the wavefunctions involved in the integrals of the computation. *Ab initio* computations are iterative in that each cycle of the computation gives an improved estimate of the energy and wavefunction that is used in the next calculation cycle. The computation is **self-consistent** in that the iteration of energy and wavefunction are repeated (after an initial approximation is made about the mathematical form of the wavefunction) until the energy and wavefunction, are unchanged to within some acceptable tolerance.

In semi-empirical methods, many of the integrals are expressed in terms of spectroscopic data or physical properties. Semi-empirical methods exist at several levels. At some levels, in order to simplify the calculations, many of the integrals are set equal to zero.

Solutions to exercises

E14.1 The valence bond description of a C—H group in a molecule is a σ bond formed from the constructive overlap of the hydrogen 1s orbital with a hybridized carbon valence orbital as shown in Figure 14.3 with the wavefunction sign indicated as a + or – in each orbital lobe. When the carbon atom is bonded with four σ bonds, it is sp^3 hybridized. If the carbon is bonded with three σ bonds and one π bond, it is sp^2 hydridized and, finally, when bonded with two σ bonds and two π bonds, it is sp hydridized.

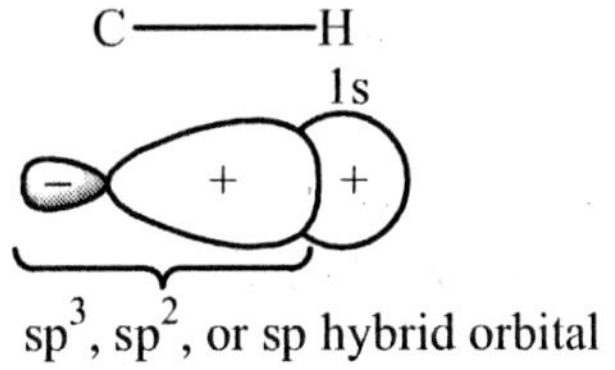

Figure 14.3

E14.2 The valence bond description of P_2 is similar to that of N_2: One $\sigma(2p_{zA}, 2p_{zB})$; one $\pi(2p_{xA}, 2p_{xB})$; and one $\pi(2p_{yA}, 2p_{yB})$ bond; along with their antibonding counterparts.

In the tetrahedral P_4 molecule, elemental white phosphorus shown in Figure 14.4, there are six single P—P bonds of roughly 200 kJ mol^{-1} bond enthalpy each. So the total bonding enthalpy is roughly 1200 kJ mol^{-1}. In the transformation

$$P_4 \rightarrow 2P_2$$

there is a loss of about 800 kJ mol^{-1} in σ bond enthalpy. This loss is not likely to be made up by the formation of 4 P—P π bonds. Period 3 atoms, such as P, are too large to get close enough to each other to form strong π bonds. Thus, P_4 is a more stable than P_2. White phosphorus is, however, unstable w/r/t to its oxides as it spontaneously inflames in air and must be stored under water.

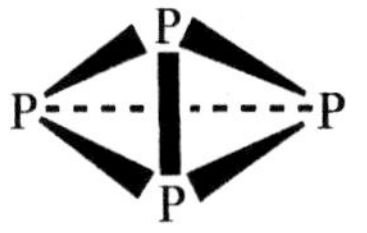

Figure 14.4

E14.3 The three valence bond wavefunctions for N_2 are of the form described by eqn 14.2.

$$\psi_1(\sigma\text{-bond}) = \psi_{2p_zA}(1)\psi_{2p_zB}(2) + \psi_{2p_zA}(2)\psi_{2p_zB}(1)$$

$$\psi_2(\pi\text{-bond}) = \psi_{2p_xA}(1)\psi_{2p_xB}(2) + \psi_{2p_xA}(2)\psi_{2p_xB}(1)$$

$$\psi_3(\pi\text{-bond}) = \psi_{2p_yA}(1)\psi_{2p_yB}(2) + \psi_{2p_yA}(2)\psi_{2p_yB}(1)$$

E14.4 We use eqn 14.3 with $R = 74.1$ pm and $Z = 1$.

$$V_{\text{nuc,nuc}} = \frac{e^2}{4\pi\varepsilon_0 R} = \frac{(1.602\times10^{-19}\ \text{C})^2}{\left(1.113\times10^{-10}\ \text{J}^{-1}\ \text{C}^2\ \text{m}^{-1}\right)\times\left(74.1\times10^{-12}\ \text{m}\right)}$$

$$= 3.11\times10^{-18}\ \text{J}$$

The molar value is

$$N_A V_{\text{nuc,nuc}} = (6.022\times10^{23}\ \text{mol}^{-1})\times(3.11\times10^{-18}\ \text{J}) = \boxed{1.87\times10^{6}\ \text{J mol}^{-1}}$$

E14.5 The carbon atom in CH_4 is sp^3 hybridized and these four equivalent hybrid orbitals are given by the LCAO of eqn 14.4. Numbering the hydrogen atoms from 1 to 4, the four valence-bond, C—H σ bonds are

$$\psi_{\sigma1} = h_1(1)\psi_{1sH1}(2) + h_1(2)\psi_{1sH1}(1)$$

$$\psi_{\sigma2} = h_2(1)\psi_{1sH2}(2) + h_2(2)\psi_{1sH2}(1)$$

$$\psi_{\sigma3} = h_3(1)\psi_{1sH3}(2) + h_3(2)\psi_{1sH3}(1)$$

$$\psi_{\sigma4} = h_4(1)\psi_{1sH4}(2) + h_4(2)\psi_{1sH4}(1)$$

E14.6 Refer to the structure shown in Figure 14.5 for the numbering of the carbon atoms in cis-retinal. Carbon atoms 5–15 each have three sp^2 hybrid atomic orbitals which form σ bonds with their neighboring atoms. There are six conjugated π bonds between these 11 C atoms and the one O atom. These six π bonds are formed from 12 p_x atomic orbitals, one on each of the 12 atoms. They are resonance hybrids all of the form:

$$\psi(\pi\text{-bond}) = \sum_{i=5}^{15} \psi_{2p_xC_i} + \psi_{2p_xO}$$

All the remaining C atoms each have four sp^3 hybrid atomic orbitals which form σ bonds with their neighboring atoms.

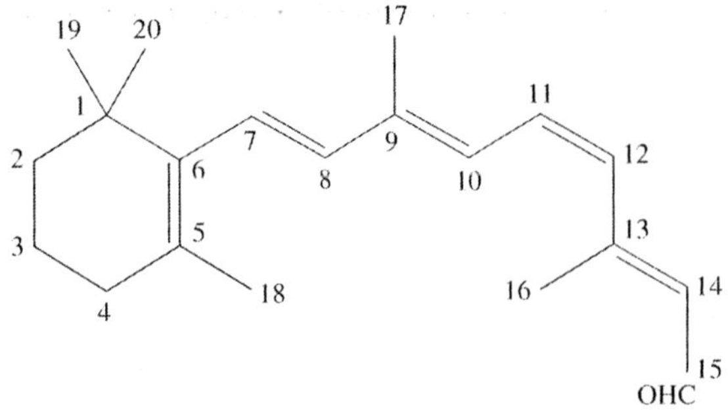

Figure 14.5

E14.7 Three resonance structure of naphthalene, $C_{10}H_8$ are shown in Figure 14.6.

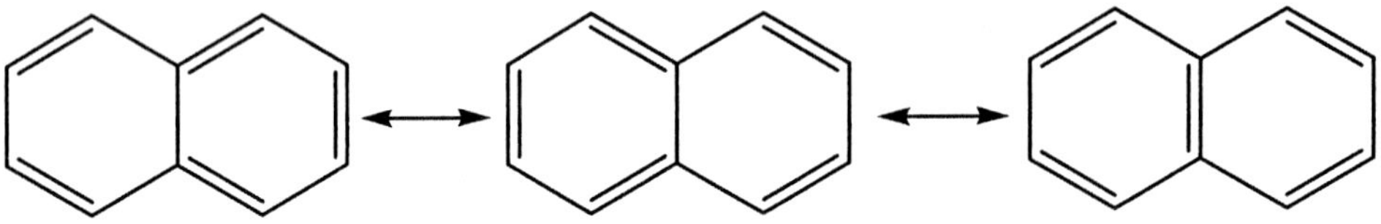

naphthalene

Figure 14.6

E14.8 In a normalized two electron wavefunction of the form $\psi = c_{\text{cov}}\psi_{\text{cov}} + c_{\text{ion}}\psi_{\text{ion}}$ where $c_{\text{cov}} = 0.889$ and $c_{\text{ion}} = 0.458$, the probability density of the electron pair is

$$\psi^2 = c_{\text{cov}}^2 \psi_{\text{cov}}^{\ 2} + c_{\text{ion}}^2 \psi_{\text{ion}}^{\ 2} + c_{\text{cov}} c_{\text{ion}} \psi_{\text{A}} \psi_{\text{B}}$$

c_{cov}^2 is the probability density if the pair were confined to ψ_{cov}, c_{ion}^2 is the probability density if the pair were confined to ψ_{ion} and $c_{\text{cov}} c_{\text{ion}}$ is an extra contribution to the density due to overlap. Thus, the ionic probability, where both electrons are confined to one atom, is

$$c_{\text{ion}}^2 = (0.458)^2 = 0.210$$

In 1000 inspections, both electrons will be observed on one atom $\boxed{\text{210 times}}$.

E14.9 Covalent structures are shown in Figure 14.7(a) while ionic structures are shown in Figure 14.7(b).

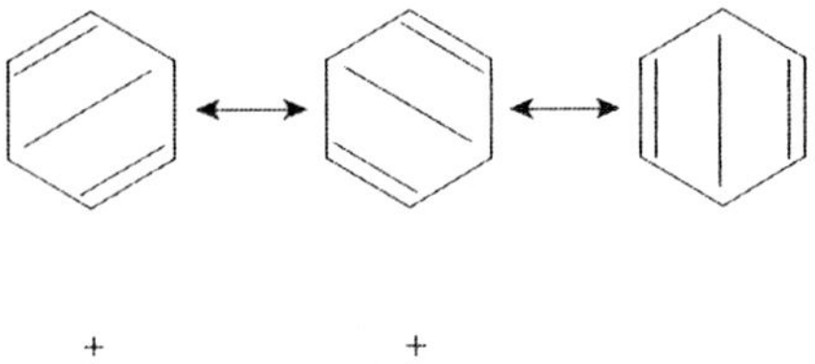

Figure 14.7(a)

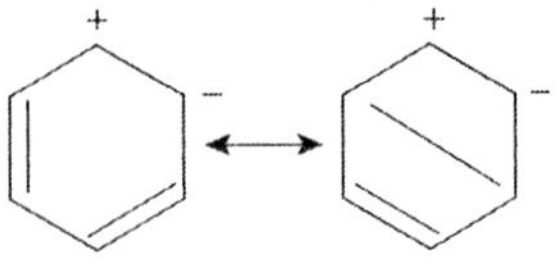

Figure 14.7(b)

In addition there are many other possible ionic structures. These structures can be safely ignored in simple descriptions of the molecule because the coefficients of the wavefunction representing these structures in the linear combination of wavefunctions for the entire resonance hybrid are very small. Benzene is a very symmetrical molecule, and we expect that all the C atoms will be equivalent.

Hence, those structures in which the C atoms are not equivalent should contribute little to the resonance hybrid.

E14.10 In general we have $\psi = c_A\ \psi_A + c_B\ \psi_B$

We need to determine the coefficients c_A and c_B.

A systematic way of finding the coefficients in the linear combinations used to build molecular orbitals is provided by the **variation principle**:

If an arbitrary wavefunction is used to calculate the energy, then the value calculated is never less than the true energy.

The arbitrary wavefunction is called the **trial wavefunction**. The principle implies that if we vary the coefficients in the trial wavefunction until we achieve the lowest energy, then those coefficients will be the best. We might get a lower energy if we use a more complicated wavefunction (for example, by taking a linear combination of several atomic orbitals on each atom), but we shall have the optimum molecular orbital that can be built from the given set of atomic orbitals.

The method can be illustrated by the trial wavefunction

$$\psi = c_A\ \psi(A) + c_B\ \psi(B)$$

This function is real but not normalized (because the coefficients can take arbitrary values), so in the following we cannot assume that $\int \psi^2 d\tau = 1$. The energy of the orbital is the expectation value of the energy operator

$$E = \frac{\int \psi H \psi d\tau}{\int \psi^2 d\tau}$$

We must search for values of the coefficients in the trial function that minimize the value of E. This is a standard problem in calculus, and is solved by finding the coefficients for which

$$\frac{\partial E}{\partial c_A} = 0 \quad \text{and} \quad \frac{\partial E}{\partial c_B} = 0$$

The first step is to express the two integrals in terms of the coefficients. The denominator is

$$\begin{aligned} \int \psi^2 d\tau &= \int \{c_A \psi(A) + c_B \psi(B)\}^2 d\tau \\ &= c_A^2 \int \psi(A)^2 d\tau + c_B^2 \int \psi(B)^2 d\tau + 2c_A c_B \int \psi(A)\psi(B) d\tau \\ &= c_A^2 + c_B^2 + 2c_A c_B S \end{aligned} \quad (1)$$

because the individual atomic orbitals are normalized and the third integral is the overlap integral S. The numerator is

$$\begin{aligned} \int \psi H \psi d\tau &= \int \{c_A \psi(A) + c_B \psi(B)\} H \{c_A \psi(A) + c_B \psi(B)\} d\tau \\ &= c_A^2 \int \psi(A) H \psi(A) d\tau + c_B^2 \int \psi(B) H \psi(B) d\tau + 2c_A c_B \int \psi(A) H \psi(B) d\tau \end{aligned}$$

There are some complicated integrals in this expression, but we can denote them by the constants

$$\alpha_A = \int \psi(A) H \psi(A) d\tau \qquad \alpha_B = \int \psi(B) H \psi(B) d\tau \qquad \beta = \int \psi(A) H \psi(B) d\tau$$

Then

$$\int \psi H \psi d\tau = c_A^2 \alpha_A + c_B^2 \alpha_B + 2c_A c_B \beta$$

α is called a **Coulomb integral**. It is negative, and can be interpreted as the energy of the electron when it occupies $\psi(A)$ (for α_A) or $\psi(B)$ (for α_B). In a homonuclear diatomic molecule, $\alpha_A = \alpha_B$. β is called a **resonance integral** (for classical reasons). It vanishes when the orbitals do not overlap, and at equilibrium bond lengths it is normally negative.

The complete expression for E is

$$E = \frac{c_A^2\alpha_A + c_B^2\alpha_B + 2c_Ac_B\beta}{c_A^2 + c_B^2 + 2c_Ac_BS}$$

Its minimum is found by differentiation with respect to the two coefficients. This involves elementary but slightly tedious work, the end result being the two **secular equations**

$$(\alpha_A - E)c_A + (\beta - ES)c_B = 0$$

$$(\beta - ES)c_A + (\alpha_B - E)c_B = 0$$

They have a solution if the determinant of the coefficients, the *secular determinant* vanishes; that is, if

$$\begin{vmatrix} \alpha_A - E & \beta - ES \\ \beta - ES & \alpha_B - E \end{vmatrix} = 0$$

This determinant expands to a quadratic equation in E, which may be solved. Its two roots give the energies of the bonding and antibonding MOs formed from the basis set and, according to the variation principle, these are the best energies for the given basis set. The corresponding values of the coefficients are then obtained by solving the secular equations using the two energies: the lower energy gives the coefficients for the bonding MO, the upper energy the coefficients for the antibonding MO. The secular equations give expressions for the *ratio* of the coefficients in each case, and so we need a further equation in order to find their individual values. This is obtained by demanding that the best wavefunction should be normalized, which means that we must also ensure (from eqn (1) above) that

$$\int \psi^2 \mathrm{d}\tau = c_A^2 + c_B^2 + 2c_Ac_BS = 1.$$

There are two cases where the roots can be written down very simply. First, when the two atoms are the same, and we can write $\alpha_A = \alpha_B = \alpha$, the solutions are

$$E_+ = \frac{\alpha + \beta}{1 + S} \qquad c_A = \left\{\frac{1}{2(1+S)}\right\}^{1/2} \qquad c_B = c_A$$

$$E_- = \frac{\alpha - \beta}{1 - S} \qquad c_A = \left\{\frac{1}{2(1-S)}\right\}^{1/2} \qquad c_B = -c_A$$

In this case, the best bonding function has the form

$$\psi_+ = \left\{\frac{1}{2(1+S)}\right\}^{1/2} \{\psi(A) + \psi(B)\}$$

and the corresponding antibonding function is

$$\psi_- = \left\{\frac{1}{2(1-S)}\right\}^{1/2} \{\psi(A) - \psi(B)\}$$

(a) When it is justifiable to neglect overlap, the secular determinant is

$$\begin{vmatrix} \alpha_A - E & \beta \\ \beta & \alpha_B - E \end{vmatrix} = 0$$

and its solutions can be expressed in terms of the parameter θ, with

$$\tan(2\theta) = \frac{2\beta}{\alpha_A - \alpha_B}$$

The solutions are:

$$E_- = \alpha_A - \beta\cot\theta \qquad \psi_- = -\sin\theta\,\psi(A) + \cos\theta\,\psi(B)$$

$$E_+ = \alpha_B + \beta\cot\theta \qquad \psi_+ = \cos\theta\,\psi(A) + \sin\theta\,\psi(B)$$

If $\theta = 0$, the wavefunction $\psi_+ = \psi(A)$; if $\theta = \pi/2$, the wavefunction $\psi_+ = \psi(B)$. So we see that this wavefunction can describe a polar covalent bond, the degree of polarity is dependent on θ. If $\theta = \pi/4$, the bond is completely covalent.

(b) We need to evaluate $\int \psi^2 d\tau$ to see if it equals 1.

$$\begin{aligned}\int(\psi_A \cos\theta + \psi_B \sin\theta)^2 d\tau &= \int(\psi_A^2 \cos^2\theta + \psi_B^2 \sin^2\theta + 2\psi_A\psi_B \sin\theta\cos\theta)d\tau \\ &= \cos^2\theta \int\psi_A^2 d\tau + \sin^2\theta\int\psi_B^2 d\tau + 2\sin\theta\cos\theta\int\psi_A\psi_B d\tau \\ &= \cos^2\theta + \sin^2\theta + 2\sin\theta\cos\theta\, S \\ &= 1\end{aligned}$$

We have used the facts that ψ_A and ψ_B are each normalized and that S is zero (i.e., ψ_A and ψ_B are orthogonal). Remember θ is a constant.

(c) In a homonuclear diatomic molecule we set $\alpha_A = \alpha_B$. Therefore, $\tan(2\theta) = \infty$ which solves to $\theta = \frac{1}{2}\arctan(\infty) = \frac{1}{2}\times 1.5708 = 0.7854$ radian $= \boxed{45°}$

E14.11 σ-bonding with a $d_{x^2-y^2}$ orbital is shown in Figure 14.8.

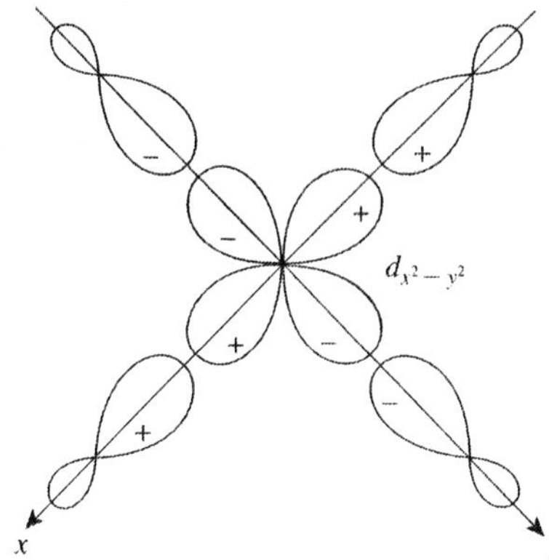

Figure 14.8

The σ-antibonding orbital looks the same but with the p-orbital lobes pointed in the opposite direction to cause destructive interference.

The σ-bonding and antibonding diagrams with the d_{xy}, d_{yz}, and d_{xz} orbitals have the same appearance as the diagram above except that the d-orbital lobes are pointed between the axes rather than along them.

σ-bonding with d_{z^2} orbital is shown in Figure 14.9.

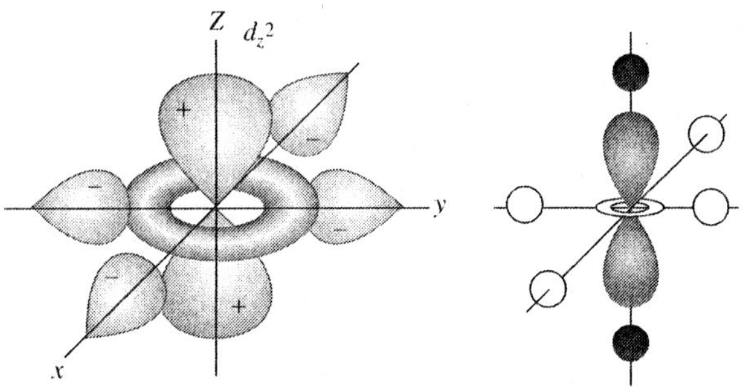

Figure 14.9

In this figure, only one of the p-orbital lobes is shown in each p-orbital. p-orbitals with positive lobes may also approach this orbital along the + and − *z*-direction as indicated in the smaller diagram on the right. The antibonding diagrams are similar, but with the signs of the p-orbital lobes reversed.

π-bonding (See Figure 14.10).

Only the d_{xy}, d_{yz}, and d_{xz} orbitals undergo π-bonding with p-orbitals on neighboring atoms. The bonding arrangement is pictured below with the d_{xy} orbital. The diagrams for the d_{yz} and d_{xz} orbitals are similar. The antibonding diagrams have the signs of the p-orbital lobes reversed.

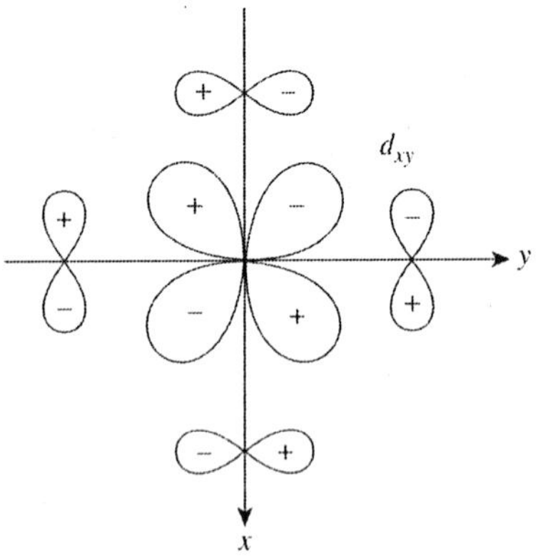

Figure 14.10

E14.12 Configurations of valence electrons are shown in Fig. 14.27 of the text.

(a) Li_2	$1\sigma_g^{\,2}$	$b=1$
(b) Be_2	$1\sigma_g^{\,2}1\sigma_u^{\,2}$	$b=0$
(c) C_2	$1\sigma_g^{\,2}1\sigma_u^{\,2}1\pi_u^{\,4}$	$b=2$

E14.13

(a) H_2^-	$1\sigma_g^{\,2}1\sigma_u^{\,1}$	$b=\frac{1}{2}$
(b) N_2	$1\sigma_g^{\,2}1\sigma_u^{\,2}1\pi_u^{\,4}2\sigma_g^{\,2}$	$b=3$
(c) O_2	$1\sigma_g^{\,2}1\sigma_u^{\,2}2\sigma_g^{\,2}1\pi_u^{\,4}1\pi_g^{\,2}$	$b=2$

E14.14

(a) CO	$1\sigma^2 2\sigma^{*2} 1\pi^4 3\sigma^2$	$b=3$
(b) NO	$1\sigma^2 2\sigma^{*2} 1\pi^4 3\sigma^2 2\pi^{*1}$	$b=5/2$
(c) CN^-	$1\sigma^2 2\sigma^{*2} 1\pi^4 3\sigma^2$	$b=3$

E14.15

B_2	$1\sigma_g^{\,2}1\sigma_u^{\,2}1\pi_u^{\,2}$	$b=1$
C_2	$1\sigma_g^{\,2}1\sigma_u^{\,2}1\pi_u^{\,4}$	$b=2$

The bond orders of B_2 and C_2 are respectively 1 and 2. Therefore, $\boxed{C_2}$ should have the greater bond dissociation enthalpy. The experimental values are approximately 4 eV and 6 eV, respectively.

E14.16 Decide whether the electron added or removed increases or decreases the bond order. The simplest procedure is to decide whether the electron occupies or is removed from a bonding or antibonding orbital. The levels for the homonuclear diatomics are shown in text Fig. 14.23 and 14.24. The level for the heteronuclear diatomics is shown in Fig. 14.33.

The following table gives the orbital involved:

	N_2	NO	O_2	C_2	F_2	CN
(a) AB^-	$1\pi_g$	2π	$1\pi_g$	$2\sigma_g$	$2\sigma_u$	3σ
Δb	$-1/2$	$-1/2$	$-1/2$	$+1/2$	$-1/2$	$+1/2$
(b) AB^+	$2\sigma_g$	2π	$1\pi_g$	$1\pi_u$	$1\pi_g$	3σ
Δb	$-1/2$	$+1/2$	$+1/2$	$-1/2$	$+1/2$	$-1/2$

Therefore,

$\boxed{\text{C}_2 \text{ and CN are stabilized by anion formation. NO, O}_2\text{, and F}_2 \text{ are stabilized by cation formation}}$.

E14.17 B_2 (6 valence electrons): $1\sigma_g^2 1\sigma_u^2 1\pi_u^2$ $b = 1$ (See Fig. 14.27)

C_2 (8 valence electrons): $1\sigma_g^2 1\sigma_u^2 1\pi_u^4$ $b = 2$

The bond orders of B_2 and C_2 are respectively 1 and 2; so $\boxed{C_2}$ should have the greater bond dissociation enthalpy. The experimental values are approximately 4eV and 6eV respectively.

E14.18 We use Figure 14.33 of the text to construct the Figure 14.11 configuration of XeF.

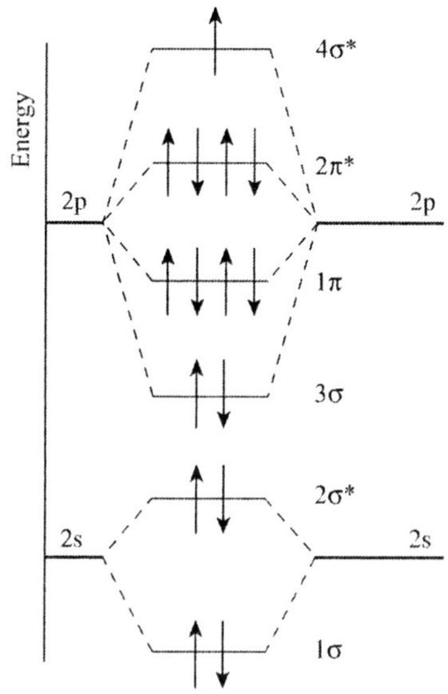

Figure 14.11

Because the bond order is increased when XeF^+ is formed from XeF (an electron is removed from an antibonding orbital), $\boxed{XeF^+ \text{ will have a shorter bond length than XeF}}$.

E14.19 (a) $1\pi^*$ is $\boxed{g}$

(b) g, u is $\boxed{\text{inapplicable}}$ to a heteronuclear molecule because it has no center of inversion.

(c) $\boxed{g}$ (See Figure 14.12(a))

(d) $\boxed{u}$ (See Figure 14.12(b))

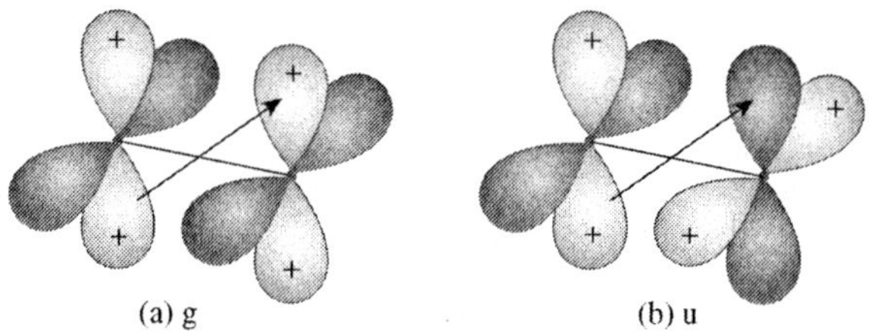

Figure 14.12

E14.20 The wavefunctions are

$$\psi_n(x) = \left(\frac{2}{L}\right)^{1/2} \sin\left(\frac{n\pi x}{L}\right) \ [12.7] \qquad \text{where } n = 1, 2, 3\ldots \ \text{ and } \ 0 \le x \le L$$

Each of these wavefunction has a centre of symmetry at $L/2$ so we invert through $x = \tfrac{1}{2}$. For $n = 1$, when $x < \tfrac{1}{2}$, $\psi_1 = +$, when $x > \tfrac{1}{2}$, $\psi_1 = +$, therefore ψ_1 is $\boxed{g}$. In a similar fashion we determine

$n = 2$, $\boxed{u}$; $n = 3$, $\boxed{g}$; $n = 4$, $\boxed{u}$

Figure 14.13, an adaptation of text Figure 12.19, illustrates the operation of inversion through the centre of symmetry. First, identify the centre of symmetry. Then, pick any non-nodal position and note the sign and magnitude of the wavefunction. Draw an arrow from that wavefunction point through the centre of symmetry to a point that is an equal distance on the opposite side of the centre of symmetry. If the wavefunction has the same magnitude and sign after the inversion, the wavefunction has gerade (g) symmetry. If the wavefunction has the same magnitude and opposite sign, the wavefunction has ungerade (u) symmetry. We quickly see that all odd-numbered (n = 1, 3, 5,...) quantum states of the particle in a box have gerade symmetry or 'parity' while all even-numbered (n = 2, 4, 6,...) states have ungerade symmetry or 'parity'.

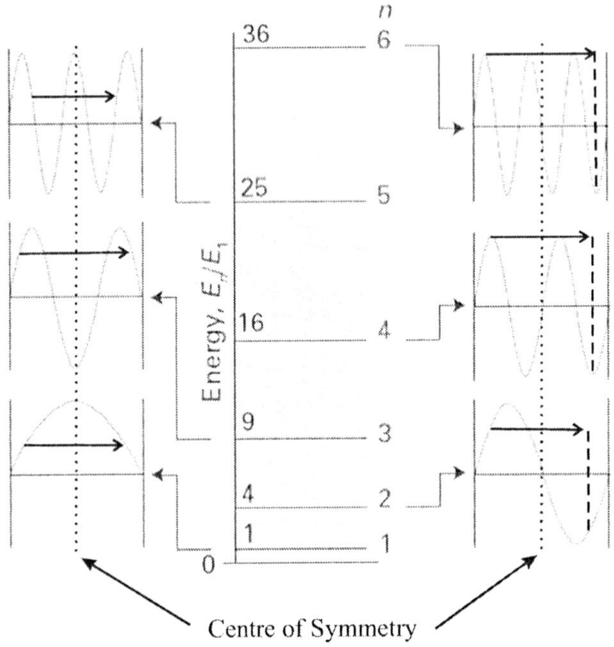

Figure 14.13

E14.21 Refer to Fig. 12.34(a) of the text. Examine the inversion through the center of these functions, i.e., replace x with $-x$.

(a) $v = 0$, $\boxed{g}$
$v = 1$, $\boxed{u}$
$v = 2$, $\boxed{g}$
$v = 3$, u (not shown in Fig. 12.34 of the text)

(b) If v is even, ψ_v is g.
If v is odd, ψ_v is u.

E14.22 The parities are given in Fig. 14.35 of the text: a_{2u}, e_{1g}, e_{2u}, and b_{2g} where e_{1g} and e_{2u} are doubly degenerate levels. The text figure shows p_z lobes on one side of the benzene molecular plane only. Figure 14.14 shows both lobes (as does text Figure 14.37) and the figure arrow shows the inversion operation that indicates the ungerade symmetry of one of the e_{2u} MOs.

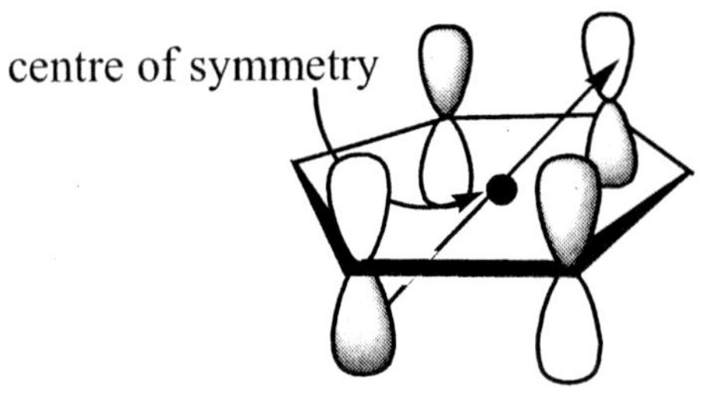

Figure 14.14

E14.23 NO $1\sigma^2 2\sigma^{*2} 1\pi^4 3\sigma^2 2\pi^{*1}$ $b = \frac{5}{2}$

N_2 $1\sigma_g^{\,2} 1\sigma_u^{\,2} 1\pi_u^{\,4} 2\sigma_g^{\,2}$ $b = 3$

Because the bond order of N_2 is greater, $\boxed{N_2}$ is likely to have the shorter bond length.

E14.24 $F_2^{\,+}$ $1\sigma_g^{\,2} 1\sigma_u^{\,2} 2\sigma_g^{\,2} 1\pi_u^{\,4} 1\pi_g^{\,3}$ $b = \frac{3}{2}$

F_2 $1\sigma_g^{\,2} 1\sigma_u^{\,2} 2\sigma_g^{\,2} 1\pi_u^{\,4} 1\pi_g^{\,4}$ $b = 1$

$F_2^{\,-}$ $1\sigma_g^{\,2} 1\sigma_u^{\,2} 2\sigma_g^{\,2} 1\pi_u^{\,4} 1\pi_g^{\,4} 2\sigma_u^{\,1}$ $b = \frac{1}{2}$

Therefore, the order of bond lengths is $\boxed{F_2^+ < F_2 < F_2^-}$.

E14.25 $F_2^{\,+}$ $1\sigma_g^{\,2} 1\sigma_u^{\,2} 2\sigma_g^{\,2} 1\pi_u^{\,4} 1\pi_g^{\,3}$ $b = \frac{3}{2}$

F_2 $1\sigma_g^{\,2} 1\sigma_u^{\,2} 2\sigma_g^{\,2} 1\pi_u^{\,4} 1\pi_g^{\,4}$ $b = 1$

$F_2^{\,-}$ $1\sigma_g^{\,2} 1\sigma_u^{\,2} 2\sigma_g^{\,2} 1\pi_u^{\,4} 1\pi_g^{\,4} 2\sigma_u^{\,1}$ $b = \frac{1}{2}$

The bond order is $\boxed{F_2^+ > F_2 > F_2^-}$.

E14.26 $O_2^{\,+}$(11 electrons): $1\sigma_g^{\,2} 1\sigma_u^{\,2} 2\sigma_g^{\,2} 1\pi_u^{\,4} 1\pi_g^{\,1}$ $b = \frac{5}{2}$

O_2(12 electrons): $1\sigma_g^{\,2} 1\sigma_u^{\,2} 2\sigma_g^{\,2} 1\pi_u^{\,4} 1\pi_g^{\,2}$ $b = 2$

$O_2^{\,-}$(13 electrons): $1\sigma_g^{\,2} 1\sigma_u^{\,2} 2\sigma_g^{\,2} 1\pi_u^{\,4} 1\pi_g^{\,3}$ $b = \frac{3}{2}$

$O_2^{\,2-}$(14 electrons): $1\sigma_g^{\,2} 1\sigma_u^{\,2} 2\sigma_g^{\,2} 1\pi_u^{\,4} 1\pi_g^{\,4}$ $b = 1$

Each electron added to O_2^+ is added to an antibonding $1\pi_g$ orbital, thus increasing the length. So the sequence $\boxed{O_2^+, O_2, O_2^-, O_2^{2-}}$ has progressively longer bonds.

E14.27 $O_2^{\,+}$(11 electrons): $1\sigma_g^{\,2} 1\sigma_u^{\,2} 2\sigma_g^{\,2} 1\pi_u^{\,4} 1\pi_g^{\,1}$ $b = \frac{5}{2}$

O_2(12 electrons): $1\sigma_g^{\,2} 1\sigma_u^{\,2} 2\sigma_g^{\,2} 1\pi_u^{\,4} 1\pi_g^{\,2}$ $b = 2$

$O_2^{\,-}$(13 electrons): $1\sigma_g^{\,2} 1\sigma_u^{\,2} 2\sigma_g^{\,2} 1\pi_u^{\,4} 1\pi_g^{\,3}$ $b = \frac{3}{2}$

$O_2^{\,2-}$(14 electrons): $1\sigma_g^{\,2} 1\sigma_u^{\,2} 2\sigma_g^{\,2} 1\pi_u^{\,4} 1\pi_g^{\,4}$ $b = 1$

Each electron added to O_2^+ is added to an antibonding $1\pi_g$ orbital, thus decreasing the bond order and the bond strength. So the bond order is $\boxed{O_2^{\,+} > O_2 > O_2^{\,-} > O_2^{\,2-}}$.

E14.28 (a) The molecular orbital energy level diagram of the hybridized CH_2 fragments and ethene, $CH_2{=}CH_2$, is shown in Figure 14.15(a).

COMMENT. Note that the π-bonding orbital must be lower in energy than the σ-antibonding orbital for π-bonding to exist in ethene.

(b) The molecular orbital energy level diagram of the hybridized CH fragments and ethyne, CH≡CH is shown in Figure 14.15(b).

Question. Would the ethyne molecule exist if the order of the energies of the π and σ* orbitals were reversed?

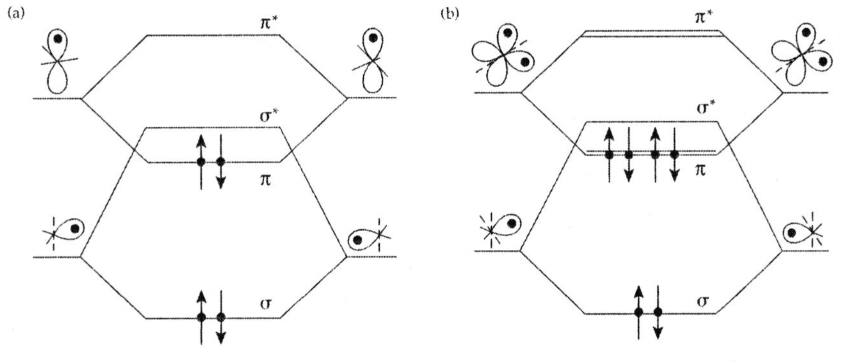

Figure 14.15

E14.29 In order to anticipate the form that the bonding and antibonding 'φ orbitals' that could be constructed from two neighbouring $f_{(l=3)}$ atomic orbitals, we first attempt a reasonable guess about the shape and parity of the f orbitals. A quick sketch of $s_{(l=0)}$, $p_{(l=1)}$, and $d_{(l=2)}$ atomic orbitals helps us recall that s, p, and d orbitals have 1, 2, and 4 lobes (except for the d_{z^2} AO, which has 3 lobes), respectively, and their parities are gerade for orbitals that have an even angular momentun quantum number and ungerade for orbitals that have an odd angular momentum quantum number. The pattern allows us to conclude that **f orbitals have ungerade parity** and 4 or more lobes. We try to imagine the possible shapes of ungerade orbitals having 4 or more lobes and arrive at those shown in Figure 14.16 with the wavefunction sign of each lobe shown as + or −. Figure 14.17 shows the possible f-f overlaps that might occur between neighbouring AO. Figure 14.17(a) shows a σ bond using the planar arrangement of f AO lobes while Figure 14.17(b) shows a π bond. There is one additional bonding orientation to consider. It is analogous to the δ bond, shown in Figure 14.18(a), which originates with d-d AO overlap. It is the sandwich-like layering of two f orbitals shown in Figure 14.18(b) as a bonding VB and in Figure 14.18(b) as an antibonding VB; broad lines are used to show lobe overlap and the internuclear axis. These latter two are the bonding and antibonding 'φ orbitals' that we set out to explore. The bond sketches show that a σ bond has no nodal plane along the internuclear axis while a π bond has one, a δ bond has two. **A φ bond has three nodal planes along the internuclear axis.**

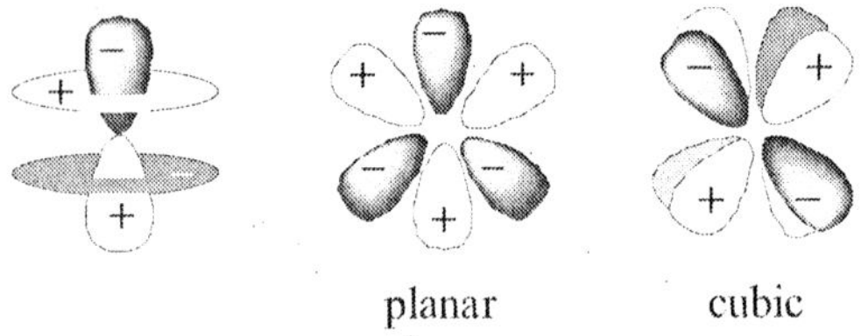

Figure 14.16

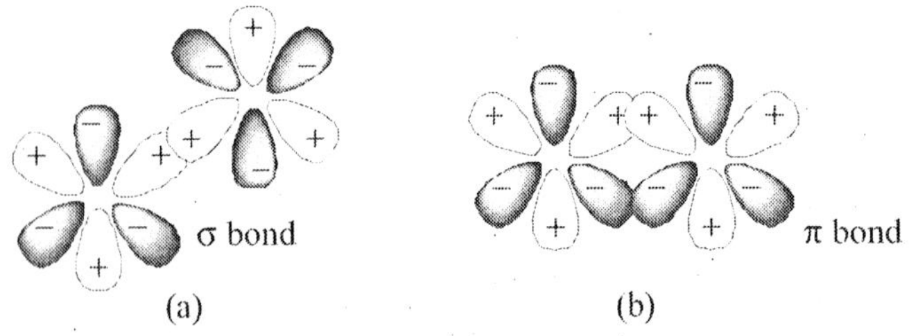

Figure 14.17

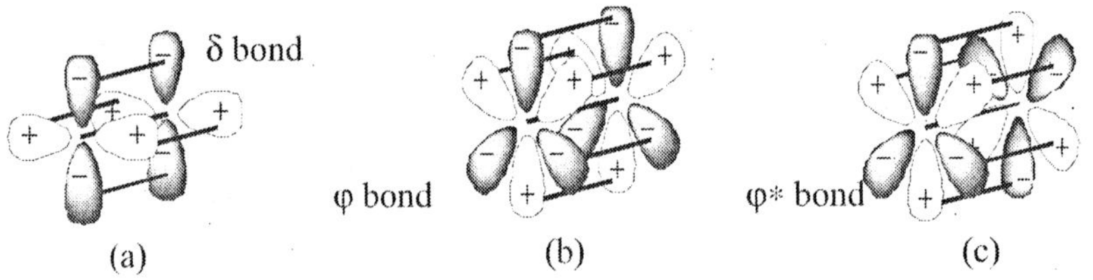

Figure 14.18

E14.30 The Pauling electronegativities (Table 14.2) are $\chi_H = 2.1$, $\chi_B = 2.0$, and $\chi_P = 2.1$.

(a) The P—H bond should be **nonpolar** as the electronegativities of P and H are identical.

(b) The B—H bond should be slightly **polarized** toward H as the electronegativity of H is slightly larger.

E14.31 In text Section 14.16 the π electron energy levels of benzene according to the Hückel method are given as $E = \alpha \pm 2\beta$ (singly degenerate levels) and $\alpha \pm \beta$ (doubly degenerate levels) where α and β are both negative.

(a) Benzene anion (7 π electrons): $\boxed{1a_{2u}^{\;2}1e_{1g}^{\;4}1e_{2u}^{\;1}}$

$$E = 2(\alpha + 2\beta) + 4(\alpha + \beta) + (\alpha - \beta) = \boxed{7\alpha + 7\beta}$$

(b) Benzene cation (5 π electrons): $1a_{2u}^{\;2}1e_{1g}^{\;3}$

$$E = 2(\alpha + 2\beta) + 3(\alpha + \beta) = \boxed{5\alpha + 7\beta}$$

E14.32 $$\psi_n(x) = \left(\frac{2}{L}\right)^{1/2} \sin\left(\frac{n\pi x}{L}\right) \quad [12.7] \qquad \text{where } n = 1,2,3... \quad \text{and} \quad 0 \le x \le L$$

$$E_n = \frac{n^2h^2}{8m_eL^2}$$

Two electrons occupy each level by the Pauli principle, and so butadiene, which has four π electrons, has two electrons in ψ_1 and two electrons in ψ_2. The occupied particle in a box wavefunctions are

$$\psi_1 = \left(\frac{2}{L}\right)^{1/2} \sin\left(\frac{\pi x}{L}\right) \quad \text{and} \quad \psi_2 = \left(\frac{2}{L}\right)^{1/2} \sin\left(\frac{2\pi x}{L}\right)$$

These FEMO wavefunctions are sketched and compared with the Hückel π model MOs in Figure 14.19.

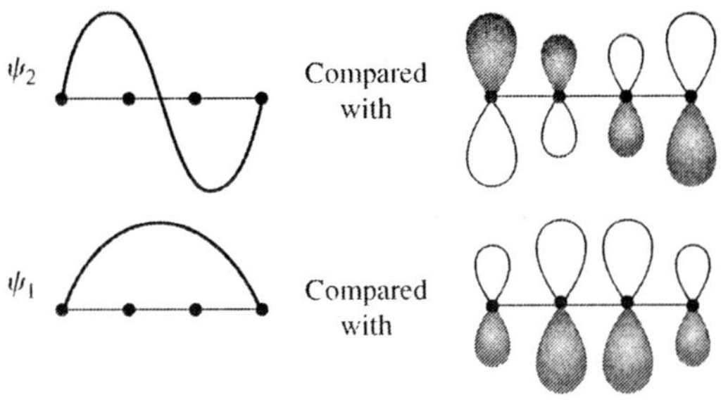

Figure 14.19

The minimum excitation energy for butadiene in the FEMO model is

$$\Delta E = E_3 - E_2 = (3^2 - 2^2)\left(\frac{h^2}{8m_eL^2}\right) = \frac{5h^2}{8m_eL^2} \qquad \text{where } L \simeq 4R = 4\times140 \text{ pm}$$

$$= \frac{5\times(6.626\times10^{-34} \text{ J s})^2}{8\times(9.109\times10^{-31} \text{ kg})\times(4\times140\times10^{-12} \text{ m})^2} = 9.61\times10^{-19} \text{ J} = 6.00 \text{ eV}$$

In $CH_2{=}CH{-}CH{=}CH{-}CH{=}CH{-}CH{=}CH_2$ there are eight π electrons to accommodate, so the HOMO will be ψ_4 and the LUMO ψ_5. The minimum excitation energy in the FEMO model is

$$\Delta E = E_5 - E_4 = (5^2 - 4^2)\left(\frac{h^2}{8m_eL^2}\right) = \frac{9h^2}{8m_eL^2} \qquad \text{where } L \simeq 8R = 8\times140 \text{ pm}$$

$$= \frac{9\times(6.626\times10^{-34} \text{ J s})^2}{8\times(9.109\times10^{-31} \text{ kg})\times(8\times140\times10^{-12} \text{ m})^2} = 4.32\times10^{-19} \text{ J} = \boxed{2.70 \text{ eV}}$$

The FEMO model ψ_4 HOMO and ψ_5 LUMO are sketched in Figure 14.20.

COMMENT. Can you identify the positions of the carbon nuclei in Figure 14.20 and superimpose the p_z orbitals of the simple Hückel π model with their wave sign upon both the HOMO and LUMO of the figure?

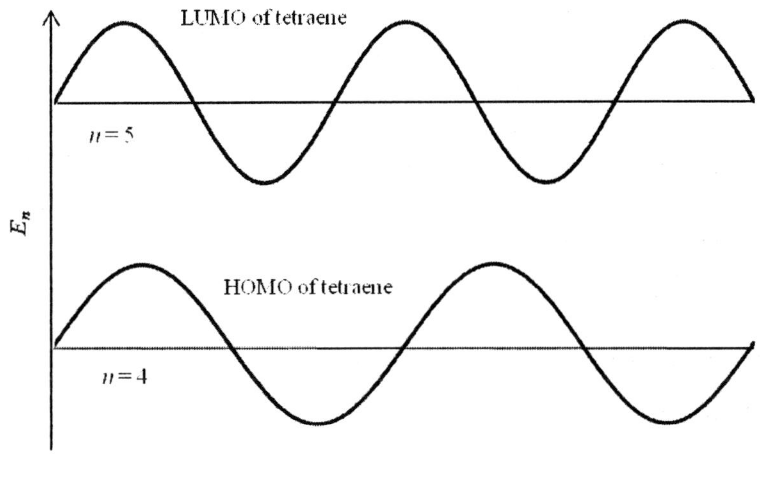

Figure 14.20

COMMENT. In the FEMO model the wavelength of the minimum excitation energy is given by

$$\lambda = \frac{hc}{\Delta E} = \frac{(6.626\times10^{-34}\ \text{J s})\times(2.998\times10^{8}\ \text{m s}^{-1})}{4.32\times10^{-19}\ \text{J}} = 4.60\times10^{-7}\ \text{m} \quad \text{or} \quad \boxed{460\ \text{nm}}$$

The wavelength 460 nm corresponds to blue light; so on the basis of this calculation alone the molecule would appear $\boxed{\text{orange}}$ in white light (because blue is subtracted). The experimental value of λ_{max} is 304 nm, and the compound is colorless. This illustrates the very approximate nature of the calculation we have performed.

E14.33 For a particle in a two dimensional rectangle the wavefunction for each state is given by the separation of variables relation $\psi_{n_X n_Y} = \psi_{n_X}\psi_{n_Y}$ where both ψ_{n_X} and ψ_{n_Y} have even parity when their respective quantum numbers are odd and odd parity when their respective quantum numbers are even (see text Figure 12.19). Thus, a state will have odd parity when one of the wavefunctions ψ_{n_X} or ψ_{n_Y} has odd parity while the other has even parity. Similarly, a state will have even parity when both of the wavefunctions ψ_{n_X} and ψ_{n_Y} have either odd parity or both have even parity (see text Figure 12.23). Having decided this important issue, we must now find the states that are occupied by the π electrons of naphthalene.

The naphthalene molecule, shown in Figure 14.21, has 10 π electrons.

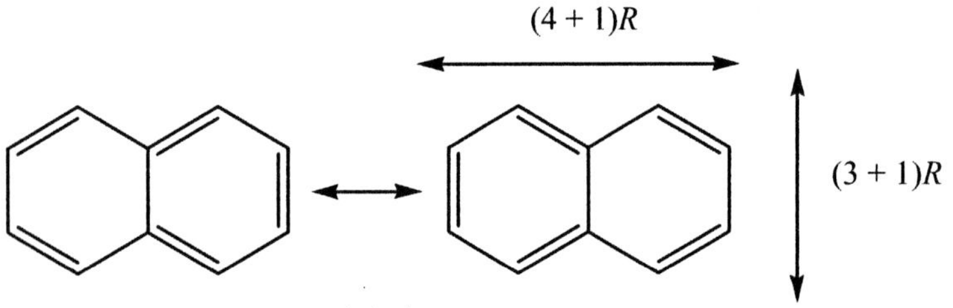

Figure 14.21

We will model these 10 π electrons with *free-electron molecular orbital* (FEMO) theory in which electrons are treated as particles in a rectangular box for which $L_X = 5R = 5L/4$ and $L_Y = 4R = L$ (an extra half bond-length is added at the end of each stick structure length shown in Figure 14.21). The question is "Which levels are occupied by the π electrons and are any of the levels accidentally degenerate?" The energies of the FEMO electrons are

$$E_{n_X n_Y} = \left(\frac{n_X^2}{L_X^2} + \frac{n_Y^2}{L_Y^2}\right)\frac{h^2}{8m_e} \quad [12.12\text{b}] \qquad \text{where } n_X, n_Y = 1,2,3,...$$

$$= \left(\frac{16n_X^2 + 25n_Y^2}{25}\right)\frac{h^2}{8m_eL^2}$$

Degeneracy occurs whenever there are multiple (n_X,n_Y) quantum number sets that give identical values of $(16n_X^2 + 25n_Y^2)/25$. We draw the following table to calculate and compare $(16n_X^2 + 25n_Y^2)/25$ values for different sets of quantum numbers.

n_X	n_Y	$(16n_X^2 + 25n_Y^2)/25$
1	1	41/25
	2	116/25
	3	241/25
	4	416/25
2	1	89/25
	2	164/25
	3	289/25
3	1	169/25
	2	244/25

Inspection of the table reveals that the five lowest energy levels have different energies so they are non-degenerate. These levels are filled with the 10 π electrons of naphthalene. From lowest energy to highest they are: $(n_X,n_Y) = (1,1), (2,1), (1,2), (2,2)$, and $(3,1)$. Thus, we find that the parities of the occupied orbitals are

$$\boxed{(n_X, n_Y, \text{parity}) = (1,1,\text{even}),\ (2,1,\text{odd}),\ (1,2,\text{odd}),\ (2,2,\text{even}),\ \text{and}\ (3,1,\text{even})}$$

E14.34 The total number of atomic orbitals in a single set of s, p, d, and f subshells is

$$1 + 3 + 5 + 7 = 16$$

In a diatomic molecule there would then be 32 AO from which $\boxed{\text{32 molecular oribtals}}$ can be constructed.

E14.35 (a) The transition occurs for photons whose energies are equal to the difference in energy between the highest occupied and lowest unoccupied orbital energies of the conjugated molecule:

$$E_{\text{photon}} = E_{\text{LUMO}} - E_{\text{HOMO}}$$

If N is the number of carbon atoms contributing an electron in a $2p_z$ orbital, then the number of π electrons is also N. These N electrons occupy the first $N/2$ π molecular orbitals, so π MO number $N/2$ is the HOMO and π MO number $1 + N/2$ is the LUMO.

The non-degenerate energy levels of FEMO theory are

$$E_k = \alpha + 2\beta \cos\frac{k\pi}{N+1} \qquad \text{where } k = 1,2,3,\ldots,N$$

Writing the photon energy in terms of the wavenumber, and recognizing that the transition is between the π HOMO and π LUMO, gives:

$$\begin{aligned} hc\tilde{\nu} &= E_{\frac{1}{2}N+1} - E_{\frac{1}{2}N} \\ &= \left(\alpha + 2\beta\cos\frac{(\frac{1}{2}N+1)\pi}{N+1}\right) - \left(\alpha + 2\beta\cos\frac{\frac{1}{2}N\pi}{N+1}\right) \\ &= 2\beta\left(\cos\frac{(\frac{1}{2}N+1)\pi}{N+1} - \cos\frac{\frac{1}{2}N\pi}{N+1}\right) \end{aligned}$$

Solving for β yields

$$\frac{\beta}{hc} = \frac{\tilde{\nu}}{2\left(\cos\frac{(\frac{1}{2}N+1)\pi}{N+1} - \cos\frac{\frac{1}{2}N\pi}{N+1}\right)}$$

Draw up the following table for the computation of β for ethene and each conjugated linear polyene.

Species	N	$\tilde{\nu}$ / cm^{-1}	β_{FEMO}/hc/cm^{-1}
ethene, C_2H_4	2	61500	–30750
butadiene, C_4H_6	4	46080	–37280
hexatriene, C_6H_8	6	39750	–44660
octatetraene C_8H_{10}	8	32900	–47370

Examination of the table computation reveals that β_{FEMO} is not a constant for the series. However, it is reasonable to estimate that $\boxed{\beta/hc \sim -40000\ \text{cm}^{-1}\ (-5.0\ \text{eV})}$ for the homologous series.

(b) The total energy of the π electron system is the sum of the energies of occupied orbitals weighted by the number of electrons that occupy them. In octatetraene, C_8H_{10}, each of the first four orbitals are doubly occupied, so

$$E_\pi = 2\sum_{k=1}^{4} E_k = 2\sum_{k=1}^{4}\left(\alpha + 2\beta\cos\frac{k\pi}{9}\right) = 8\alpha + 4\beta\sum_{k=1}^{4}\cos\frac{k\pi}{9} = 8\alpha + 9.518\beta$$

The delocalization energy is the difference between this quantity and that of four isolated double bonds:

$$E_{\text{deloc}} = E_{\pi} - 8(\alpha+\beta) = 8\alpha + 9.518\beta - 8(\alpha+\beta) = \boxed{1.518\beta}$$

Using the estimate of β from the above table for octatetraene yields

$$\boxed{E_{\text{deloc}}/hc = 60720\ \text{cm}^{-1}\ (7.35\ \text{eV})}$$

E14.36 (a) In the absence of numerical values for α and β, we express orbital energies as $(E_k - \alpha)/\beta$ for the purpose of comparison. Recall that β is negative (as is α for that matter), so the orbital with the greatest value of $(E_k - \alpha)/\beta$ has the lowest energy. Draw up the following table, evaluating the Hückel theory expression

$$\frac{E_k - \alpha}{\beta} = 2\cos\frac{2k\pi}{N}$$

$$k = 0, \pm1, \pm2, \ldots, \pm N/2 \text{ for even } N \quad \text{or} \quad k = 0, \pm1, \pm2, \ldots, \pm(N-1)/2 \text{ for odd } N$$

where N is the number of carbon atoms contributing an electron in a $2p_z$ orbital to the π system of a monocyclic conjugated polyene.

	Energy $(E_k - \alpha)/\beta$	
Orbital, k	Benzene, C_6H_6 $N = 6$	Octatriene, C_8H_8 $N = 8$
±4		−2.000
±3	−2.000	−1.414
±2	−1.000	0
±1	1.000	1.414
0	2.000	2.000

In each case, the lowest and highest energy levels are non-degenerate, while the other energy levels are doubly degenerate. The degeneracy is clear for all energy levels except, perhaps, the highest: each value of the quantum number k corresponds to a separate MO, and positive and negative values of k therefore give rise to a pair of MOs of the same energy. This is not the case for the highest energy level, though, because there are only as many MOs as there were AOs input to the calculation, which is the same as the number of carbon atoms; having a doubly-degenerate top energy level would yield one extra MO.

(b) The total energy of the π electron system is the sum of the energies of occupied orbitals weighted by the number of electrons that occupy them.

In C_6H_6, each of the first three orbitals are doubly occupied, but the second level ($k = \pm1$) is doubly degenerate, so

$$E_{\pi} = 2E_0 + 4E_1 = 2(\alpha + 2\beta\cos 0) + 4\left(\alpha + 2\beta\cos\frac{2\pi}{6}\right) = 6\alpha + 8\beta$$

The delocalization energy is the difference between this quantity and that of three isolated double bonds:

$$E_{\text{deloc}} = E_{\pi} - 6(\alpha+\beta) = 6\alpha + 8\beta - 6(\alpha+\beta) = \boxed{2\beta}$$

For linear hexatriene, $E_{\text{deloc}} = 0.988\beta$, so benzene has considerably more delocalization energy (assuming that β is similar in the two molecules). This extra stabilization is an example of the special stability of $\boxed{\text{aromatic}}$ compounds.

(c) In C_8H_8, each of the first three orbitals are doubly occupied, but the second level ($k = \pm1$) is doubly degenerate. The next level is also doubly degenerate, with a single electron occupying each orbital. So the energy is

$$E_\pi = 2E_0 + 4E_1 + 2E_2$$
$$= 2(\alpha + 2\beta\cos 0) + 4\left(\alpha + 2\beta\cos\frac{2\pi}{8}\right) + 2\left(\alpha + 2\beta\cos\frac{4\pi}{8}\right)$$
$$= 8\alpha + 9.657\beta$$

The delocalization energy is the difference between this quantity and that of four isolated double bonds:

$$E_{\text{deloc}} = E_\pi - 8(\alpha+\beta) = 8\alpha + 9.657\beta - 8(\alpha+\beta) = \boxed{1.657\beta}$$

This delocalization energy is not much different from that of linear octatetraene (1.518β), so cycloocatetraene does not have much additional stabilitzation over the linear structure. Once again, though, we do see that the delocalization energy stabilizes the π orbitals of the closed ring conjugated system to a greater extent than what is observed in the open chain conjugated system. However, the benzene/hexatriene comparison shows a much greater stabilization than does the cyclooctatetraene/octatetraene system. This is a demonstration of the Hückel $4n + 2$ rule, which states that any planar, cyclic, conjugated system exhibits unusual aromatic stabilization if it contains $4n + 2$ π electrons where "n" is an integer. Benzene with its 6 π electrons has this aromatic stabilization whereas cycloctatetraene with 8 π electrons doesn't have this unusual stabilization. We can say that it is $\boxed{\text{not aromatic}}$, consistent with indicators of aromaticity such as the Hückel $4n + 2$ rule.

Answers to projects

P14.37 The atomic orbital φ is normalized when $\int \varphi^2 \mathrm{d}\tau = 1$ where the integral over τ represents an integral over all possible values of x, y, and z. Two atomic orbitals φ_1 and φ_2 are orthogonal if $\int \varphi_1\varphi_2 \mathrm{d}\tau = 0$. Each AO of the set s, p_x, p_y, and p_z is both normalized and orthogonal to other members of the set so they are said to be an **orthonormal** set.

(a) $h_1 = s + p_x + p_y + p_z$ and $h_2 = s - p_x - p_y + p_z$

h_1 and h_2 are orthogonal providing that $\int h_1 h_2 \mathrm{d}\tau = 0$. We check that this condition is satisfied.

$$\int h_1h_2\mathrm{d}\tau = \int (s + p_x + p_y + p_z)(s - p_x - p_y + p_z)\mathrm{d}\tau$$
$$= \int (s^2 - sp_x - sp_y + sp_z)\mathrm{d}\tau + \int (sp_x - p_x^2 - p_xp_y + p_xp_z)\mathrm{d}\tau$$
$$+ \int (sp_y - p_xp_y - p_y^2 + p_yp_z)\mathrm{d}\tau + \int (sp_z - p_xp_z - p_yp_z + p_z^2)\mathrm{d}\tau$$

All of the above integrals of the type $\int (sp_x)\mathrm{d}\tau$ and $\int (p_xp_y)\mathrm{d}\tau$ vanish because the integrand AOs are orthogonal. Therefore,

$$\int h_1h_2\mathrm{d}\tau = \int (s^2)\mathrm{d}\tau - \int (p_x^2)\mathrm{d}\tau - \int (p_y^2)\mathrm{d}\tau + \int (p_z^2)\mathrm{d}\tau$$
$$= 1 - 1 - 1 + 1 \text{ because the AO are normalized}$$
$$= 0$$

Thus, h_1 and h_2 are orthogonal.

(b) We need to demonstrate that $\int h_{sp^2}{}^2\mathrm{d}\tau = 1$, where $h_{sp^2} = \dfrac{s + \sqrt{2}p}{\sqrt{3}}$.

$$\int h_{sp^2}{}^2\mathrm{d}\tau = \tfrac{1}{3}\int \left(s + \sqrt{2}p\right)^2 \mathrm{d}\tau$$
$$= \tfrac{1}{3}\int \left(s^2 + 2p^2 + 2\sqrt{2}\ sp\right)\mathrm{d}\tau$$
$$= \tfrac{1}{3}(1 + 2 + 0) \quad \text{as } \int s^2\mathrm{d}\tau = 1,\ \int p^2\mathrm{d}\tau = 1, \text{ and } \int sp\ \mathrm{d}\tau = 0 \text{ (orthonormality)}$$
$$= 1$$

Thus, this hybrid orbital is normalized to 1.

(c) Rewrite the normalized sp^2 hybrid orbital of part (b) as $h_1 = \frac{s+\sqrt{2}p_x}{\sqrt{3}}$. This hybrid orbital points along the line l_1 on the x-axis as shown in Figure 14.22. There are two additional sp^2 hybrid orbitals, h_2 and h_3; they point along the lines l_2 and l_3. These three orbitals have the same size and shape. They differ only in the direction to which they point. To construct h_2, we appropriately weigh the s, p_x, and p_y AOs so that the sum points along l_2 while simultaneously being normalized and orthogonal to h_1. In order to point along l_2, the weight of p_y must be positive but the weight of p_x must be negative. Furthermore, as shown in Figure 14.22, the weight of p_y must be $3^{1/2}$ times the weight of p_x. Thus,

$$h_2 = as - bp_x + 3^{1/2}bp_y \quad \text{where the weights } a \text{ and } b \text{ are positive numbers}$$

The values of the weights a and b are found with the orthonormal conditions. The orthogonality of h_1 and h_2 provides a useful relation:

$$\int h_1 h_2 \mathrm{d}\tau = \int \left(\frac{s+2^{1/2}p_x}{3^{1/2}}\right)\left(as - bp_x + 3^{1/2}bp_y\right)\mathrm{d}\tau$$

$$= \frac{a}{3^{1/2}}\int s^2\mathrm{d}\tau - \frac{2^{1/2}b}{3^{1/2}}\int p_x^2\mathrm{d}\tau \quad \text{because terms like } \int sp_x\mathrm{d}\tau \text{ equal zero (orthgonality)}$$

$$= \frac{a}{3^{1/2}} - \frac{2^{1/2}b}{3^{1/2}} \quad \text{because the orbitals are normalized}$$

The above expression equals zero when $a = 2^{1/2}b$ so $h_2 = b\left(2^{1/2}s - p_x + 3^{1/2}p_y\right)$. Now we use the normalization condition $\int h_2^2\mathrm{d}\tau = 1$ to determine the value of b.

$$\int h_2^2\mathrm{d}\tau = b^2\int\left(2^{1/2}s - p_x + 3^{1/2}p_y\right)^2\mathrm{d}\tau$$

$$= b^2\left\{2\int s^2\mathrm{d}\tau + \int p_x^2\mathrm{d}\tau + 3\int p_y^2\mathrm{d}\tau\right\} \quad \text{because terms like } \int sp_x\mathrm{d}\tau \text{ equal zero (orthgonality)}$$

$$= 6b^2 \quad \text{because the orbitals are normalized}$$

Thus, $b = 6^{-1/2}$ and substitution gives

$$\boxed{h_2 = \sqrt{\frac{1}{3}}\,s - \sqrt{\frac{1}{6}}\,p_x + \sqrt{\frac{1}{2}}\,p_y}$$

The sp^2 hybrid h_3 is a reflection of h_2 through the y-axis so we need only change the weight of p_y by changing the weight sign, thereby, giving the last of the three sp^2 hybrids.

$$\boxed{h_3 = \sqrt{\frac{1}{3}}\,s - \sqrt{\frac{1}{6}}\,p_x - \sqrt{\frac{1}{2}}\,p_y}$$

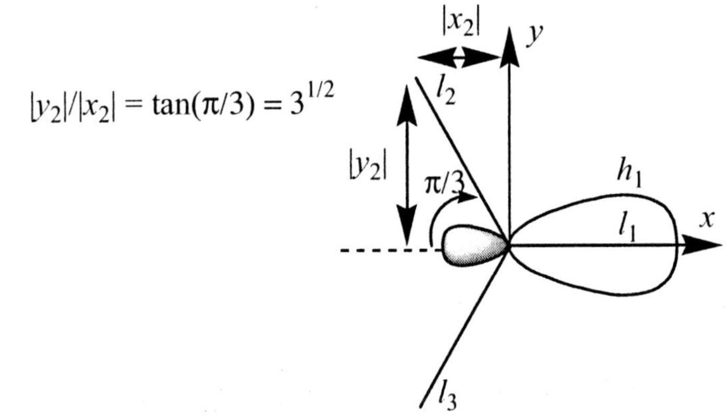

Figure 14.22

P14.38 (a) When the s orbital shares a common centre with the p orbital, the constructive interference between the s orbital and one of the p orbital lobe is exactly balanced with the destructive interference between the s orbital and the other p orbital lobe (see Figure 14.23). Thus, when $R = 0$, $S = 0$. As the two orbitals separate the overlap increases to a maximum after which larger separation results in an ever decreasing overlap, which is sketched in Figure 14.24.

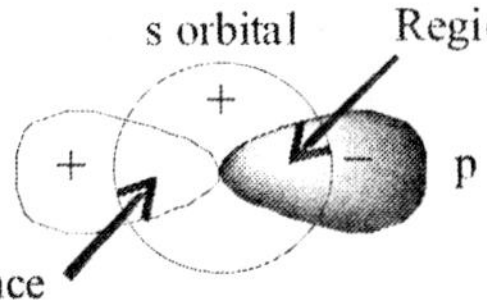

Figure 14.23

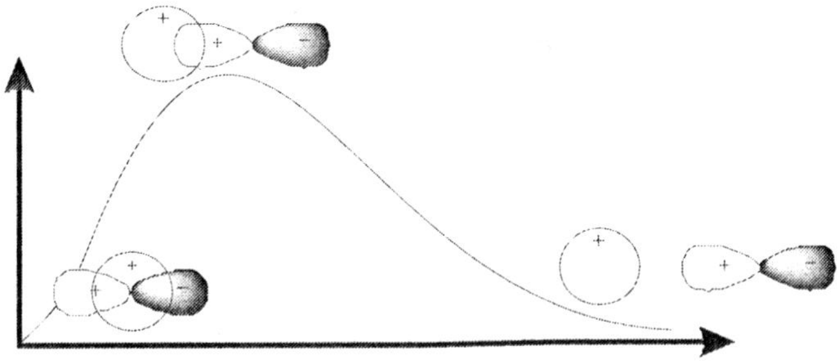

Figure 14.24

(b) $S = (R/a_0)\{1 + (R/a_0) + \frac{1}{3}(R/a_0)^2\}e^{-R/a_0}$

Draw up the following table and plot S against R/a_0 as shown in Figure 14.25. A maximum occurs at $\boxed{R/a_0 = 2.11}$.

R/a_0	0	1	2	3	4	5	6	7	8	9	10
S	0	0.858	1.173	1.046	0.757	0.483	0.283	0.155	0.081	0.041	0.02

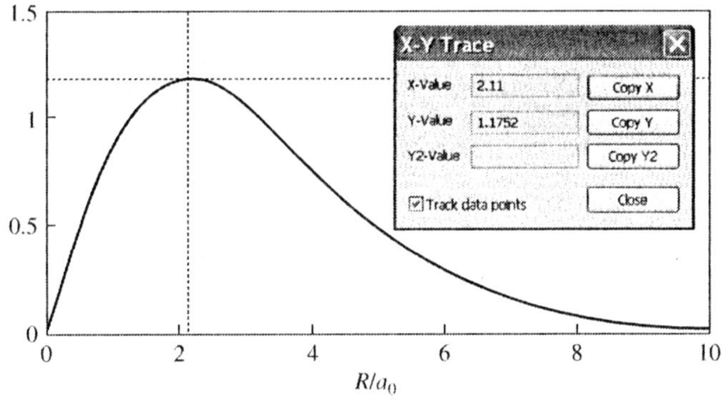

Figure 14.25

(c) We seek an orbital of the form $aA + bB$, where a and b are constants, which is orthogonal to the orbital $N(0.245A + 0.644B)$. Orthogonality implies

$$\int (aA + bB)N(0.245A + 0.644B)\,d\tau = 0$$

$$N\int [0.245aA^2 + (0.245b + 0.644a)AB + 0.644bB^2]\,d\tau = 0$$

The integrals of squares of orbitals are 1 and the integral $\int AB\,d\tau$ is the overlap integral S, so

$$0 = (0.245 + 0.644\,S)a + (0.245\,S + 0.644)b \qquad \text{so } a = -\frac{0.245\,S + 0.644}{0.245 + 0.644\,S}b$$

This would make the orbitals orthogonal, but not necessarily normalized. If $S=0$, the expression simplifies to

$$a=-\frac{0.644}{0.245}b$$

and the new orbital would be normalized if $a=0.644N$ and $b=-0.245N$. That is

$$\boxed{N(0.644A-0.245B)}$$

(d) Assuming that ψ_{cov} and ψ_{ion} are individually normalized and writing $S=\int\psi_{\text{cov}}\psi_{\text{ion}}\mathrm{d}\tau$:

$$\begin{aligned}\int\psi^2\mathrm{d}\tau &= N^2\int(\psi_{\text{cov}}+\lambda\psi_{\text{ion}})^2\mathrm{d}\tau=1\\ &= N^2\int(\psi_{\text{cov}}^2+\lambda^2\psi_{\text{ion}}^2+2\lambda\psi_{\text{cov}}\psi_{\text{ion}})\mathrm{d}\tau=1\\ &= N^2(1+\lambda^2+2\lambda S)\end{aligned}$$

This is normalized to 1 when $\boxed{N=\left(\frac{1}{1+2\lambda S+\lambda^2}\right)^{1/2}}$.

15 Molecular interactions

Answers to discussion questions

D15.1 Molecules with a permanent separation of electric charge have a **permanent dipole moment** μ. In molecules containing atoms of differing electronegativity, the bonding electrons may be displaced in such a way as to produce a net separation of charge in the molecule. Separation of charge may also arise from a difference in atomic radii of the bonded atoms. The separation of charges in the bonds is usually, though not always, in the direction of the more electronegative atom but depends on the precise bonding situation in the molecule as described in Section 15.2. A heteronuclear diatomic molecule necessarily has a dipole moment if there is a difference in electronegativity between the atoms, but the situation in polyatomic molecules is more complex. A polyatomic molecule has a permanent dipole moment only if at least one of its μ_x, μ_y, μ_z components is non-zero (see Eq. 15.4a). Thus, the tetrahedral CCl_4 molecule has polar bonds but the sum of the polar components balance so as to cancel and give $\mu = 0$. Molecular symmetry is reduced in $CHCl_3$, a molecule that has a permanent dipole moment because the C—H bond does not balance the polarity of the C—Cl bonds. Similarly, 1,4-dichlorobenzene is nonpolar while 1,2-dichlorobenzene is a polar molecule.

Question. Why does a molecule of trans-1*e*,4*e*-dichlorocyclohexane have no permanent dipole while a molecule of cis-1*a*,4*e*-dichlorocyclohexane does have a permanent dipole?

See the discussion of ozone and carbon dioxide in Section 15.2 for further examples of the importance of molecular symmetry and electronegativities in deciding whether a polyatomic molecule is polar or not. The discussion of carbon monoxide is a very important example of a molecule that has a dipole moment in the opposite direction to that expected from electronegativity considerations alone because the polarity of the HOMO antibonding orbital, which is reversed from the electronegativity expectation, provides a large contribution to the observed polarity.

Both nonpolar and polar molecules may acquire a temporary **induced dipole moment** μ^* as a result of the influence of an **electric field** $\mathcal{E}$ generated by a nearby ion or polar molecule. The field distorts the electron distribution of the molecule, and gives rise to an electric dipole. The induced dipole moment is proportional to the field [15.8] and the constant of proportionality is called the **polarizability** α. See Section 15.4 for a bullet list discussion of the molecular structure features that affect polarizabilty. They include: molecular size, nuclear control, ionization energy, and the relative orientation of the molecule with the external electric field.

D15.2 The progression from a $1/r$ separation dependence for the interaction energy between two point charges (the Coulomb potential) to a more rapidly declining $1/r^6$ dependence observed for many attractive interactions between molecules is a very important feature of molecular science. Theoretical analysis, beginning with the Coulomb potential, concludes that the potential energy of dipole μ_1 in the presence of a point charge Q_2 is given by

$$V = -\frac{\mu_1 Q_2 \cos\theta}{4\pi\varepsilon_0 r^2} \quad [15.5b]$$

From the viewpoint of the point charge, the partial charges of the dipole seem to merge and cancel as the distance r increases, thereby, causing the observed potential to vary as $1/r^2$, a more rapid decline than the Coulomb $1/r$ potential.

The theoretical analysis of the interaction energy between two dipoles μ_1 and μ_2 in a stationary orientation, which is not time-averaged as the molecules rotate, is given by

$$V = -\frac{\mu_1\mu_2\left(1-3\cos^2\theta\right)}{4\pi\varepsilon_0 r^3} \quad [15.6]$$

The potential energy decreases even more rapidly (as $1/r^3$ instead of $1/r^2$) because the charges of both dipoles seem to merge as the separation of the dipoles increases.

When a dipole rotates under the influence of a neighboring dipole (either permanent or induced), lower energy orientations are favoured and the average orientation yields an average interaction potential given by the Keesom interaction:

$$V_{\text{mean}} = -\frac{2\mu_1^2\mu_2^2}{3\left(4\pi\varepsilon_0\right)^2 kTr^6} \quad [15.7]$$

The $1/r^6$ dependence originates from the product of the dipole-dipole potential energy of interaction that goes as $1/r^3$ and a weighting factor for favoured orientations that also goes as $1/r^3$ (because it depends upon the dipole-dipole energy). Additional van der Waals type interactions that depend upon distance as $1/r^6$ include the permanent-dipole–induced-dipole-interaction, and the induced-dipole–induced-dipole, or London dispersion, interaction. In each case, we can visualize the distance dependence of the potential energy as arising from the $1/r^3$ dependence of the orienting field (and hence the magnitude of the induced dipole) and the $1/r^3$ dependence of the potential energy of interaction of the dipoles (either permanent or induced).

D15.3 Molecular energy interactions that are proportional to the inverse sixth power of the separation are called **van der Waals interactions**. They include the rotating permanent-dipole-permanent-dipole interaction, the permanent-dipole–induced-dipole-interaction, and the induced-dipole–induced-dipole, or London dispersion, interaction. Each depends upon the structure of the molecule through the questions of whether the molecule has a dipole and whether the electronic structure is polarizable. See Discussion question 15.1 for some molecular structure considerations.

D15.4 A **hydrogen bond** ($\cdots$) is an attractive interaction between two species that arises from a link of the form A—H$\cdots$B, where A and B are highly electronegative elements (usually nitrogen, oxygen, or fluorine) and B possesses a lone pair of electrons. It is a contact-like attraction that requires AH to touch B. Experimental evidences supports a linear or near-linear structural arrangement and a bond strength of about 20 kJ mol^{-1}. The hydrogen bond strength is considerably weaker than a covalent bond but it is larger than, and dominates, other intermolecular attractions such as dipole-dipole attractions. Its formation can be understood in terms of either the (a) electrostatic interaction model or with (b) molecular orbital calculations (Ch. 14).

(a) A and B, being highly electronegative, are viewed as having partial negative charges (δ^-) in the electrostatic interaction model of the hydrogen bond. Hydrogen, being less electronegative than A, is viewed as having a partial positive (δ^+). The linear structure maximizes the electrostatic attraction between H and B:

$$\overset{\delta^-}{\text{A}}\text{——}\overset{\delta^+}{\text{H}}\text{---------:}\overset{\delta^-}{\text{B}}$$

This model is conceptually very useful. However, it is impossible to exactly calculate the interaction strength with this model because the partial atomic charges cannot be precisely defined. There is no way to define which fraction of the electrons of the AB covalent bond should be assigned to one or the other nucleus.

(b) *Ab initio* MO quantum calculations are needed in order to explore questions about the linear structure, the role of the lone pair, the shape of the potential energy surface, and the extent to which the hydrogen bond has covalent sigma bond character. Yes, the hydrogen bond appears to have some sigma bond character. This was initially suggested by Linus Pauling in the 1930's and more recent experiments with Compton scattering of x-rays and NMR techniques indicate that the covalent character may provide as much as 20% of the hydrogen bond strength. A three-center molecular orbital model provides a degree of insight. A linear combination of an appropriate sigma orbital on A, the 1s hydrogen orbital, and an appropriate orbital for the lone pair on B yields a total of three molecular orbitals:

$$\psi = c_1\psi_A + c_2\psi_H + c_3\psi_B$$

One of the MOs is bonding, one is almost nonbonding, and the third is antibonding (see text Figure 15.7). Both the bonding MO and the almost nonbonding orbital are occupied by two electrons (the sigma bonding electrons of A—H and the lone pair of B). The antibonding MO is empty. Thus, depending on the precise location of the almost nonbonding orbital, the nonbonding orbital may lower the total energy and account for the hydrogen bond.

D15.5 The increase in entropy of a solution when hydrophobic molecules or groups in molecules cluster together and reduce their structural demands on the solvent (water) is the origin of the hydrophobic interaction that tends to stabilize clustering of hydrophobic groups in solution. A manifestation of the hydrophobic interaction is the clustering together of hydrophobic groups in biological macromolecules. For example, the side chains of amino acids that are used to form the polypeptide chains of proteins are hydrophobic, and the hydrophobic interaction is a major contributor to the tertiary structure of polypeptides. At first thought, this clustering would seem to be a nonspontaneous process as the clustering of the solute results in a decrease in entropy of the solute. However, the clustering of the solute results in greater freedom of movement of the solvent molecules and an accompanying increase in disorder and entropy of the solvent. The total entropy of the system has increased and the process is spontaneous.

D15.6 In the Monte Carlo method, the particles in the box are moved through small but otherwise random distances, and the change in total potential energy of the N particles in the box, ΔV_N, is calculated using one of the intermolecular potentials discussed in this chapter. Whether this new configuration is accepted is then judged from the following rules:

1. If the potential energy is not greater than before the change, then the configuration is accepted.
2. If the potential energy is greater than before the change, the Boltzmann factor $e^{-\Delta V_N/kT}$ is compared with a random number between 0 and 1; if the factor is larger than the random number, the configuration is accepted; if the factor is not larger, the configuration is rejected. This procedure ensures that at equilibrium the probability of occurrence of any configuration is proportional to the Boltzmann factor.

In the molecular dynamics approach, the history of an initial arrangement is followed by calculating the trajectories of all the particles under the influence of the intermolecular potentials. Newton's laws are used to predict where each particle will be after a short time interval (about 1 fs which is shorter than the average time between collisions), and then the calculation is repeated for tens of thousands of such steps. The time-consuming part of the calculation is the evaluation of the net force on the molecule arising from all the other molecules present in the system. The calculation gives a series of snapshots of the liquid.

Solutions to exercises

E15.1 $\chi(\mathrm{H}) = 2.1 \qquad \chi(\mathrm{F}) = 4.0 \qquad \Delta\chi = 1.9$

We use

$$\mu = \Delta\chi\ \mathrm{D}\ [15.2] = \boxed{1.9\ \mathrm{D}}$$

$$\mu = 1.9 \times (3.33564\times10^{-30}\ \mathrm{C\ m}) = \boxed{6.3\times10^{-30}\ \mathrm{C\ m}}$$

The experimental value is 1.82 D.

E15.2 PCl_5

Lewis structure

(lone pair on the Cl atoms are not shown; there is no lone pair on the P atom)

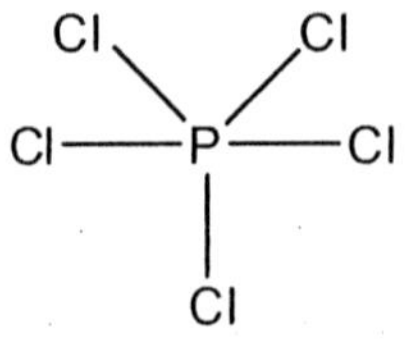

Orientations caused by repulsions between five bonding pair on P atom (no lone pair)

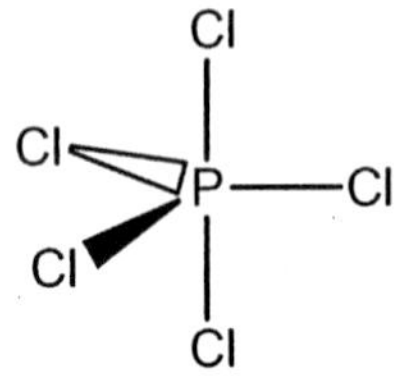

Molecular shape: trigonal bipyramidal with equatorial bond angles of 120° and axial bond angles of 90°. The polarity of the axial P—Cl bonds cancel as they point in opposite directions. Also, the vector sum of the three equatorial bond polarities cancels to zero. Thus, the molecule is $\boxed{\text{nonpolar}}$.

E15.3 Refer to diagram **4** of the text and Eqn 15.3. Here $\mu_1 = \mu_2 = 0.40$ D.

$$\mu_{\text{res}} = (\mu_1^2 + \mu_2^2 + 2\mu_1\mu_2 \cos\theta)^{1/2} = (2\mu_1^2 + 2\mu_1^2 \cos\theta)^{1/2} = \sqrt{2}(1+\cos\theta)^{1/2}\mu_1$$

(a) 1,2-dimethylbenzene, $\theta = 60° = \pi/3$ radians

$$\mu_{\text{res}} = \sqrt{2}\,(1+\cos 60°)^{1/2}\mu_1 = \sqrt{3}\,\mu_1 = \sqrt{3}\times 0.4\ \text{D} = \boxed{0.7\ \text{D}}$$

(b) 1,3-dimethylbenzene, $\theta = 120° = 2\pi/3$ radians

$$\mu_{\text{res}} = \sqrt{2}\,(1+\cos 120°)^{1/2}\mu_1 = \mu_1 = \boxed{0.4\ \text{D}}$$

(c) 1,4-dimethylbenzene, $\theta = 180° = \pi$ radians

$\mu_{\text{res}} = \boxed{0}$. The polarities of all equivalent bonds are opposed and cancel.

We can be sure about the answer to part (c) because by symmetry the polarities of all equivalent bonds are opposed and cancel.

E15.4

$$\mu = (\mu_1^2 + \mu_2^2 + 2\mu_1\mu_2 \cos\theta)^{1/2} \quad [17.3]$$

$$= \left\{(1.20)^2 + (0.60)^2 + 2\times 1.20\times 0.60\times(\cos 107°)\right\}^{1/2}\ \text{D} = \boxed{1.26\ \text{D}}$$

E15.5 Add the dipole moments vectorially using the dipole moments calculated in Exercise 15.3.

(a) 1,2,3-trimethylbenzene, $\mu_{\text{res}} = m\text{-xylene} + \text{toluene} = 0.4\ \text{D} + 0.4\ \text{D} = \boxed{0.8\ \text{D}}$

(b) 1,2,4-trimethylbenzene, $\mu_{\text{res}} = p\text{-xylene} + \text{toluene} = 0 + 0.4\ \text{D} = \boxed{0.4\ \text{D}}$

(c) 1,3,5-trimethylbenzene, $\mu_{\text{res}} = m\text{-xylene} - \text{toluene} = 0.4\ \text{D} - 0.4\ \text{D} = \boxed{0}$

We can be sure about the answer to part (c) because by symmetry the polarities of all equivalent bonds are opposed and cancel.

E15.6 Place the z axis along the C—C bond of the 1,2-dichloroethane molecule. Place the x axis along the horizontal line of staggered conformations (**13**), (**14**), and (**15**); place the y axis along the vertical line of these Newman projections. The z components of each conformation cancel by symmetry so we can ignore this component. It is a helpful simplification to analyze the dipole moment x and y

components (μ_x and μ_y) of each conformation in units of the maximum value of the bond dipole projection upon the y axis, $\mu_{y,\text{max}}$. Assume that the carbon atoms are sp^3 hybridized with 109.5° bond angles and assume that the "bond dipole" (1.50 D) is along the C—Cl bond, then, as shown in Figure 15.1, $\mu_{y,\text{max}}$ is given by

$$\mu_{y,\text{max}} = \mu_{\text{bond}} \sin(180° - 109.5°) = (1.50\text{ D}) \times \sin(180° - 109.5°) = 1.414\text{ D}$$

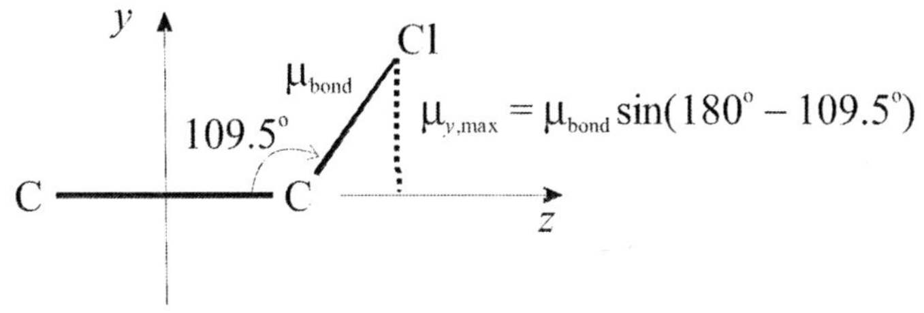

Figure 15.1

This means that, if the Newman projection shows the C—Cl bond to be at the angle φ from the y axis in the xy plane of the projection, the x and y components of the bond dipole are given by

$$\mu_{x,\text{bond}} = \mu_{y,\text{max}} \sin(\varphi) \quad \text{and} \quad \mu_{y,\text{bond}} = \mu_{y,\text{max}} \cos(\varphi)$$

The Newman projections show the C—Cl bond to be at the angles $\varphi = -60°$, 0°, 60°, and 180°. There are two such bonds in each conformation and addition of the x and y components of the two bond dipoles gives the x and y components of the molecular dipole moment.

Conformation (**13**)

$$\mu_x = \mu_{y,\text{max}} \sin(0°) + \mu_{y,\text{max}} \sin(180°) = 0$$

$$\mu_y = \mu_{y,\text{max}} \cos(0°) + \mu_{y,\text{max}} \cos(180°) = 0$$

$$\boldsymbol{\mu} = \mathbf{0}$$

Conformation (**14**)

$$\mu_x = \mu_{y,\text{max}} \sin(0°) + \mu_{y,\text{max}} \sin(60°) = \sqrt{3}/2 \times \mu_{y,\text{max}}$$

$$\mu_y = \mu_{y,\text{max}} \cos(0°) + \mu_{y,\text{max}} \cos(60°) = 3/2 \times \mu_{y,\text{max}}$$

$$\boldsymbol{\mu} = \sqrt{3}/2 \times \mu_{y,\text{max}}\mathbf{i} + 3/2 \times \mu_{y,\text{max}}\mathbf{j}$$

$$\mu = (\boldsymbol{\mu}\cdot\boldsymbol{\mu})^{1/2} = \left\{\left(\sqrt{3}/2 \times \mu_{y,\text{max}}\right)^2 + \left(3/2 \times \mu_{y,\text{max}}\right)^2\right\}^{1/2} = \sqrt{3}\mu_{y,\text{max}}$$

Conformation (**15**)

$$\mu_x = \mu_{y,\text{max}} \sin(0°) + \mu_{y,\text{max}} \sin(-60°) = -\sqrt{3}/2 \times \mu_{y,\text{max}}$$

$$\mu_y = \mu_{y,\text{max}} \cos(0°) + \mu_{y,\text{max}} \cos(-60°) = 3/2 \times \mu_{y,\text{max}}$$

$$\boldsymbol{\mu} = -\sqrt{3}/2 \times \mu_{y,\text{max}}\mathbf{i} + 3/2 \times \mu_{y,\text{max}}\mathbf{j}$$

$$\mu = (\boldsymbol{\mu}\cdot\boldsymbol{\mu})^{1/2} = \left\{\left(\sqrt{3}/2 \times \mu_{y,\text{max}}\right)^2 + \left(3/2 \times \mu_{y,\text{max}}\right)^2\right\}^{1/2} = \sqrt{3}\mu_{y,\text{max}}$$

(a) The probabilities of confirmations (**13**), (**14**), and (**15**) are ⅓, ⅓, and ⅓, respectively. Conformation (**13**) has no dipole moment so it can be ignored.

$$\mu_x = \tfrac{1}{3}\{\mu_x(\mathbf{14}) + \mu_x(\mathbf{15})\} = 0$$

$$\mu_y = \tfrac{1}{3}\{\mu_y(\mathbf{14}) + \mu_y(\mathbf{15})\} = \tfrac{1}{3}\left\{3/2 \times \mu_{y,\text{max}} + 3/2 \times \mu_{y,\text{max}}\right\} = \mu_{y,\text{max}}$$

$$\mu_{\text{mean}} = \{\mu_x^2 + \mu_y^2\}^{1/2} = \mu_{y,\text{max}} = \boxed{1.414\text{D}}$$

(b) The probabilities of confirmations (**13**), (**14**), and (**15**) are 0, 1, and 0, respectively.

$$\mu_x = \mu_x(\mathbf{14}) = \sqrt{3}/2\times\mu_{y,\max}$$

$$\mu_y = \mu_y(\mathbf{14}) = 3/2\times\mu_{y,\max}$$

$$\mu_{\text{mean}} = \{\mu_x^2 + \mu_y^2\}^{1/2} = \sqrt{3}\mu_{y,\max} = \boxed{2.45\text{ D}}$$

(c) The probabilities of confirmations (**13**), (**14**), and (**15**) are ½, ¼, and ¼, respectively.

$$\mu_x = \tfrac{1}{4}\{\mu_x(\mathbf{14}) + \mu_x(\mathbf{15})\} = 0$$

$$\mu_y = \tfrac{1}{4}\{\mu_y(\mathbf{14}) + \mu_y(\mathbf{15})\} = \tfrac{1}{4}\{3/2\times\mu_{y,\max} + 3/2\times\mu_{y,\max}\} = \tfrac{3}{4}\mu_{y,\max}$$

$$\mu_{\text{mean}} = \{\mu_x^2 + \mu_y^2\}^{1/2} = \tfrac{3}{4}\mu_{y,\max} = \boxed{1.06\text{ D}}$$

(d) The probabilities of confirmations (**13**), (**14**), and (**15**) are $\tfrac{1}{5}$, $\tfrac{2}{5}$, and $\tfrac{2}{5}$, respectively.

$$\mu_x = \tfrac{2}{5}\{\mu_x(\mathbf{14}) + \mu_x(\mathbf{15})\} = 0$$

$$\mu_y = \tfrac{2}{5}\{\mu_y(\mathbf{14}) + \mu_y(\mathbf{15})\} = \tfrac{2}{5}\{3/2\times\mu_{y,\max} + 3/2\times\mu_{y,\max}\} = \tfrac{6}{5}\mu_{y,\max}$$

$$\mu_{\text{mean}} = \{\mu_x^2 + \mu_y^2\}^{1/2} = \tfrac{6}{5}\mu_{y,\max} = \boxed{1.70\text{ D}}$$

E15.7 The dipole moment is the vector sum (see Fig. 15.2)

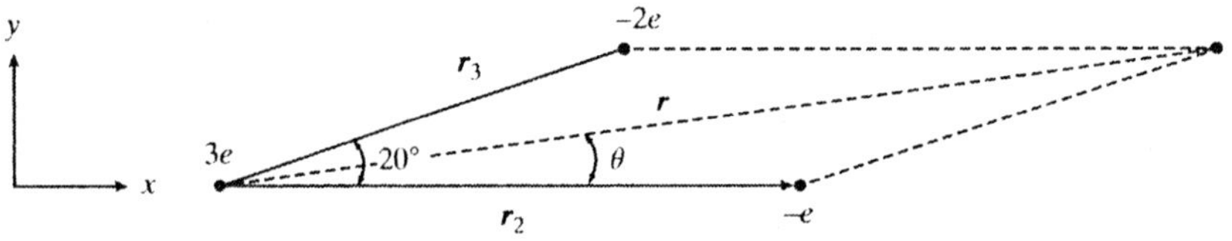

Figure 15.2

$$\mu = \sum_i Q_i r_i = 3e(0) - er_2 - 2er_3 = -er_2 - 2er_3$$

$$r_2 = \mathbf{i}x_2, \qquad r_3 = \mathbf{i}x_3 + \mathbf{j}y_3$$

$$x_2 = +0.32\text{ nm}$$

$$x_3 = r_3\cos 20° = (+0.23\text{ nm})\times(0.940) = 0.21\bar{6}\text{ nm}$$

$$y_3 = r_3\sin 20° = (+0.23\text{ nm})\times(0.342) = 0.078\bar{7}\text{ nm}$$

The components of the vector resultant are the sums of the components. That is (with all distances in nm)

$$\mu_x = -ex_2 - 2ex_3 = -(e)\times\{(0.32) + (2)\times(0.21\bar{6})\} = -(e)\times(0.752\text{ nm})$$

$$\mu_y = -2ey_3 = -(e)\times(2)\times(0.078\bar{7}) = -(e)\times(0.1574\text{ nm})$$

$$\mu = (\mu_x^2 + \mu_y^2)^{1/2} = (e)\times(0.76\bar{8}\text{ nm}) = (1.602\times10^{-19}\text{ C})\times(0.76\bar{8}\times10^{-9}\text{ m})$$

$$= 1.2\bar{3}\times10^{-28}\text{ C m} = \boxed{37\text{D}}$$

The angle that μ makes with the x-axis is given by

$$\cos\theta = \frac{|\mu_x|}{\mu} = \frac{0.752}{0.768}; \quad \boxed{\theta = 11.7°}$$

E15.8 This solution has the O—H oxygen atom at the position (49, −107, 88) while the nitrogen atom is at the position (−101, −11, −126). Other positions, which we assume to be in the picometre unit, are shown in text (**16**).

$$\begin{aligned}\mu_x &= 0.02e\times(-86\text{ pm})+0.02e\times(34\text{ pm})+0.06e\times(-195\text{ pm})\\&\quad+0.18e\times(-199\text{ pm})-0.36e\times(-101\text{ pm})+0.45e\times(82\text{ pm})\\&\quad-0.38e\times(199\text{ pm})+0.18e\times(-80\text{ pm})-0.38e\times(49\text{ pm})\\&\quad+0.42e\times(129\text{ pm})\\&= -29.76e\text{ pm}\end{aligned}$$

$$\begin{aligned}\mu_y &= 0.02e\times(118\text{ pm})+0.02e\times(146\text{ pm})+0.06e\times(70\text{ pm})\\&\quad+0.18e\times(-1\text{ pm})-0.36e\times(-11\text{ pm})+0.45e\times(-15\text{ pm})\\&\quad-0.38e\times(16\text{ pm})+0.18e\times(-110\text{ pm})-0.38e\times(-107\text{ pm})\\&\quad+0.42e\times(-146\text{ pm})\\&= -40.03e\text{ pm}\end{aligned}$$

$$\begin{aligned}\mu_z &= 0.02e\times(37\text{ pm})+0.02e\times(-98\text{ pm})+0.06e\times(-38\text{ pm})\\&\quad+0.18e\times(-100\text{ pm})-0.36e\times(-126\text{ pm})+0.45e\times(34\text{ pm})\\&\quad-0.38e\times(-38\text{ pm})+0.18e\times(-111\text{ pm})-0.38e\times(88\text{ pm})\\&\quad+0.42e\times(126\text{ pm})\\&= +53.1e\text{ pm}\end{aligned}$$

The magnitude of the dipole moment is given by

$$\begin{aligned}\mu &= \left(\mu_x^2+\mu_y^2+\mu_z^2\right)^{1/2} \qquad [15.4\text{a}]\\&= \left[(-29.76)^2+(-40.03)^2+(53.1)^2\right]^{1/2} e\text{ pm}\\&= 72.8\overline{5}\,e\text{ pm}\\&= \left(72.8\overline{5}\right)\times\left(1.6022\times10^{-19}\text{ C}\right)\times\left(10^{-12}\text{ m}\right)\\&= 1.17\times10^{-29}\text{ C m}\\&= \boxed{3.50\text{ D}}\end{aligned}$$

E15.9 First, determine the dipole moment of the OH fragment, $\mu_{\text{O-H}}$, of the H_2O molecule by considering the total dipole moment of the molecule, $\mu_{\text{H-O-H}} = 1.85$ D, to be the resultant of the dipoles of two identical OH fragments at an angle θ equal to 104.5° with respect to each other. This is the bond angle in water. Use Eqn 15.3 with $\mu_1 = \mu_2 = \mu_{\text{O-H}}$. When this is the case, eqn 15.3 simplifies to

$$\mu_{\text{H-O-H}} = \left(2\mu_{\text{O-H}}{}^2 + 2\mu_{\text{O-H}}{}^2\cos\theta\right)^{1/2} = \sqrt{2}\mu_{\text{O-H}}\left(1+\cos\theta\right)^{1/2} = 2\mu_{\text{O-H}}\cos(\theta/2)$$

$$\mu_{\text{O-H}} = \frac{\mu_{\text{H-O-H}}}{2\cos(\theta/2)} = \frac{1.85\text{ D}}{2\cos(52.25°)} = 1.51\text{ D}$$

Then, using μ to represent the angle between the two O—H bond dipoles in H_2O_2, we have

$$\mu_{\text{H-O-O-H}} = \boxed{2\mu_{\text{O-H}}\cos(\phi/2)}$$

(a) $\mu_{\text{H-O-O-H}}$ is plotted as function of ϕ in the Figure 15.3. At 90°, $\mu_{\text{H-O-O-H}}$ is $\boxed{2.13\text{ D}}$, which is the experimental value.

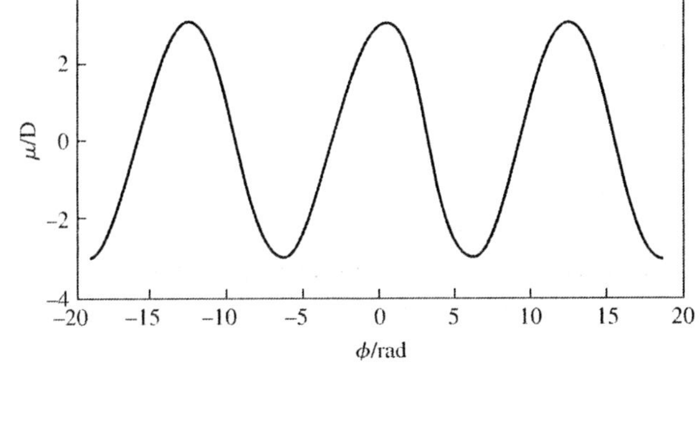

Figure 15.3

(b) The angle can be related to the dipole moment as follows

$$\phi = 2\arccos(\mu_{\text{H-O-O-H}}/2\mu_{\text{O-H}}) = \boxed{2\arccos(\mu_{\text{H-O-O-H}}/3.02\ \text{D})}$$

E15.10 We assume that the dipole of the water molecule and the Li^+ ion are collinear and that the separation of charges in the dipole is smaller than the distance to the ion. With these assumptions we can use eqn 1.5a of the text. To flip the water molecule over requires twice the energy of interaction given by eqn 15.5a.

$$E = \left(-\frac{Q_2(-\mu_1)}{4\pi\varepsilon_0 r^2}\right) - \left(-\frac{Q_2(+\mu_1)}{4\pi\varepsilon_0 r^2}\right) = \frac{2Q_2\mu_1}{4\pi\varepsilon_0 r^2} \qquad Q_2 = e = 1.602\times10^{-19}\,\text{C}$$

$$= \frac{2\times(1.602\times10^{-19}\ \text{C})\times(1.85\ \text{D})\times(3.336\times10^{-30}\,\text{C m D}^{-1})}{(4\pi\times8.854\times10^{-12}\,\text{J}^{-1}\ \text{C}^2\ \text{m}^{-1})\times r^2}$$

$$= \frac{1.777\times10^{-38}\,\text{J m}^2}{r^2}$$

(a) $$E = \frac{1.777\times10^{-38}\,\text{J m}^2}{(150\times10^{-12}\,\text{m})^2} = 7.90\times10^{-19}\,\text{J}$$

$$\text{molar energy} = N_A \times E = \boxed{476\ \text{kJ mol}^{-1}}$$

(b) $$E = \frac{1.777\times10^{-38}\,\text{J m}^2}{(350\times10^{-12}\,\text{m})^2} = 1.45\times10^{-19}\,\text{J}$$

$$\text{molar energy} = N_A \times E = \boxed{87.4\ \text{kJ mol}^{-1}}$$

E15.11 Figure 15.4 defines symbols for distances.

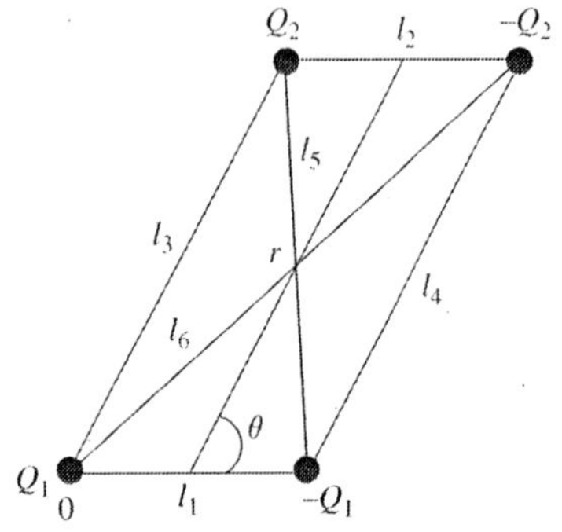

Figure 15.4

To simplify the analysis let us assume that $l_1 = l_2 = l$. Then $l_3 = l_4 = r$. Take the origin to be 0. Then the coordinates of the charges Q_2, $-Q_2$, Q_1, and $-Q_1$ are

Q_2: $x = r\cos\theta,\ y = r\sin\theta$

$-Q_2$: $x = r\cos\theta + l,\ y = r\sin\theta$

Q_1: $x = 0,\ y = 0$

$-Q_1$: $x = l,\ y = 0$

Then $l_6 = [(r\cos\theta + l)^2 + (r\sin\theta)^2]^{1/2}$

$l_5 = [(r\cos\theta - l)^2 + (r\sin\theta)^2]^{1/2}$

Let $\lambda = l/r$, then $l = \lambda r$ and we can write

$$(r\cos\theta + l)^2 = (r\cos\theta + \lambda r)^2 = r^2(\cos\theta + \lambda)^2$$
$$(r\cos\theta - l)^2 = (r\cos\theta - \lambda r)^2 = r^2(\cos\theta - \lambda)^2$$
$$l_6 = [r^2(\cos\theta + \lambda)^2 + r^2\sin^2\theta]^{1/2} = r\,[(\cos\theta + \lambda)^2 + \sin^2\theta]^{1/2}$$
$$= r\,[1 + 2\lambda\cos\theta + \lambda^2]^{1/2}$$
$$l_5 = r\left(1 - 2\lambda\cos\theta + \lambda^2\right)^{1/2}$$

Let $4\pi\varepsilon_0 = K$, then the coulomb interactions are

$$V = \frac{Q_1Q_2}{Kr} + \frac{Q_1Q_2}{Kr} - \frac{Q_1Q_2}{Kr\left(1 - 2\lambda\cos\theta + \lambda^2\right)^{1/2}} - \frac{Q_1Q_2}{Kr\left(1 + 2\lambda\cos\theta + \lambda^2\right)^{1/2}}$$

$$= -\frac{Q_1Q_2}{Kr}\left(-2 + \frac{1}{\left(1 - 2\lambda\cos\theta + \lambda^2\right)^{1/2}} + \frac{1}{\left(1 + 2\lambda\cos\theta + \lambda^2\right)^{1/2}}\right)$$

We will assume that $2\lambda\cos\theta + \lambda^2 \ll 1$, then we can expand the denominator in a Taylor's series expansion.

$$\frac{1}{\left(1 + 2(\cos\theta)\lambda + \lambda^2\right)^{1/2}} = 1 - (\cos\theta)\lambda + \left(\frac{3\cos^2\theta}{2} - \frac{1}{2}\right)\lambda^2 + \cdots$$

$$\frac{1}{\left(1 - 2(\cos\theta)\lambda + \lambda^2\right)^{1/2}} = 1 + (\cos\theta)\lambda + \left(\frac{3\cos^2\theta}{2} - \frac{1}{2}\right)\lambda^2 + \cdots$$

$$V = \frac{-Q_1Q_2}{Kr}\left[-2 + 1 + (\cos\theta)\lambda + \left(\frac{3\cos^2\theta - 1}{2}\right)\lambda^2 + 1 - (\cos\theta)\lambda + \left(\frac{3\cos^2\theta - 1}{2}\right)\lambda^2\right]$$

$$V = \frac{-Q_1Q_2}{Kr}\left[\left(3\cos^2\theta - 1\right)\lambda^2\right] = -\frac{Q_1Q_2}{Kr}\left[\left(3\cos^2\theta - 1\right)\left(\frac{l}{r}\right)^2\right]$$

Now use $\mu_1 = Q_1 l$, $\mu_2 = Q_2 l$, $K = 4\pi\varepsilon_0$, and rearrange to get

$$V = \frac{\mu_1\mu_2\left(1 - 3\cos^2\theta\right)}{4\pi\varepsilon_0 r^3} \quad \text{Q.E.D.}$$

E15.12 (a) The kinetic energy per mole is given by

$$E = \tfrac{3}{2}RT = \boxed{3.7\ \text{kJ mol}^{-1}} \text{ at 298 K}$$

(b) For 0.50 mol of molecules in a 1.0 dm^3 volume, the volume occupied per molecule is on average

$$v = \frac{2.00\times10^{-3}\ \text{m}^3\ \text{mol}^{-1}}{6.022\times10^{23}\ \text{mol}^{-1}} = 3.32\times10^{-27}\ \text{m}^3$$

This places the molecules at an average distance of $r = v^{1/3}$ with respect to each other.

$$r = (3.32\times10^{-27}\ \text{m}^3)^{1/3} = 1.49\times10^{-9}\text{m} = 1.49\ \text{nm}$$

Then, the average potential energy of interaction per pair of HCl molecules ($\mu = 1.08$ D, Table 15.2) that rotate is given by the Keesom interaction of eqn 15.7.

$$V = -\frac{2\mu_1^{\,2}\mu_2^{\,2}}{3(4\pi\varepsilon_0)^2\, kTr^6} \quad [15.7]$$

$$= -\frac{2\times(1.08\text{ D})^4\times(3.336\times10^{-30}\text{ C m D}^{-1})^4}{3\times(4\pi\times8.854\times10^{-12}\text{ J}^{-1}\text{ C}^2\text{ m}^{-1})^2\times(1.381\times10^{-23}\text{ J K}^{-1})\times(298\text{ K})\times(1.49\times0^{-9}\text{m})^6}$$

$$= -2.02\times10^{-25}\text{J}$$

Each molecule has an average of 6 nearest neighbors, but we count a pair interaction only once. Then the total potential energy per mole in this sample is

$$V = -3\times N_A\times(2.02\times10^{-25}\text{ J}) = \boxed{-0.365\text{ J mol}^{-1}}$$

This potential energy is exceedingly small compared to 3.7 kJ mol^{-1}, so the kinetic theory of gases is justifiable for this sample.

E15.13 (a) Polarizability, $\alpha = \mu^*/\mathcal{E}$ [15.8] where μ^*, the induced dipole moment has the SI unit "C m". The electric field strength, $\mathcal{E}$, is the force per unit charge experienced by a charge. It has the unit "N C^{-1}" or "J C^{-1} m^{-1}". Consequently, polarizability has the SI unit

$$(\text{C m}) / (\text{J C}^{-1}\text{ m}^{-1}) = \text{C}^2\text{ m}^2\text{ J}^{-1}.$$

(b) Polarizability volume, $\alpha' = \alpha / 4\pi\varepsilon_0$ [15.9] where α has the SI unit "C^2 m^2 J^{-1}" and ε_0 has the "C^2 J^{-1} m^{-1}" unit. Consequently, polarizability volume has the SI unit

$$(\text{C}^2\text{ m}^2\text{ J}^{-1}) / (\text{C}^2\text{ J}^{-1}\text{ m}^{-1}) = \text{m}^3.$$

E15.14

$$\mu^* = \alpha\mathcal{E}\ [15.8] = 4\pi\varepsilon_0\alpha'\mathcal{E}\ [15.9] = 4\pi\varepsilon_0\alpha'\left(\frac{e}{4\pi\varepsilon_0 r^2}\right) = 1.85\text{ D}\quad [Q = e\text{ for a proton}]$$

$$\frac{\alpha' e}{r^2} = 1.85\text{ D}$$

Solve for r,

$$r = \left(\frac{\alpha' e}{1.85\text{ D}}\right)^{1/2}$$

$$= \left(\frac{1.48\times10^{-30}\text{ m}^3\times1.602\times10^{-19}\text{ C}}{(1.85\text{ D})\times(3.336\times10^{-30}\text{ C m D}^{-1})}\right)^{1/2}$$

$$= 1.96\times10^{-10}\text{ m} = \boxed{196\text{ pm}}$$

E15.15 For argon: $\alpha'_{Ar} = 1.66\times10^{-30}\text{ m}^3$ [Table 15.2] and $I_{Ar} = 15.76\text{ eV} = 2.525\times10^{-18}\text{ J}$ [Table 13.2]

The London formula gives a reasonable estimate of the dispersion interaction. Use eq. 15.11 with $\alpha'_1 = \alpha'_2 = \alpha'_{Ar}$ and $I_1 = I_2 = I_{Ar}$.

$$V = -\tfrac{2}{3}\times\frac{\alpha'_1\alpha'_2}{r^6}\times\frac{I_1 I_2}{I_1+I_2}\ [15.11] = -\tfrac{1}{3}\times\frac{(\alpha'_{Ar})^2 I_{Ar}}{r^6}$$

$$= -\tfrac{1}{3}\times\frac{(1.66\times10^{-30}\text{ m}^3)^2(2.525\times10^{-18}\text{ J})}{(1.00\times10^{-9}\text{ m})^6} = \boxed{-2.32\times10^{-24}\text{ J}}$$

E15.16 The interaction is a dipole-induced-dipole interaction. The energy is given by eqn 15.10:

$$V = -\frac{\mu_1{}^2\alpha_2'}{4\pi\varepsilon_0 r^6} \ [15.10]$$

$$= -\frac{[(2.7\,\text{D})\times(3.336\times10^{-30}\,\text{C m D}^{-1})]^2\times(1.04\times10^{-29}\,\text{m}^3)}{4\pi(8.854\times10^{-12}\,\text{J}^{-1}\,\text{C}^2\,\text{m}^{-1})\times(4.0\times10^{-9}\,\text{m})^6} \ [\text{Table 15.2}]$$

$$V = \boxed{-1.8\times10^{-27}\,\text{J} = -1.1\times10^{-3}\,\text{J mol}^{-1}}.$$

E15.17

$$V = -\tfrac{2}{3}\times\frac{\alpha_1'\alpha_2'}{r^6}\times\frac{I_1 I_2}{I_1+I_2} \ [15.11] = -\tfrac{1}{3}\,\frac{(\alpha')^2 I}{r^6}$$

$$= -\tfrac{1}{3}\times\frac{\left(10.4\times10^{-30}\,\text{m}^3\right)^2\times(5.0\,\text{eV})}{\left(4.0\times10^{-9}\,\text{m}\right)^6}$$

$$= -4.4\times10^{-8}\,\text{eV} = 7.1\times10^{-27}\,\text{J}$$

$$= \boxed{-4.2\times10^{-3}\,\text{J mol}^{-1}}$$ The negative sign indicates that the interaction is attractive.

E15.18 The geometry at the hydrogen bond is linear (see Figure 15.5 and text Figure 15.6), that is, θ in structure (**19**) is 0°. The partial charges are as given in Table 15.1. Distances in structure (**19**) are not given, but we may take $r_{\text{O–H}}$ to be 100 pm and $R_{\text{O}\cdots\text{N}}$ to be 300 pm as a typical value.

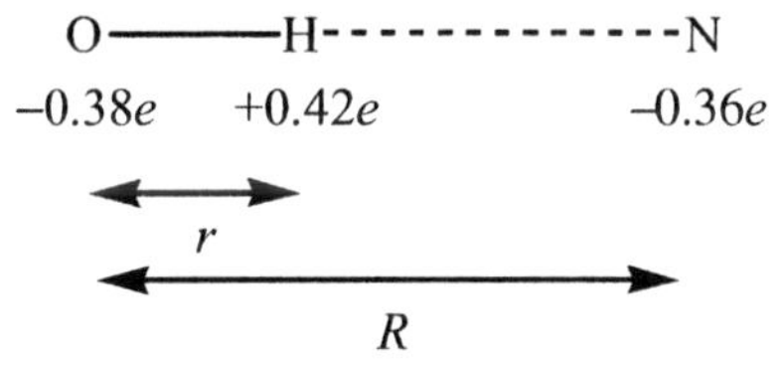

Figure 15.5

Then

$$V_{\text{O}\cdots\text{N}} = \frac{N_A\delta_O\delta_N e^2}{4\pi\varepsilon_0 R} \ [15.1\text{a}] \quad \text{and} \quad V_{\text{N}\cdots\text{H}} = \frac{N_A\delta_N\delta_H e^2}{4\pi\varepsilon_0 (R-\text{r})} \ [15.1\text{a}]$$

$$\frac{N_A e^2}{4\pi\varepsilon_0} = 1.389\times10^{-4}\,\text{J m mol}^{-1}$$

$$V_{\text{hydrogen bond}} = V_{\text{O}\cdots\text{N}} + V_{\text{N}\cdots\text{H}} = \left(1.389\times10^{-4}\,\text{J m mol}^{-1}\right)\times\left(\frac{\delta_O\delta_N}{R} + \frac{\delta_N\delta_H}{(R-r)}\right)$$

$$= \left(1.389\times10^{-4}\,\text{J m mol}^{-1}\right)\times\left(\frac{(-0.38)\times(-0.36)}{300\times10^{-12}\,\text{m}} + \frac{(-0.36)\times(+0.42)}{200\times10^{-12}\,\text{m}}\right)$$

$$= \boxed{-42\,\text{kJ mol}^{-1}}$$

E15.19 Individual acetic acid molecules have a non-zero dipole moment but the two dipoles of the dimer are exactly opposed and cancel. At low temperature a significant fraction of molecules are a part of a dimer, which means that their individual dipole moments will not be observed. However, as temperature is increased, hydrogen bonds of the dimer are broken, thereby, releasing individual molecules and the apparent dipole moment increases.

E15.20 The distances that appear in the denominator of the Coulombic interaction are shown in Figure 15.6. They are deduced with the standard formula $d = \left\{(x_1-x_2)^2 + (y_1-y_2)^2\right\}^{1/2}$ where the nth atom has the coordinate (x_n, y_n). The value of the vertical height above the C—C axis (y = 107.4 pm) is based upon the assumption that the O—C—O angle equals 120°. The magnitude of the partial charges is assumed to be identical for all relevant atoms (±δ).

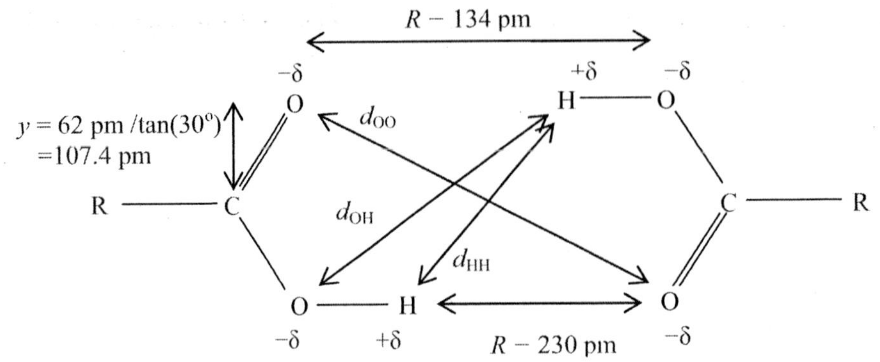

$d_{CC} = R$ $\quad d_{OO} = \left\{(R-124)^2 + 4y^2\right\}^{1/2}$ pm

$d_{OH} = \left\{(R-240)^2 + 4y^2\right\}^{1/2}$ pm $\quad d_{HH} = \left\{(R-336)^2 + 4y^2\right\}^{1/2}$ pm

Figure 15.6

Each Coulombic interaction has the form $V_{1,2} = \pm\dfrac{\delta^2}{4\pi\varepsilon_0 d_{1,2}}$ where the positive sign designates repulsion between atoms 1 and 2 and the negative designates attraction. The total potential is the sum of all such terms. It is most conveniently analyzed as the function

$$f(R) = \frac{4\pi\varepsilon_0 V_{\text{total}}}{\delta^2}$$

$$= -\frac{2}{R-230\text{ pm}} + \frac{2}{R-134\text{ pm}}$$

$$+\frac{2}{\left\{(R-124\text{ pm})^2+4y^2\right\}^{1/2}} - \frac{2}{\left\{(R-240\text{ pm})^2+4y^2\right\}^{1/2}} + \frac{1}{\left\{(R-336\text{ pm})^2+4y^2\right\}^{1/2}}$$

V_{total} becomes attractive at the value of R for which $f(R) = 0$. This value may be determined either with the numeric solve function on a scientific calculator or it may be determined graphically. The Figure 15.7 plot yields $\boxed{R = 461.2\text{ pm}}$.

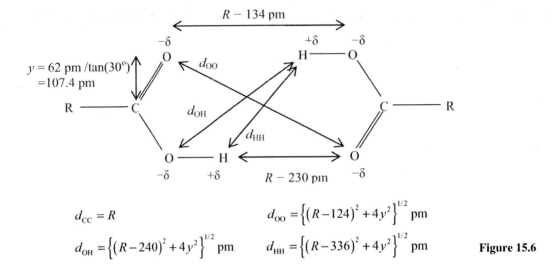

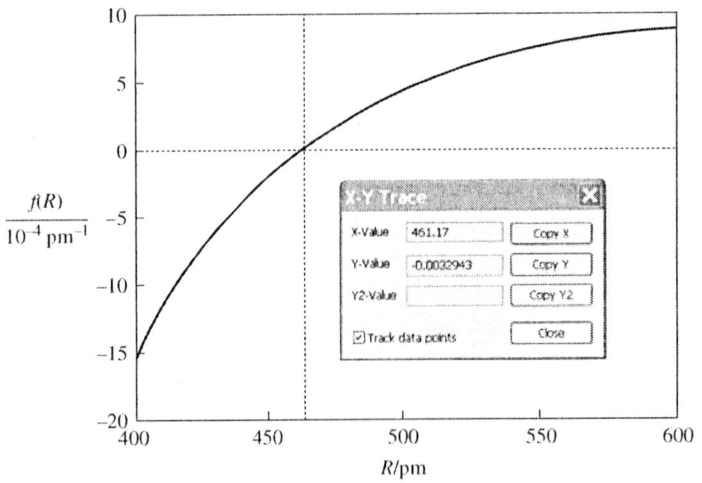

Figure 15.7

E15.21 (a) $\varphi = 0$ in the maximum potential eclipsed conformation. In the minimum potential staggered conformation, $\varphi = (2n + 1)\pi/3$ where $n = 0,1,2,3,\ldots$. For convenience, use the potential minimum that occurs at $\varphi = \pi/3$.

$$\begin{aligned}\Delta V &= V_{\text{eclipsed}} - V_{\text{staggered}} = V(0) - V(\pi/3)\\ &= \tfrac{1}{2}V_0\{1+\cos(3\times 0)\} - \tfrac{1}{2}V_0\{1+\cos(3\times\pi/3)\}\\ &= \tfrac{1}{2}V_0\{1+1\} - \tfrac{1}{2}V_0\{1+(-1)\}\\ &= \boxed{V_0}\end{aligned}$$

(b) The harmonic potential has a potential minimum at a displacement of $x = 0$ and an increasing potential that is given by $V_{\text{harmonic}} = \tfrac{1}{2}kx^2$ [12.20b] for both very small negative and positive values of displacement x. The frequency of the harmonic oscillator is given by $\nu = (2\pi)^{-1}(k/m)^{1/2}$ [12.21] where m is the particle mass. To aid the comparison with the harmonic oscillator, let $\varphi = (\chi + \pi)/3$. φ is the displacement angle from the eclipsed conformation; χ is the displacement angle from the equilibrium, staggered conformation. The potential energy of a CH_3 group in ethane is a minimum when $\chi = 0$ and by writing the potential in terms of χ we can more readily compare it to the harmonic oscillator potential.

$$\begin{aligned}V &= \tfrac{1}{2}V_0\{1+\cos(3\varphi)\} = \tfrac{1}{2}V_0\{1+\cos(\chi+\pi)\} = \tfrac{1}{2}V_0\{1+\cos(\chi)\cos(\pi)-\sin(\chi)\sin(\pi)\}\\ &= \tfrac{1}{2}V_0\{1+\cos(\chi)(-1)-\sin(\chi)\times 0\} = \tfrac{1}{2}V_0\{1-\cos(\chi)\}\end{aligned}$$

For small values of χ the function $\cos(\chi)$ may be expanded in a Taylor series around $\chi = 0$, and the series may be truncated at fourth order terms because the higher order terms are negligibly small when $\chi \ll 1$.

$$\begin{aligned}V &= \tfrac{1}{2}V_0\{1-[1-\chi^2/2!+\chi^4/4!-\chi^6/6!+\cdots]\} = \tfrac{1}{2}V_0\{1-[1-\chi^2/2]\}\\ &= \tfrac{1}{4}V_0\chi^2\end{aligned}$$

The square dependence on χ clearly shows that the torsional motion around the C—C bond of ethane has the harmonic oscillator behavior around the equilibrium azimuthal angle.

(c) We can estimate the frequency of the torsional oscillation by comparing a reasonable total energy for this motion with the total energy and frequency of the harmonic oscillator. Since the rotational motion around the C—C bond involves the kinetic energy of six hydrogen atoms traveling in circles of radius r perpendicular to the axis of the C—C bond, we estimate that each hydrogen atom has kinetic energy provided by its rotational energy (Section 12.8). Adding the potential energy gives:

$$E_{\text{rotation}} = 6\times\left(\frac{J^2_{\text{C-C}}}{2m_{\text{H}}r^2}\right) + \tfrac{1}{4}V_0\chi^2$$

where $r = R_{\text{C-H}}\sin(180° - \theta_{\text{H-C-C}}) = (110.7\text{ pm})\sin(180° - 109.5°) = 104.4\text{ pm}$

The total energy of the harmonic oscillator is

$$E_{\text{vibration}} = \frac{p_x}{2m} + \tfrac{1}{2}kx^2$$

Comparison of the two relationships shows that the harmonic oscillator energy transforms into the rotational energy using $m \to m_{\text{H}}r^2/6$ and $k \to V_0/2$. Using these transformations in the eqn. 12.21 expression for frequency, yields an estimate of the frequency of the torsional oscillation.

$$\begin{aligned}\nu &= \frac{1}{2\pi}\left(\frac{V_0/2}{m_{\text{H}}r^2/6}\right)^{1/2} = \frac{1}{2\pi}\left(\frac{3V_0}{m_{\text{H}}r^2}\right)^{1/2}\\ &= \frac{1}{2\pi}\left(\frac{3(11.6\times 10^3\text{ J mol}^{-1})}{(0.00100\text{ kg mol}^{-1})\times(104.4\times 10^{-12}\text{ m})^2}\right)^{1/2}\\ &= \boxed{8.99\times 10^{12}\text{ Hz}}\end{aligned}$$

Expressing this as a wavenumber gives

$$\tilde{\nu} = \frac{\nu}{c} = \frac{8.99\times10^{12}\ \text{s}^{-1}}{3.00\times10^{8}\ \text{m s}^{-1}} = \boxed{300.\ \text{cm}^{-1}}$$

E15.22 By the law of cosines $r_{\text{O-H}}^2 = r_{\text{O-H}}^2 + r_{\text{O-O}}^2 - 2r_{\text{O-H}}r_{\text{O-O}}\cos\theta$. Therefore,

$$r_{\text{O-H}} = f(\theta) = \left(r_{\text{O-H}}^2 + r_{\text{O-O}}^2 - 2r_{\text{O-H}}r_{\text{O-O}}\cos\theta\right)^{1/2}$$

$$V_{\text{m}} = \frac{N_{\text{A}}e^2}{4\pi\varepsilon_0}\left\{\frac{\delta_{\text{O}}\delta_{\text{H}}}{r_{\text{O-H}}} + \frac{\delta_{\text{O}}\delta_{\text{H}}}{r_{\text{O-H}}} + \frac{\delta_{\text{O}}^2}{r_{\text{O-O}}}\right\} = \frac{N_{\text{A}}e^2}{4\pi\varepsilon_0}\left\{\delta_{\text{O}}\delta_{\text{H}}\left(\frac{1}{r_{\text{O-H}}} + \frac{1}{f(\theta)}\right) + \frac{\delta_{\text{O}}^2}{r_{\text{O-O}}}\right\}$$

$$= \frac{N_{\text{A}}e^2}{4\pi\varepsilon_0\times\left(10^{-12}\ \text{m}\right)}\left\{\delta_{\text{O}}\delta_{\text{H}}\left(\frac{1}{r_{\text{O-H}}/\text{pm}} + \frac{1}{f(\theta)/\text{pm}}\right) + \frac{\delta_{\text{O}}^2}{r_{\text{O-O}}/\text{pm}}\right\}$$

$$= \left(139\ \text{MJ mol}^{-1}\right)\times\left\{\delta_{\text{O}}\delta_{\text{H}}\left(\frac{1}{r_{\text{O-H}}/\text{pm}} + \frac{1}{f(\theta)/\text{pm}}\right) + \frac{\delta_{\text{O}}^2}{r_{\text{O-O}}/\text{pm}}\right\}$$

With $\delta_{\text{O}} = -0.83$, $\delta_{\text{H}} = 0.45$, $r_{\text{O-H}} = 95.7$ pm, and $r_{\text{O-O}} = 200$ pm we draw up a tabular computation of $f(\theta)$ and $V_{\text{m}}(\theta)$ over the range $0 \leq \theta \leq 2\pi$ and plot $V_{\text{m}}(\theta)$ in Figure 15.8. As expected, the potential is a minimum when $\theta = 0$ because at that angle the hydrogen lies directly between the two oxygen atoms, which repel.

θ/deg	θ/radian	$f(\theta)$	V_{m} / kJ/mol^{-1}
0	0	104.30	−561
15	0.261799	110.38	−534
30	0.523599	126.52	−474
45	0.785398	148.63	−413
60	1.047198	173.26	−363
75	1.308997	198.12	−326
90	1.570796	221.72	−298
105	1.832596	243.04	−277
120	2.094395	261.34	−262
135	2.356194	276.09	−252
150	2.617994	286.90	−245
165	2.879793	293.49	−241
180	3.141593	295.70	−239

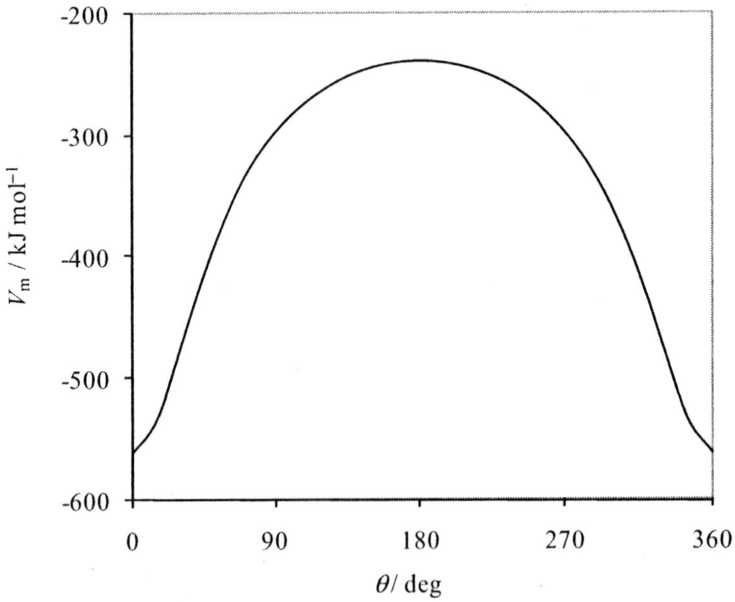

Figure 15.8

Alternatively, if you interpret the problem as an exploration of the electrostatic model of hydrogen bond alone, omit the electrostatic attraction between the bonded O and H atoms. This gives

$$V_{\text{hydrogen bond}} = \frac{e^2 N_A}{4\pi\varepsilon_0}\left\{\frac{\delta_O^2}{R} - \frac{\delta_O \delta_H}{\left(R^2 + r^2 - 2rR\cos\theta\right)^{1/2}}\right\}$$

where $R = r_{O\text{-}O}$ and $r = r_{O\text{-}H}$. The Figure 15.9 plot of $V_{\text{hydrogen bond}}$ shows a minimum of −19.0 kJ mol^{-1} at $\theta = 0°$. Thus, the hydrogen bond strength is 19.0 kJ mol^{-1} in this model. Furthermore, the model shows that small deviations from $\theta = 0°$ destroys the hydrogen bond because the potential becomes greater than zero.

Question. The hydrogen bond model of Exercise 15.18 found a hydrogen bond strength of 42 kJ mol^{-1}. Which of these two models has more reasonable characteristics?

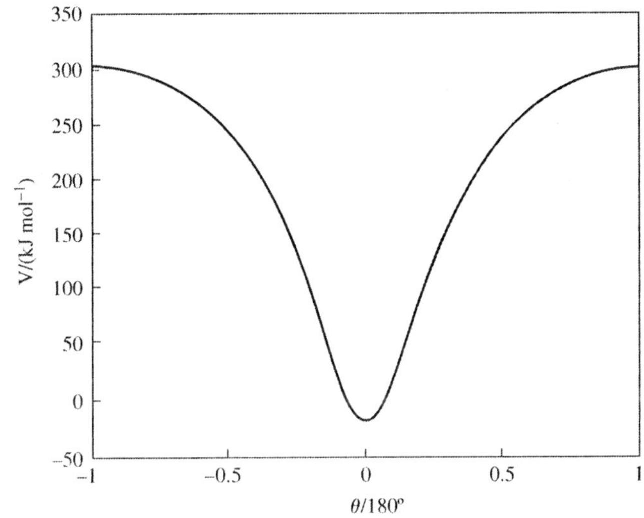

Figure 15.9

Answers to projects

P15.23 (a) We assume that the Lennard-Jones potential, eqn [15.15], adequately represents the potential energy in this case.

$$V(R) = 4\varepsilon\left\{\left(\frac{\sigma}{R}\right)^{12} - \left(\frac{\sigma}{R}\right)^{6}\right\} = \frac{4\varepsilon\sigma^6}{R^6}\left\{\left(\frac{\sigma}{R}\right)^{6} - 1\right\}$$

$$V(R+\delta R) = 4\varepsilon\left\{\left(\frac{\sigma}{R+\delta R}\right)^{12} - \left(\frac{\sigma}{R+\delta R}\right)^{6}\right\} = \frac{4\varepsilon\sigma^6}{R^6}\left\{\frac{\sigma^6}{R^6}\left(\frac{1}{1+\frac{\delta R}{R}}\right)^{12} - \left(\frac{1}{1+\frac{\delta R}{R}}\right)^{6}\right\}$$

Since $(1+x)^{-12} \simeq 1-12x$ and $(1+x)^{-6} \simeq 1-6x$ when $x = 1$, and since $\delta R / R = 1$,

$$V(R+\delta R) = \frac{4\varepsilon\sigma^6}{R^6}\left\{\frac{\sigma^6}{R^6}\left(1-12\frac{\delta R}{R}\right) - \left(1-6\frac{\delta R}{R}\right)\right\}$$

$$\Delta V = V(R+\delta R) - V(R) = \frac{4\varepsilon\sigma^6}{R^6}\left\{\frac{\sigma^6}{R^6}\left(1-12\frac{\delta R}{R}\right) - \left(1-6\frac{\delta R}{R}\right)\right\} - \frac{4\varepsilon\sigma^6}{R^6}\left\{\left(\frac{\sigma}{R}\right)^{6} - 1\right\}$$

$$= \frac{24\varepsilon\sigma^6}{R^6}\left(\frac{\delta R}{R}\right)\left\{1 - 2\left(\frac{\sigma}{R}\right)^{6}\right\}$$

$$F = -\frac{\Delta V}{\Delta r} = -\left[\frac{\frac{24\varepsilon\sigma^6}{R^6}\left(\frac{\delta R}{R}\right)\left\{1-2\left(\frac{\sigma}{R}\right)^6\right\}}{(R+\delta R)-R}\right] = -\frac{24\varepsilon\sigma^6}{R^7}\left\{1-2\left(\frac{\sigma}{R}\right)^6\right\}$$

Thus, F equals zero when the factor $1-2\left(\frac{\sigma}{R}\right)^6$ equals zero or $\boxed{R = 2^{1/6}\sigma}$.

(b) $$F = -\frac{\mathrm{d}V}{\mathrm{d}r} = -4\varepsilon\frac{\mathrm{d}}{\mathrm{d}r}\left\{\left(\frac{\sigma}{r}\right)^{12} - \left(\frac{\sigma}{r}\right)^6\right\} = -4\varepsilon\left\{\frac{12\sigma^{12}}{r^{13}} - \frac{6\sigma^6}{r^7}\right\} = -\frac{24\varepsilon\sigma^6}{r^7}\left\{\frac{2\sigma^6}{r^6} - 1\right\}$$

The minimum occurs when

$\frac{2\sigma^6}{r^6} - 1 = 0$ which solves to $\boxed{r = 2^{1/6}\sigma}$.

P15.24 (a) We want to construct the exponential-6 potential so that $V(r_0) = 0$ and the depth of the potential well is $-\varepsilon$ when $r = r_e$. Consequently, we write

$$V(r) = A\varepsilon\left[Be^{-r/r_0} - \left(\frac{r_0}{r}\right)^6\right]$$

where the constants A and B are chosen to satisfy the conditions. For example, examine $V(r_0)$ to find an expression for B.

$$V(r_0) = A\varepsilon\left[Be^{-r_0/r_0} - \left(\frac{r_0}{r_0}\right)^6\right] = A\varepsilon\left[Be^{-1} - 1\right] = 0$$

So, $B = e$ and the potential becomes $V(r) = A\varepsilon\left[e^{1-r/r_0} - \left(\frac{r_0}{r}\right)^6\right]$.

The constant A is evaluated at the equilibrium distance r_e.

$$-\varepsilon = A\varepsilon\left[e^{1-r_e/r_0} - \left(\frac{r_0}{r_e}\right)^6\right]$$

So, $$A = -\left[e^{1-r_e/r_0} - \left(\frac{r_0}{r_e}\right)^6\right]^{-1}$$ where $r_e > r_0$ and $A < 0$

The repulsive exponential term, the attractive term, and the potential (in the unit $|A\varepsilon|$) are sketched in Fig. 15.10. The point at which the potential is a minimum is labeled as $x_e = r_e/r_0$.

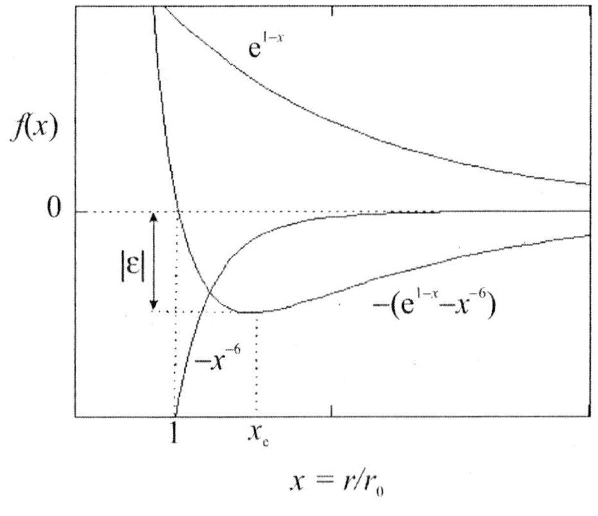

Figure 15.10

(b) Let $x = r/r_0$, then $V(x) = A\varepsilon\left[e^{1-x} - \frac{1}{x^6}\right]$ and the potential minimum occurs when

$$\left.\frac{dV(x)}{dx}\right|_{x=x_e} = 0$$

$$\frac{dV(x)}{dx} = A\varepsilon\left[-e^{1-x} + \frac{6}{x^7}\right]$$

$$\left.\frac{dV(x)}{dx}\right|_{x=x_e} = A\varepsilon\left[-e^{1-x_e} + \frac{6}{x_e^7}\right] = 0$$

Thus, the solution of the transcendental equation

$$e^{1-x_e} = \frac{6}{x_e^7}$$

gives the value x_e at which V is a minimum.

x_e may be found as the intersection of the curves e^{1-x} and $6/x^7$ or it may be found using the numeric solver of a scientific calculator. Here is a short Mathcad worksheet solution for x_e.

$x_e := 1$ Estimate of x_e for Given / Find solve block. ($x_e = r_e / r_0$)

Given $e^{1-x_e} = 6 \cdot x_e^{-7}$ $x_e :=$ Find (x_e)

$x_e = 1.3598$

Thus, $\boxed{r_e = 1.3598\, r_0}$.

$$A = -\left[e^{1-r_e/r_0} - \left(\frac{r_0}{r_e}\right)^6\right]^{-1} = -\left[e^{-0.3598} - \left(\frac{1}{1.3598}\right)^6\right]^{-1} = -1.8531$$

P15.25 (a) The left side molecule in Figure 15.11 is methyladenine. Please note that we have taken the liberty of placing the methyl group in the position that would be occupied by a sugar in RNA and DNA. The wavefunction, structure, and atomic electrostatic charges (shown in the figure) calculation was performed with Spartan '06™ using a Hartree-Fock procedure with a 6-31G* basis set. The atomic electrostatic charge (ESP) numerical method generates charges that reproduce the electrostatic field from the entire wavefunction. The right side molecule in Figure 15.11 is methylthymine.

(b) The two molecules will hydrogen bond into a stable dimer in an orientation for which hydrogen bonding is linear, maximized, and steric hindrance is avoided. We expect hydrogen bonds of the type N−H·····O and N−H·····N with the N and O atoms having large negative electrostatic charges and the H atoms having large positive charges. These atoms are evident in the figure.

(c) Figure 15.11 shows one of three arrangements of hydrogen bonding between the two molecules. Another can be drawn by rotating methylthymine over the top of methyladenine and a third involves rotation to the bottom. The dashed lines show the alignments of two strong hydrogen bonds between the molecules.

(d) The A-to-T base pairing shown in Figure 15.11 has the largest charges in the most favorable positions for strong hydrogen bonding. Also the N-to-O distance of one hydrogen bond equals the N-to-N distance of the other, a favorable feature in RNA and DNA polymers where this pairing and alignment is observed naturally.

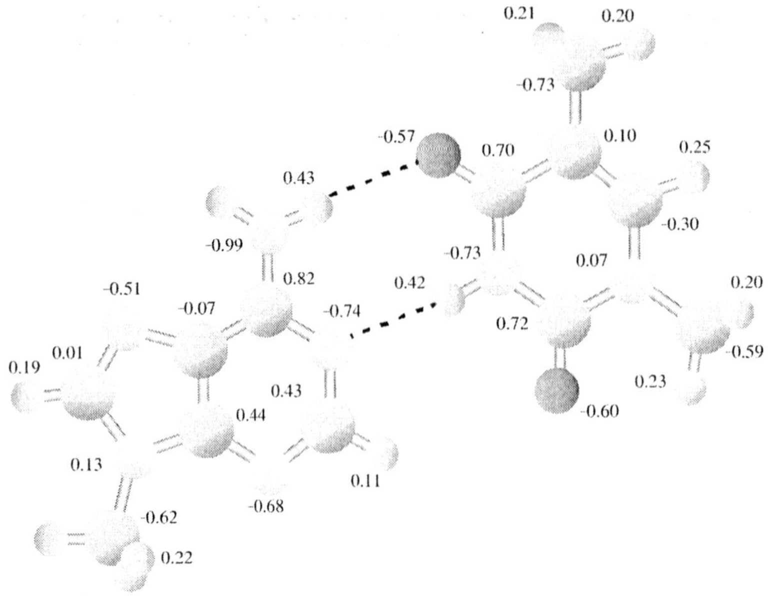

Figure 15.11

P15.26 The drug Crixivan, shown in Figure 15.12 with some of its hydrogen-bonding interactions with HIV protease, is a competitive inhibitor of HIV protease and has several molecular features that optimize binding to the enzyme's active site. First, the highlighted hydroxyl group displaces a H_2O molecule that acts as the nucleophile in the hydrolysis of the substrate. Second, the carbon atom to which the key -OH group is bound has a tetrahedral geometry that mimics the structure of the transition state of the peptide hydrolysis reaction. However, the tetrahedral moiety in the drug is not cleaved by the enzyme. Third, the inhibitor is anchored firmly to the active site via a network of hydrogen bonds involving the carbonyl groups of the drug, a water molecule, and peptide NH groups from the enzyme.

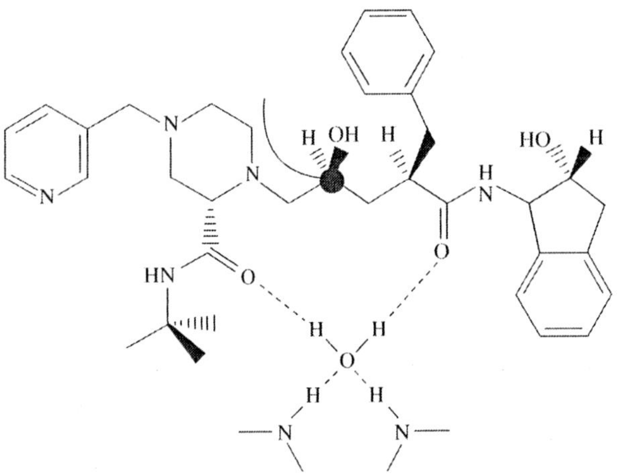

Figure 15.12

16 Materials: macromolecules and aggregates

Answers to discussion questions

D16.1 **Number-average molar mass,** $\bar{M}_n$, is the value obtained by weighting each molar mass by the number of molecules of that mass.

$$\bar{M}_n = \frac{1}{N}\sum_i N_i M_i = \frac{1}{n}\sum_i n_i M_i \quad [16.1a, 16.1b]$$

In this expression, N_i is the number of molecules of molar mass M_i and N is the total number of molecules and, since $N_i = N_A n_i$ and $N = N_A n$, N_i / N can be replaced by n_i/n. Measurements of the osmotic pressures of macromolecular solutions yield the number-average molar mass.

Weight-average molar mass, $\bar{M}_w$, is the value obtained by weighting each molar mass by the mass of each one present.

$$\bar{M}_w = \frac{1}{m}\sum_i m_i M_i \quad [16.1c]$$

Since $m_i = n_i M_i$ and $m = \sum_j n_j M_j$, the ratio m_i / m in the weight average molar mass expression can be replaced by either of the equivalent ratios $n_i M_i / \sum_j n_j M_j$ and $N_i M_i / \sum_j N_j M_j$.

$$\bar{M}_w = \frac{\sum_i n_i M_i^2}{\sum_j n_j M_j} = \frac{\sum_i N_i M_i^2}{\sum_j N_j M_j}$$

Laser light scattering measures $\bar{M}_w$.

$\bar{M}_n$ and $\bar{M}_w$ are identical for molar mass distributions that are very narrow or 'monodisperse'. However, $\bar{M}_w$ is greater than $\bar{M}_n$ for a polydisperse sample and the difference between the two becomes greater for very wide molar mass distributions. The **heterogeneity index**, $\bar{M}_w / \bar{M}_n$, is an indicator of the distribution width. If the index is less than 1.1, the polymer sample is 'monodisperse'. Larger values indicate a wide, or polydisperse, molar mass distribution, which is considered to be heterogeneous.

D16.2 **Contour length,** R_c: the length of the macromolecule measured along its backbone, the length of all its monomer units placed end to end. This is the stretched-out length of the macromolecule with bond angles maintained within the monomer units and 180° angles at unit links. It is proportional to the number of monomer units, N, and to the length of each unit (eqn 16.3a).

Root mean square separation, R_{rms}: one measure of the average separation of the ends of a random coil. It is the square root of the mean value of R^2, where R is the separation of the two ends of the coil. R_{rms} is proportional to $N^{1/2}$ and the length of each unit (eqn 16.3b).

Radius of gyration, R_G: the radius of a thin hollow spherical shell of the same mass and moment of inertia as the macromolecule. In general, it is not easy to visualize this distance geometrically.

However, for the simple case of a molecule consisting of a chain of identical atoms this quantity can be visualized as the root mean square distance of the atoms from the centre of mass. It also depends on $N^{1/2}$, but is smaller than the root mean square separation by a factor of $(1/6)^{1/2}$ (eqn 16.3c).

D16.3 The freely jointed random coil model of a polymer chain of 'units' or 'residues' gives the simplest possibility for the conformation of the polymer that is not capable of forming hydrogen bonds or any other type of non-linkage bond. In this model, a bond that links adjacent units in the chain is free to make any angle with respect to the preceding one (Fig. 16.4 of the text). We assume that the residues occupy zero volume, so different parts of the chain can occupy the same region of space. We also assume in the derivation of the expression for the probability of the ends of the chain being a distance nl apart, that the chain is compact in the sense that $n \ll N$. This model is obviously an oversimplification because a bond is actually constrained to a cone of angles around a direction defined by its neighbor. In a hypothetical one-dimensional freely jointed chain all the residues lie in a straight line, and the angle between neighbors is either 0° or 180°. The residues in a three-dimensional freely jointed chain are not restricted to lie in a line or a plane.

The random coil model ignores the role of the solvent: a poor solvent will tend to cause the coil to tighten; a good solvent does the opposite. Therefore, calculations based on this model are best regarded as lower bounds to the dimensions of a polymer in a good solvent and as an upper bound for a polymer in a poor solvent. The model is most reliable for a polymer in a bulk solid sample, where the coil is likely to have its natural dimensions.

D16.4 **Matrix-assisted laser desorption/ionization (MALDI)** embeds the macromolecule sample in a solid matrix composed of an organic material and inorganic salt. This is irradiated with a pulsed laser, which ejects excited matrix ions, cations, and neutral macromolecules into a dense gas plume above the matrix surface. The macromolecule is ionized by collisions and complexation with small cations and the masses of the resulting ions are determined in a mass spectrometer. Molar mass averages are calculated from the data of the mass spectrum.

The **ultracentrifuge** technique spins a column of macromolecule solution to produce accelerations equivalent to about 10^5 g. The rate at which the macromolecules recede is related to the number-average molar mass.

Laser light scattering by macromolecules in solution is a convenient method for determination of the weight-average molar mass of a polydisperse sample. It is used to characterize polymers, colloids, and biological systems from proteins to viruses. The method involves measuring the dependence of the intensity of scattered light on the angle between the incident and scattered beams.

A **viscometer** is often used to measure the viscosity, which is proportional to the drainage time through a capillary in the basic instrument, of a macromolecule solution over a range of concentrations in order to deduce the viscosity-average molar mass.

D16.5 **Elastomers** are polymers with numerous crosslinks that pull them back into their original shape when a stress is removed. The internal energy of a **perfect elastomer** is independent of the extension of the random coil; it acts like a coil spring because for small displacements from the random coil, it spontaneously returns to a random coil conformation with a force that is proportional to the displacement. This is driven by an increase in entropy of the perfect elastomer while the entropy of the surroundings remains unchanged as no energy has been released, or absorbed, by the coil.

D16.6 Molecular interactions of polymeric materials include hydrogen bonding (Section 15.6), ionic interactions (Section 15.1), van der Waals interactions (Sections 15.2 and 15.3), hydrophobic interactions (Section 15.5), and disulfide links. Physical entanglement of long polymeric chains also plays an important role in determining the thermal stability and mechanical strength of these materials.

D16.7 A surfactant is a species that is active at the interface of two phases or substances, such as the interface between hydrophilic and hydrophobic phases. A surfactant accumulates at the interface and modifies the properties of the surface, in particular, decreasing its surface tension. A typical surfactant consists of a long hydrocarbon tail and other non-polar materials, and a hydrophilic head group, such as the carboxylate group, $—CO_2^-$, that dissolves in a polar solvent, typically water. In other words, a surfactant is an amphipathic substance, meaning that it has both hydrophobic and hydrophilic regions.

How does the surfactant decrease the surface tension? Surface tension is a result of cohesive forces and the solute molecules must weaken the attractive forces between solvent molecules. Thus, molecules with bulky hydrophobic regions such as fatty acids can decrease the surface tension because they attract solvent molecules less strongly than solvent molecules attract each other.

D16.8 The surface of a disperse, colloidal particle shows two distinctive regions of charge. First, there is a fairly immobile layer of solvated ions and water molecules that stick tightly to the surface. The radius of the sphere that captures this rigid layer is call the **radius of shear**, and it is the major factor determining the mobility of the particle. The electric potential at the radius of shear relative to its value in the distant, bulk medium is called the **electrokinetic potential,** ζ. The charged unit attracts an oppositely charged ionic atmosphere. The inner shell of charge and outer atmosphere jointly constitute the **electric double layer**.

The electric double layer is the major source of colloidal kinetic non-lability and physical stability. Colliding colloidal particles break through the double layer and coalesce only if the collision is sufficiently energetic to disrupt the layers of ions and solvating molecules, or if thermal motion has stirred away the surface accumulation of charge.

Solutions to exercises

E16.1 Equal amounts imply equal numbers of molecules $N_1/N = N_2/N = \frac{1}{2}$. Hence the number-average molar mass is

$$\bar{M}_n = \frac{N_1M_1 + N_2M_2}{N} \; [16.1a] = \tfrac{1}{2}(M_1 + M_2)$$
$$= \tfrac{1}{2}(82+108)\,\text{kg mol}^{-1} = \boxed{95\text{ kg mol}^{-1}}$$

and, since $n_1 = n_2$, the weight-average molar mass is

$$\bar{M}_w = \frac{m_1M_1 + m_2M_2}{m} \; [16.1c] = \frac{n_1M_1^2 + n_2M_2^2}{n_1M_1 + n_2M_2} = \frac{M_1^2 + M_2^2}{M_1 + M_2}$$
$$= \frac{82^2 + 108^2}{82+108}\,\text{kg mol}^{-1} = \boxed{97\text{ kg mol}^{-1}}$$

E16.2 (a) Osmometry gives the number-average molar mass and, since the concentration is in terms of mass percentage, it is convenient to reference enough solution to give 100 g total of monomer (70%, 15 kg mol^{-1}) and dimer (30%, 30 kg mol^{-1}).

$$\bar{M}_n = \frac{n_1M_1 + n_2M_2}{n_1 + n_2} \; [16.1b] = \frac{\left(\frac{m_1}{M_1}\right)M_1 + \left(\frac{m_2}{M_2}\right)M_2}{\left(\frac{m_1}{M_1}\right) + \left(\frac{m_2}{M_2}\right)} = \frac{m_1 + m_2}{\left(\frac{m_1}{M_1}\right) + \left(\frac{m_2}{M_2}\right)}$$
$$= \frac{100\text{ g}}{\left(\frac{30\text{ g}}{30\text{ kg mol}^{-1}}\right) + \left(\frac{70\text{ g}}{15\text{ kg mol}^{-1}}\right)}$$
$$= \boxed{18\text{ kg mol}^{-1}}$$

(b) Light-scattering gives the weight-average molar mass.

$$\bar{M}_{\mathrm{w}}=\frac{m_1M_1+m_2M_2}{m_1+m_2}\ [16.1\mathrm{c}]=\frac{(30)\times(30)+(70)\times(15)}{100}\ \mathrm{kg\ mol^{-1}}=\boxed{20\ \mathrm{kg\ mol^{-1}}}$$

E16.3 Text Example 16.1 illustrates the methodology for calculating the heterogeneity index ($\bar{M}_{\mathrm{W}}/\bar{M}_{\mathrm{n}}$).

Interval/(kg mol^{-1})	5–10	10–15	15–20	20–25	25–30	30–35	Total
Molar mass/(kg mol^{-1})	6.5	11.5	19.5	23.5	28.5	35.5	
Mass in interval/g	16.0	27.1	29.5	13.4	8.7	3.5	98.2
Amount/mmol	2.46	2.36	1.51	0.570	0.31	0.099	7.30
Mass × Molar mass/(kg^2 mol^{-1})	104	312	575	315	248	124	1678

$$\bar{M}_{\mathrm{n}}=\frac{1}{n}\sum_i n_iM_i\ [16.1\mathrm{b}]=\frac{1}{n}\sum_i m_i=\frac{m}{n}$$

$$=\frac{98.2}{7.30}\ \mathrm{kg\ mol^{-1}}$$

$$=13.5\ \mathrm{kg\ mol^{-1}}$$

$$\bar{M}_{\mathrm{W}}=\frac{1}{m}\sum_i m_iM_i\ [16.1\mathrm{c}]$$

$$=\frac{1678}{98.2}\ \mathrm{kg\ mol^{-1}}$$

$$=17.1\ \mathrm{kg\ mol^{-1}}$$

$$\text{Heterogeneity index}=\bar{M}_{\mathrm{W}}/\bar{M}_{\mathrm{n}}=\frac{17.1}{13.5}=\boxed{1.27}$$

E16.4 The peaks are separated by $104\ \mathrm{g\ mol^{-1}}$, so this is the molar mass of the repeating unit of the polymer. This peak separation is consistent with the identification of the polymer as polystyrene, for the repeating group of $CH_2CH(C_6H_5)$ (8 C atoms and 8 H atoms) has a molar mass of $8\times(12+1)\ \mathrm{g\ mol^{-1}}=104\ \mathrm{g\ mol^{-1}}$. A consistent difference between peaks suggests a pure system and points away from different numbers of subunits of different molecular weight (such as the *t*-butyl initiators) being incorporated into the polymer molecules. The most intense peak has a molar mass equal to that of *n* repeating groups plus that of a silver cation plus that of terminal groups:

$$M(\text{peak})=nM(\text{repeat})+M(\mathrm{Ag^+})+M(\text{terminal})$$

If both ends of the polymer have terminal *t*-butyl groups, then

$$M(\text{terminal})=2M(t\text{-butyl})=2(4\times12+9)\ \mathrm{g\ mol^{-1}}=114\ \mathrm{g\ mol^{-1}}.$$

and $$n=\frac{M(\text{peak})-M(\mathrm{Ag^+})-M(\text{terminal})}{M(\text{repeat})}=\frac{25578-108-114}{104}=\boxed{244}$$

E16.5 $$\bar{M}_{\mathrm{W}}=\frac{2RT}{(r_2^2-r_1^2)b\omega^2}\ln\frac{c_2}{c_1}\ [16.2]\ \text{rearranges to}\ \ln c_2=\frac{\bar{M}_{\mathrm{W}}(r_2^2-r_1^2)b\omega^2}{2RT}+\ln c_1$$

Let the pair (c_2,r_2) be any unspecified pair (c,r) and let the pair (c_1,r_1) be the reference pair $(c_{\mathrm{ref}},r_{\mathrm{ref}})$ and the working equation can be written in the linear form:

$$\ln c=slope\times r^2+intercept$$

where $slope=\dfrac{\bar{M}_{\mathrm{W}}b\omega^2}{2RT}$ and $intercept=\ln c_{\mathrm{ref}}-\dfrac{\bar{M}_{\mathrm{W}}b\omega^2r_{\mathrm{ref}}^2}{2RT}$

Thus, the slope of a ln c against r^2 data plot determines $\bar{M}_W$.

$$\bar{M}_W = \frac{2RT \times slope}{b\omega^2}$$

$$= \frac{(2)\times(8.3145\ \mathrm{J\ K^{-1}\ mol^{-1}})\times(300\ \mathrm{K})\times(659\times10^4\ \mathrm{m^{-2}})}{(1-0.996\times0.61)\times\left(\frac{(2\pi\ \mathrm{rad\ revolution^{-1}})\times(50000\ \mathrm{revolutions\ min^{-1}})}{60\ \mathrm{s\ min^{-1}}}\right)^2}$$

$$= \boxed{3.1\times10^3\ \mathrm{kg\ mol^{-1}}}$$

E16.6 (a) $R_c = Nl$ [16.3a] $= (800)\times(1.10\ \mathrm{nm}) = \boxed{880\ \mathrm{nm}}$

(b) $R_{rms} = N^{1/2}l$ [16.3b] $= (800)^{1/2}\times(1.10\ \mathrm{nm}) = \boxed{31.1\ \mathrm{nm}}$

E16.7 The repeating unit (monomer) of polyethylene is CH_2—CH_2 which has a molar mass of 28 g mol^{-1}. The number of repeating units, N, is therefore

$$N = \frac{250\ 000\ \mathrm{g\ mol^{-1}}}{28\ \mathrm{g\ mol^{-1}}} = 8.93\times10^3$$

and $l = 2R$(C—C) [Add half a bond-length on either side of monomer.]

Apply the formulas for the contour length (16.3a) and rms separation (16.3b):

$$R_c = Nl = 2\times(8.93\times10^3)(154\ \mathrm{pm}) = 2.75\times10^6\ \mathrm{pm} = \boxed{2.75\ \mu\mathrm{m}}$$

$$R_{rms} = N^{1/2}l = 2\times(8.93\times10^3)^{1/2}\times(154\ \mathrm{pm}) = 2.91\times10^4\ \mathrm{pm} = \boxed{29.1\ \mathrm{nm}}$$

E16.8 For a random coil, the radius of gyration is (16.3c)

$$R_G = \left(\frac{N}{6}\right)^{1/2} l \qquad \text{so } N = 6\left(\frac{R_G}{l}\right)^2 = (6)\times\left(\frac{7.3\ \mathrm{nm}}{0.154\ \mathrm{nm}}\right)^2 = \boxed{1.3\times10^4}$$

E16.9 A simple procedure is to generate numbers in the range 1 to 8, and to step north for a 1 or 2, east for 3 or 4, south for 5 or 6, and west for 7 or 8 on a uniform grid. One such walk is shown in Figure 16.1. Roughly, the mean and most probable separation of the ends appear to vary as $N^{1/2}$. (See Exercise/Project 16.19.)

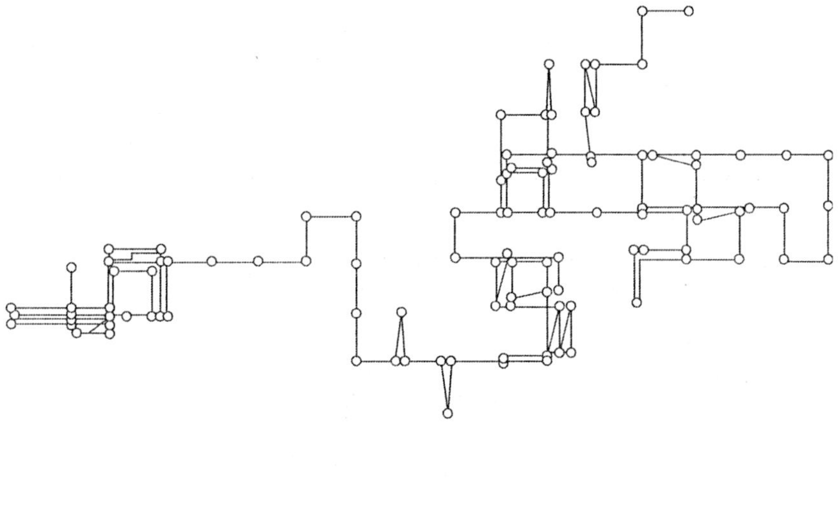

Figure 16.1

E16.10 Classical rotation of a solid object of uniform density is an important aspect of physics. One application involves using its mathematical formulas to model protein shapes in solution so as to estimate whether the shape is spherical or ellipsoidal. In this exercise we have an interest in estimating whether the shape is spherical-like or rod-like or disk-like so we make that lovely trip to the physics stacks within the library and find that the mathematics of rotation is simplified by expressing the motion in terms of a point particle that has circular motion, the same mass and moment of inertia as the object of interest. The radius of the circular motion is called the **radius of gyration, R_G** and it is related to the moment of inertia of the object by the expression

$$R_G = \left(\frac{I}{m}\right)^{1/2} \quad \text{where } I \text{ and } m \text{ are the moment of inertia and mass of the object}$$

So far, so good. But, what are the formulas for the moment of inertia of the different objects? Well, this is a purely mathematical problem so we expect that the formulas can be found in any basic handbook of mathematics and sure enough we find the formulas of the following table.

Object	I	Motion
solid sphere, radius r	$2mr^2/5$	about axis through center
thin rod, length l	$ml^2/12$	about axis perpendicular to rod and through centre of mass
thin disk, radius r	$mr^2/2$	about axis perpendicular to disk and through centre
thin ring, radius r	mr^2	about axis perpendicular to ring and through centre

We need only examine the expression for the solid sphere. Substitution gives

$$R_{G,\text{sphere}} = \left(\frac{2mr^2}{5m}\right)^{1/2} = \left(\frac{2}{5}\right)^{1/2} r$$

Two relations are used to find an expression for r in terms of data provided in the exercise. They are

$$V_{\text{sphere}} = \tfrac{4}{3}\pi r^3 \qquad \text{and} \qquad V_{\text{sphere}} = mv_s = Mv_s / N_A$$

Solving them for r gives

$$r = \left(\frac{3Mv_s}{4N_A}\right)^{1/3}$$

and the working equation for the radius of gyrations becomes

$$R_{G,\text{sphere}} = \left(\frac{2}{5}\right)^{1/2}\left(\frac{3Mv_s}{4N_A}\right)^{1/3}$$

Using the data for serum albumin, we find

$$R_{G,\text{sphere}} = \left(\frac{2}{5}\right)^{1/2}\left(\frac{3\times(66\times10^3\ \text{g mol}^{-1})\times(0.752\times10^{-6}\ \text{m}^3\ \text{g}^{-1})}{4\times(6.022\times10^{23}\ \text{mol}^{-1})}\right)^{1/3} = 2.5\ \text{nm}$$

This radius of gyration value and those calculated for bushy stunt virus and DNA are summarized along with the experimental data in the following table. The values calculated for serum albumin and bushy stunt virus are in reasonable agreement with the experiment but that of DNA does not agree. Therefore, serum albumin and bushy stunt virus resemble solid spheres, but DNA does not.

	$M/(\text{g mol}^{-1})$	$v_s/(\text{cm}^3\ \text{g}^{-1})$	$(R_g/\text{nm})_{\text{expt}}$	$(R_G/\text{nm})_{\text{sphere}}$
Serum albumin	66×10^3	0.752	2.98	2.5
Bushy stunt virus	10.6×10^6	0.741	12.0	13.5
DNA	4×10^6	0.556	117.0	9

E16.11 $\nu = 0.1$

$$\Delta S / kN = -\tfrac{1}{2}\ln\left\{(1+\nu)^{1+\nu}(1-\nu)^{1-\nu}\right\} \text{ [16.4]}$$

$$= -\tfrac{1}{2}\ln\left\{(1.1)^{1.1}\times(0.90)^{0.9}\right\} = \boxed{-5.0\times10^{-3}}$$

E16.12 The glass transition temperature T_g is the temperature at which internal bond rotations freeze. In effect, the easier such rotations are, the lower T_g. Internal rotations are more difficult for polymers that have bulky side chains than for polymers without such chains because the side chains of neighboring molecules can impede each others' motion. Of the four polymers in this problem, polystyrene has the largest side chain (phenyl) and the largest T_g. The chlorine atoms in poly(vinyl chloride) interfere with each other's motion more than the smaller hydrogen atoms that hang from the carbon backbone of polyethylene. Poly(oxymethylene), like polyethylene, has only hydrogen atoms protruding from its backbone; however, poly(oxymethylene) has fewer hydrogen protrusions and a still lower T_g than polyethylene.

E16.13 $N\,\mathrm{S} \rightleftharpoons \mathrm{S}_N \quad K = \dfrac{[\mathrm{S}_N]}{[\mathrm{S}]^N}$ [16.6a] and $[\mathrm{S}]_{\text{total}} = [\mathrm{S}] + N[\mathrm{S}_N]$ by the conservation of mass

There are times when the choice of a problem's 'unknown' leads to mathematical difficulties. Before we show an example of this difficulty, let us solve the above equilibrium expression for [S] using the substitution $[\mathrm{S}_N] = \dfrac{1}{N}\left\{[\mathrm{S}]_{\text{total}} - [\mathrm{S}]\right\}$ in the case for which $N = 2$.

$$NK = \frac{\left\{[\mathrm{S}]_{\text{total}} - [\mathrm{S}]\right\}}{[\mathrm{S}]^N}$$

$$NK[\mathrm{S}]^N + [\mathrm{S}] - [\mathrm{S}]_{\text{total}} = 0$$

$$2K[\mathrm{S}]^2 + [\mathrm{S}] - [\mathrm{S}]_{\text{total}} = 0 \text{ (use solution for a quadratic equation)}$$

$$[\mathrm{S}] = \frac{-1 \pm \sqrt{1 + 8K[\mathrm{S}]_{\text{total}}}}{4K}$$

[S] must be positive, so we immediately drop the negative from the ± possibilities giving us the unambiguous solution:

$$\boxed{[\mathrm{S}] = \frac{-1 + \sqrt{1 + 8K[\mathrm{S}]_{\text{total}}}}{4K} \quad \text{and} \quad [\mathrm{S}_2] = \frac{1}{2}\left\{[\mathrm{S}]_{\text{total}} - \frac{-1 + \sqrt{1 + 8K[\mathrm{S}]_{\text{total}}}}{4K}\right\}}$$

This time we create some difficulty by choosing to solve the equilibrium expression for $[\mathrm{S}_N]$ using the substitution $[\mathrm{S}] = [\mathrm{S}]_{\text{total}} - N[\mathrm{S}_N]$ in the case for which $N = 2$.

$$K = \frac{[\mathrm{S}_N]}{\left([\mathrm{S}]_{\text{total}} - N[\mathrm{S}_N]\right)^N} \quad \text{[16.6b]}$$

$$K\left([\mathrm{S}]_{\text{total}} - N[\mathrm{S}_N]\right)^N - [\mathrm{S}_N] = 0$$

$$K\left([\mathrm{S}]_{\text{total}} - 2[\mathrm{S}_2]\right)^2 - [\mathrm{S}_2] = 0$$

$$4K[\mathrm{S}_2]^2 - \left(4K[\mathrm{S}]_{\text{total}} + 1\right)[\mathrm{S}_2] + K[\mathrm{S}]_{\text{total}}^2 = 0 \text{ (use solution for a quadratic equation)}$$

$$[\mathrm{S}_2] = \frac{\left(4K[\mathrm{S}]_{\text{total}} + 1\right) \pm \sqrt{\left(4K[\mathrm{S}]_{\text{total}} + 1\right)^2 - 16K^2[\mathrm{S}]_{\text{total}}^2}}{8K}$$

It is not at all immediately apparent which of the ± possibilities gives a physically plausible answer. A good deal of additional work is needed to find the solution from this expression.

E16.14 We begin by rearranging eqn 16.6b into the form

$$K\left([\mathrm{S}]_{\text{total}} - N[\mathrm{S}_N]\right)^N - [\mathrm{S}_N] = 0$$

Divide by $K^{1/(1-N)}$ and rearrange slightly to give

$$(Y - NX)^N - X = 0$$

$$\text{where } X = [\mathrm{S}_N]/K^{1/(1-N)} \quad \text{and} \quad Y = [\mathrm{S}]_{\text{total}}/K^{1/(1-N)}$$

When using this simple looking expression, we consider X to be the 'unknown'. X can also be considered to be a function of both Y and N. Since this expression is transcendental, for a given N we must numerically solve the expression at each value of Y using the numeric solver of a scientific calculator or using appropriate computer software. (Mathcad is useful.) Here is a tabular set of calculations for $N = 3$:

Y	0.01	0.08	0.15	0.22	0.29	0.36	0.43	0.50
X	1.0×10^{-6}	4.8×10^{-4}	2.8×10^{-3}	7.7×10^{-3}	0.015	0.024	0.035	0.047

By plotting X against Y for a range of N values, we find that

(i) the plots are concave up at small values of Y,

(ii) the plots are linear for large values of Y

(iii) for large values of N, X is zero for all practical purposes when $Y < 1$ but X increases linearly when $Y > 1$.

These things happen because Y is small when K is small and a low value for the equilibrium constant implies that both $[\mathrm{S}_N]$ and X must be small. Conversely, Y is large when K is large and a high value for the equilibrium constant implies that both $[\mathrm{S}_N]$ and X must be large. Thus, our major discovery is finding that there is virtually no significant amount of $[\mathrm{S}_N]$ unless $[\mathrm{S}]_{\text{total}} > K^{1/(1-N)}$ when N is very large. For large values of N there is literally a transition from no significant $[\mathrm{S}_N]$ to significant $[\mathrm{S}_N]$ at the point $[\mathrm{S}]_{\text{total}} = K^{1/(1-N)}$.

Another verification of these observations is provided by the examination of the ratio X/Y. The above definitions indicate that the ratio equals $[\mathrm{S}_N]/[\mathrm{S}]_{\text{total}}$. Calculation of this ratio for $N = 3$ and $Y = 0.01$ (see the above table) gives

$$[\mathrm{S}_N]/[\mathrm{S}]_{\text{total}} = X/Y = (1.0 \times 10^{-6})/0.01 = 1.0 \times 10^{-8}$$

and we see that $[\mathrm{S}_N]$ is a negligible fraction of the whole.

In contrast for $N = 50$ and $Y = 2.0$ we find that $X = 0.021$ (see plot on opposite page). Thus,

$$[\mathrm{S}_N]/[\mathrm{S}]_{\text{total}} = X/Y = (0.021)/2.0 = 0.01$$

and we see that $[\mathrm{S}_N]$ has now become a significant fraction (1%) of the whole.

The following presents a powerful Mathcad worksheet for numerically solving for X over a range of Y values at any specified N. Mathcad plots are included in Figures 16.2 and 16.3 to demonstrate the observations outlined above.

$\mathrm{Y_{min}} := 0.01$ $\quad\quad$ $\mathrm{Y_{max}} := 2$

$\mathrm{M} := 50$ $\quad\quad$ $\mathrm{i} := 0..\mathrm{M}$ $\quad\quad$ $\mathrm{Y_i} := \mathrm{Y_{min}} + (\mathrm{Y_{max}} - \mathrm{Y_{min}}) \cdot \mathrm{i/M}$

$\mathrm{Z} := 0.01 \cdot \mathrm{Y}$ $\quad\quad$ Convenient estimate of X for the solve block.

$\mathrm{f(Z, N)} := (\mathrm{Y} - \mathrm{N} \cdot \mathrm{Z})^{\mathrm{N}}$ $\quad\quad$ Convenient function for the solve block.

Given

$\mathrm{f(Z, N)} - \mathrm{Z} = 0$

$\mathrm{X(N)} := \mathrm{Find(Z)}$ $\quad\quad$ Given/Find solve block finds X as a function of N.

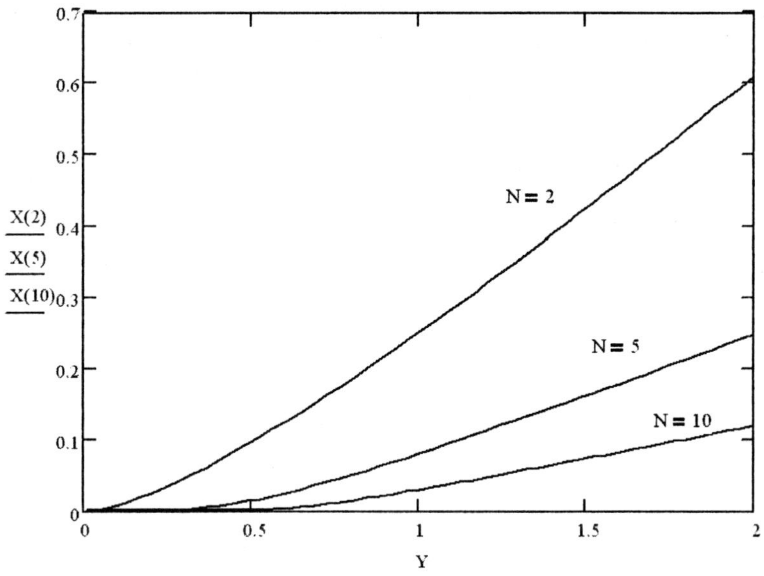

Figure 16.2

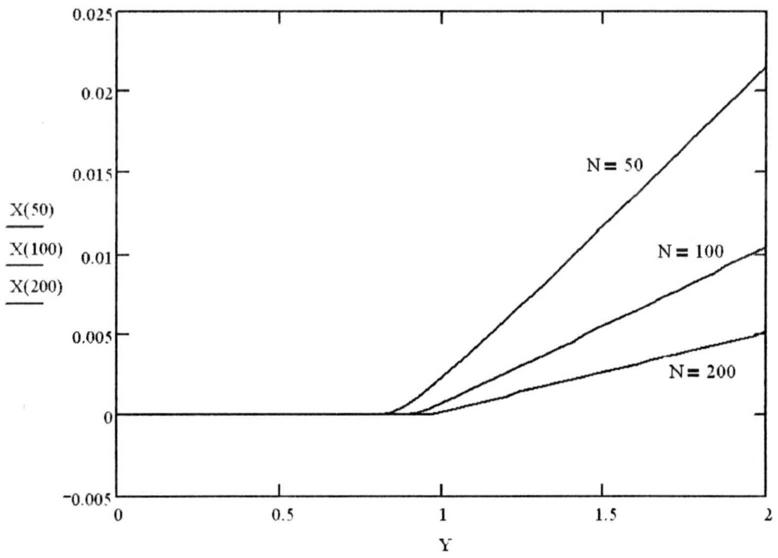

Figure 16.3

E16.15 $\Delta p = p_{\text{concave}} - p_{\text{convex}} = \dfrac{2\gamma}{r}$ [16.8]

(a) $\Delta p = \dfrac{2\times(72\times10^{-3}\ \text{N m}^{-1})}{0.10\times10^{-3}\ \text{m}} = 1.4\times10^{3}\ \text{Pa} = \boxed{1.4\ \text{kPa}}$

(b) $\Delta p = \dfrac{2\times(72\times10^{-3}\ \text{N m}^{-1})}{1.0\times10^{-3}\ \text{m}} = 1.4\times10^{2}\ \text{Pa} = \boxed{0.14\ \text{kPa}}$

E16.16 $h = \dfrac{2\gamma}{\rho g r}$ [16.9]

$= \dfrac{2\times(22.39\times10^{-3}\ \text{N m}^{-1})}{(789\ \text{kg m}^{-3})\times(9.80665\ \text{m s}^{-2})\times(0.10\times10^{-3}\ \text{m})} = 5.8\times10^{-2}\ \text{m}$

$= \boxed{5.8\ \text{cm}}$

E16.17 $h = \dfrac{2\gamma}{\rho g r}$ [16.9] Solve for γ. $\gamma = \frac{1}{2}\rho g r h$

$\gamma = \frac{1}{2}(791\ \text{kg m}^{-3})\times(9.80665\ \text{m s}^{-2})\times(0.20\times10^{-3}\ \text{m})\times(5.8\times10^{-2}\ \text{m})$

$= 4.5\times10^{-2}\ \text{kg s}^{-2} = \boxed{45\ \text{mN m}^{-1}}$ The accepted value is 22.6 mN m^{-1} at 293 K.

E16.18 $\Gamma = \dfrac{n_{\text{total}} - n_{\text{solution}}}{\sigma}$ [16.10]

$= \dfrac{(c_{\text{total}} - c_{\text{solution}})V}{\pi r_{\text{beaker}}^2}$ where c is molar concentration

$$= \frac{(0.100\times10^3\ \text{mol m}^{-3} - 0.0981\times10^3\ \text{mol m}^{-3})\times(100\times10^{-6}\ \text{m}^3)}{\pi(2.5\times10^{-2}\ \text{m})^2}$$

$= 9.7\times10^{-2}\ \text{mol m}^{-2} = \boxed{97\ \text{mmol m}^{-2}}$

Solutions to projects

P16.19 The general expression for the mean nth power of the end-to-end separation of a three-dimensional random coil is

$$\langle R^n\rangle = \int_0^\infty R^n f(R)\,\text{d}R \qquad \text{where } f(R) = 4\pi\left(\frac{a}{\pi^{1/2}}\right)^3 R^2\text{e}^{-a^2R^2} \quad \text{and} \quad a = \left(\frac{3}{2Nl^2}\right)^{1/2}$$

(a) $R_{\text{rms}} = \langle R^2\rangle^{1/2}$

$$\langle R^2\rangle = \int_0^\infty R^2 f(R)\,\text{d}R$$

$$= 4\pi\left(\frac{a}{\pi^{1/2}}\right)^3 \int_0^\infty R^4\text{e}^{-a^2R^2}\,\text{d}R \text{ (This standard integral is found in handbooks of mathematics.)}$$

$$= 4\pi\left(\frac{a}{\pi^{1/2}}\right)^3 \times \left(\frac{3\pi^{1/2}}{8a^5}\right)$$

$$= \frac{3}{2a^2} = Nl^2$$

Hence, $R_{\text{rms}} = \boxed{N^{1/2}l}$. This confirms eqn 16.3b.

(b) The mean separation is $R_{\text{mean}} = \langle R\rangle$.

$$R_{\text{mean}} = \int_0^\infty Rf(R)\,\text{d}R$$

$$= 4\pi\left(\frac{a}{\pi^{1/2}}\right)^3 \int_0^\infty R^3\text{e}^{-a^2R^2}\,\text{d}R \text{ (This standard integral is found in handbooks of mathematics.)}$$

$$= 4\pi\left(\frac{a}{\pi^{1/2}}\right)^3 \times \left(\frac{1}{2a^4}\right) = \frac{2}{a\pi^{1/2}}$$

$$= \boxed{\left(\frac{8N}{3\pi}\right)^{1/2} l}$$

(c) The most probable separation is the value of R for which f is a maximum, so set $\frac{\mathrm{d}f}{\mathrm{d}R}=0$ and solve for R.

$$\frac{\mathrm{d}f}{\mathrm{d}R}=4\pi\left(\frac{a}{\pi^{1/2}}\right)^3\{2R-2a^2R^3\}\mathrm{e}^{-a^2R^2}=0 \qquad \text{when } a^2R^2=1$$

Therefore, the most probable separation, R^*, is

$$R^*=\frac{1}{a}=\boxed{\left(\frac{2}{3}N\right)^{1/2}l}$$

When $N = 5000$ and $l = 154$ pm,

(a) $R_{\text{rms}}=\boxed{10.9\text{ nm}}$ (b) $R_{\text{mean}}=\boxed{10.0\text{ nm}}$ (c) $R^*=\boxed{8.89\text{ nm}}$

P16.20 (a) We estimate the force required for a 10% expansion from the coiled state with the one-dimensional perfect elastomer model ($\nu = 0.1$ in eqn 16.5a). Assume $l = 154$ pm, which is the C—C bond length in a typical polymer.

$$F=\frac{kT}{2l}\ln\left(\frac{1+\nu}{1-\nu}\right) \quad [16.5\text{a}]$$

$$=\frac{(1.381\times10^{-23}\text{ J K}^{-1})\times(300\text{ K})}{2\times(154\times10^{-12}\text{ m})}\ln\left(\frac{1.1}{0.9}\right)=2.70\times10^{-12}\text{ J}$$

$$=\boxed{2.70\text{ pJ}}$$

Having finished the estimate with the one-dimensional model, let us now see what a three-dimensional perfect elastomer model looks like with $\nu = 0.1$. In this case we consider ν to result from the vector displacements ν_x, ν_y, ν_z in an isotropic media in which $\nu_x = \nu_y = \nu_z$. Then,

$$\nu^2=\nu_x^2+\nu_y^2+\nu_z^2=3\nu_x^2 \text{ and, consequently, } \nu_x=\nu/\sqrt{3}=0.10/\sqrt{3}=0.057\overline{7}$$

The restoring force F also results from independent vector components and each component is given by the one-dimensional expression of eqn 16.5a.

$$F^2=F_x^2+F_y^2+F_z^2=3F_x^2$$

$$F=\sqrt{3}F_x=\sqrt{3}\frac{kT}{2l}\ln\left(\frac{1+\nu_x}{1-\nu_x}\right)$$

$$=\sqrt{3}\frac{(1.381\times10^{-23}\text{ J K}^{-1})\times(300\text{ K})}{2\times(154\times10^{-12}\text{ m})}\ln\left(\frac{1+0.057\overline{7}}{1-0.057\overline{7}}\right)=2.69\times10^{-12}\text{ J}$$

$$=\boxed{2.69\text{ pJ}}$$

The two estimates are not significantly different.

(b) $$F=-T\frac{\mathrm{d}S}{\mathrm{d}x} \qquad \text{where } \Delta S=S-S_{\mathrm{e}}=-\tfrac{1}{2}kN\ln\left\{(1+\nu)^{1+\nu}(1-\nu)^{1-\nu}\right\} \quad [16.4]$$

The math is facilitated by transforming the expression for the restoring force to an expression in ν.

$$\nu=\frac{n}{N}=\frac{x-x_{\mathrm{e}}}{Nl}$$

Thus, $\dfrac{\mathrm{d}v}{\mathrm{d}x}=\dfrac{1}{Nl}$.

$$F=-T\frac{\mathrm{d}S}{\mathrm{d}x}=-T\frac{\mathrm{d}\Delta S}{\mathrm{d}x}$$

$$=-T\frac{\mathrm{d}v}{\mathrm{d}x}\frac{\mathrm{d}\Delta S}{\mathrm{d}v}=-\frac{T}{Nl}\frac{\mathrm{d}\Delta S}{\mathrm{d}v}$$

$$=-\frac{T}{Nl}\frac{\mathrm{d}}{\mathrm{d}v}\left[-\tfrac{1}{2}kN\ln\left\{(1+v)^{1+v}(1-v)^{1-v}\right\}\right]$$

$$=\frac{kT}{2l}\frac{\mathrm{d}}{\mathrm{d}v}\left[\ln\left\{(1+v)^{1+v}(1-v)^{1-v}\right\}\right]$$

$$=\frac{kT}{2l}\frac{\mathrm{d}}{\mathrm{d}v}\left[(1+v)\ln(1+v)+(1-v)\ln(1-v)\right]\ \text{(Use } \ln(ab)=\ln a+\ln b \text{ and } \ln a^b=b\ln a\text{)}$$

$$=\frac{kT}{2l}\left[\ln(1+v)+\frac{1+v}{1+v}-\ln(1-v)-\frac{1-v}{1-v}\right]\ \left(\text{Use } \frac{\mathrm{d}fg}{\mathrm{d}v}=g\frac{\mathrm{d}f}{\mathrm{d}v}+f\frac{\mathrm{d}g}{\mathrm{d}v}\right)$$

$$=\frac{kT}{2l}\ln\left(\frac{1+v}{1-v}\right)\ \text{(Use } \ln a-\ln b=\ln(a/b)\text{)}$$

This confirms eqn 16.5a. The derivation of eqn 16.5b is found in Section 16.4 of the text.

P16.21 (a) $V_{\text{bend}}=\tfrac{1}{2}k_{\text{bend}}(\theta-\theta_{\text{e}})^2$ where $V_{\text{bend}}=8.5\text{ kJ mol}^{-1}/N_{\text{A}}$ and $\theta-\theta_{\text{e}}=(30-15)°=15°$

$$k_{\text{bend}}=\frac{2V_{\text{bend}}}{(\theta-\theta_{\text{e}})^2}=\frac{2\times(8.5\times10^3\text{ J mol}^{-1})}{(6.022\times10^{23}\text{ mol}^{-1})\times(15°)^2}$$

$$=\boxed{1.3\times10^{-22}\text{ J deg}^{-2}}$$

(b) $V_{\text{stretch}}(\text{per mole})=N_A\times\left\{\tfrac{1}{2}k_{\text{stretch}}(R-R_{\text{e}})^2\right\}$

$$=6.022\times10^{23}\text{ mol}^{-1}\times\left\{\tfrac{1}{2}\times(400\text{ N m}^{-1})\times[(165-152)\times10^{-12}\text{ m}]^2\right\}$$

$$=2.04\times10^4\text{ J mol}^{-1}=\boxed{20.4\text{ kJ mol}^{-1}}$$

P16.22 (a) Consider the motion within the confines of a membrane to be two-dimensional. Eqn 11.20 tells us that the root-mean-square distance traveled during a one-dimensional random walk of duration t is given by $d_x=(2Dt)^{1/2}$. In two-dimensions the root-mean-square distance traveled is that given by the vector addition of motion in both the x and y directions. Therefore,

$$d^2=d_x^2+d_y^2=2(2Dt)=4Dt$$

and

$$t=\frac{d^2}{4D}$$

In a cell plasma membrane: $t=\dfrac{(10\times10^{-7}\text{ cm})^2}{4\times(1.0\times10^{-8}\text{ cm}^2\text{ s}^{-1})}=\boxed{2.5\times10^{-5}\text{ s}}$

In a lipid bilayer: $t=\boxed{2.5\times10^{-6}\text{ s}}$

(b) Unsaturated lipids have lower melting points than comparable saturated lipid because of the alkene-alkene π repulsions between adjacent molecules and because the cis-conformation at each alkene group bends the lipid in a manner that reduces stacking and reduces the net strength of intermolecular attractions. Thus, cells produce a greater degree of unsaturated lipid at lower temperatures to maintain membrane-melting temperatures close to ambient temperature.

17 Metallic, ionic, and covalent solids

Answers to discussion questions

D17.1 A **metallic conductor** is a substance with a conductivity that decreases as the temperature is raised. A **semiconductor** is a substance with a conductivity that increases as the temperature is raised. A semiconductor generally has a lower conductivity than that typical of metals, but the magnitude of the conductivity is not the criterion of the distinction. It is conventional to classify semiconductors with very low electrical conductivities, such as most synthetic polymers and diamond, as **insulators**. We shall use this term. But it should be appreciated that it is one of convenience rather than one of fundamental significance.

The conductivity of these three kinds of materials is explained by **band theory**. When each of N atoms of a metallic element contributes one atomic orbital to the formation of molecular orbitals, the resulting N molecular orbitals form an almost continuous band of levels. The orbital at the bottom of the band is fully bonding between all neighbours, and the orbital at the top of the band is fully antibonding between all immediate neighbours. If the atomic orbitals are s-orbitals, then the resulting band is called an s-band; if the original orbitals are p-orbitals, then they form a p band. In a typical case, there is so large an energy difference between the s and p atomic orbitals that the resulting s and p bands are separated by a region of energy in which there are no orbitals. This region is called the band gap, and its width is denoted E_g.

When electrons occupy the orbitals in the bands, they do so in accord with the Pauli principle. If insufficient electrons are present to fill the band, the electrons close to the top of the band are mobile and the solid is a metallic conductor. An unfilled band is called a **conduction band** and the energy of the highest occupied orbital at $T = 0$ K is called the **Fermi level**. Only the electrons close to the Fermi level can contribute to conduction and to the heat capacity of a metal. If the band is full, then the electrons cannot transport a current readily, and the solid is an insulator; more formally, it is a species of semiconductor with a large band gap. A full band is called a **valence band**. The detailed population of the levels in a band taking into account the role of temperature is expressed by the Fermi–Dirac distribution.

The distinction between metallic conductors and semiconductors can be traced to their band structure: a metallic conductor has an incomplete band, its conductance band, and a semiconductor has full bands, and hence lacks a conductance band. The decreasing conductance of a metallic conductor with temperature stems from the scattering of electrons by the vibrating atoms of the metal lattice. The increasing conductance of a semiconductor arises from the increasing population of an upper empty band as the temperature is increased. Many substances, however, have such large band gaps that their ability to conduct an electric current remains very low at all temperatures: it is conventional to refer to such solids as insulators. The ability of a semiconductor to transport charge is enhanced by doping it, or adding substances in controlled quantities. If the dopant provides additional electrons, then the semiconductor is classified as n-type. If it removes electrons from the valence band and thereby increases the number of positive holes, it is classified as p-type.

Diamond and graphite are **covalent solids** in which covalent bonds in a definite spatial orientation link the atoms in a network extending through the crystal. In diamond each sp^3 hybridized carbon is covalently bonded tetrahedrally to its four neighbours. Each bond involves a low energy, localized σ electron pair. This means that there is a large gap between the filled σ band and the empty π band above it; it is a large band gap semiconductor, an insulator. In graphite, σ bonds between sp^2

hybridized carbon atoms form hexagonal rings that repeat throughout a graphene sheet to which each carbon contributes one p_z orbital and one electron. Neighbouring p_z orbitals overlap to produce a π band that extends throughout the graphene sheet and is half full; it is, therefore, a conduction band that lies within the graphene sheets alone.

D17.2 Lattice planes are labelled by their Miller indices h, k, and l, where h, k, and l refer respectively to the reciprocals of the smallest intersection distances (in units of the lengths of the unit cell, a, b, and c) of the plane with the $\boldsymbol{a}$, $\boldsymbol{b}$, and $\boldsymbol{c}$ axes of the unit cell. The axes may be, but are not always, orthogonal (see text Figure 17.24).

D17.3 The phase problem arises with the analysis of X-ray diffraction data to determine a crystal structure. The analysis requires knowledge of the so called **structure factor**, F_{hkl}, which is related to the measured intensity, I_{hkl}, of the diffracted radiation by the expression $I_{hkl} \propto |F_{hkl}|^2$ (i.e., the modulus square $F_{hkl}{}^* F_{hkl}$). In the simplest case, even though we have measured I_{hkl}, we do not know whether F_{hkl} is positive or negative. Thus, when we attempt to compute the electron density distribution, $\rho(\boldsymbol{r})$, with the **Fourier synthesis**, which has the form

$$\rho(x) = \frac{1}{V}\left\{F_0 + 2\sum_{h=1}^{\infty} F_h \cos(2h\pi x)\right\} \quad [17.9]$$

in a simple case, we do not know the sign of each term in the sum. Things are even more difficult in the general case because the structure factor has the general form $F_{hkl} = |F_{hkl}|e^{i\alpha}$ where $i = \sqrt{-1}$ and α is the **phase** of F_{hkl}. Upon taking the modulus square of F_{hkl}, information about the phase is completely lost. Thus, measurement of radiation intensity gives no information about the phase and it is not available for use in the Fourier synthesis. This is the phase problem.

The phase problem may be evaded by the use of a **Patterson synthesis** or tackled directly by using the so-called **direct methods of phase allocation**. The Patterson synthesis is a technique of data analysis in X-ray diffraction which helps to circumvent the phase problem. In it, a function P is formed by calculating the Fourier transform of the squares of the structure factors (which are proportional to the intensities):

$$P(r) = \frac{1}{V}\sum_{hkl} |F_{hkl}|^2 e^{-2\pi i(hx+ky+lz)}$$

The outcome is a map of the *separations* of the atoms in the unit cell of the crystal. If some atoms are heavy (perhaps because they have been introduced by isomorphous replacement), they dominate the Patterson function, and their locations can be deduced quite simply. Their locations can then be used in the determination of the locations of lighter atoms. Direct methods dominate modern X-ray diffraction analysis. These methods use statistical techniques, and the considerable computational capacity of the modern computer, to compute the probabilities that the phases have a particular value.

D17.4 The majority of metals crystallize in structures that can be interpreted as the closest packing arrangements of hard spheres. These are the cubic close-packed (ccp) and hexagonal close-packed (hcp) structures. In these models, 74% of the volume of the unit cell is occupied by the atoms (packing fraction = 0.74). Most of the remaining metallic elements crystallize in the body-centered cubic (bcc) arrangement, which is not too much different from the close-packed structures in terms of the efficiency of the use of space (packing fraction 0.68 in the hard sphere model). Polonium is an exception; it crystallizes in the simple cubic structure, which has a packing fraction of 0.52. See the solution to Exercise 17.23 for a derivation of the packing fraction in ccp systems. If atoms were truly hard spheres, we would expect that all metals would crystallize in either the ccp or hcp close-packed structures. The fact that a significant number crystallize in other structures is proof that a simple hard sphere model is an inaccurate representation of the interactions between the atoms. Covalent bonding between the atoms may influence the structure. Text Figures 17.34–36 illustrate the difference between ccp and hcp structures.

D17.5 The caesium-chloride and rock-salt structures are illustrated and described in text Figures 17.38 and 17.39. The **radius ratio**, $\gamma = r_{\text{smaller ion}} / r_{\text{larger ion}}$ [17.11], is an aid in the classification of ionic compound structure. If $\gamma > 0.732$, the caesium chloride structure is indicated. If $0.414 < \gamma < 0.732$, the rock salt structure is indicated. If $\gamma < 0.414$, the zinc blend (ZnS) structure is indicated (see text Figure 17.40).

The structures can also be described in terms of the occupation of holes in expanded closed-packed lattices. In a face-centred cubic close-packed lattice, there is an octahedral hole in the centre. The rock-salt structure can be thought of as being derived from an fcc structure of Cl^- ions in which Na^+ ions have filled the octahedral holes.

The caesium-chloride structure can be considered to be derived from the ccp structure by having Cl^- ions occupy all the primitive lattice points and octahedral sites, with all tetrahedral sites occupied by Cs^+ ions. This is exceedingly difficult to visualize and describe without carefully constructed figures or models. Refer to S.-M. Ho and B. E. Douglas, *J. Chem. Educ.* **46**, 208, 1969, for the appropriate diagrams.

D17.6 **Diamagnetic** substances have negative magnetic susceptibilities and they tend to move out of a magnetic field. Most molecules with no unpaired electron spins are diamagnetic because the magnetic field induces the circulation of electronic currents, which give rise to a magnetic field that opposes the applied field, in the substance. **Paramagnetic** substances have positive magnetic susceptibilities and they tend to move into a magnetic field. Molecules with unpaired electrons are paramagnetic because the spin of the electron has an associated magnetic field that lines up in parallel with the applied field so as to lower energy. Some paramagnetic solid substances, like iron and cobalt, have microscopic domains in which many unpaired electrons are aligned parallel. They can undergo a transition in which the spin of many domains align to give rise to the strong magnetism called **ferromagnetism**. Some paramagnetic substances have structures that cause domains to orient with alternating up-down spins; this low-magnetization arrangement is called **anti-ferromagnetism**. Chromium exhibits this behavior.

Solutions to exercises

E17.1 (a) P(group V) has one more valence electron than Ge(group IV); therefore Ge doped with P forms an $\boxed{\text{n} - \text{type}}$ semiconductor.

(b) In(group III) has one less valence electron than Ge(group IV); therefore Ge doped with In forms an $\boxed{\text{p} - \text{type}}$ semiconductor.

E17.2 The resistance of metals increases with increasing temperature; the opposite is true of semiconductors; therefore, this substance is a $\boxed{\text{metallic conductor}}$.

E17.3 As discussed in Section 12.5, the wavefunction must:

1. Be continuous.
2. Have a continuous slope, which is indicated as ψ' in our discussion.
3. Be single-valued and cannot become infinite over a finite region of space.
4. Be square-integrable.

This means that, when a line of N tight-bonding identical atoms are wrapped into a ring, the wavefunction must satisfy the conditions: ψ(left end) = ψ(right end) and ψ'(left end) = ψ'(right end). These boundary conditions are more restrictive than those required for the line of atoms: ψ(left end) = $\pm\psi$(right end) and ψ'(left end) = $\pm\psi'$(right end). As the line is wrapped into a ring, the states for which ψ(left end) = $-\psi$(right end) become forbidden and only alternate quantum numbers for which ψ(left end) = ψ(right end) are allowed. This explains the appearance of the factor of 2 within the cosine term of the allowed ring energies. Additionally, the quantum number for the electron moving around the ring of atoms may be either positive of negative because the angular momentum may be in either direction around the axis of the ring.

E17.4 Figure 17.1(a) shows a dark univalent probe cation in a vacancy within a two-dimensional square ionic lattice of grey univalent cations and white univalent anions. Let $d_0 = 200$ pm be the distance between nearest neighbors and let V_0 be the absolute value of the Coulombic interaction between nearest neighbors.

$$V_0 = \frac{e^2}{4\pi\varepsilon_0 d_0} = \frac{(1.602\times10^{-19}\ \text{C})^2}{(1.113\times10^{-10}\ \text{J}^{-1}\ \text{C}^2\ \text{m}^{-1})\times(200\times10^{-12}\ \text{m})} = 1.153\times10^{-18}\ \text{J}$$

The symmetry of the lattice around the probe cation consists of four regions like that of Figure 17.1(b) so we calculate the total Coulombic interaction of the probe within the lattice quadrant of Figure 17.1(b) and multiply by 4. The calculation is pursued one column at a time and the column interactions are summed.

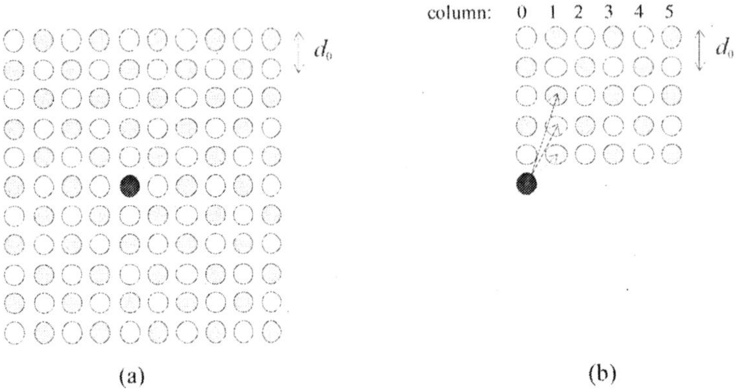

Figure 17.1

Probe-to-column 0 interaction:

$$V_{\text{column 0}} = -V_0 \times \left(1 - \frac{1}{2} + \frac{1}{3} - \frac{1}{4} + \frac{1}{5} \cdots\right) = -V_0 \ln 2 = V_0 \sum_{n=1}^{\infty} \frac{(-1)^{n+0}}{(n^2+0)^{1/2}} \quad \text{(useful form)}$$

Probe-to-column 1 interaction using the Pythagorean theorem for the probe-ion distance:

$$V_{\text{column 1}} = V_0 \times \left(\frac{1}{2^{1/2}} - \frac{1}{5^{1/2}} + \frac{1}{10^{1/2}} - \frac{1}{17^{1/2}} \cdots\right) = V_0 \sum_{n=1}^{\infty} \frac{(-1)^{n+1}}{(n^2+1)^{1/2}}$$

Similarly, the probe-to-column m interaction, using the Pythagorean theorem for the probe-ion distance, is

$$V_{\text{column } m} = V_0 \sum_{n=1}^{\infty} \frac{(-1)^{n+m}}{(n^2+m)^{1/2}} \quad \text{for } 0 \le m < \infty \text{ (the sum is performed with a calculator or software)}$$

The total interaction for the region shown in Fig. 17.1(b) is the sum of the above expression over all columns

$$\begin{aligned} V_{\text{Fig. 17.1(b)}} &= V_0 \sum_{m=0}^{\infty} \sum_{n=1}^{\infty} \frac{(-1)^{n+m}}{(n^2+m)^{1/2}} \\ &= -0.442 V_0 = (-0.442)\times(1.153\times10^{-18}\ \text{J}) \\ &= -5.096\times10^{-19}\ \text{J} \end{aligned}$$

The total Coulombic interaction of the probe cation with the lattice is

$$V_{\text{total}} = 4V_{\text{Fig. 17.1(b)}} = 4(-5.096\times10^{-19}\ \text{J}) = \boxed{-2.039\times10^{-18}\ \text{J}}$$

where the negative value indicates a net attraction.

COMMENT. Suppose that you interpret the exercise to involve placement of the probe cation at the foot of the two-dimensional step shown in Fig. 17.2, you calculate the probe potential with

$$V_{\text{total}} = 3V_{\text{Fig. 17.1(b)}} - V_{\text{column 0}}$$
$$= 3\times(-0.442V_0) - (-V_0 \ln 2)$$
$$= -0.633V_0$$
$$= (-0.633)\times(1.153\times10^{-18}\ \text{J}) = \boxed{-7.30\times10^{-19}\ \text{J}}$$

Once again the negative value indicates a net attraction.

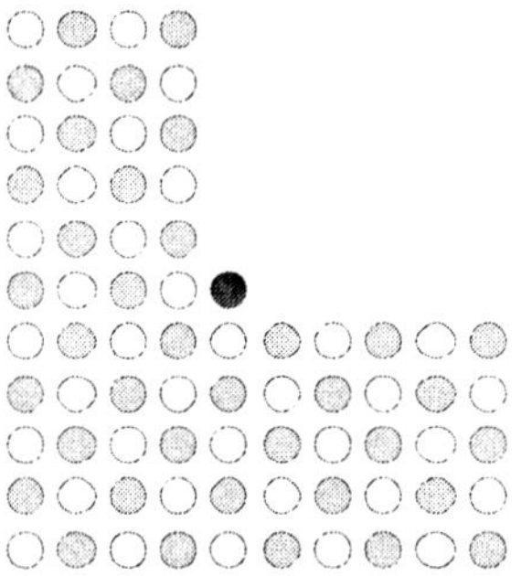

Figure 17.2

If you interpret the exercise to involve placement of the probe cation at the top of the two-dimensional surface shown in Fig. 17.3, you calculate the probe potential with

$$V_{\text{total}} = 2V_{\text{Fig. 17.1(b)}} - V_{\text{column 0}}$$
$$= 2\times(-0.442V_0) - (-V_0 \ln 2)$$
$$= -0.191V_0$$
$$= (-0.191)\times(1.153\times10^{-18}\ \text{J}) = \boxed{-2.20\times10^{-19}\ \text{J}}$$

Once again the negative value indicates a net attraction.

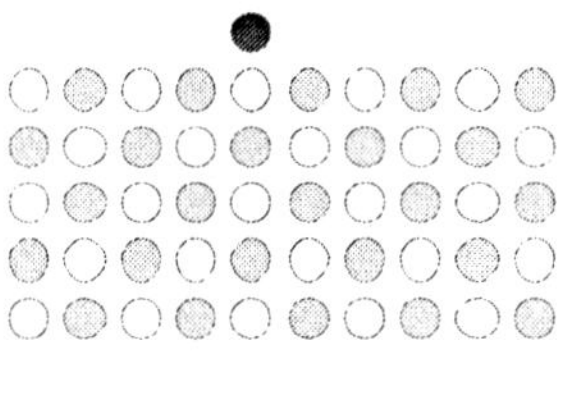

Figure 17.3

E17.5 The valence electrons of Ca(Group 2) are 4s electrons, those of O(Group 16) are 2p electrons. We expect that the O2p electrons will be much lower in energy than the Ca4s electrons. The energy band diagram of CaO is expected to look like that sketched in Figure 17.4.

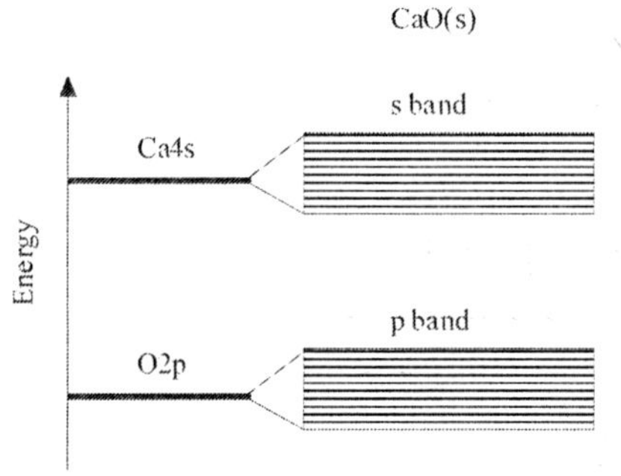

Figure 17.4

There are three degenerate O2p orbitals available for band formation for each of N oxygen atoms, which results in a p band that can hold $3 \times 2 \times N = 6N$ electrons. Each oxygen atom contributes 4 electrons to the p band while each calcium atom also contributes 2 electrons. Thus, the p band is filled with $6N$ electrons leaving no electrons to occupy the s band. There will very likely be a large band gap so CaO is a large gap semiconductor, an insulator. Effectively electrons have been transferred from Ca to O as in the ionic model.

E17.6 The lattice enthalpy is the difference in enthalpy between an ionic solid and the corresponding isolated ions. In this exercise, it is the enthalpy corresponding to the process

$$\text{CaO(s)} \rightarrow \text{Ca}^{2+}(\text{g}) + \text{O}^{2-}(\text{g})$$

The standard lattice enthalpy can be computed from the standard enthalpies given in the exercise by considering the formation of CaO(s) from its elements as occurring through the following steps: sublimation of Ca(s), removing two electrons from Ca(g), atomization of O_2(g), two-electron attachment to O(g), and formation of CaO(s) lattice from gaseous ions. The formation reaction of CaO(s) is

$$\text{Ca(s)} + \tfrac{1}{2}\,\text{O}_2(\text{g}) \rightarrow \text{CaO(s)}$$

$$\Delta_f H^{\ominus}(\text{CaO,s}) = \Delta_{\text{sub}}H^{\ominus}(\text{Ca,s}) + \Delta_{\text{ion}}H^{\ominus}(\text{Ca,g}) + \tfrac{1}{2}\Delta_{\text{bond diss}}H^{\ominus}(\text{O}_2,\text{g}) + \Delta_{\text{eg}}H^{\ominus}(\text{O,g}) + \Delta_{\text{eg}}H^{\ominus}(\text{O}^-,\text{g}) - \Delta_L H^{\ominus}(\text{CaO,s})$$

So the lattice enthalpy is

$$\Delta_L H^{\ominus}(\text{CaO,s}) = \Delta_{\text{sub}}H^{\ominus}(\text{Ca,s}) + \Delta_{\text{ion}}H^{\ominus}(\text{Ca,g}) + \tfrac{1}{2}\Delta_{\text{bond diss}}H^{\ominus}(\text{O}_2,\text{g}) + \Delta_{\text{eg}}H^{\ominus}(\text{O,g}) + \Delta_{\text{eg}}H^{\ominus}(\text{O}^-,\text{g}) - \Delta_f H^{\ominus}(\text{CaO,s})$$

$$\Delta_L H^{\ominus}(\text{CaO,s}) = [178 + 1735 + \tfrac{1}{2}(497) - 141 + 844 + 635]\ \text{kJ mol}^{-1} = \boxed{3500.\ \text{kJ mol}^{-1}}$$

E17.7 The lattice enthalpy is the difference in enthalpy between an ionic solid and the corresponding isolated ions. In this exercise, it is the enthalpy corresponding to the process

$$\text{SrI}_2(\text{s}) \rightarrow \text{Sr}^{2+}(\text{g}) + 2\,\text{I}^-(\text{g})$$

The standard lattice enthalpy can be computed from the standard enthalpies given in the exercise by considering the formation of SrI_2(s) from its elements as occurring through the following steps: sublimation of Sr(s), removing two electrons from Sr(g), sublimation of I_2(s), atomization of I_2(g), electron attachment to I(g), and formation of the solid SrI_2(s) lattice from gaseous ions.

$$\Delta_f H^{\ominus}(\text{SrI}_2,\text{s}) = \Delta_{\text{sub}}H^{\ominus}(\text{Sr,s}) + \Delta_{\text{ion}}H^{\ominus}(\text{Sr,g}) + \Delta_{\text{sub}}H^{\ominus}(\text{I}_2,\text{s}) + \Delta_{\text{bond diss}}H^{\ominus}(\text{I}_2,\text{g}) + 2\Delta_{\text{eg}}H^{\ominus}(\text{I,g}) - \Delta_L H^{\ominus}(\text{SrI}_2,\text{s})$$

So the lattice enthalpy is

$$\Delta_L H^{\ominus}(\text{SrI}_2,\text{s}) = \Delta_{\text{sub}}H^{\ominus}(\text{Sr,s}) + \Delta_{\text{ion}}H^{\ominus}(\text{Sr,g}) + \Delta_{\text{sub}}H^{\ominus}(\text{I}_2,\text{s}) + \Delta_{\text{bond diss}}H^{\ominus}(\text{I}_2,\text{g}) + 2\Delta_{\text{eg}}H^{\ominus}(\text{I,g}) - \Delta_f H^{\ominus}(\text{SrI}_2,\text{s})$$

$$= [164.4 + 1626.1 + 62.4 + 75.6 - 2(303.8) + 828.9]\ \text{kJ mol}^{-1} = \boxed{2149.8\ \text{kJ mol}^{-1}}$$

E17.8 Let there be N charges ze on the first sphere, of radius d; then there will by $N/2$ of charge $-ze$ on the second sphere of radius $2d$, $N/3$ of charge $+ze$ on the third sphere of radius $3d$, and so on. The total potential energy for the interaction of point charge Q at the center with the alternating charges in the spherical layers is therefore the sum

$$V=\frac{Q}{4\pi\varepsilon_0}\left(\frac{Nze}{d}-\frac{Nze}{2\times 2d}+\frac{Nze}{3\times 3d}-\frac{Nze}{4\times 4d}+\cdots\right)$$
$$=\frac{Q\,Nze}{4\pi\varepsilon_0 d}\left(1-\frac{1}{2^2}+\frac{1}{3^2}-\frac{1}{4^2}+\cdots\right)$$
$$=\frac{Q\,Nze}{4\,\pi\varepsilon_0 d}\left(\frac{\pi^2}{12}\right)$$
$$=\boxed{\frac{Q\,Nze\,\pi}{48\,\varepsilon_0 d}}$$

E17.9 We need to evaluate the ratio of the factors that depend on $d = r_{\text{cation}} + r_{\text{anion}}$ for the two compounds using $d_{\text{CaO}} = (100 + 140)\text{pm} = 240$ pm and $d_{\text{SrO}} = (116 + 140)\text{pm} = 256$ pm (Table 17.5).

$$\frac{\Delta H_{\text{L}}^{\ominus}(\text{CaO})}{\Delta H_{\text{L}}^{\ominus}(\text{SrO})}=\frac{\frac{1}{d_{\text{CaO}}}\times\left(1-\frac{d^*}{d_{\text{CaO}}}\right)}{\frac{1}{d_{\text{SrO}}}\times\left(1-\frac{d^*}{d_{\text{SrO}}}\right)}\ [17.5]\text{ where } d^*=34.5\text{ pm}$$
$$=\frac{\frac{1}{240}\times\left(1-\frac{34.5}{240}\right)}{\frac{1}{256}\times\left(1-\frac{34.5}{256}\right)}$$
$$=\boxed{1.06}$$

E17.10 The relationship between critical temperature and critical magnetic field is given by

$$\mathcal{H}_{\text{c}}(T)=\mathcal{H}_{\text{c}}(0)\left(1-\frac{T^2}{T_{\text{c}}^2}\right)$$

Solving for T gives the critical temperature for a given magnetic field:

$$T=T_{\text{c}}\left(1-\frac{\mathcal{H}_{\text{c}}(T)}{\mathcal{H}_{\text{c}}(0)}\right)^{1/2}=(7.19\text{ K})\times\left(1-\frac{20\text{ kA m}^{-1}}{63.9\text{ kA m}^{-1}}\right)^{1/2}=\boxed{6.0\text{ K}}$$

E17.11 The points and planes are shown in Figure 17.5.

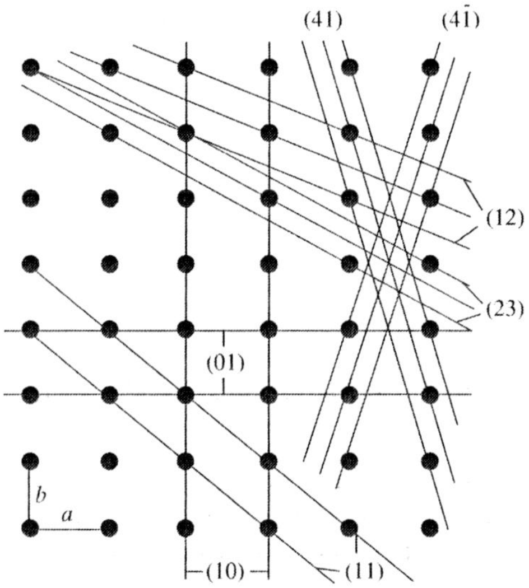

Figure 17.5

E17.12 The points and planes are shown in Figure 17.6.

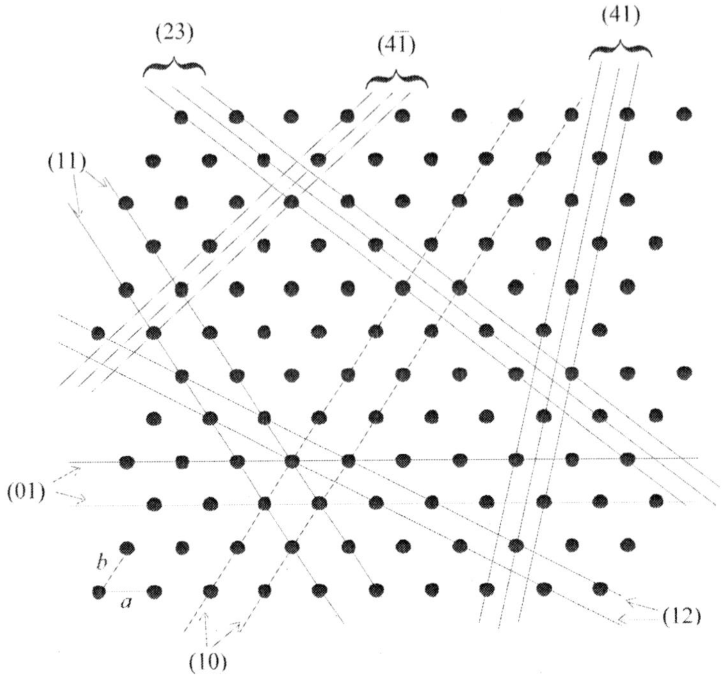

Figure 17.6

E17.13 Draw up the following table, using the procedure set out in Section 17.11.

Original	Reciprocal	Clear fractions	Miller indices
$(2a, 3b, c)$ or (2, 3, 1)	$(\frac{1}{2}, \frac{1}{3}, 1)$	(3, 2, 6)	(326)
(a, b, c) or (1, 1, 1)	(1, 1, 1)	(1, 1, 1)	(111)
$(6a, 3b, 3c)$ or (6, 3, 3)	$(\frac{1}{6}, \frac{1}{3}, \frac{1}{3})$	(1, 2, 2)	(122)
$(2a, -3b, -3c)$ or (2, –3, –3)	$(\frac{1}{2}, -\frac{1}{3}, -\frac{1}{3})$	(3, –2, –2)	$(3\bar{2}\bar{2})$

E17.14 See Figure 17.7(a) for the (100), (010), (001), (011), (101), and (111) planes. See Figure 17.7(b) for the 101 plane.

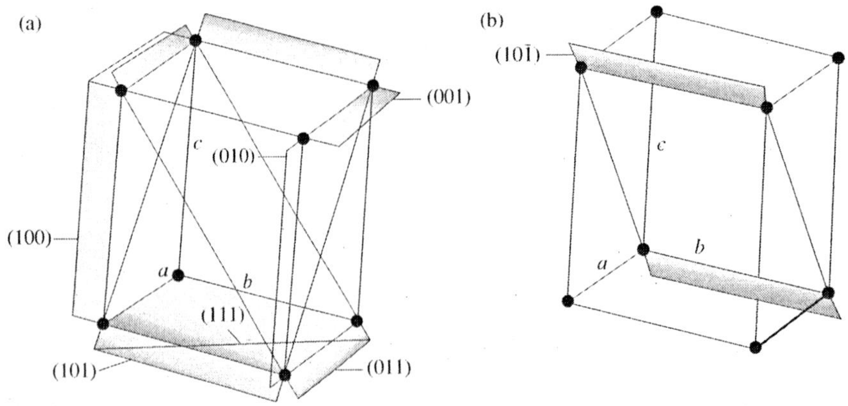

Figure 17.7

E17.15 See Figure 17.8(a) for the (100), (010), (001), (011), (101), and (111) planes. See Figure 17.8(b) for the 101 plane.

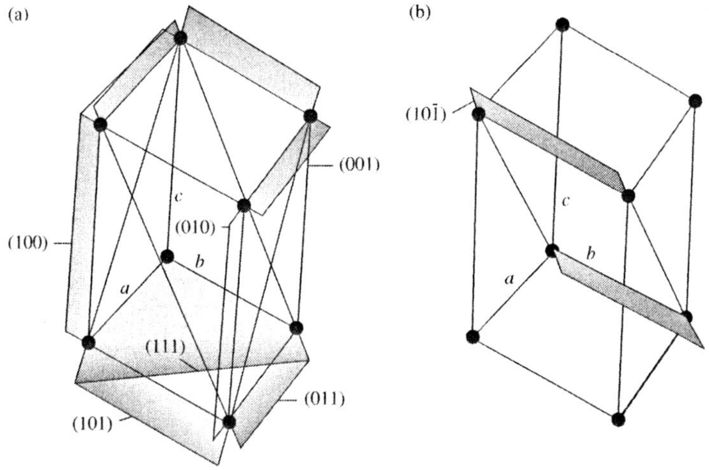

Figure 17.8

E17.16

$$\frac{1}{d_{hkl}^{2}}=\frac{h^2}{a^2}+\frac{k^2}{b^2}+\frac{l^2}{c^2} \quad [17.7]$$

For a cubic unit cell in which $a=b=c$

$$d_{hkl}=\frac{a}{\left(h^2+k^2+l^2\right)^{1/2}}$$

Therefore,

$$d_{111}=\frac{a}{3^{1/2}}=\frac{572\text{ pm}}{3^{1/2}}=\boxed{330\text{ pm}}$$

$$d_{211}=\frac{a}{6^{1/2}}=\frac{572\text{ pm}}{6^{1/2}}=\boxed{234\text{ pm}}$$

$$d_{100}=a=\boxed{572\text{ pm}}$$

E17.17

$$\frac{1}{d_{hkl}^{2}}=\frac{h^2}{a^2}+\frac{k^2}{b^2}+\frac{l^2}{c^2} \quad [17.7] \qquad \text{or} \qquad d_{hkl}=\left(\frac{h^2}{a^2}+\frac{k^2}{b^2}+\frac{l^2}{c^2}\right)^{-1/2}$$

$$d_{123}=\left(\frac{1^2}{(784\text{ pm})^2}+\frac{2^2}{(633\text{ pm})^2}+\frac{3^2}{(454\text{ pm})^2}\right)^{-1/2}=\boxed{135\text{ pm}}$$

$$d_{236}=\left(\frac{2^2}{(784\text{ pm})^2}+\frac{3^2}{(633\text{ pm})^2}+\frac{6^2}{(454\text{ pm})^2}\right)^{-1/2}=\boxed{70.1\text{ pm}}$$

E17.18

$$\lambda=2d\sin\theta \ [17.8]=2\times(97.3\text{ pm})\times(\sin 19.85°)=\boxed{66.1\text{ pm}}$$

E17.19 $d_{100}=a=b=c=350$ pm in both the fcc and bcc cases. Let N be the number of Li atoms in the unit cell and let V be the volume of the unit cell. Then,

$$\rho=\frac{NM}{VN_A}$$

Solve for N.

$$N=\frac{\rho VN_A}{M}=\frac{(0.53\times10^{6}\text{ g m}^{-3})\times(350\times10^{-12}\text{ m})^{3}\times(6.022\times10^{23}\text{ mol}^{-1})}{6.94\text{ g mol}^{-1}}=1.97$$

An fcc cubic cell has $N=4$ and a bcc unit cell has $N=2$. Therefore, lithium has a $\boxed{\text{bcc unit cell}}$.

E17.20 (a) $a = b = c = 361$ pm for the fcc copper unit cell. Thus,

$$\frac{1}{d_{hkl}^{2}} = \frac{h^2}{a^2} + \frac{k^2}{b^2} + \frac{l^2}{c^2} \ [17.7] \quad \text{or} \quad d_{hkl} = \left(\frac{h^2}{a^2} + \frac{k^2}{b^2} + \frac{l^2}{c^2}\right)^{-1/2} = a \times \left(h^2 + k^2 + l^2\right)^{-1/2}$$

Solve Bragg's law, $\lambda = 2d\sin\theta$ [17.8], for the glancing angle.

$$\theta_{khl} = \arcsin\left\{\frac{\lambda}{2a}\left(h^2 + k^2 + l^2\right)^{1/2}\right\}$$

Calculate successively larger glancing angles in the powder diffraction pattern using $\lambda = 154$ pm.

(hkl)	100	110	111	200	210	220
$\theta_{hkl}/°$	12.3	17.6	21.7	25.3	28.5	37.1

(hkl)	300	310	311	222	320	321	400
$\theta_{hkl}/°$	39.8	42.4	45.0	47.6	50.3	52.9	58.6

COMMENT. Many of the diffraction pattern 'spots' that are predicted in the above table for the fcc unit cell are not observed. This includes reflection from the (100), (110), (210), (300), (310), (320), and (321) planes. In fact, fcc reflections are observed only from those planes for which the Miller indices, *(hkl)*, are either all even or all odd because the structure factors for all other combinations equal zero.

(b) $V = a^3$ and $N = 4$ for the fcc unit cell. Therefore, the density is calculate from

$$\rho = \frac{NM}{VN_A} = \frac{4\times(63.55 \text{ g mol}^{-1})}{(3.61\times10^{-8} \text{ cm})^3 \times (6.022\times10^{23} \text{ mol}^{-1})} = \boxed{8.97 \text{ g cm}^{-3}}$$

E17.21

$$V\rho(x) = F_0 + 2\sum_{h=1}^{\infty} F_h \cos(2h\pi x) \ [17.9]$$

$$= 30 + 16.4 \cos(2\pi x) + 13.0 \cos(4\pi x) + 8.2 \cos(6\pi x)$$
$$+ 11 \cos(8\pi x) - 4.8 \cos(10\pi x) + 10.8 \cos(12\pi x)$$
$$+ 6.4 \cos(14\pi x) - 2.0 \cos(16\pi x) + 2.2 \cos(18\pi x)$$
$$+ 13.0 \cos(20\pi x) + 10.4 \cos(22\pi x) - 8.6 \cos(24\pi x)$$
$$- 2.4 \cos(26\pi x) + 0.2 \cos(28\pi x) + 4.2 \cos(30\pi x)$$

A plot of $V\rho(x)$ is shown in Figure 17.9.

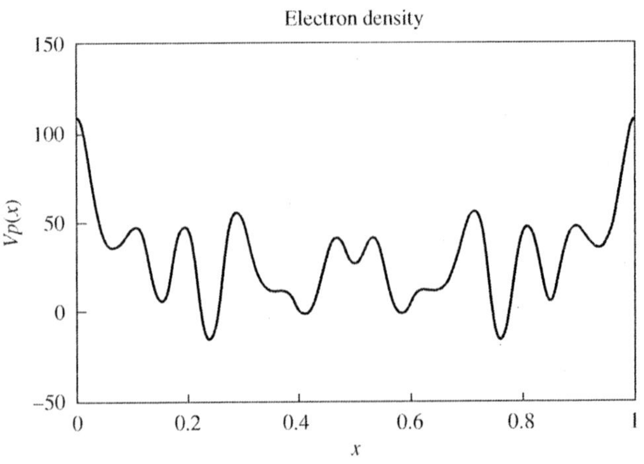

Figure 17.9

E17.22 The hatched area shown in Figure 17.10 is $h\times 2R = 3^{1/2}R\times 2R = 2\sqrt{3}R^2$ where $h = 2R\cos 30°$. The net number of cylinders in a hatched area is 1, and the area of the cylinder's base is πR^2. The volume of the prism (of which the hatched area is the base) is $2\sqrt{3}R^2L$, and the volume occupied by the cylinders is $\pi R^2 L$. Hence, the packing fraction is

$$f = \frac{\pi R^2 L}{2\sqrt{3}R^2 L} = \frac{\pi}{2\sqrt{3}} = \boxed{0.9069}$$

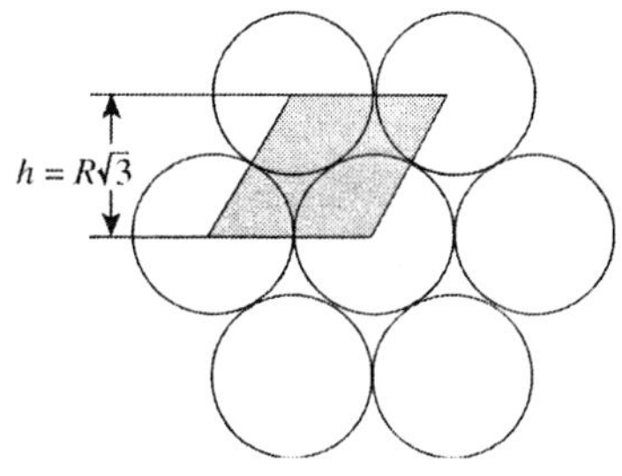

Figure 17.10

E17.23 To calculate a packing fraction of a ccp structure, we first calculate the volume of a unit cell shown in Figure 17.11 (Fig. 20.36 From p. 716 of P. Atkins and J. de Paula, *Physical Chemistry*, 8th edn, W.H. Freeman & Co., New York (2006)), and then calculate the total volume of the spheres that fully or partial occupy it. The first part of the calculation is a straightforward exercise in geometry. The second part involves counting the fraction of spheres that occupy the cell. Refer to the figure. Because a diagonal of any face passes completely through one sphere and halfway through two other spheres, its length is $4R$. The length of a side is therefore $8^{1/2}R$ and the volume of the unit cell is $8^{3/2}R^3$. Because each cell contains the equivalent of $6\times\frac{1}{2} + 8\times\frac{1}{8} = 4$ spheres, and the volume of each sphere is $\frac{4}{3}\pi R^3$ the total occupied volume is $8^{3/2}R^3$. The fraction of space occupied is therefore $\left(\frac{16}{3}\pi R^3\right)/\left(8^{3/2}R^3\right)$, or $\boxed{0.740}$. Because an hcp structure has the same coordination number, its packing fraction is the same.

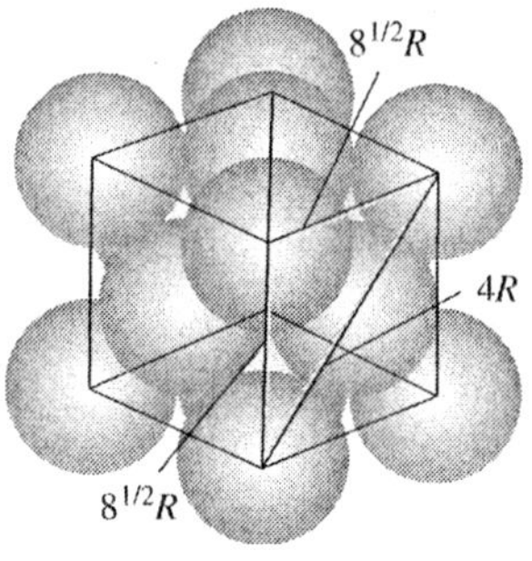

Figure 17.11

E17.24 We need the packing fraction for hexagonal close-packing, which is 0.740. Therefore the density of the solid virus sample is

$$0.740\times 1.00\ \text{g cm}^{-3} = \boxed{0.740\ \text{g cm}^{-3}}$$

E17.25 Refer to Fig. 17.37 of the text.

(a) eight nearest neighbors

(b) six next-nearest neighbors

Nearest neighbors touch each along the body diagonal of the cube. If the side of the cube is a, then $\sqrt{3}\,a$ is the length of the body diagonal. As there are 2 atoms along the body diagonal, the distance between nearest neighbors is given by

$$2d = \sqrt{3}a$$

$$d = \frac{\sqrt{3}}{2}a = \frac{\sqrt{3}}{2}\times 600\ \text{nm} = \boxed{520\ \text{nm}}$$

The next nearest neighbors are the length of a side away, $\boxed{600\ \text{nm}}$.

E17.26 Refer to Figures 17.34 and 17.36 of the text.

(a) $\boxed{12}$ nearest neighbors

(b) $\boxed{6}$ next nearest neighbors

Nearest neighbors touch each other along the diagonal of a face; therefore with a = length of side of unit cell.

$$(2d)^2 = 2a^2 \qquad d = \text{distance between neighbors}$$

$$2d = \sqrt{2}a$$

$$d = a/\sqrt{2} = 600\ \text{nm}/\sqrt{2} = \boxed{424\ \text{nm}}.$$

For next nearest neighbors, $d = a = \boxed{600\ \text{nm}}$.

E17.27 The packing fraction for cubic close-packing is 0.74, that for body-centered structures is 0.68. Therefore, (a) $\boxed{\text{less dense}}$, and (b) its density would decrease to 0.68/0.74 = 0.92, or $\boxed{92\%}$ of its former value.

E17.28
$$V = 651\ \text{pm}\times 651\ \text{pm}\times 934\ \text{pm} = \boxed{3.96\times 10^{-28}\ \text{m}^3}$$

If we assume a simple tetragonal unit cell, then there is one formula unit per unit cell.

$$574.79\ \text{g mol}^{-1}\times(1\ \text{mol}/6.02214\times 10^{23}\ \text{unit}) = 9.5446\times 10^{-22}\ \text{g unit}^{-1}$$

$$d = m/V = \left(9.5446\times 10^{-22}\ \text{g}\right)/\left(3.96\times 10^{-28}\ \text{m}^3\right) = \boxed{2.41\times 10^{6}\ \text{g m}^{-3}}$$

E17.29
$$\rho = \frac{\text{mass of unit cell}}{\text{volume of unit cell}} = \frac{m}{V}$$

$$m = nM = NM/N_A \quad [N \text{ is the number of formula units per unit cell}]$$

Then, $$\rho = \frac{NM}{VN_A}$$

and $$N = \frac{\rho V N_A}{M}$$

$$= \frac{(3.9\times 10^{6}\ \text{g m}^{-3})\times\left(634\times 10^{-12}\ \text{m}\right)\times\left(784\times 10^{-12}\ \text{m}\right)\times\left(516\times 10^{-12}\ \text{m}\right)\times\left(6.022\times 10^{23}\ \text{mol}^{-1}\right)}{154.77\ \text{g mol}^{-1}}$$

$$= 3.9$$

Therefore, $\boxed{N=4}$ and the true calculated density (in the absence of defects) is

$$\rho = \frac{4\times(154.77\ \mathrm{g\ mol^{-1}})}{(634\times10^{-10}\ \mathrm{cm})\times(784\times10^{-10}\ \mathrm{cm})\times(516\times10^{-10}\ \mathrm{cm})\times(6.022\times10^{23})\ \mathrm{mol^{-1}}}$$

$$= \boxed{4.01\ \mathrm{g\ cm^{-3}}}$$

E17.30 $\frac{1}{d_{hkl}^2} = \frac{h^2}{a^2}+\frac{k^2}{b^2}+\frac{l^2}{c^2}$ [17.7] or $d_{hkl} = \left(\frac{h^2}{a^2}+\frac{k^2}{b^2}+\frac{l^2}{c^2}\right)^{-1/2}$

(a) $d_{321} = \left\{\left(\frac{3}{812\ \mathrm{pm}}\right)^2+\left(\frac{2}{947\ \mathrm{pm}}\right)^2+\left(\frac{1}{637\ \mathrm{pm}}\right)^2\right\}^{-1/2} = \boxed{220\ \mathrm{pm}}$

(b) $d_{642} = \tfrac{1}{2}d_{321} = \boxed{110\ \mathrm{pm}}$

E17.31 Radius ratio $= \frac{72\ \mathrm{pm}}{140\ \mathrm{pm}} = 0.51$ [Table 17.5]

$0.41 < 0.51 < 0.73$

Therefore, the $\boxed{\text{rock salt structure}}$ is predicted.

Answers to projects

P17.32 (a) $E_k = \alpha + 2\beta\cos\left(\frac{k\pi}{N+1}\right) \quad k = 1,2,\ldots,N$ [17.1]

$$E_{k+1}-E_k = \left[\alpha+2\beta\cos\left(\frac{(k+1)\pi}{N+1}\right)\right]-\left[\alpha+2\beta\cos\left(\frac{k\pi}{N+1}\right)\right]$$

$$= 2\beta\left\{\cos\left(\frac{(k+1)\pi}{N+1}\right)-\cos\left(\frac{k\pi}{N+1}\right)\right\}$$

In the limit as $N\to\infty$, the denominator of each cosine term becomes infinite and, therefore, each cosine term becomes cos(0), which equals 1. Thus,

$$\lim_{N\to\infty}(E_{k+1}-E_k) = 2\beta\lim_{N\to\infty}\left\{\cos\left(\frac{(k+1)\pi}{N+1}\right)-\cos\left(\frac{k\pi}{N+1}\right)\right\}$$

$$= 2\beta\left\{\cos\left(\frac{(k+1)\pi}{\infty}\right)-\cos\left(\frac{k\pi}{\infty}\right)\right\} = 2\beta\{\cos(0)-\cos(0)\} = 2\beta\{1-1\}$$

$$= \boxed{0}$$

This means that the separation between neighbouring levels goes to zero as N increases to infinity.

(b) The density of energy levels is:

$$\rho = \frac{\mathrm{d}E}{\mathrm{d}k}$$

where $\frac{dE}{dk} = \frac{d}{dk}\left(\alpha + 2\beta\cos\frac{k\pi}{N+1}\right)$ [17.1]

$$= -\frac{2\pi\beta}{N+1}\sin\frac{k\pi}{N+1}$$

so $$\rho(k) = -\frac{2\pi\beta}{N+1}\sin\frac{k\pi}{N+1}$$

A plot of $\rho(k)$ against k is shown in Figure 17.12.

The sine factor is eliminated from the energy density expression by application of the trigonometric identity

$$\sin^2\theta + \cos^2\theta = 1 \quad \text{or} \quad \sin\theta = \left(1 - \cos^2\theta\right)^{1/2}.$$

$$\rho(k) = -\frac{2\pi\beta}{N+1}\sin\frac{k\pi}{N+1} = -\frac{2\pi\beta}{N+1}\left(1 - \cos^2\frac{k\pi}{N+1}\right)^{1/2}$$

But from equation [17.1], $\cos\frac{k\pi}{N+1} = \frac{E_k - \alpha}{2\beta}$, so

$$\boxed{\rho(E) = -\frac{2\pi\beta}{N+1}\left\{1 - \left(\frac{E-\alpha}{2\beta}\right)^2\right\}^{1/2}}$$

The density of states is greatest when $E = \alpha$ by this definition of density of states.

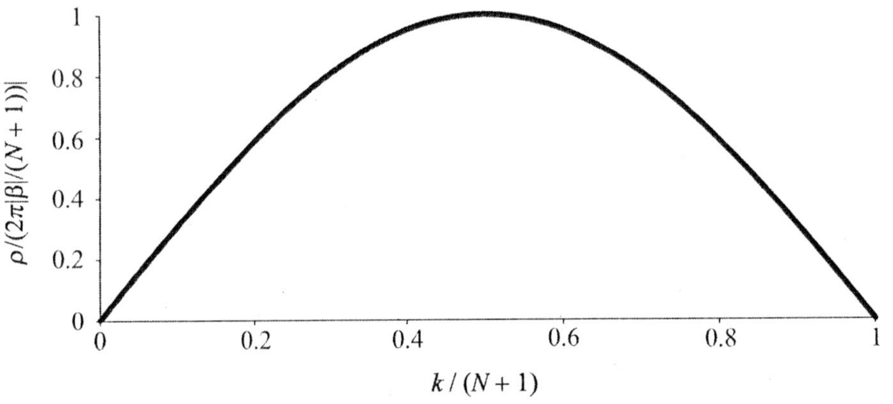

Figure 17.12

(c) Permitted states at the low energy edge of the band must have a relatively long characteristic wavelength while the permitted states at the high energy edge of the band must have a relatively short characteristic wavelength. There are few wavefunctions that have these characteristics so the density of states is lowest at the edges. This is analogous to the MO picture that shows a few bonding MOs that lack nodes and a few antibonding MOs that have the maximum number of nodes.

Another insightful view is provided by consideration of the spatially periodic potential that the electron experiences within a crystal. The periodicity demands that the electron wavefunction be a periodic function of the position vector $\vec{r}$. We can approximate it with a Bloch wave: $\psi \propto e^{i\vec{k}\cdot\vec{r}}$

where $\vec{k} = k_x\hat{i} + k_y\hat{j} + k_z\hat{k}$ is called the wave number vector. This is a bold, "free" electron approximation and in the spirit of searching for a conceptual explanation, not an accurate solution, suppose that the wavefunction satisfies a Hamiltonian in which the potential can be neglected: $\hat{H} = -(\hbar^2/2m)\nabla^2$. The eigenvalues of the Bloch wave are: $E = \hbar^2 |\vec{k}|^2 /2m$. The Bloch wave is periodic when the components of the wave number vector are multiples of a basic repeating unit. Writing the repeating unit as $2\pi/L$ where L is a length that depends upon the structure of the unit cell, we find: $k_x = 2n_x\,\pi/L$ where $n_x = 0, \pm1, \pm2,\ldots.$ Similar equations can be written for k_y and k_z and with substitution the eigenvalues become: $E = (1/2m)(2\pi\hbar/L)^2\left(n_x^2 + n_y^2 + n_z^2\right)$. This equation suggests that the density of states for energy level E can be visually evaluated by looking at a plot of permitted n_x, n_y, n_z values as shown in the following graph of Fig.17.13. The number of n_x, n_y, n_z values within a thin, spherical shell around the origin equals the density of states which have energy E. Three shells, labeled 1, 2, and 3, are shown in the graph. All have the same width but their energies increase with their distance from the origin. It is obvious that the low energy shell 1 has a much lower density of states than the intermediate energy shell 2. The sphere of shell 3 has been cut into the shape determined by the periodic potential pattern of the crystal and, because of this phenomena, it also has a lower density of states than the intermediate energy shell 2. The general concept is that the low energy and high energy edges of a band have lower density of states than that of the band center.

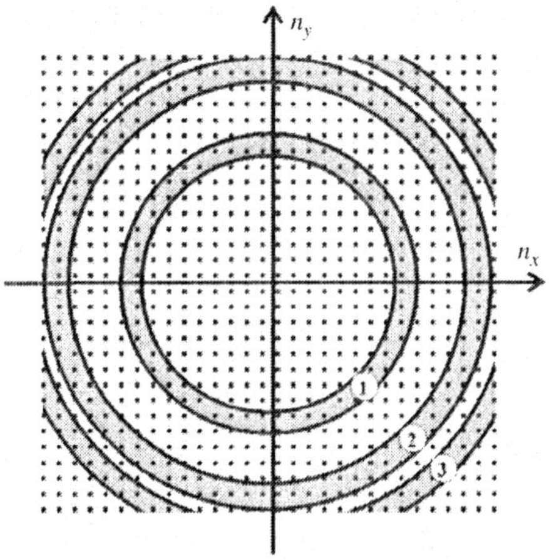

Figure 17.13

P17.33 Special time-resolved X-ray diffraction techniques have become available in recent years and it is now possible to make exquisitely detailed measurements of atomic motions during chemical and biochemical reactions. Time-resolved X-ray diffraction techniques make use of synchrotron sources, which can emit intense polychromatic pulses of X-ray radiation with pulse widths varying from 100 ps to 200 ps. Instead of the Bragg method, the Laue method is used because many reflections can be collected simultaneously, rotation of the sample is not required, and data acquisition times are short. However, good diffraction data cannot be obtained from a single X-ray pulse and reflections from several pulses must be averaged together. In practice, this averaging dictates the time resolution of the experiment, which is commonly tens of microseconds or less.

During your literature search of X-ray studies of photoactive yellow protein (PYP) it may be helpful to review the research article by H. Ihee (Ihee H. et al., *PNAS* 2005;102:7145–7150) in the *Proceedings of the National Academy of Sciences of the United States* and references therein. Figure 17.14 (http://cars9.uchicago.edu/biocars/pages/timeresolved.shtml) is a summary of the time-resolved X-ray data and its analysis. For details of data collection and analysis see M. Schmidt, H. Ihee, R. Pahl and V. Srajer, "Protein-ligand interactions probed by time-resolved crystallography," in *Protein-Ligand Interactions: Methods and Protocols* (U. Nienhaus, eds.), Humana Press, Totawa, NJ, pp. 115–154 (2005).

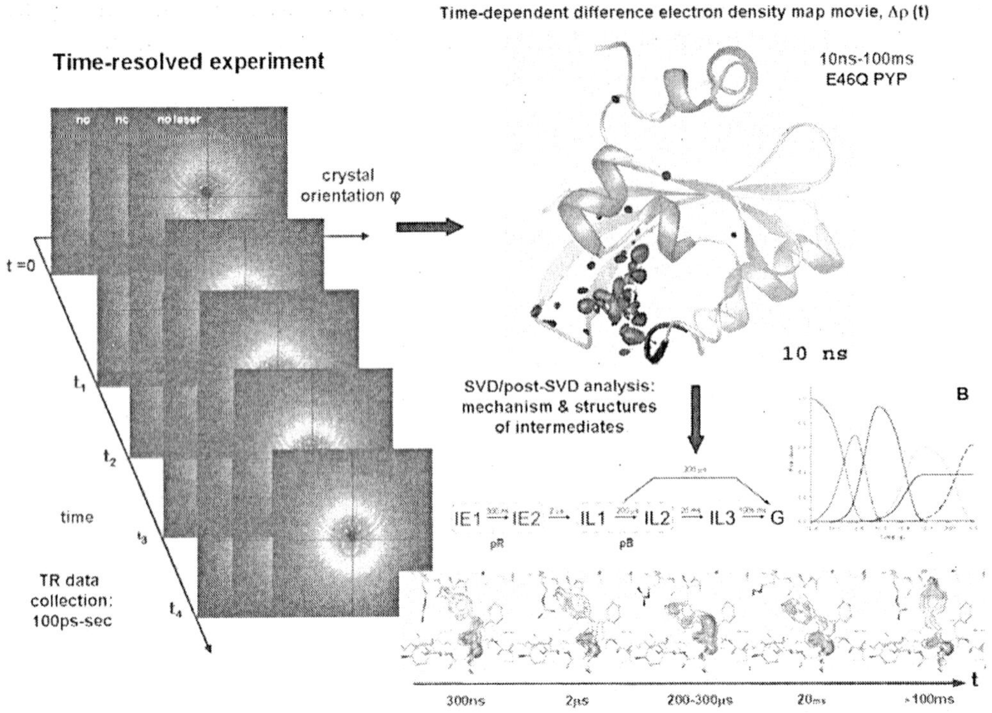

Figure 17.14

P17.34 Tans and coworkers (S.J. Tans et al., *Nature*, **393**, 49 (1998)) have draped a semiconducting carbon nanotube (CNT) over metal (gold in Fig. 17.15) electrodes that are 400 nm apart atop a silicon surface coated with silicon dioxide. A bias voltage between the electrodes provides the source and drain of the molecular field-effect transistor(FET). The silicon serves as a gate electrode and the thin silicon oxide layer (at least 100 nm thick) insulates the gate from the CNT circuit. By adjusting the magnitude of an electric field applied to the gate, current flow across the CNT may be turned on and off.

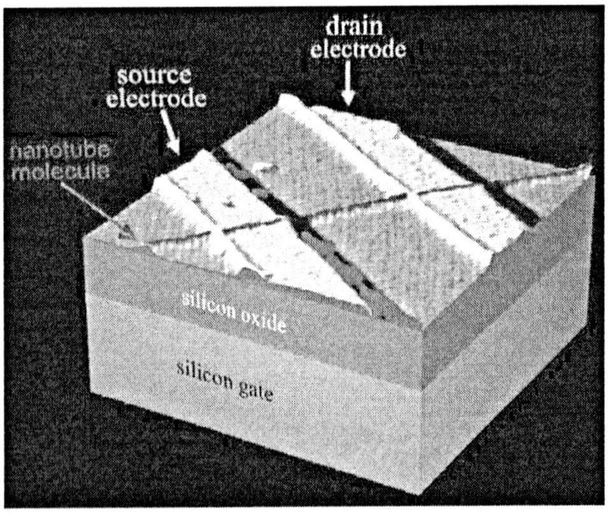

Figure 17.15

Wind and coworkers (S.J. Wind et al., *Applied Physics Letters*, **80**(20, May 20), 3817 (2002)) have designed (Fig. 17.16) a CNT-FET of improved current carrying capability. The gate electrode is above the conduction channel and separated from the channel by a thin oxide dielectric. In this manner the CNT-to-air contact is eliminated, an arrangement that prevents the circuit from acting like a p-type transistor. This arrangement also reduces the gate oxide thickness to about 15 nm, allowing for much smaller gate voltages and a steeper subthreshold slope, which is a measure of how well a transistor turns on or off.

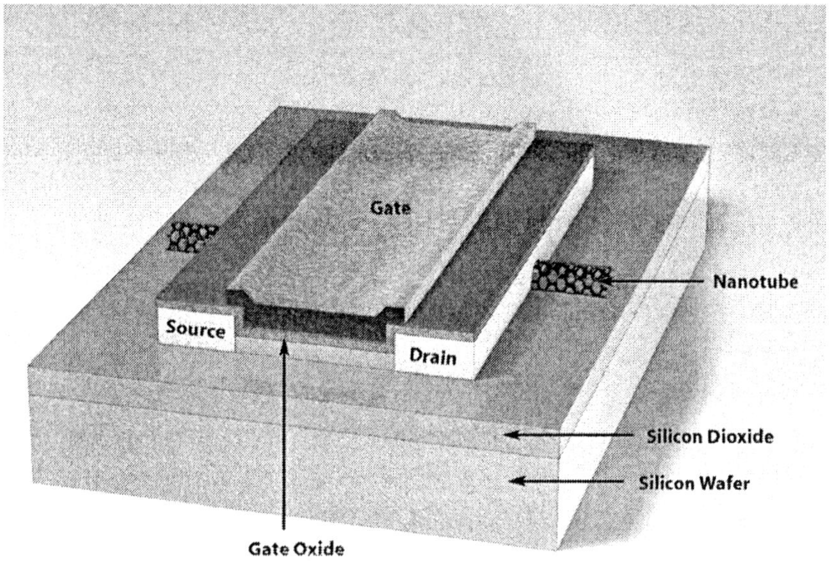

Figure 17.16

A single-electron transistor (SET) has been prepared by Cees Dekker and coworkers (*Science*, **293**, 76, (2001)) with a CNT. The SET is prepared by putting two bends in a CNT with the tip of an AFM (Fig. 17.17). Bending causes two buckles that, at a distance of 20 nm, serves as a conductance barrier. When an appropriate voltage is applied to the gate below the barrier, electrons tunnel one at a time across the barrier.

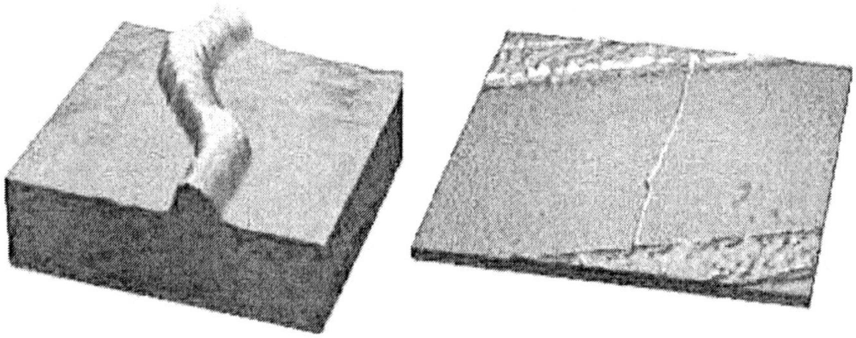

Figure 17.17

Weitz et al. (Phys. Stat. Sol. (b) **243**, 13, 3394 (2006)) report on the construction of a single-wall CNT using a silane-based organic self-assembled monolayer (SAM) as a gate dielectric on top of a highly doped silicon wafer. The organic SAM is made of 18-phenoxyoctadecyltrichlorosilane. This ultrathin layer (Fig. 17.18) ensures strong gate coupling and therefore low operation voltages. Single-electron transistors (SETs) were obtained from individual metallic SWCNTs. Field-effect transistors made from individual semiconducting SWCNTs operate with gate-source voltages of −2 V, show good saturation, small hysteresis (200 mV) as well as a low subthreshold swing (290 mV/dec).

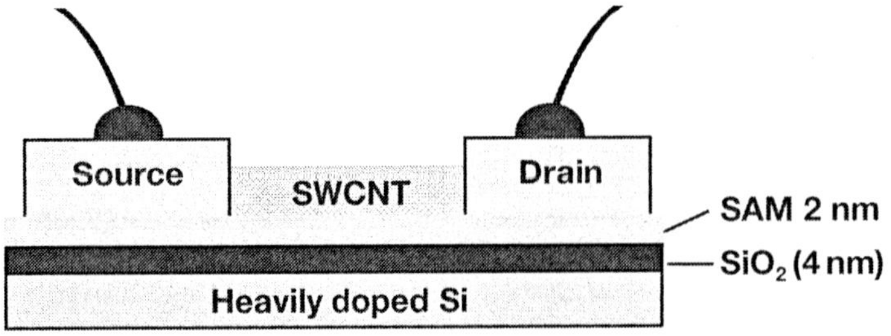

Figure 17.18

John Rodgers and researchers at the University of Illinois have reported a technique for producing near perfect alignment of CNT transistors (Fig. 17.19). The array is prepared by patterning thin strips of an iron catalyst on quartz crystals and then growing nanometer-wide CNTs along those strips using conventional carbon vapor deposition. The quartz crystal aligns the nanotubes. Transistor development then includes depositing source, drain, and gate electrodes using conventional photolithography. Transistors made with about 2,000 nanotubes can carry currents of one ampere, which is several orders of magnitude larger than the current possible with single nanotubes. The research group also developed a technique for transferring the nanotube arrays onto any substrate, including silicon, plastic, and glass. See Coskun Kocabas, Seong Jun Kang, Taner Ozel, Moonsub Shim, and John A. Rogers, *J. Phys. Chem. C* **2007**, *111*, 17879, Improved Synthesis of Aligned Arrays of Single-Walled Carbon Nanotubes and Their Implementation in Thin Film Type Transistors.

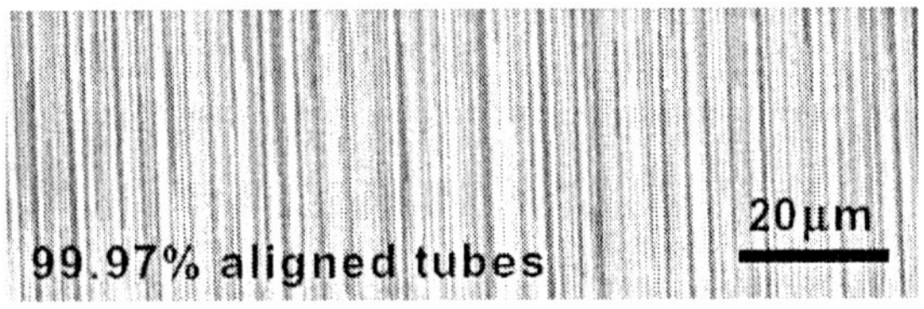

Figure 17.19

Also, see the review A.K. Feldman, et. al., *Acc. Chem. Res.*, 2008, 41 (12), pp 1731–1741; "Molecular Electronic Devices Based on Single-Walled Carbon Nanotube Electrodes".

18 Solid surfaces

Answers to discussion questions

D18.1 Characterizing the composition and structure of a surface requires that the cleanliness of a freshly prepared surface be assured with **ultrahigh vacuum** (UHV) techniques, which provide low pressures in the range of 1 μPa to 1 nPa. At these pressures collision fluxes may be as low as once a day.

One technique that may be used to characterize a solid surface is **photoemission spectroscopy**, in which X-rays (for XPS) or hard (short-wavelength) ultraviolet (for UPS) ionizing radiation is used to eject electrons from adsorbed species. The kinetic energies of the electrons ejected from their orbitals are measured and the pattern of energies is used to identify the material present. UPS, which examines electrons ejected from valence shells, is also used to establish the bonding characteristics and the details of electronic structures of substances on the surface. Its usefulness is its ability to reveal which orbitals of the adsorbate are involved in the bond to the substrate.

A technique, which is widely used in the microelectronics industry, is **Auger electron spectroscopy** (AES). The **Auger effect** (pronounced oh-zhey) is the emission of a second electron after high-energy radiation has expelled another electron. The first electron to depart leaves a hole in a low lying orbital, and an upper electron falls into it. The energy released in this transition may result either in the generation of radiation, which is called **X-ray fluorescence** or in the ejection of another electron. The latter is the secondary electron of the Auger effect. The energies of the secondary electrons are characteristic of the material present. In practice, the Auger spectrum is normally obtained by irradiating the sample with an electron beam rather than electromagnetic radiation. In **scanning Auger electron microscopy** (SAM), the finely focused electron beam is scanned over the surface and a map of composition is compiled; the resolution can reach to below about 50 nm.

A technique for determining the arrangement of the atoms close to and adsorbed on the surface is **low-energy electron diffraction** (LEED). This technique is like X-ray diffraction but uses the wave character of electrons. The use of low-energy electrons (with energies in the range 10–200 eV, corresponding to wavelengths in the range 100–400 pm) ensures that the diffraction is caused only by atoms on and close to the surface. The presence of terraces, steps, and kinks in a surface shows up in LEED patterns, and their surface density (the number of defects in a region divided by the area of the region) can be estimated.

Terraces, steps, kinks, and dislocations on a surface may be observed by **scanning tunnelling microscopy** (STM), and **atomic force microscopy** (AFM). In scanning tunnelling microscopy a platinum–rhodium or tungsten needle is scanned across the surface of a conducting solid. When the tip of the needle is brought very close to the surface, electrons tunnel across the intervening space. In the *constant-current mode* of operation, the stylus moves up and down corresponding to the form of the surface, and the topography of the surface, including any adsorbates, can be mapped on an atomic scale. The vertical motion of the stylus is achieved by fixing it to a piezoelectric cylinder, which contracts or expands according to the potential difference it experiences. In the *constant-z mode*, the vertical position of the stylus is held constant and the current is monitored. Because the tunnelling probability is very sensitive to the size of the gap, the microscope can detect tiny, atom-scale variations in the height of the surface.

In *atomic force microscopy* (AFM) a sharpened stylus attached to a beam is scanned across the surface. The force exerted by the surface and any adsorbate pushes or pulls on the stylus and deflects the beam. The deflection is monitored by using a laser beam. Because no current is needed between the sample and the probe, the technique can be applied to nonconducting surfaces too.

D18.2 The **Langmuir isotherm** applies under the following conditions:

1. Adsorption cannot proceed beyond monolayer coverage.
2. All sites are equivalent and the surface is uniform.
3. The ability of a molecule to adsorb at a given site is independent of the occupation of neighbouring sites.

Assumption 1 is justified when the initial adsorbed layer cannot act as a substrate for the adsorption of multilayers because the pressure is much lower than the vapor pressure of the adsorbate. Assumption 2 and 3 imply, respectively, that the enthalpy of adsorption is the same for all sites and is independent of the extent of surface coverage. The incidence of steps, kinks, and dislocations on a surface must be minimal.

D18.3 Text Figure 18.21 demonstrates that the BET isotherm describes multilayer adsorption. For a given c, which equals K_0/K_1 where K_0 is the equilibrium constant for adsorption on to the substrate and K_1 is the equilibrium constant for physisorption on to the overlaying layers already present, the ratio V/V_{mon} becomes greater than 1 at a pressure that is less than the vapour pressure of the liquid, p^*. Ratios for which $V/V_{\text{mon}} > 1$ imply multilayer adsorption. To find a relationship for $z = p/p^*$ at the point for which $V/V_{\text{mon}} = 1$ on the BET isotherm, set $V/V_{\text{mon}} = 1$ in eqn 18.7 and solve for z. The mathematics involves a second-order polynomial in x and application of the solution of the quadratic equation gives

$$\text{at } V/V_{\text{mon}} = 1, \qquad z = \frac{1 \pm \sqrt{c}}{1-c} \qquad \text{where } c \neq 1$$

z must be in the range $0 < z < 1$ while the range of c is $0 < c < \infty$. These ranges imply that of the two possible solutions to the above equation (±) only the negative sign in the numerator provides a physically reasonable value. Thus,

$$\text{at } V/V_{\text{mon}} = 1, \qquad z = \frac{1 - \sqrt{c}}{1-c} \qquad \text{where } c \neq 1$$

Direct examination of the BET isotherm when $V/V_{\text{mon}} = 1$ and $c = 1$, reveals that $z = 0.50$ under these conditions. Draw a table summarizing the value of z at $V/V_{\text{mon}} = 1$ on the BET isotherm over a range of c values.

c	0.001	0.01	0.1	1	10	100	1000
z	0.9693	0.9090	0.7597	0.5000	0.2403	0.09091	0.03065

The table indicates that low values of c require relatively large values of z before a second layer begins to form. Very large values of c cause multilayer formation to occur at low pressure (i.e., low values of z).

D18.4 In the **Langmuir–Hinshelwood mechanism** (LH mechanism) of surface catalysed reactions, the reaction takes place by encounters between molecular fragments and atoms already adsorbed on the surface. We therefore expect the rate law to be second-order in the extent of surface coverage:

$$\mathrm{A + B \rightarrow P} \qquad v = k_r \theta_A \theta_B$$

Insertion of the appropriate isotherms for A and B then gives the reaction rate in terms of the partial pressures of the reactants. For example, if A and B follow Langmuir isotherms (eqn 18.5), and adsorb without dissociation, then it follows that the rate law is

$$v = \frac{k_r K_A K_B p_A p_B}{(1 + K_A p_A + K_B p_B)^2} \qquad \text{[18.17, competition for the same surface site].}$$

The parameters K in the isotherms and the rate constant k_r are all temperature dependent, so the overall temperature dependence of the rate may be strongly non-Arrhenius (in the sense that the reaction rate is unlikely to be proportional to $\exp(-E_a/RT)$.

In the **Eley-Rideal mechanism** (ER mechanism) of a surface-catalysed reaction, a gas phase molecule collides with another molecule already adsorbed on the surface. The rate of formation of product is expected to be proportional to the partial pressure, p_B of the non-adsorbed gas B and the extent of surface coverage, θ_A, of the adsorbed gas A. It follows that the rate law should be

$$A + B \rightarrow P \qquad v = k_r p_B \theta_A .$$

The rate constant, k_r, might be much larger than for the uncatalysed gas-phase reaction because the reaction on the surface has a low activation energy and the adsorption itself is often not activated.

If we know the adsorption isotherm for A, we can express the rate law in terms of its partial pressure, p_A. For example, if the adsorption of A follows a Langmuir isotherm in the pressure range of interest, then the rate law would be

$$v = \frac{k_r K p_A p_B}{1 + K p_A} \quad [18.18]$$

According to eqn 18.18, when the partial pressure of A is high (in the sense $Kp_A \gg 1$, there is almost complete surface coverage, and the rate is equal to $k_r p_B$. Now the rate-determining step is the collision of B with the adsorbed fragments. When the pressure of A is low ($Kp_A \ll 1$), perhaps because of its reaction, the rate is equal to $k_r K p_A p_B$; and now the extent of surface coverage is important in the determination of the rate.

The LH and ER mechanism have distinctive isotherms that can be distinguished with plots of measured rates against a full range of p_A at constant p_B. The prediction of the LH mechanism with competition for the same surface site is shown in Figure 18.1 and the ER prediction is shown in Figure 18.2. In the LH mechanism the rate is limited by the low fractional coverage of A (θ_A) at low values of $K_A p_A$ while the rate is limited by low fractional coverage of B (θ_B) at high values of $K_A p_A$. Thus, the LH isotherm goes through a maximum when $K_A p_A = 1 + K_B p_B$. In contrast, the ER rate is limited only by the low fractional coverage of A (θ_A) at low values of Kp_A while the rate reaches a plateau, equal to $k_r^{-1} p_B$, as Kp_A becomes large. Almost all thermal surface-catalysed reactions are thought to take place by the LH mechanism.

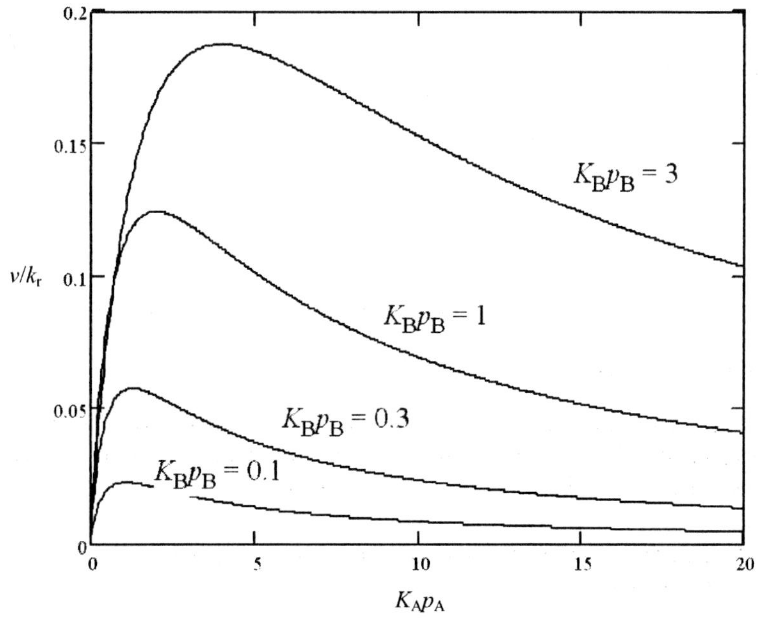

Figure 18.1

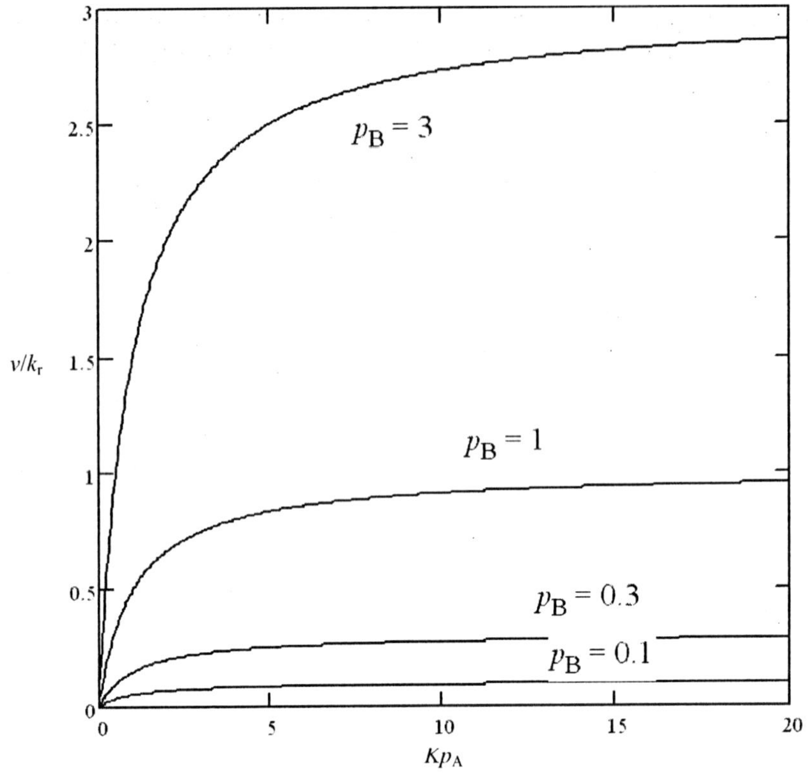

Figure 18.2

D18.5 In the **Mars van Krevelen mechanism of catalytic oxidation** (MvK), for example in the partial oxidation of propene to propenal, the first stage is the adsorption of the propene molecule with loss of a hydrogen to form the allyl radical, $CH_2{=}CHCH_2$. An O atom in the surface can now transfer to this radical, leading to the formation of acrolein (propenal, $CH_2{=}CHCHO$) and its desorption from the surface. The H atom also escapes with a surface O atom, and goes on to form H_2O, which leaves the surface. The surface is left with vacancies and metal ions in lower oxidation states. These vacancies are attacked by O_2 molecules in the overlying gas, which then chemisorb as O_2^- ions, so reforming the catalyst. This sequence of events involves great upheavals of the surface, and some materials break up under the stress.

The MvK mechanism can be identified by finding that a product leaves the catalytic surface with a constituent of the surface lattice. The lattice contributes an atom, or molecular fragment, to formation of an intermediate product and is more than a substrate alone. For example an oxygen atom of the lattice may be carried away as product and the resulting oxygen vacancies are refilled by gas phase oxygen in a separate reaction. Changes in the surface structure may be observed along with changes in both the coordination numbers and oxidation states of transition metals on the catalytic surface.

D18.6 The **electrical double layer** model of the electrode–electrolyte interface consists of a sheet of positive charge at the surface of the electrode and a sheet of negative charge next to it in the solutions (or vice versa). This creates the **Galvani potential difference** between the bulk of the electrode and the bulk of the solution. In the **Helmholtz layer model** of the interface the solvated ions arrange themselves along the surface of the electrode but are held away from it by their hydration spheres. The plane running through the solvated ions is called the outer Helmholtz plane (OHP). The model results in an electrical potential that changes linearly between the electrode and the OHP. In a refinement of this model, ions that have discarded their solvating molecules and have become attached to the electrode surface by chemical bonds are regarded as forming the **inner Helmholtz plane** (IHP). In the **Gouy–Chapman model** of the **diffuse double layer**, the disordering effect of thermal motion is taken into account by replacing the OHP with a diffuse counter ionic atmosphere.

D18.7 In cyclic voltammetry, the current at a working electrode is monitored as the applied potential difference is changed back and forth at a constant rate between pre-set limits (Figs 18.32). As the potential difference approaches $E^{\ominus}(\text{Ox, Red})$ for a solution that contains the reduced component (Red), current begins to flow as Red is oxidized. When the potential difference is swept beyond $E^{\ominus}(\text{Ox, Red})$, the current passes through a maximum and then falls as all the Red near the electrode is consumed and converted to Ox, the oxidized form. When the direction of the sweep is reversed and the potential difference passes through $E^{\ominus}(\text{Ox, Red})$, current flows in the reverse direction. This current is caused by the reduction of the Ox formed near the electrode on the forward sweep. It passes through the maximum as Ox near the electrode is consumed. The forward and reverse current maxima bracket $E^{\ominus}(\text{Ox, Red})$, so the species present can be identified. Furthermore, the forward and reverse peak currents are proportional to the concentration of the couple in the solution, and vary with the sweep rate. If the electron transfer at the electrode is rapid, so that the ratio of the concentrations of Ox and Red at the electrode surface have their equilibrium values for the applied potential (that is, their relative concentrations are given by the Nernst equation), the voltammetry is said to be *reversible*. In this case, the peak separation is independent of the sweep rate and equal to $(59\,\text{mV})/n$ at room temperature, where n is the number of electrons transferred. If the rate of electron transfer is low, the voltammetry is said to be *irreversible*. Now, the peak separation is greater than $(59\,\text{mV})/n$ and increases with increasing sweep rate. If homogeneous chemical reactions accompany the oxidation or reduction of the couple at the electrode, the shape of the voltammogram changes, and the observed changes give valuable information about the kinetics of the reactions as well as the identities of the species present.

Solutions to exercises

E18.1

$$Z_W = \frac{p}{(2\pi mkT)^{1/2}}\ [18.1] = \frac{p}{(2\pi MkT/N_A)^{1/2}}$$

$$= \left\{4.825\times10^{21}\times(\text{kg mol})^{-1/2}\times(\text{s/m})\right\}\frac{p}{M^{1/2}} \quad \text{at } 25°\text{C}$$

$$= \left\{4.825\times10^{17}\ \text{cm}^{-2}\ \text{s}^{-1}\right\}\frac{p/\text{Pa}}{(M/\text{kg mol}^{-1})^{1/2}} \quad \text{or} \quad \left\{6.433\times10^{19}\ \text{cm}^{-2}\ \text{s}^{-1}\right\}\frac{p/\text{Torr}}{(M/\text{kg mol}^{-1})^{1/2}}$$

Hence, we can draw up the following table.

	H_2	C_3H_8
$M/(\text{kg mol}^{-1})$	0.002016	0.04409
$Z_W/(\text{cm}^{-2}\ \text{s}^{-1})$		
(i) 100 Pa	1.075×10^{21}	2.298×10^{20}
(ii) 10^{-7} Torr	1.433×10^{14}	3.064×10^{13}

E18.2

$$A = \tfrac{1}{4}\pi d^2 = \tfrac{1}{4}\pi(2.5\times10^{-3}\ \text{m})^2 = 4.9\bar{1}\times10^{-6}\ \text{m}^2 \quad \text{and} \quad Z_W A = 8.5\times10^{20}\ \text{s}^{-1} \text{ at } 450\ \text{K}$$

$$Z_W = \frac{p}{(2\pi mkT)^{1/2}}\ [18.1] = \frac{p}{(2\pi MkT/N_A)^{1/2}}$$

Solve for p.

$$p = Z_W A\times(2\pi MkT/N_A)^{1/2}/A$$

$$= (8.5\times10^{20}\ \text{s}^{-1})$$

$$\times\left\{\begin{array}{l}2\pi(0.03995\ \text{kg mol}^{-1})\\ \times(1.381\times10^{-23}\ \text{J K}^{-1})\times(450\ \text{K})/(6.022\times10^{23}\ \text{mol}^{-1})\end{array}\right\}^{1/2} /(4.9\bar{1}\times10^{-6}\ \text{m}^2)$$

$$= 8.8\times10^3\ \text{Pa} = \boxed{0.088\ \text{bar}}$$

E18.3

$$Z_W = \frac{p}{(2\pi mkT)^{1/2}}\ [18.1] = \frac{p}{(2\pi MkT/N_A)^{1/2}}$$

$$= 3.2\bar{9}\times10^{24}\ \text{m}^{-2}\ \text{s}^{-1} \text{ for He at 100 K and 25 Pa}$$

The area occupied by a Cu atom is $\frac{1}{2}\times(361\times10^{-12}\ \text{m})^2 = 6.52\times10^{-20}\ \text{m}^2$ (in an fcc unit cell, there is the equivalent of two Cu atoms per face). Therefore,

$$\text{rate per Cu atom} = (3.2\bar{9}\times10^{24}\ \text{m}^{-2}\ \text{s}^{-1})\times(6.52\times10^{-20}\ \text{m}^2) = \boxed{2.1\times10^5\ \text{s}^{-1}}$$

E18.4

$$Z_W = \frac{p}{(2\pi mkT)^{1/2}}\ [18.1] = \frac{nRT}{(2\pi mkT)^{1/2}V} = constant\times T^{1/2}$$

$$\frac{Z_W(T')}{Z_W(T)} = \left(\frac{T'}{T}\right)^{1/2}$$

$$= \left(\frac{400\ \text{K}}{300\ \text{K}}\right)^{1/2} = \boxed{1.15}$$

E18.5 (a) $$v = Ae^{-d/l} = (5\times10^{14}\ \text{s}^{-1})e^{-(750\ \text{pm})/(70\ \text{pm})} = \boxed{1.\bar{1}\times10^{10}\ \text{s}^{-1}}$$

(b) $$\frac{v(d_2)}{v(d_1)} = \frac{Ae^{-d_2/l}}{Ae^{-d_2/l}} = e^{-(d_2-d_1)/l} = e^{-(850\ \text{pm}-750\ \text{pm})/(70\ \text{pm})} = \boxed{0.24}$$

E18.6 The number of CO molecules adsorbed on the catalyst is

$$N = nN_A = \frac{pVN_A}{RT} = \frac{(1.00\ \text{bar})\times(4.25\times10^{-3}\ \text{dm}^3)\times(6.022\times10^{23}\ \text{mol}^{-1})}{(0.083145\ \text{dm}^3\ \text{bar K}^{-1}\ \text{mol}^{-1})\times(273.15\ \text{K})}$$

$$= 1.13\times10^{20}$$

The area of the surface must be the same as that of the molecules spread into a monolayer, namely, the number of molecules times each one's effective area a, which we take to be about the same as the collision cross-section of a dinitrogen molecule (0.43 nm^2, Table 1.3).

$$A = Na = (1.13\times10^{20})\times(0.43\times10^{-18}\,\text{m}^2) = \boxed{49\ \text{m}^2}$$

E18.7

$$\theta = \frac{Kp}{1+Kp}\ [18.3], \text{ which implies that } p = \left(\frac{\theta}{1-\theta}\right)\frac{1}{K}$$

(a) $$p = \left(\frac{0.10}{0.90}\right)\times\left(\frac{1}{1.85\ \text{kPa}^{-1}}\right) = \boxed{0.060\ \text{kPa}}$$

(b) $$p = \left(\frac{0.90}{0.10}\right)\times\left(\frac{1}{1.85\ \text{kPa}^{-1}}\right) = \boxed{4.9\ \text{kPa}}$$

E18.8

$$s = \frac{\text{rate of adsorption of particles by the surface}}{\text{rate of collision of particles with the surface}}\ [18.12]$$

Suppose that the sticking probability, s, has the form $s = (1 - \theta)s_0$ where s_0 is the sticking probability on a perfectly clean surface and θ is the fractional coverage. Then, using the rate of adsorption for a Langmuir isotherm and a collision rate that equals $Z_W \times N$, where N is the number of adsorption sites, we may write

$$(1-\theta)s_0 = \frac{k_a N(1-\theta)p}{Z_W N}$$

Solve for k_a.

$$k_a = \frac{s_0 Z_W}{p} = \frac{s_0}{(2\pi mkT)^{1/2}} \quad [18.1]$$

The eqn for k_a along with the eqn 18.13 expression for k_d allow us to estimate the equilibrium constant of the Langmuir isotherm as

$$K = \frac{k_a}{k_d} = \boxed{\frac{s_0 e^{E_d/RT}}{(2\pi mkT)^{1/2} A}}$$ where A is the pre-exponential factor for desorption

E18.9 We follow Example 18.2 of the text, where it is shown that for a Langmuir isotherm

$$\frac{p}{V} = \frac{p}{V_\infty} + \frac{1}{KV_\infty},$$

and draw up the following table.

p/Pa	25	129	253	540	1000	1593
p/V/Pa cm^{-3}	595	791	1145	1682	2433	3382

p/V is plotted against p in Fig. 18.3. The plot is observed to be linear so we conclude that the data fits the Langmuir isotherm for these low pressures and, therefore, low coverage. The linear regression slope equals $1/V_\infty$; the regression intercept equals $1/KV_\infty$. Thus,

$$V_\infty = 1/slope = 1/(1.77\text{ cm}^{-3}) = \boxed{0.565\text{ cm}^3}$$

and

$$K = 1/(V_\infty \times intercept) = 1/(0.565\text{ cm}^3 \times 629\text{ Pa cm}^{-3}) = \boxed{2.81 \times 10^{-3}\text{ Pa}^{-1}}$$

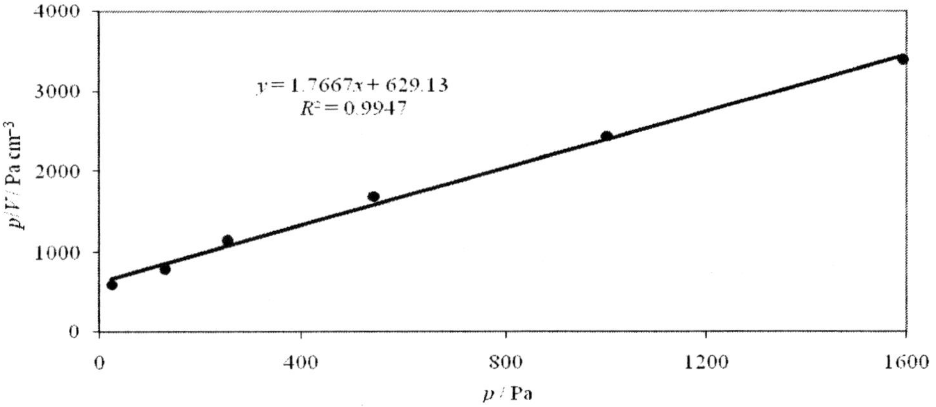

Figure 18.3

E18.10 $$\ln K' - \ln K = \frac{\Delta_{ads}H^\ominus}{R}\left(\frac{1}{T} - \frac{1}{T'}\right) \quad \text{van't Hoff equation [18.4]}$$

$$\Delta_{ads}H^\ominus = \frac{R(\ln K' - \ln K)}{\left(\frac{1}{T} - \frac{1}{T'}\right)} = \frac{R\ln\left(\frac{K'}{K}\right)}{\left(\frac{1}{T} - \frac{1}{T'}\right)}$$

$$= \frac{(8.3145\text{ J mol}^{-1}\text{ K}^{-1})\ln\left(\frac{1.0\times10^{-3}\text{ Torr}^{-1}}{2.7\times10^{-3}\text{ Torr}^{-1}}\right)}{\left(\frac{1}{250\text{ K}} - \frac{1}{273\text{ K}}\right)} = \boxed{-2\bar{5}\text{ kJ mol}^{-1}}$$

E18.11 Text Example 18.3 (the isosteric enthalpy of adsorption) demonstrates that a plot of ln p against $1/T$ should be a straight line of slope $\Delta_{\text{ads}}H^{\ominus}/R$ so we draw up the following table, prepare a plot to check linearity (Figure 18.4), and perform a linear least squares fit in order to acquire the slope.

T/K	200	210	220	230	240	250
p/kPa	4.32	5.59	7.07	8.80	10.67	12.80
$10^3/(T/\text{K})$	5.00	4.76	4.55	4.35	4.17	4.00
ln(p/kPa)	1.46	1.72	1.96	2.17	2.37	2.55

We find that

$$\Delta_{\text{ads}}H^{\ominus}/R = slope = -1.08\bar{7}\times10^3\ \text{K}$$

$$\Delta_{\text{ads}}H^{\ominus} = (slope)\times R = \left(-1.08\bar{7}\times10^3\ \text{K}\right)\times\left(8.3145\ \text{J mol}^{-1}\ \text{K}^{-1}\right) = \boxed{-9.04\ \text{kJ mol}^{-1}}$$

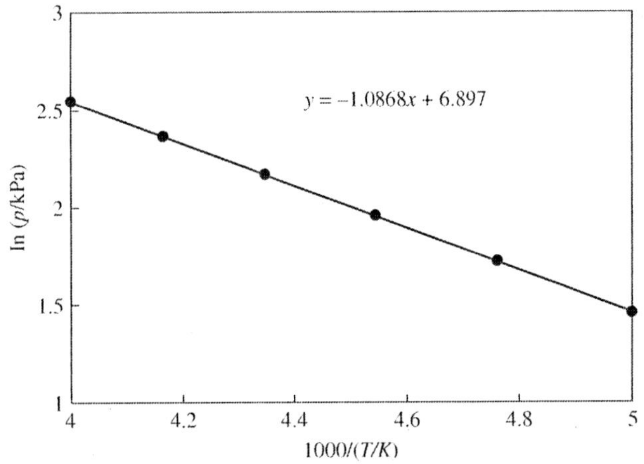

Figure 18.4

E18.12

$$\theta = \frac{(Kp)^{1/2}}{1+(Kp)^{1/2}} \quad [18.5]$$

$p(\theta)$ is found by first solving the above expression for $(Kp)^{1/2}$.

$$\theta\left\{1+(Kp)^{1/2}\right\} = (Kp)^{1/2} \text{ or } (Kp)^{1/2} = \frac{\theta}{1-\theta}. \text{ Therefore, } \boxed{p = \frac{1}{K}\left(\frac{\theta}{1-\theta}\right)^2}.$$

E18.13 $O_3(g) + M(\text{surface}) \rightleftharpoons 3\ O\text{—}M(\text{surface})$ $\quad K = k_a/k_d$

For adsorption with dissociation, the rate of adsorption is proportional to the pressure and to the probability that all three atoms of ozone will find sites, which is proportional to the cube of the number of vacant sites:

$$\text{Rate of adsorption} = k_a p\{N(1-\theta)\}^3$$

The rate of desorption is proportional to the frequency of encounters of three atoms on the surface, and is therefore third-order in the number of atoms present:

$$\text{Rate of desorption} = k_d(N\theta)^3$$

At equilibrium, the two rates are equal, so we can write

$$k_{\mathrm{a}}p\{N(1-\theta)\}^3 = k_{\mathrm{d}}(N\theta)^3$$

$$Kp\left(\frac{1-\theta}{\theta}\right)^3 = 1$$

Solving for θ gives the isotherm

$$\boxed{\theta = \frac{(Kp)^{1/3}}{1+(Kp)^{1/3}}}$$

A plot of θ against p at low pressures (where the denominator is approximately 1) shows a weaker pressure dependence for the dissociation than that shown by a Langmuir isotherm without dissociation [18.3].

E18.14 Derivation 18.1 provides a model for deriving equations [18.6]. However, with A and B competing for N adsorption sites, the number of sites not occupied equals $(1 - \theta_{\mathrm{A}} - \theta_{\mathrm{B}})N$ where θ_{A} and θ_{B} are the fraction of sites occupied by A and B, respectively. The rate at which A adsorbs to the surface is proportional to the partial pressure of A, p_{A}, and to $(1 - \theta_{\mathrm{A}} - \theta_{\mathrm{B}})N$.

$$\text{Rate of adsorption of A} = k_{\mathrm{a,A}}N(1 - \theta_{\mathrm{A}} - \theta_{\mathrm{B}})p_{\mathrm{A}}$$

The rate at which adsorbed A molecules leave the surface is proportional to the number currently on the surface, $N\theta_{\mathrm{A}}$.

$$\text{Rate of desorption of A} = k_{\mathrm{d,A}}N\theta_{\mathrm{A}}$$

At equilibrium the two rates are equal:

$$k_{\mathrm{a,A}}N(1 - \theta_{\mathrm{A}} - \theta_{\mathrm{B}})p_{\mathrm{A}} = k_{\mathrm{d,A}}N\theta_{\mathrm{A}}.$$

Using $K_{\mathrm{A}} = k_{\mathrm{a,A}}/k_{\mathrm{d,A}}$, this rearranges to

$$K_{\mathrm{A}}(1 - \theta_{\mathrm{A}} - \theta_{\mathrm{B}})p_{\mathrm{A}} = \theta_{\mathrm{A}}. \qquad \text{(i)}$$

Similarly,

$$K_{\mathrm{B}}(1 - \theta_{\mathrm{A}} - \theta_{\mathrm{B}})p_{\mathrm{B}} = \theta_{\mathrm{B}}. \qquad \text{(ii)}$$

Division of eqn (ii) by eqn (i) yields

$$\theta_{\mathrm{B}} = (K_{\mathrm{B}}p_{\mathrm{B}}/K_{\mathrm{A}}p_{\mathrm{A}})\theta_{\mathrm{A}} \qquad \text{(iii)}$$

and, upon substitution of eqn (iii) into eqn (i) and solving for θ_{A}, it is found that

$$\boxed{\theta_{\mathrm{A}} = \frac{K_{\mathrm{A}}p_{\mathrm{A}}}{1+K_{\mathrm{A}}p_{\mathrm{A}}+K_{\mathrm{B}}p_{\mathrm{B}}}}.$$

Substitution of $\theta_{\mathrm{A}} = (K_{\mathrm{A}}p_{\mathrm{A}}/K_{\mathrm{B}}p_{\mathrm{B}})\,\theta_{\mathrm{B}}$ into eqn (ii) and solving for θ_{B}, yields

$$\boxed{\theta_{\mathrm{B}} = \frac{K_{\mathrm{B}}p_{\mathrm{B}}}{1+K_{\mathrm{A}}p_{\mathrm{A}}+K_{\mathrm{B}}p_{\mathrm{B}}}}$$

E18.15

$$\frac{V}{V_{\mathrm{mon}}} = \frac{cz}{(1-z)\{1-(1-c)z\}} \quad \left[18.7\text{a, BET isotherm, } z = \frac{p}{p^*}\right]$$

This rearranges to

$$\frac{z}{(1-z)V} = \frac{1}{cV_{\mathrm{mon}}} + \frac{(c-1)z}{cV_{\mathrm{mon}}}$$

Therefore, a plot of the left-hand side against z should result in a straight line if the data obeys the BET isotherm. We draw up the following table.

0°C, $p^* = 429.6$ kPa

p / kPa	14.0	37.6	65.6	79.2	82.7	100.7	106.4
V/cm^3	11.1	13.5	14.9	16.0	15.5	17.3	16.5
$10^3 z$	32.6	87.5	152.7	184.4	192.4	234.3	247.7
$\dfrac{10^3 z}{(1-z)(V/\text{cm}^3)}$	3.04	7.10	12.1	14.1	15.4	17.7	20.0

The points are plotted in Fig. 18.5 and, since the plot is linear, we analyze the data by a linear least squares procedure.

The intercept is 4.64×10^{-4}. Hence,

$$\frac{1}{cV_{\text{mon}}} = 4.64\times10^{-4}\,\text{cm}^{-3}$$

The slope is 0.07612. Hence

$$\frac{c-1}{cV_{\text{mon}}} = .07612\ \text{cm}^{-3}$$

Solving the equations gives

$$c-1 = 164.\bar{1}$$

and hence

$$c = \boxed{165.\bar{1}} \quad \text{and} \quad V_{\text{mon}} = \boxed{13.1\ \text{cm}^3}$$

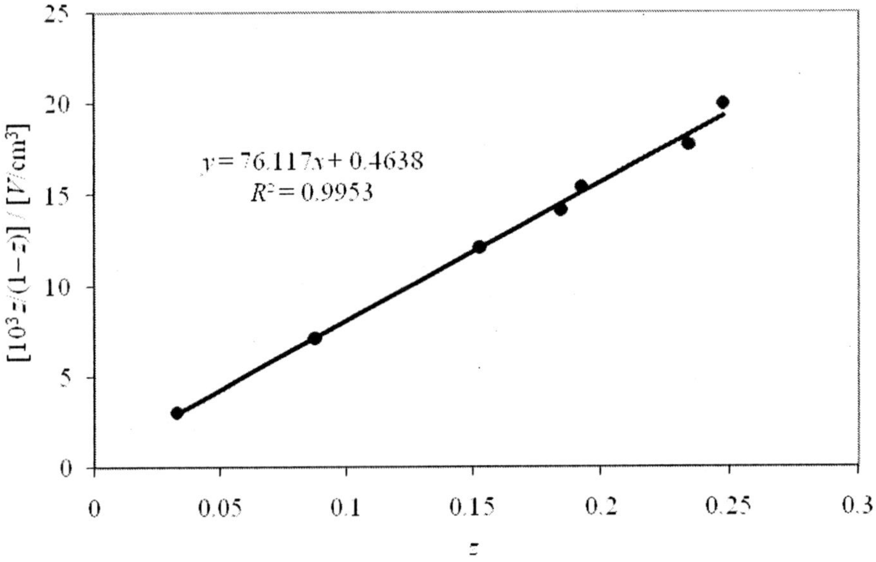

Figure 18.5

E18.16 The probability that a molecule will escape the surface at time t is proportional to $\theta(t)/\theta(0)$, which equals $e^{-k_d t}$ for first-order rate of desorption. Consequently, the mean lifetime for the presence of a molecule on a surface is given by

$$\langle t\rangle = (\text{normalization constant})\times\int_0^\infty t\times\left(\frac{\theta(t)}{\theta(0)}\right)\text{d}t = \frac{\int_0^\infty t\times\left(\frac{\theta(t)}{\theta(0)}\right)\text{d}t}{\int_0^\infty\left(\frac{\theta(t)}{\theta(0)}\right)\text{d}t} = \frac{\int_0^\infty t\,e^{-k_d t}\,\text{d}t}{\int_0^\infty e^{-k_d t}\,\text{d}t}.$$

Standard methods of integration yield the relationships

$$\int_0^\infty e^{-k_d t}\,dt = \frac{1}{k_d} \qquad \text{and} \qquad \int_0^\infty t\,e^{-k_d t}\,dt = \frac{1}{k_d^2}.$$

Thus, $\langle t\rangle = \frac{1}{k_d} = \frac{t_{1/2}}{\ln 2}[18.14] = \frac{\tau_0\,e^{E_d/RT}}{\ln 2}[18.14] \approx \frac{\tau_0\,e^{|\Delta_{ads}H^{\ominus}|/RT}}{\ln 2}$ (see Section 18.5)

With the guess $\tau_0 \approx 10^{-12}$ s, which is about the inverse of the vibrational frequency of a weak molecule–surface bond, the estimated mean lifetime is

$$\langle t\rangle \approx \frac{(10^{-12}\,\text{s})\times\exp\left(\dfrac{155\times10^3\ \text{J mol}^{-1}}{(8.3145\ \text{J K}^{-1}\,\text{mol}^{-1})\times(600\,\text{K})}\right)}{\ln 2} = \boxed{4\bar{5}\ \text{s}}$$

E18.17 The average residence time, $\langle t\rangle$, for particle adsorbed on the surface is

$$\langle t\rangle = \frac{1}{k_d} = \frac{t_{1/2}}{\ln 2}[18.14] = \frac{\tau_0\,e^{E_d/RT}}{\ln 2}[18.14]\ \text{(see exercise 18.16)}$$

$$\ln\langle t\rangle - \ln\langle t\rangle' = \frac{E_d}{R}\times\left(\frac{1}{T}-\frac{1}{T'}\right)$$

$$E_d = \frac{R\times\left(\ln\langle t\rangle - \ln\langle t\rangle'\right)}{\left(\dfrac{1}{T}-\dfrac{1}{T'}\right)} = \frac{R\times\ln\left(\dfrac{\langle t\rangle}{\langle t\rangle'}\right)}{\left(\dfrac{1}{T}-\dfrac{1}{T'}\right)}$$

(a) $$E_d = \frac{(8.3145\ \text{J K}^{-1}\,\text{mol}^{-1})\times\ln(0.36/3.49)}{\dfrac{1}{2548\ \text{K}}-\dfrac{1}{2362\ \text{K}}} = \boxed{61\bar{1}\ \text{kJ mol}^{-1}}$$

(b) $$\langle t\rangle = \frac{\tau_0\,e^{E_d/RT}}{\ln 2}\ \text{(see exercise 18.16)} = A^{-1}\,e^{E_d/RT}\ [18.14]$$

$$A = \frac{e^{E_d/RT}}{\langle t\rangle} = \left(\frac{1}{3.49\ \text{s}}\right)e^{61\bar{1}\times10^3\ \text{J mol}^{-1}/(8.3145\ \text{J K}^{-1}\ \text{mol}^{-1}\times2362\ \text{K})} = \boxed{9.3\times10^{12}\ \text{s}^{-1}}$$

E18.18 For first order desorption: $\theta(t) = \theta(0)\,e^{-k_d t}$

Let $k_d t$ and $k_d' t'$ be values at T and T', respectively. Then, we see from the desorption eqn that, if $\theta(t) = \theta(t')$ then $k_d t = k_d' t'$

We now use eqn 18.13 as a substitution for k_d and k_d' and solve for E_d.

$$k_d t = k_d' t'$$

$$Ae^{-E_d/RT}t = Ae^{-E_d/RT'}t' \quad [18.13]$$

$e^{E_d\left\{\frac{1}{T'}-\frac{1}{T}\right\}/R} = \frac{t'}{t}$ [This eqn in the form $t = t'e^{E_d\left\{\frac{1}{T}-\frac{1}{T'}\right\}/R}$ is used in parts (a) and (b).]

$$E_d\left\{\frac{1}{T'}-\frac{1}{T}\right\}/R = \ln\left(\frac{t'}{t}\right)$$

$$E_d = R\left\{\frac{1}{T'}-\frac{1}{T}\right\}^{-1}\ln\left(\frac{t'}{t}\right)$$

$$= (8.3145\ \text{J K}^{-1}\ \text{mol}^{-1})\times\left\{\frac{1}{1856\ \text{K}}-\frac{1}{1978\ \text{K}}\right\}^{-1}\ln\left(\frac{27\ \text{min}}{2.0\ \text{min}}\right) = \boxed{65\bar{1}\ \text{kJ mol}^{-1}}$$

(a) The above equation for t is used to calculate the length of time for the same amount to desorb at any temperature T. Let the primed values be either of the values provided in the exercise.

$t' = 27$ min and $T' = 1856$ K

$$t = t'\mathrm{e}^{E_d\left\{\frac{1}{T}-\frac{1}{T'}\right\}/R}$$

$$= (27\ \text{min})\times\left(\mathrm{e}^{(651\times10^3\ \text{J mol}^{-1})\times\left\{\frac{1}{T}-\frac{1}{1856\ \text{K}}\right\}/(8.3145\ \text{J K}^{-1}\ \text{mol}^{-1})}\right)$$

$$= (1.2\bar{9}\times10^{-17}\ \text{min})\mathrm{e}^{(7.83\times10^4\ \text{K})/T}$$

At 298 K,

$$t = (1.2\bar{9}\times10^{-17}\ \text{min})\mathrm{e}^{(7.83\times10^4)/298.15} = 1.4\times10^{97}\ \text{min} = \boxed{2.7\times10^{91}\ \text{a}}$$

This is incredibly longer than the age of the universe. There is really no desorption at this temperature.

(b) At 3000 K,

$$t = (1.2\bar{9}\times10^{-17}\ \text{min})\mathrm{e}^{(7.83\times10^4)/3000} = 2.8\times10^{-6}\ \text{min} = \boxed{0.17\ \text{ms}}$$

E18.19

$$s = \frac{\text{rate of adsorption of particles by the surface}}{\text{rate of collision of particles with the surface}}\quad[18.12]$$

$$= \frac{\text{rate of adsorption of particles by the surface}}{Z_W A}$$

$$= \frac{\text{rate of adsorption of particles by the surface}}{pA(2\pi MRT)^{-1/2}}\quad[18.1]$$

$$= \frac{0.33\times10^{-3}\ \text{mol s}^{-1}}{(10.0\ \text{Pa})\times(10\times10^{-4}\ \text{m}^2)\times\{2\pi(0.01703\ \text{kg mol}^{-1})\times(8.3145\ \text{J K}^{-1}\ \text{mol}^{-1})\times(210\ \text{K})\}^{-1/2}}$$

$$= \boxed{0.45}$$

E18.20

$$\text{Rate of desorption} = k_d\theta = \frac{k_dKp}{1+Kp}\quad[18.3]$$

Since HI is very strongly adsorbed on Au, we surmise that $Kp \gg 1$. Thus, the rate of desorption is $\boxed{\text{zero order on gold}}$ because

$$\text{Rate of desorption} = \frac{k_dKp}{1+Kp} = \frac{k_dKp}{Kp} = k_d$$

Since HI is very weakly adsorbed on Pt, we surmise that $Kp \ll 1$. Thus, the rate of desorption is $\boxed{\text{first order in pressure on platinum}}$ because

$$\text{Rate of desorption} = \frac{k_dKp}{1+Kp} = \frac{k_dKp}{1} = k_dKp$$

E18.21 (a) $$v = \frac{k_r K_A K_B p_A p_B}{(1+K_A p_A + K_B p_B)^2} \quad [18.17]$$

(b) $$\lim_{p_A, p_B \to 0} \left(\frac{k_r K_A K_B p_A p_B}{(1+K_A p_A + K_B p_B)^2} \right) = \boxed{k_r K_A K_B p_A p_B}$$

(c) Eqn 18.17 is highly non-linear so from the mathematical point of view the rate is not zero'th order over any appreciable ranges of partial pressures. Admittedly, if $K_A p_A = K_B p_B$ and $K_A p_A >> 1$, then $v/k_r = ¼$, which is zero'th order but this is a very limited situation. It is more likely that the experimentalist will study the isotherms over a wide range of partial pressures and observe a strong pressure dependence. To be sure of this last assertion, it is worth taking a moment to prepare a surface plot of the rate expressed by eqn 18.17. Software programs are abundantly available for such plots and that of Figure 18.6 is prepared with the following Mathcad worksheet.

$$X_{max} := 50$$

$$N := 50 \qquad i := 0 \ldots N \qquad X_{A_i} := \frac{X_{max} i}{N} \qquad j := 0 \ldots N \qquad X_{B_j} := \frac{X_{max} j}{N}$$

$$Rate_{i,j} := \frac{X_{A_i} \cdot X_{B_j}}{(1 + X_{A_i} + X_{B_j})^2} \qquad max(Rate) = 0.245 \qquad Rate_{N,N} = 0.245$$

The vertical axis of Figure 18.6 is v/k_r (called "Rate" in the above worksheet where X_A represents $K_A p_A$). The two base axes are $K_A p_A$ and $K_B p_B$. The surface plot shows a strong pressure dependence when either $K_A p_A$ or $K_B p_B$ are small but v/k_r quickly grows toward the limiting value of 0.25 for other values. Furthermore, the top of the surface appears rather flat for a large range of intermediate to large Kp values. This flatness can be important as it means that from the experimental point of view imprecise rate measurements made in these regimes can potentially appear to be zero'th order (independent of partial pressures) even though the Langmuir-Hinshelwood mechanism is appropriate. The contour map of Figure 18.7 helps identification of this "zero'th order" regime. The axes of Figure 18.7 are $K_A p_A$ and $K_B p_B$ and rate measurements made within the 0.23 contour will appear to be zero'th order if there is a measurement imprecision of around 8%. Binding to the surface is so strong in this regime that the rate determining step for the reaction depends only upon breaking the bond; the rate is nearly independent of partial pressures.

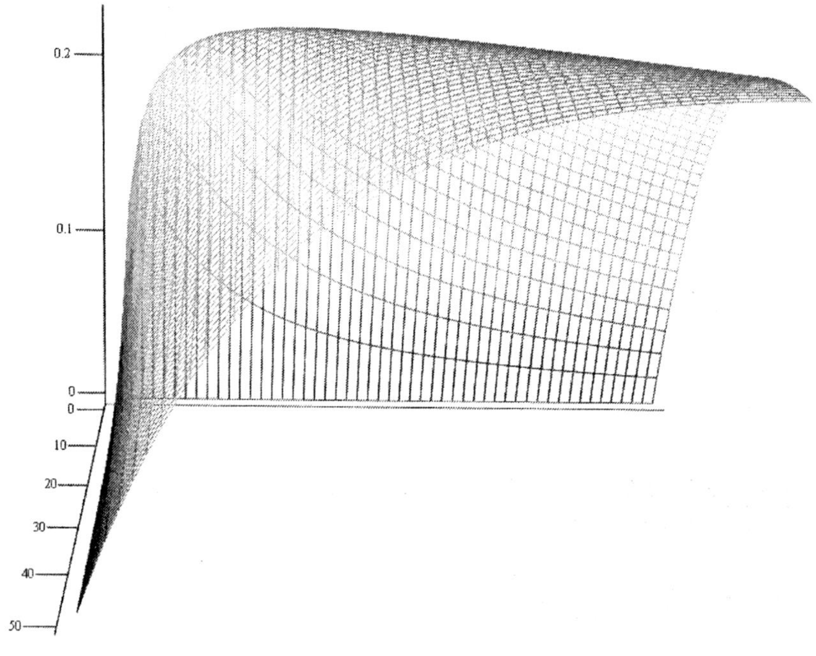

Figure 18.6

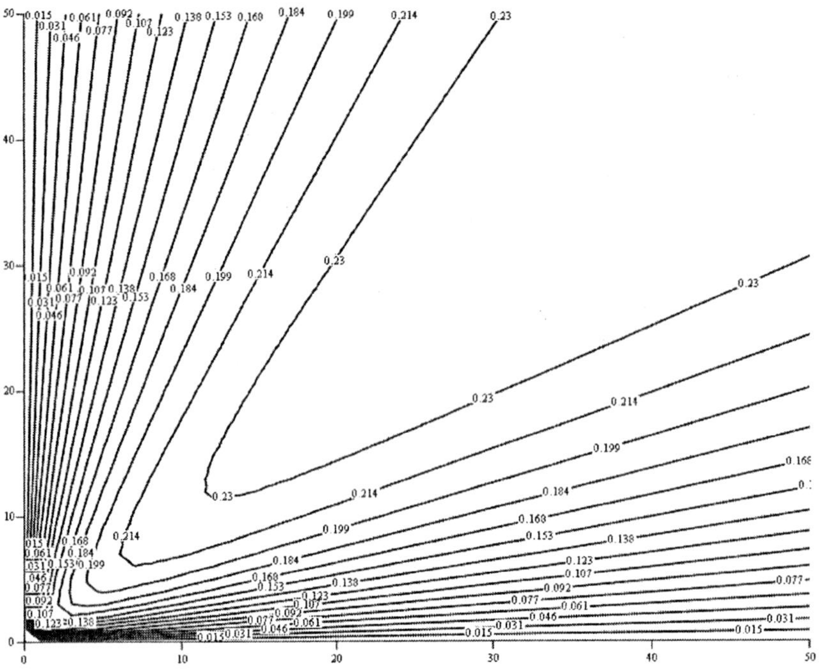

Figure 18.7

E18.22 Butler–Volmer eqn: $j = j_0\left\{e^{(1-\alpha)f\eta} - e^{-\alpha f\eta}\right\}$ [18.20] where $f = F/RT = 38.92\ \text{V}^{-1}$ at 25°C

First we examine the two exponential terms using the data point $j = 17\ \text{mA cm}^{-2}$ when $\eta = 115$ mV:

$$e^{(1-\alpha)f\eta} - e^{-\alpha f\eta} = e^{(1-0.48)\times(38.92\ \text{V}^{-1})\times(0.115\ \text{V})} - e^{-(0.48)\times(38.92\ \text{V}^{-1})\times(0.115\ \text{V})} = 10.25 - 0.12$$

This shows that the second term in the Butler–Volmer eqn, which is about 1% of the first term, is negligibly small and we conclude that the large, positive overpotential limit of the eqn adequately describes the data.

$$j = j_0 e^{(1-\alpha)f\eta} \qquad \text{where } j_0 = j/e^{(1-\alpha)f\eta} = \left(17.0\ \text{mA cm}^{-2}\right)/\left(10.25\right) = 1.66\ \text{mA cm}^{-2}$$

Taking the natural logarithm of this eqn gives eqn 18.23a, which we use to find the desired overpotential.

$$\ln j = \ln j_0 + (1-\alpha)f\eta$$

$$\eta = \frac{\ln(j/j_0)}{(1-\alpha)f}$$

$$= \frac{\ln\left(38\ \text{mA cm}^{-2}/1.66\ \text{mA cm}^{-2}\right)}{(1-0.48)\times\left(38.92\ \text{V}^{-1}\right)} = \boxed{155\ \text{mV}}$$

E18.23 Butler–Volmer eqn: $j = j_0\left\{e^{(1-\alpha)f\eta} - e^{-\alpha f\eta}\right\}$ [18.20] where $f = F/RT = 38.92\ \text{V}^{-1}$ at 25°C

$$j_0 = j/\left\{e^{(1-\alpha)f\eta} - e^{-\alpha f\eta}\right\}$$

$$= \left(17.0\ \text{mA cm}^{-2}\right)/\left\{e^{(1-0.48)\times(38.92\ \text{V}^{-1})\times(0.115\ \text{V})} - e^{-(0.48)\times(38.92\ \text{V}^{-1})\times(0.115\ \text{V})}\right\}$$

$$= \boxed{1.68\ \text{mA cm}^{-2}}$$

This is almost identical to the estimate of Exercise 18.22.

E18.24

$$\frac{j}{j_0} = e^{(1-\alpha)f\eta} - e^{-\alpha f\eta} \; [18.20] = e^{\frac{1}{2}f\eta} - e^{-\frac{1}{2}f\eta} \qquad [\alpha = 0.5]$$

$$= 2\sinh\left(\tfrac{1}{2}f\eta\right) \quad \left[\sinh x = \tfrac{1}{2}\left(e^{x} - e^{-x}\right)\right] \qquad \text{where } f = F/RT = 38.92\ \text{V}^{-1} \text{ at } 25°\text{C}$$

and we use $\frac{1}{2}f\eta = \frac{1}{2}\times\left(38.92\ \text{V}^{-1}\right)\times\eta = 0.01946(\eta/\text{mV})$.

Thus, $j = 2j_0 \sinh\left(\frac{1}{2}f\eta\right) = (1.58\,\text{mA cm}^{-2})\times\sinh\left(0.01946\eta/\text{mV}\right)$

(a) $\eta = 10\,\text{mV}$

$$j = (1.58\,\text{mA cm}^{-2})\times(\sinh 0.1946) = \boxed{0.31\,\text{mA cm}^{-2}}$$

(b) $\eta = 100\,\text{mV}$

$$j = (1.58\,\text{mA cm}^{-2})\times(\sinh 1.946) = \boxed{5.41\,\text{mA cm}^{-2}}$$

(c) $\eta = -5.0\,\text{V}$

$$j = (1.58\,\text{mA cm}^{-2})\times(\sinh -97.3) \approx \boxed{-1.43\times10^{39}\ \text{A cm}^{-2}}$$

E18.25 The current density of electrons is j_0/e because each one carries a charge of magnitude e and the number of electrons passing through the electrode surface each second is j_0A/e where A is the surface area of the electrode. Therefore,

(a) $\text{Pt}\,|\,\text{H}_2\,|\,\text{H}^+;\ j_0 = 0.79\ \text{mA cm}^{-2}$ [Table 18.5]

$$\frac{j_0A}{e} = \frac{\left(0.79\ \text{mA cm}^{-2}\right)\times\left(1.0\ \text{cm}^2\right)}{1.602\times10^{-19}\ \text{C}} = \boxed{4.9\times10^{15}\,\text{s}^{-1}}$$

(b) $\text{Pt}|\text{Fe}^{3+},\ \text{Fe}^{2+};\ j_0 = 2.5\ \text{mA cm}^{-2}$

$$\frac{j_0A}{e} = \frac{\left(2.5\,\text{mA cm}^{-2}\right)\times\left(1.0\ \text{cm}^2\right)}{1.602\times10^{-19}\ \text{C}} = \boxed{1.6\times10^{16}\,\text{s}^{-1}}$$

(c) $\text{Pb}\,|\,\text{H}_2\,|\,\text{H}^+;\ j_0 = 5.0\times10^{-12}\ \text{A cm}^{-2}$

$$\frac{j_0A}{e} = \frac{\left(5.0\times10^{-12}\ \text{A cm}^{-2}\right)\times\left(1.0\ \text{cm}^2\right)}{1.602\times10^{-19}\ \text{C}} = \boxed{3.1\times10^{7}\ \text{s}^{-1}}$$

There are approximately

$$\frac{1.0\ \text{cm}^2}{(280\,\text{pm})^2} = 1.2\bar{8}\times10^{15} \text{ atoms in each square centimeter of surface.}$$

The numbers of electrons transferred through each atom per second is given by $\dfrac{j_0A}{\left(1.2\bar{8}\times10^{15}\right)e}$. For these electrodes the numbers are: $\boxed{3.8\ \text{s}^{-1}}$, $\boxed{13\ \text{s}^{-1}}$, and $\boxed{2.4\times10^{-8}\ \text{s}^{-1}}$, respectively. The last corresponds to less than one event per year.

E18.26

$$\ln j = \ln j_0 + (1-\alpha)f\eta \quad [18.23\text{a}] \qquad \text{where } f = F/RT = 38.92\ \text{V}^{-1} \text{ at } 25°\text{C}$$

Draw up the following table

η / mV	50	100	150	200	250
j/mA cm^{-2}	2.66	8.91	29.9	100	335
ln(j/mA cm^{-2})	0.978	2.19	3.40	4.61	5.81

The points are plotted in Fig. 18.8.

The intercept is at −0.230, and so $j_0/(\text{mA cm}^{-2}) = e^{-0.230} = \boxed{0.795}$. The slope is 0.0242, and so $(1-\alpha)f = 0.0242\ \text{mV}^{-1}$. It follows that $1-\alpha = 0.622$, and so $\alpha = \boxed{0.378}$.

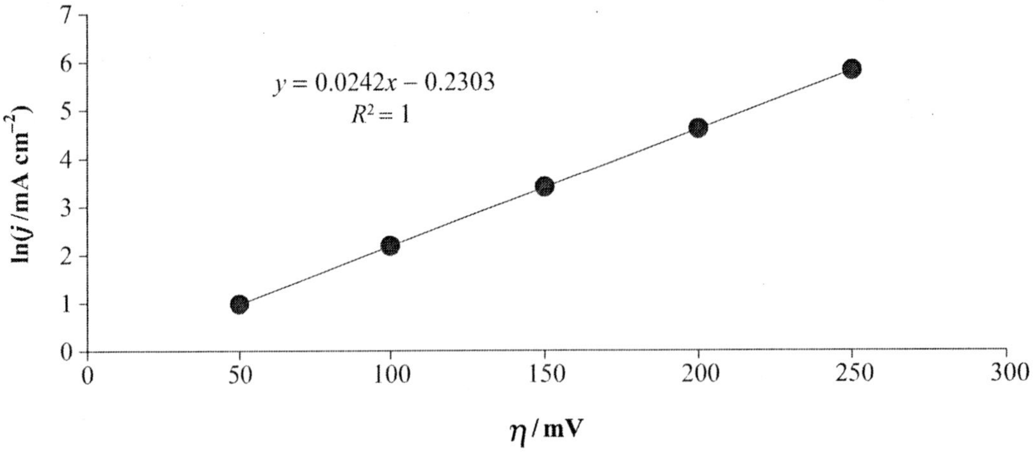

Figure 18.8

If η were negative,

$$j = j_0\left\{e^{(1-\alpha)f\eta} - e^{-\alpha f\eta}\right\}[18.20] \qquad \text{where } \alpha = 0.378 \quad \text{and} \quad j_0 = 0.795\ \text{mA cm}^{-2}$$

and we draw up the following table of computed current densities at this electrode.

η/mV	−50	−100	−150	−200	−250
$-j/(\text{mA cm}^{-2})$	1.42	3.39	7.20	15.1	31.5

E18.27 This problem differs somewhat from the simpler one-electron transfers considered in the text. In place of $\text{Ox} + e^- \rightarrow \text{Red}$ we have here

$$In^{3+} + 3\ e^- \rightarrow In$$

namely, a three-electron transfer. Therefore, the Butler-Volmer equation [18.20] and the Tafel large overpotential equations [18.23a and 18.23b] need to be modified by including the factor z, the number of electrons that appear in the half-reaction. Thus,

$$\ln j = \ln j_0 + z(1-\alpha)f\eta \qquad \text{anode; } f = F/RT = 38.92\ \text{V}^{-1} \text{ at } 25°\text{C and } z = 3$$

$$\ln(-j) = \ln j_0 - z\alpha f\eta \qquad \text{cathode}$$

We draw up the following table using $\eta = E' + 0.388\ \text{V}$ [18.21].

$j/(\text{A m}^{-2})$	$-E$/V	η/V	$\ln(j/\text{A m}^{-2})$
0	0.388	0	
0.590	0.365	0.023	−0.5276
1.438	0.350	0.038	0.3633
3.507	0.335	0.053	1.255

A plot of $\ln j$ against η is shown in Figure 18.9. It is seen to be linear so we perform a linear least squares regression fit. The regression intercept equals $\ln j_0$; the regression slope equals $z(1-\alpha)f$.

$$\ln\left(j_0/\text{A m}^{-2}\right) = \textit{intercept} = -1.894$$

$$j_0 = e^{-1.894}\ \text{A m}^{-2} = \boxed{0.150\ \text{A m}^{-2}}$$

$$z(1-\alpha)f = slope = 59.41\ \text{V}^{-1}$$

$$\alpha = 1 - \frac{slope}{zf}$$

$$= 1 - \frac{59.41\ \text{V}^{-1}}{3\times(38.92\ \text{V}^{-1})} = \boxed{0.491}$$

The cathodic current density is obtained from

$$\ln(-j_c/\text{A m}^{-2}) = \ln(j_0/\text{A m}^{-2}) - z\alpha f\eta \quad \text{where } \eta = 0.023\,\text{V at } E = -0.365\ \text{V}$$

$$= -1.894 - 3\times(0.491)\times(38.92\ \text{V}^{-1})\times(0.023\ \text{V})$$

$$= -3.21$$

$$j_c = -e^{-3.21}\ \text{A m}^{-2} = \boxed{-0.0404\ \text{A m}^{-2}}$$

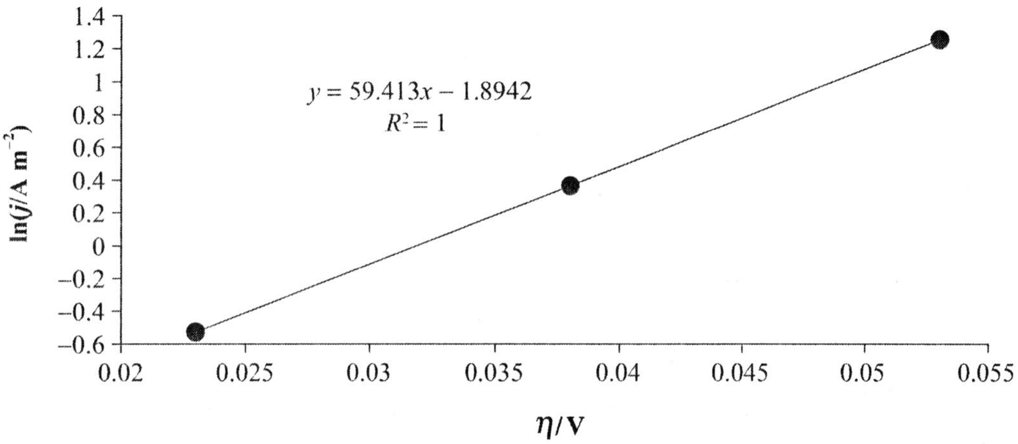

Figure 18.9

E18.28 $H^+(aq) + e^- \rightarrow H(Hg)$

$$2\ H(Hg) \rightarrow H_2(g)$$

At large positive values of the overpotential the current density is anodic.

$$\ln j = \ln j_0 + z(1-\alpha)f\eta \ \ [18.23a] \qquad \text{where } f = F/RT = 38.92\ \text{V}^{-1} \text{ at } 25°\text{C and } z = 1$$

A plot of $\ln j$ against η is shown in Figure 18.10. It is seen to be linear so we perform a linear least squares regression fit. The regression intercept equals $\ln j_0$; the regression slope equals $z(1-\alpha)f$.

$$\ln(j_0/\ \text{mA m}^{-2}) = intercept = -10.83$$

$$j_0 = e^{-10.83}\ \text{mA m}^{-2} = \boxed{19.80\ \text{nA m}^{-2}}$$

$$z(1-\alpha)f = slope = 19.55\ \text{V}^{-1}$$

$$\alpha = 1 - \frac{slope}{zf}$$

$$= 1 - \frac{19.55\ \text{V}^{-1}}{1\times(38.92\ \text{V}^{-1})} = \boxed{0.498}$$

The Tafel plot appears to have no non-linear character and the linear regression explains 99.8% of the variation in the plot with the remainder appearing to be random experimental uncertainty. There are $\boxed{\text{no}}$ deviations from the Tafel equation for large positive overpotentials.

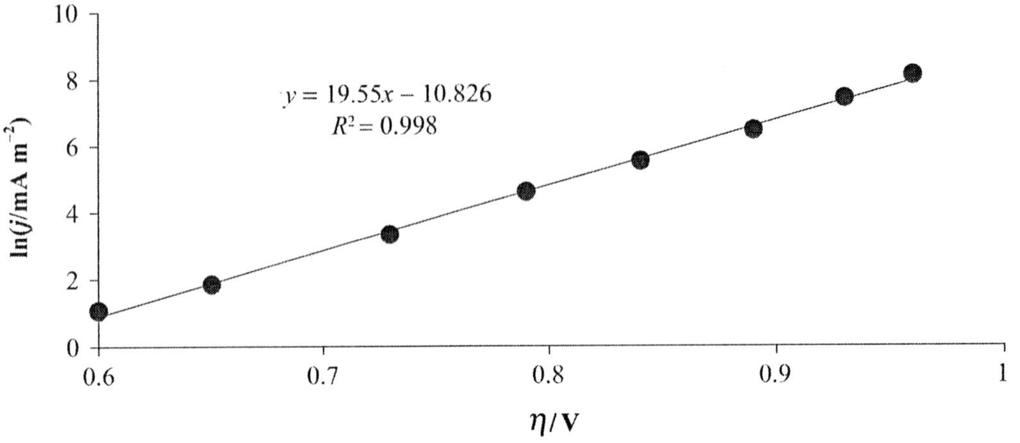

Figure 18.10

E18.29 (a) The electroactive species is reduced during the forward sweep of potential. The process is reversible because the reduced species shows oxidization upon reversing the direction of the potential sweep. The peak reduction current and peak oxidation current lie symmetrically about the standard reduction potential.

(b) The electroactive species experiences a second reduction at high potential during the forward sweep of potential. Both reductions are reversible and, consequently, show oxidation currents upon reversing the potential scan toward low potentials.

(c) Reduction of the electroactive species is observed during the forward potential sweep. However, the process is not reversible. No oxidation current is observed upon reversing the sweep.

(d) The electroactive species experiences a second reduction at high potential during the forward sweep of potential. Upon reversing the potential sweep, the highly reduced species is reversibly oxidized with the loss of one electron but loss of a second electron is not observed at low potential. The complete reoxidation of the species is irreversible.

Answers to projects

P18.30 (a) $$F = -\frac{\mathrm{d}V}{\mathrm{d}r} = -\frac{\mathrm{d}}{\mathrm{d}r}\left(\frac{Q_1Q_2}{4\pi\varepsilon_0 r}\right) = \frac{Q_1Q_2}{4\pi\varepsilon_0 r^2} = \frac{e^2}{4\pi\varepsilon_0 r^2}$$ for the repulsion of two electrons

$$F_{r=0.50\text{ nm}} = \frac{(1.602\times10^{-19}\text{ C})^2}{4\pi(8.854\times10^{-12}\text{ J}^{-1}\text{ C}^2\text{ m}^{-1})\times(0.50\times10^{-9}\text{ m})^2} = \boxed{9.2\times10^{-10}\text{ N}}$$

$$F_{r=0.60\text{ nm}}/F_{r=0.50\text{ nm}} = \left(\frac{0.50\text{ nm}}{0.60\text{ nm}}\right)^2 = \boxed{0.69}$$

(b) We assume that the Lennard-Jones potential, eqn [15.15], adequately represents the potential energy in this case.

$$V(r) = 4\varepsilon \left\{\left(\frac{\sigma}{r}\right)^{12} - \left(\frac{\sigma}{r}\right)^{6}\right\}$$

$$F = -\frac{dV}{dr} = -4\varepsilon\frac{d}{dr}\left\{\left(\frac{\sigma}{r}\right)^{12} - \left(\frac{\sigma}{r}\right)^{6}\right\} = \boxed{\frac{24\varepsilon}{\sigma}\left\{\frac{2\sigma^{13}}{r^{13}} - \frac{\sigma^{7}}{r^{7}}\right\}}$$

Furthermore, the force has its minimal value of zero when

$$\frac{2\sigma^{13}}{r^{13}} = \frac{\sigma^{7}}{r^{7}} \text{ which solves to } r = 2^{1/6}\sigma.$$

P18.31 The van't Hoff equation for adsorption equilibrium is $\frac{d\ln K}{dT} = \frac{\Delta_{ads}H^{\ominus}}{RT^2}$ at fixed surface coverage, θ. $\Delta_{ads}H^{\ominus}$ is called the isosteric enthalpy of adsorption. Solving the Langmuir isotherm for ln K and taking the derivative w/r/t T at fixed surface coverage, we find

$$\theta = \frac{Kp}{1+Kp} \quad \text{or} \quad \theta(1+Kp) = Kp$$

$$\theta = (1-\theta)Kp$$

$$Kp = \frac{\theta}{1-\theta}$$

$$\ln(Kp) = \ln\left(\frac{\theta}{1-\theta}\right)$$

$$\ln K + \ln p = \ln\left(\frac{\theta}{1-\theta}\right)$$

$$\frac{d\ln K}{dT} + \frac{d\ln p}{dT} = \frac{d\ln\left(\frac{\theta}{1-\theta}\right)}{dT} = 0 \quad \text{because } \theta \text{ is constant.}$$

$$\frac{d\ln K}{dT} = -\frac{d\ln p}{dT}$$

Substitution into the van't Hoff equation yields

$$\boxed{\frac{d\ln p}{dT} = -\frac{\Delta_{ads}H^{\ominus}}{RT^2}} \text{ at fixed surface coverage, } \theta.$$

With $d(1/T)/dT = -1/T^2$, this expression rearranges to $\boxed{\frac{d\ln p}{d(1/T)} = \frac{\Delta_{ads}H^{\ominus}}{R}}$.

Therefore, a plot of ln p against $1/T$ should be a straight line of slope $\Delta_{ads}H^{\ominus}/R$.

P18.32

$$E^{\ominus} = \frac{-\Delta_r G^{\ominus}}{\nu F}$$

(a) (i) $H_2 + \frac{1}{2} O_2 \rightarrow H_2O$; $\Delta_r G^{\ominus} = -237 \text{ kJ mol}^{-1}$

Since $\nu = 2$,

$$E^{\ominus} = \frac{-(-237 \text{ kJ mol}^{-1})}{(2)\times(96.48 \text{ kC mol}^{-1})} = \boxed{+1.23 \text{ V}}$$

(ii) $CH_4 + 2\ O_2 \rightarrow CO_2 + 2\ H_2O$

$$\Delta_r G^{\ominus} = 2\Delta_f G^{\ominus}(H_2O) + \Delta_f G^{\ominus}(CO_2) - \Delta_f G^{\ominus}(CH_4)$$
$$= [(2)\times(-237.1) + (-394.4) - (-50.7)]\,\text{kJ mol}^{-1} = -817.9\ \text{kJ mol}^{-1}$$

As written, the reaction corresponds to the transfer of eight electrons. It follows that, for the species in their standard states,

$$E^{\ominus} = \frac{-(-817.9\,\text{kJ mol}^{-1})}{(8)\times(96.48\,\text{kC mol}^{-1})} = \boxed{+1.06\ \text{V}}$$

(b) $j_0 = 6.3\times10^{-6}\ \text{A cm}^{-2}$, $\alpha = 0.58$ [Table 18.5]

(i) Butler–Volmer eqn:

$$j = j_0\{e^{(1-\alpha)f\eta} - e^{-\alpha f\eta}\}\ [18.20] \qquad \text{where } f = F/RT = 38.92\ \text{V}^{-1} \text{ at } 25°\text{C}$$

$$j = (6.3\times10^{-6}\ \text{A cm}^{-2})\times\left(e^{\{(1-0.58)\times(38.92\ \text{V}^{-1})\times(0.20\ \text{V})\}} - e^{\{-(0.58)\times(38.92\ \text{V}^{-1})\times(0.20\ \text{V})\}}\right)$$
$$= \boxed{0.17\ \text{mA cm}^{-2}}$$

(ii) Tafel eqn for large, positive overpotential:

$$j = j_0 e^{(1-\alpha)f\eta}\ [18.23a]$$

$$j = (6.3\times10^{-6}\ \text{A cm}^{-2})\times e^{\{(1-0.58)\times(38.92\ \text{V}^{-1})\times(0.20\ \text{V})\}}$$
$$= \boxed{0.17\ \text{mA cm}^{-2}}$$

Comparison of (i) with (ii) shows that there is negligible difference between the Butler–Volmer eqn and the Tafel eqn when $\eta = 0.20$ V. The validity of the Tafel is unaffected by overpotentials that are greater than 0.20 V.

19 Spectroscopy: molecular rotations and vibrations

Answers to discussion questions

D19.1 (a) For *microwave rotational spectroscopy*, the allowed transitions depend on the existence of an oscillating dipole moment, which can stir the electromagnetic field into oscillation (and vice versa for absorption). This implies that the molecule must have a permanent dipole moment, which is equivalent to an oscillating dipole when the molecule is rotating. See Figure 19.11 of the text.

(b) The gross selection rule for *rotational Raman spectroscopy* is that the molecule must be anisotropically polarizable, which is to say that its polarizability, α, depends upon the direction of the electric field relative to the molecule. Non-spherical rotors satisfy this condition. Therefore, linear and symmetric rotors are rotationally Raman active.

D19.2 *Doppler broadening*. This contribution to the linewidth is due to the Doppler effect which shifts the frequency of the radiation emitted or absorbed when the atoms or molecules involved are moving towards or away from the detecting device. Molecules have a wide range of speeds in all directions in a gas and the detected spectral line is the absorption or emission profile arising from all the resulting Doppler shifts. The shape of a Doppler-broadened spectral line reflects the Maxwell distribution of speeds in the sample at the temperature of the experiment; hence the line broadens as the temperature is increased because the molecules acquire a wider range of speeds. Therefore, to decrease the linewidth, the temperature of the sample should be decreased.

Lifetime broadening. The Doppler broadening is significant in gas-phase samples, but lifetime broadening occurs in all states of matter. This kind of broadening is a quantum mechanical effect related to the uncertainty principle in the form of eqn 19.14a and is due to the finite lifetimes of the states involved in the transition. When τ is finite, the energy of the states is smeared out and hence the transition frequency is broadened as shown in eqn 19.14b.

Pressure broadening or collisional broadening. The actual mechanism affecting the lifetime of energy states depends on various processes, one of which is collisional deactivation and another is spontaneous emission. Lowering the pressure can reduce the first of these contributions; the second cannot be changed and results in a natural linewidth. The rate of spontaneous emission cannot be changed; hence it is a natural limit to the breadth of a spectral line.

D19.3 The answer to this question depends precisely on what is meant by equilibrium bond length. The angular velocity increases with the quantum number J. Thus, centrifugal distortion and bond length r_c is greater for higher rotational energy levels. But the equilibrium bond length r_e remains constant if by that term one means the value of r corresponding to a vibrating non-rotating molecule with $J = 0$.

D19.4 The rotational constants in vibrationally excited states are smaller than in the vibrational ground state and continue to get smaller as the vibrational level increases. Any anharmonicity in the vibration causes a slight extension of the bond length in the excited state. This results in an increase in the moment of inertia, and a consequent decrease in the rotational constant. The equation that describes how the rotational constant changes with increasing vibrational levels is $\tilde{B}_v = \tilde{B}_e - a(v+\frac{1}{2})$, where $\tilde{B}_e$ and a are constants and $\tilde{B}_v$ is the rotational constant in level v. The constant a is related to the anharmonicity of the potential energy and is zero for a strictly parabolic (harmonic) potential; hence the rotational constant is independent of the vibrational state for the strictly parabolic potential. (See Atkins, P. W., de Paula, J. C., and Friedman, R. S. (2009). *Quanta, Matter, and Change*. Oxford: Oxford University Press for further discussion of this effect and equation.)

D19.5 Refer to fig. 19.18 for a view of the placement in space of both linear and non-linear polyatomic molecules. A total of $3N$ coordinates are necessary to specify the location in space of a molecule of N atoms. Three of these coordinates are necessarily translational coordinates, corresponding to the motion of the center of mass of the molecule, whether the molecule is linear or not. Of the remaining $3N - 3$ coordinates, as can be seen from figure 19.18, only two angles are necessary to specify the orientation of a linear molecule in space, whereas three are necessary for a nonlinear molecule. The specification of the third angle in the nonlinear molecule effectively eliminates one of the two perpendicular bending motions the molecule would have if it were linear. This is most easily pictured for the linear triatomic, CO_2 which has two perpendicular bending motions. If the molecule were permanently bent into an angular molecule only one bending motion remains, a bending motion that changes the dihedral angle.

D19.6 (a) In the case of *infrared vibrational spectroscopy*, the physical basis of the gross selection rule is that the molecule must have a structure that allows for the existence of an oscillating dipole moment when the molecule vibrates. Polar molecules necessarily satisfy this requirement, but non-polar molecules may also have a fluctuating dipole moment upon vibration. See Figure 19.17 of the text.

The gross selection rule for *vibrational Raman spectroscopy* is that the polarizability of the molecule must change as the molecule vibrates. All diatomic molecules satisfy this condition as the molecules swell and contract during a vibration, the control of the nuclei over the electrons varies, and the molecular polarizability changes. Hence both homonuclear and heteronuclear diatomics are vibrationally Raman active. In polyatomic molecules it is usually quite difficult to judge by inspection whether or not the molecular polarizability changes upon vibration; hence group theoretical methods are relied on for judging the Raman activity of the various normal modes of vibration. The procedure is an advanced topic but the **exclusion rule** can be useful. It states that, if the molecule has a center of inversion, then no modes can be both infrared and Raman active.

(b) The exclusion rule applies to the benzene molecule because it has a center of inversion. Consequently, none of the normal modes of vibration of benzene can be both infrared and Raman active. If we wish to characterize all the normal modes, we must obtain both kinds of spectra.

D19.7 In general, vibrational frequencies are determined by the effective masses of the group of atoms participating in the mode of vibration. Since the mass of ^{13}C is greater than the mass of ^{12}C, in general we expect that vibrational frequencies would be different in $^{13}CO_2$ than in $^{12}CO_2$. However, in the symmetric stretch of CO_2, the C atom is stationary, and the effective mass of the mode depends only on the O atoms. Consequently we expect that the vibrational frequency of this mode would be independent of the mass of the carbon atom.

D19.8 See Section 19.12 of the text. For a more complete discussion see Atkins, P. W., de Paula, J. C. (2006). *Physical Chemistry* (8th ed.). Oxford: Oxford University Press.

The vibration-rotation energy levels are given by eqn 19.20 which is

$$E_{v,J} = \left(v + \frac{1}{2}\right)h\nu + hBJ(J+1)$$

In general, transitions corresponding to $\Delta J = \pm 1$ and $\Delta J = 0$ are allowed. Therefore, as can be seen from eqn 19.20, three sets of lines may occur in the vibration-rotation spectrum. One set, the P branch corresponds to $\Delta J = -1$, another set, the R branch to $\Delta J = +1$, and a third set, the Q branch to $\Delta J = 0$. The Q branch, however, is not always seen. The selection rule $\Delta J = 0$ corresponds to an allowed transition only when the molecule posses angular momentum about its axis, as in the case of the electronic orbital angular momentum of the paramagnetic molecule NO.

Solutions to exercises

E19.1 (a) $$\nu = \frac{c}{\lambda} = \frac{2.998\times10^{8}\ \text{m s}^{-1}}{4.42\times10^{-7}\ \text{m}} = \boxed{6.78\times10^{14}\ \text{s}^{-1} = 6.78\times10^{14}\ \text{Hz}}.$$

(b) $$\tilde{\nu} = \frac{1}{\lambda} = \frac{1}{4.42\times10^{-7}\ \text{m}} = 2.26\times10^{6}\ \text{m}^{-1} = \boxed{2.26\times10^{4}\ \text{cm}^{-1}}.$$

E19.2 (a) $$\tilde{\nu} = \frac{\nu}{c} = \frac{88.0\times10^{6}\ \text{s}^{-1}}{2.998\times10^{8}\ \text{m s}^{-1}} = 0.294\ \text{m}^{-1} = \boxed{2.94\times10^{-3}\ \text{cm}^{-1}}.$$

(b) $$\lambda = \frac{1}{\tilde{\nu}} = \frac{1}{0.294\ \text{m}^{-1}} = \boxed{3.40\ \text{m}}.$$

E19.3 $E_J = hBJ(J+1) \approx hBJ^2$ for large J.

$$J = \left(\frac{E_J}{hB}\right)^{1/2},\quad B = \frac{\hbar}{4\pi I},\quad I = mR^2.$$

$$J = \frac{\pi R}{h}(8mE_J)^{1/2}$$

$$= \frac{\pi\times0.70\ \text{m}}{6.63\times10^{-34}\ \text{J s}}\times(8\times0.75\ \text{kg}\times0.2\ \text{J})^{1/2}$$

$$\approx \boxed{4\times10^{33}}.$$

E19.4 (a) $$I = \mu R^2 \qquad \mu = \frac{m_A m_B}{m_A + m_B} = \frac{m}{2}\ \text{for H}_2$$

$$I = \frac{m}{2}R^2 = \frac{1.0078\times1.661\times10^{-27}\ \text{kg}\times(74.14\times10^{-12}\ \text{m})^2}{2}$$

$$= \boxed{4.601\times10^{-48}\ \text{kg m}^2}$$

(b) $$I = \frac{2.0140\times1.661\times10^{-27}\ \text{kg}\times(74.15\times10^{-12}\ \text{m})^2}{2}$$

$$= \boxed{9.196\times10^{-48}\ \text{kg m}^2}$$

(c) For a linear BA_2 molecule, $I = 2\,m_A R^2$ (Table 19.1)

$$I = 2\times15.9949\times1.661\times10^{-27}\ \text{kg}\times(116.00\times10^{-12}\ \text{m})^2$$

$$= \boxed{7.15\times10^{-46}\ \text{kg m}^2}$$

(d) Assuming the same internuclear distance, $I = \boxed{7.15\times10^{-46}\ \text{kg m}^2}$

E19.5 $$\text{unit of } B = \text{unit of}\left(\frac{\hbar}{4\pi I}\right) = \frac{\text{J s}}{\text{kg m}^2} = \frac{\text{kg m}^2\ \text{s}^{-2}\ \text{s}}{\text{kg m}^2} = \text{s}^{-1}$$

So we see that the unit of B is s^{-1} = Hz

(a) $$B = \frac{1.0546\times10^{-34}\ \text{J s}}{4\pi\times4.601\times10^{-48}\ \text{kg m}^2} = \boxed{1.824\times10^{12}\ \text{Hz}}$$

(b) $$B = \frac{1.0546\times10^{-34}\ \text{J s}}{4\pi\times9.196\times10^{-48}\ \text{kg m}^2} = \boxed{9.126\times10^{11}\ \text{Hz}}$$

(c) $B = \dfrac{1.0546\times10^{-34}\ \text{J s}}{4\pi\times7.15\times10^{-46}\ \text{kg m}^2} = \boxed{1.17\times10^{10}\ \text{Hz}}$

(d) $B = \boxed{1.17\times10^{10}\ \text{Hz}}$

E19.6 (a) An octahedral AB_6 molecule is a spherical rotor and all moments of inertia are equal.

$$I = 4m_B R^2 \ \text{[Table 19.1]}$$

(b) $B = \dfrac{\hbar}{4\pi I} = \dfrac{1.0546\times10^{-34}\ \text{J s}}{4\pi\times4\times19.00\times1.661\times10^{-27}\ \text{kg}\times(158\times10^{-12}\ \text{m})^2}$

$= \boxed{2.663\times10^{9}\ \text{Hz}}$

E19.7 $\boxed{I_{\parallel} = 4mR^2 \ \text{and}\ I_{\perp} = 2mR^2}$ See the definitions in Figure 19.1.

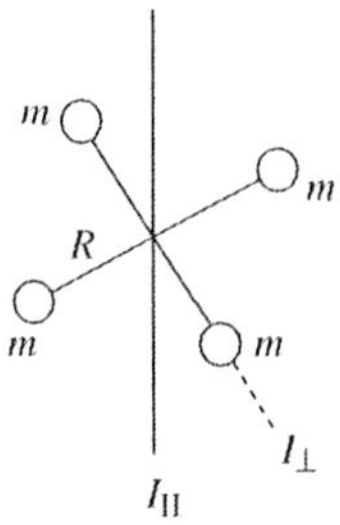

Figure 19.1

E19.8 (a) SO_3 is a trigonal planar molecule. $I_{\parallel}$ is perpendicular to the plane of the molecule and along the vertical axis that passes through the S atom. $I_{\perp}$ is shown in the Figure 19.2. The bond angles are 120°.

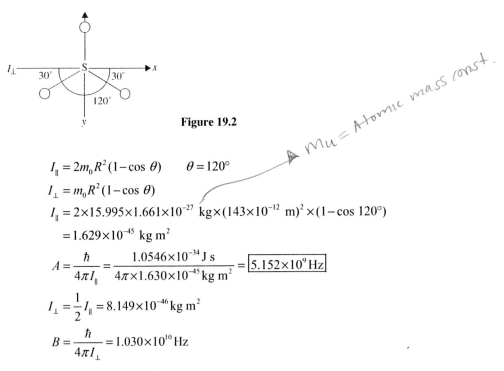

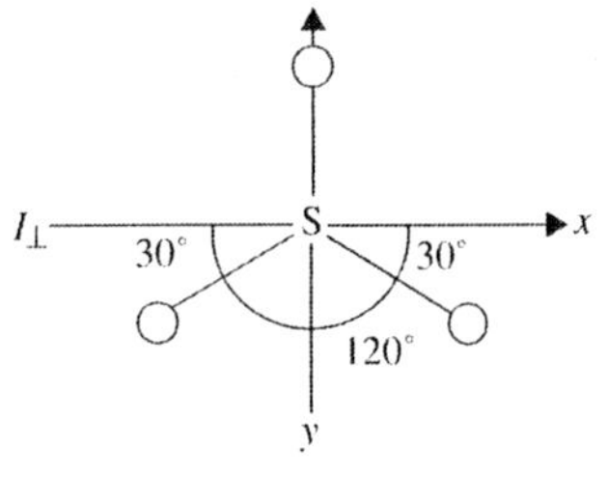

Figure 19.2

Mu = Atomic mass const.

$$I_{\parallel} = 2m_0R^2(1-\cos\theta) \qquad \theta = 120°$$

$$I_{\perp} = m_0R^2(1-\cos\theta)$$

$$I_{\parallel} = 2\times15.995\times1.661\times10^{-27}\ \text{kg}\times(143\times10^{-12}\ \text{m})^2\times(1-\cos 120°)$$

$$= 1.629\times10^{-45}\ \text{kg m}^2$$

$$A = \frac{\hbar}{4\pi I_{\parallel}} = \frac{1.0546\times10^{-34}\ \text{J s}}{4\pi\times1.630\times10^{-45}\ \text{kg m}^2} = \boxed{5.152\times10^{9}\ \text{Hz}}$$

$$I_{\perp} = \frac{1}{2}I_{\parallel} = 8.149\times10^{-46}\ \text{kg m}^2$$

$$B = \frac{\hbar}{4\pi I_{\perp}} = 1.030\times10^{10}\ \text{Hz}$$

(b) The mass of the sulfur atom does not affect the rotational constants, so microwave spectroscopy $\boxed{\text{could not be used}}$ to distinguish between $^{32}S^{16}O_3$ and $^{33}S^{16}O_3$.

E19.9 Polar molecules show a pure rotational spectrum. Therefore, select the polar molecules based on their well-known structures.

$\boxed{\text{(a), (b), (c), and (d)}}$ have dipole moments and will have a pure rotational spectrum. $\boxed{\text{(e) will not.}}$

E19.10 None of the molecules listed are spherical rotors; therefore $\boxed{\text{all will show}}$ a rotational Raman spectrum.

E19.11 Methane is a spherical rotor; hence $I_{\parallel} = I_{\perp}$ and $A = B$ and

$$E_{J,K} = hBJ(J+1)$$

and there is $\boxed{\text{one energy level with } J = 8}$.

There is a degeneracy of $(2J + 1)^2 = 289$ associated with this level, so there are a total of $\boxed{289}$ quantum states. [See Project 19.34 for the formula for the degeneracy of a spherical rotor.]

E19.12

$$E_{J,K} = hBJ(J+1) + h(A-B)K^2 \qquad [19.6]$$

$$J = 0, 1, 2.... \qquad K = J, J-1, \ldots, -J \quad (2J+1 \text{ possible } K \text{ values for a given } J)$$

There are $\boxed{\text{17 states}}$ for $J = 8$, but as K^2 determines the energy, there are only $\boxed{\text{9 energy levels}}$, and only one, $K = 0$, will be at energy $hBJ(J + 1)$.

E19.13

$$E_J = hBJ(J+1)$$

$$\Delta E = E_{J+1} - E_J = hB(J+1)(J+2) - hBJ(J+1)$$

$$= 2hB(J+1)$$

The separations of the lines are therefore

$$2hB, 4hB, 6hB,$$

(a) The frequencies of the lines are

$$2B, 4B, 6B, \ldots = \boxed{636\text{ GHz}, 1272\text{ GHz}, 1908\text{ GHz}\ldots}$$

(b) The wavenumbers of the lines are given by

$$\tilde{\nu} = \frac{\nu}{c} = \frac{636\times10^9\,\text{s}^{-1}}{2.998\times10^8\,\text{m s}^{-1}} = 2121\text{ m}^{-1} = \boxed{21.21\text{ cm}^{-1}}$$

So the set of lines corresponds to

$$\boxed{21.21\text{ cm}^{-1}, 42.42\text{ cm}^{-1}, 63.63\text{ cm}^{-1}, \ldots}$$

E19.14

$$B = \frac{\hbar}{4\pi cI}[19.4 \text{ in units of cm}^{-1}], \text{ implying that } I = \frac{\hbar}{4\pi cB}$$

Then, with $I = \mu R^2$, $\quad R = \left(\frac{\hbar}{4\pi\mu cB}\right)^{1/2}$

We use $\mu = \dfrac{m_1 m_2}{m_1 + m_2} = \dfrac{(126.904)\times(34.9688)}{(126.904)+(34.9688)}\text{u} = 27.4146\,\text{u}$

and hence obtain

$$R = \left(\frac{1.05457\times10^{-34}\,\text{Js}}{(4\pi)\times(27.4146)\times(1.66054\times10^{-27}\,\text{kg})\times(2.99792\times10^{10}\,\text{cm s}^{-1})\times(0.1142\,\text{cm}^{-1})}\right)^{1/2}$$

$$= \boxed{232.1\text{ pm}}$$

E19.15 $$B = \frac{\hbar}{4\pi I} \qquad \nu = 2B \qquad \tilde{\nu} = \frac{2B}{c}$$

$$I = \mu R^2 \qquad \mu = \frac{m_H m_{Cl}}{m_H + m_{Cl}}$$

If m_H increase as it does when ^{2}H replaces ^{1}H, then μ increases. When μ increases, I increases. When I increases, B decreases, and hence $\tilde{\nu}$ decreases (shifts to lower wavenumber).

E19.16 The wavenumber of a Stokes line in rotational Raman is

$$\tilde{\nu}_{\text{Stokes}} = \tilde{\nu}_i - 2B(2J+3) \text{ [19.15 with } B \text{ in cm}^{-1}\text{]}$$

where J is the initial (lower) rotational state. So

$$\tilde{\nu}_{\text{Stokes}} = 20623 \text{ cm}^{-1} - 2(1.4457 \text{ cm}^{-1}) \times [2(2)+3] = \boxed{20603 \text{ cm}^{-1}}$$

E19.17 $$B = \frac{\hbar}{4\pi I} = 11.70 \text{ GHz} = 11.70 \times 10^9 \text{ s}^{-1}$$

$$I = 2m_O R^2$$

Solve for R^2 in terms of B

$$R^2 = \frac{\hbar}{8\pi m_O B}$$

$$R^2 = \frac{1.0546 \times 10^{-34} \text{ J s}}{8\pi \times 15.995 \times 1.66054 \times 10^{-27} \text{ kg} \times 11.70 \times 10^9 \text{ s}^{-1}}$$
$$= 1.350 \times 10^{-20} \text{ m}^2$$

$$R = 1.162 \times 10^{-10} \text{ m} = \boxed{116.2 \text{ pm}}$$

E19.18 $$2B = 384 \text{ GHz} = \text{separation of lines}$$

$$B = 192 \text{ GHz}$$

$$B = \frac{\hbar}{4\pi I} = \frac{\hbar}{4\pi\mu R^2}$$

$$R^2 = \frac{\hbar}{4\pi\mu B}$$

$$\mu(^1\text{H}^{127}\text{I}) = \frac{1.0078 \times 126.9045}{1.0078 + 126.9045} \times 1.66054 \times 10^{-27} \text{ kg}$$
$$= 1.6603 \times 10^{-27} \text{ kg}$$

$$R^2 = \frac{1.05457 \times 10^{-34} \text{ J s}}{4\pi \times 1.6603 \times 10^{-27} \text{ kg} \times 192 \times 10^9 \text{ s}^{-1}}$$
$$= 2.632 \times 10^{-20} \text{ m}^2$$

$$R = 1.622 \times 10^{-10} \text{ m} = \boxed{162.2 \text{ pm}}$$

$$\mu(^2\text{H}^{127}\text{I}) = 3.2921 \times 10^{-27} \text{ kg}$$

$$B = \frac{\hbar}{4\pi\mu R^2} = \frac{1.05457 \times 10^{-34} \text{ J s}}{4\pi \times 3.2921 \times 10^{-27} \text{ kg} \times 2.632 \times 10^{-20} \text{ m}^2}$$
$$= 96.9 \times 10^9 \text{ s}^{-1} = 96.9 \text{ GHz}$$

$$2B = \boxed{194 \text{ GHz}} = \text{separation of lines}$$

E19.19 Equation [19.12] indicates that a plot of $\frac{\tilde{\nu}_J}{J+1}$ against $(J+1)^2$ should be linear with intercept equal to $2\tilde{B}$ and slope equal to $-4\tilde{D}$. $\tilde{B}$ and $\tilde{D}$ are the values of the constants B and D expressed in units of wavenumber, cm^{-1}.

The transition assignments are determined by guessing assignments and checking that the plot is linear. For example, guessing the lowest energy observed line has the assignment $J = 0 \rightarrow 1$ and subsequent lines have the assignments $J = 1 \rightarrow 2$, $J = 2 \rightarrow 3$, and $J = 3 \rightarrow 4$ yields a very non-linear plot. The assignments $J = 2 \rightarrow 3$, $J = 3 \rightarrow 4$, $J = 4 \rightarrow 5$, and $J = 5 \rightarrow 6$ also yield a non-linear plot. Only the assignments $J = 1 \rightarrow 2$, $J = 2 \rightarrow 3$, $J = 3 \rightarrow 4$, and $J = 4 \rightarrow 5$ yield a highly linear plot (see Figure 19.3). The intercept of the linear plot equals $0.405704\ cm^{-1}$ and the slope equals $-2.493 \times 10^{-7}\ cm^{-1}$.

$$\tilde{B} = \frac{\text{intercept}}{2} = \frac{0.405704\ cm^{-1}}{2} = \boxed{0.202852\ cm^{-1}}$$

$$\tilde{D} = -\frac{\text{slope}}{4} = -\frac{(-2.493\times10^{-7}\ cm^{-1})}{4} = \boxed{6.2\times10^{-8}\ cm^{-1}}$$

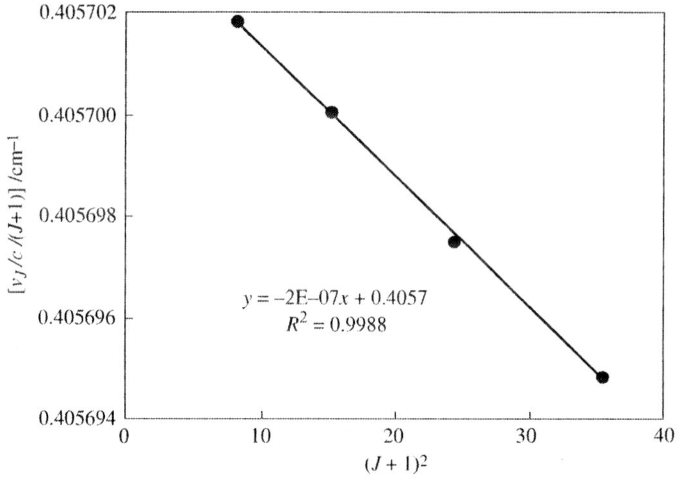

Figure 19.3

E19.20 From the equation for a linear rotor in Table 19.1 it is possible to show that $Im = m_a m_c (R+R')^2 + m_a m_b R^2 + m_b m_c R'^2$.

Thus, $$I(^{16}O^{12}C^{32}S) = \left(\frac{m(^{16}O)m(^{32}S)}{m(^{16}O^{12}C^{32}S)}\right)\times(R+R')^2 + \left(\frac{m(^{12}C)\{m(^{16}O)R^2 + m(^{32}S)R'^2\}}{m(^{16}O^{12}C^{32}S)}\right)$$

$$I(^{16}O^{12}C^{34}S) = \left(\frac{m(^{16}O)m(^{34}S)}{m(^{16}O^{12}C^{34}S)}\right)\times(R+R')^2 + \left(\frac{m(^{12}C)\{m(^{16}O)R^2 + m(^{34}S)R'^2\}}{m(^{16}O^{12}C^{34}S)}\right)$$

$m(^{16}O) = 15.9949\ u$, $m(^{12}C) = 12.0000\ u$, $m(^{32}S) = 31.9721\ u$, and $m(^{34}S) = 33.9679\ u$. Hence,

$$I(^{16}O^{12}C^{32}S)/u = (8.5279)\times(R+R')^2 + (0.20011)\times(15.9949R^2 + 31.9721R'^2)$$

$$I(^{16}O^{12}C^{34}S)/u = (8.7684)\times(R+R')^2 + (0.19366)\times(15.9949R^2 + 33.9679R'^2)$$

The spectral data provides the experimental values of the moments of inertia based on the relation $\nu = 2B(J+1)$ [19.3] with $B = \frac{\hbar}{4\pi I}$ [19.4]. These values are set equal to the above equations which are then solved for R and R'. The mean values of I obtained from the data are

$$I(^{16}O^{12}C^{32}S) = 1.37998\times10^{-45}\ kg\ m^2$$

$$I(^{16}O^{12}C^{34}S) = 1.41460\times10^{-45}\ kg\ m^2$$

Therefore, after conversion of the atomic mass units to kg, the equations we must solve are

$$1.37998\times10^{-45}\ \text{m}^2 = (1.4161\times10^{-26})\times(R+R')^2 + (5.3150\times10^{-27}R^2) + (1.0624\times10^{-26}R'^2)$$

$$1.41460\times10^{-45}\ \text{m}^2 = (1.4560\times10^{-26})\times(R+R')^2 + (5.1437\times10^{-27}R^2) + (1.0923\times10^{-26}R'^2)$$

These two equations may be solved for R and R'. They are tedious to solve by hand, but straightforward. Readily available mathematical software can be used to quickly give the result. The outcome is $R = \boxed{116.28\ \text{pm}}$ and $R' = \boxed{155.97\ \text{pm}}$. These values may be checked by direct substitution into the equations.

E19.21 65 m.p.h. = 29.1 m s^{-1}

We use $\nu' = \left(\dfrac{1+s/c}{1-s/c}\right)^{1/2}\nu$ [eqn 19.5a of the 4th edition of *Elements of Physical Chemistry*]

or $\lambda' = \left(\dfrac{1-s/c}{1+s/c}\right)^{1/2}\lambda$

$$= \left(\frac{1-29.1/2.998\times10^8}{1+29.1/2.998\times10^8}\right)^{1/2}\lambda$$

$$= \boxed{0.999\,999\,9029\times660\ \text{nm}}$$

The formula for λ' can be rewritten after expanding the numerator and denominator

$$(1\pm x)^{1/2} = 1\pm\frac{x}{2}+\cdots$$

Then

$$\lambda' \approx (1-s/c)\lambda$$

Now solve for s such that $\lambda' = 520$ nm

$$s = \left(1-\frac{\lambda'}{\lambda}\right)c$$

$$= \left(1-\frac{520}{660}\right)\times2.998\times10^8\ \text{m s}^{-1}$$

$$= \boxed{6.36\times10^7\ \text{m s}^{-1}} = 2.3\times10^8\ \text{kph} = 1.4\times10^8\ \text{mph}$$

E19.22 The star is receding so we use

$\nu' = \left(\dfrac{1-s/c}{1+s/c}\right)^{1/2}\nu$ [eqn 19.5a of the 4th edition of *Elements of Physical Chemistry*]

or

$$\lambda' = \left(\frac{1+s/c}{1-s/c}\right)^{1/2}\lambda$$

or $\lambda' \approx (1+s/c)\lambda$

Solve for s

$$s = \left(\frac{\lambda'}{\lambda} - 1\right)c$$

$$= \left(\frac{706.5}{654.2} - 1\right) \times 2.998\times10^{8}\ \mathrm{m\ s^{-1}} = \boxed{2.397\times10^{7}\ \mathrm{m\ s^{-1}}}$$

$$\delta\lambda = \frac{2\lambda}{c}\left(\frac{2RT}{M}\ln 2\right)^{1/2} \qquad \left[19.13,\ \text{note } \frac{R}{M} = \frac{k}{m}\right]$$

Solve for T to obtain

$$T = \frac{m}{2k\ln 2}\left(\frac{c\delta\lambda}{2\lambda}\right)^{2}$$

$$= \frac{48\times(1.6605\times10^{-27}\ \mathrm{kg})}{2\times(1.381\times10^{-23}\ \mathrm{J\ K^{-1}})\times\ln 2}\left(\frac{(2.998\times10^{8}\ \mathrm{m\ s^{-1}})\times(61.8\times10^{-12}\ \mathrm{m})}{2\times(654.2\times10^{-9}\ \mathrm{m^{2}})}\right)^{2}$$

$$= \boxed{8.4\times10^{5}\ \mathrm{K}}$$

E19.23

$$\tilde{\nu} = \frac{5.3\ \mathrm{cm^{-1}}}{\tau/\mathrm{ps}} \text{ implying that } \tau = \frac{5.3\ \mathrm{ps}}{\tilde{\nu}/\mathrm{cm^{-1}}}$$

(a) $\tau = \dfrac{5.3\ \mathrm{ps}}{0.10} = \boxed{53\ \mathrm{ps}}$

(b) $\tau = \dfrac{5.3\ \mathrm{ps}}{1.0} = \boxed{5.3\ \mathrm{ps}}$

(c) $\lambda = \dfrac{2.998\times10^{8}\ \mathrm{m\ s^{-1}}}{1.0\times10^{9}\ \mathrm{s^{-1}}} = 0.300\ \mathrm{m} = 30.0\ \mathrm{cm}$

$$\tilde{\nu} = \frac{1}{\lambda} = 0.0333\ \mathrm{cm^{-1}}$$

$$\tau = \frac{5.3}{0.0333} = \boxed{1.6\times10^{2}\ \mathrm{ps}}$$

E19.24

$$\tilde{\nu} = \frac{5.3\ \mathrm{cm^{-1}}}{\tau/\mathrm{ps}}$$

(a) $\tau \approx 1\times10^{-13}\ \mathrm{s} = 0.1\ \mathrm{ps}$, implying that $\boxed{\tilde{\nu} = 53\ \mathrm{cm^{-1}}}$.

(b) $\tau \approx 200\times(1\times10^{-13}\ \mathrm{s}) = 20\ \mathrm{ps}$, implying that $\boxed{\tilde{\nu} = 0.27\ \mathrm{cm^{-1}}}$.

E19.25

$$\nu = \frac{1}{2\pi}\left(\frac{k}{\mu}\right)^{1/2} \quad [19.17\mathrm{b}]$$

(a) $\mu = \dfrac{m_{\mathrm{c}}m_{\mathrm{o}}}{m_{\mathrm{c}}+m_{\mathrm{o}}} = \dfrac{12.0000\times15.9949}{12.0000+15.9949}\times1.66054\times10^{-27}\ \mathrm{kg}$

$$= 1.1385\times10^{-26}\ \mathrm{kg}$$

$$\nu = \frac{1}{2\pi}\left(\frac{908\ \mathrm{Nm^{-1}}}{1.1385\times10^{-26}\ \mathrm{kg}}\right)^{1/2}$$

$$= 4.49\times10^{13}\ \mathrm{s^{-1}} = \boxed{4.49\times10^{13}\ \mathrm{Hz}}$$

(b) $\mu = 1.1910 \times 10^{-26}$ kg

$$\nu = 4.39 \times 10^{13}\ \mathrm{s}^{-1} = \boxed{4.39 \times 10^{13}\ \mathrm{Hz}}$$

E19.26 $$\tilde{\nu} = \frac{\nu}{c} = \frac{1}{2\pi c}\left(\frac{k}{\mu}\right)^{1/2} \qquad \mu = \frac{1}{2}m(^{35}\mathrm{Cl})$$

Solve the above for k

$$k = (2\pi c\tilde{\nu})^2 \mu = (2\pi c\tilde{\nu})^2 \times \frac{1}{2}\ m\ (^{35}\mathrm{Cl})$$

$$= (2\pi \times 2.998 \times 10^8\ \mathrm{m\ s^{-1}} \times 5.65 \times 10^4\ \mathrm{m^{-1}})^2 \times \frac{1}{2} \times 34.9688 \times 1.66054 \times 10^{-27}\ \mathrm{kg}$$

$$= \boxed{329\ \mathrm{N\ m^{-1}}}$$

E19.27 As shown in the solution to Exercise 19.26

$$k = (2\pi c\tilde{\nu})^2\ \mu$$

$$\mu(\mathrm{HF}) = \frac{1.0078 \times 18.998}{1.0078 + 18.998}\mathrm{u} = 0.9570\ \mathrm{u}$$

$$\mu(\mathrm{H}^{35}\mathrm{Cl}) = \frac{1.0078 \times 34.9688}{1.0078 + 34.9688}\mathrm{u} = 0.9796\ \mathrm{u}$$

$$\mu(\mathrm{H}^{81}\mathrm{Br}) = \frac{1.0078 \times 80.9163}{1.0078 + 80.9163}\mathrm{u} = 0.9954\ \mathrm{u}$$

$$\mu(\mathrm{H}^{127}\mathrm{I}) = \frac{1.0078 \times 126.9045}{1.0078 + 126.9045}\mathrm{u} = 0.9999\ \mathrm{u}$$

Using the above equation draw up the following table:

	HF	HCl	HBr	HI
$\tilde{\nu}$	4141.3	2988.9	2649.7	2309.5
μ/u	0.9570	0.9796	0.9954	0.9999
k/(N m^{-1})	967.1	515.6	411.8	314.2

E19.28 Form $\tilde{\nu} = \frac{1}{2\pi c}\left(\frac{k}{\mu}\right)^{1/2}$ with the values of k from Exercise 19.27 and the reduced masses, for example:

$$\mu(^2\mathrm{HF}) = \frac{2.0141 \times 18.9908}{2.0141 + 18.9908}\mathrm{u} = 1.8210\ \mathrm{u}$$ and similarly for the other halides.

Draw up the following table:

	^{2}HF	^{2}HCl	^{2}HBr	^{2}HI
$\tilde{\nu}$	3002	2144	1886	1640
μ/u	1.8210	1.9044	1.9652	1.9826
k/(N m^{-1})	967.1	515.6	411.8	314.2

E19.29 The R branch obeys the relation

$$\tilde{\nu}_R(J) = \tilde{\nu} + 2\tilde{B}(J+1) \quad \text{(See section 19.12)}$$

$$\tilde{\nu} = 2649.7 \text{ cm}^{-1} \quad \text{(See Exercise 19.27)}$$

$$I = \mu R^2 \text{ [Table 19.1]}$$

$$= \left(\frac{1.0078\times(126.9045) \text{ g mol}^{-1}}{1.0078+126.9045}\right)\times\left(\frac{1 \text{ kg}}{10^3 \text{ g}}\right)\times\left(\frac{1 \text{ mol}}{6.022\times10^{23}}\right)\times\left(141\times10^{-12} \text{ m}\right)^2 \quad \text{(Table 19.2)}$$

$$= 3.30\times10^{-47} \text{ kg m}^2$$

$$\tilde{B} = \frac{\hbar}{4\pi c I} \text{ [19.4 with } \tilde{B} \text{ in units of cm}^{-1}\text{]}$$

$$= \frac{1.054\times10^{-34} \text{ J s}}{4\pi\left(2.998\times10^{10} \text{ cm s}^{-1}\right)\times\left(3.30\times10^{-47} \text{ kg m}^2\right)} = 8.48 \text{ cm}^{-1}$$

Hence, $\tilde{\nu}_R(2) = \tilde{\nu} + 6\tilde{B} = (2649.7) + (6)\times(8.48 \text{ cm}^{-1}) = \boxed{2700.6 \text{ cm}^{-1}}$

E19.30 Select those molecules in which a vibration gives rise to a change in dipole moment. It is helpful to write down the structural formulas of the compounds.

The molecules that show infrared absorption are:

$$\boxed{\text{(b) HCl, (c) } CO_2\text{, (d) } H_2O\text{, (e) } CH_3CH_3\text{, (f) } CH_4\text{, and (g) } CH_3Cl}$$

E19.31 For nonlinear molecules $3N - 6$

For linear molecules $3N - 5$

We need to establish the linearity of the molecules listed. (c) and (d) are clearly non-linear. From the Lewis structures of (a) and (b) and VSEPR we decide (a) is non-linear, and (b) is linear

(a) $3N - 6 = 9 - 6 = \boxed{3}$

(b) $3N - 5 = 9 - 5 = \boxed{4}$

(c) C_6H_{12}, $3N - 6 = 3 \times 18 - 6 = \boxed{48}$

(d) C_6H_{14}, $3N - 6 = 3 \times 20 - 6 = \boxed{54}$

E19.32 The uniform expansion mode is depicted in Figure 19.4.

Benzene is centrosymmetric and it has a centre of inversion. Consequently, the exclusion rule applies (Section 19.13). The mode is $\boxed{\text{infrared inactive}}$ (symmetric breathing leaves the molelcular dipole moment unchanged at zero), and therefore the mode may be $\boxed{\text{Raman active}}$ (and is).

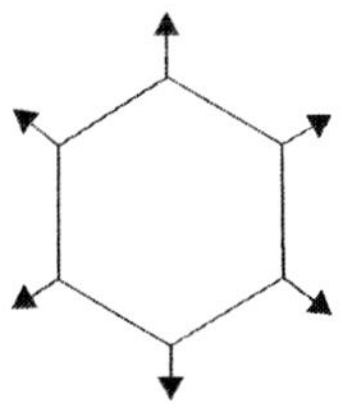

Figure 19.4

E19.33 Summarize the six observed vibrations according to their wavenumbers ($\tilde{\nu}/\text{cm}^{-1}$):

IR	870	1370	2869	3417
Raman	877	1408	1435	3407

(a) If H_2O_2 were linear, it would have $3N-5=\boxed{7}$ vibrational modes.

(b) The exclusion rule applies to structure **1** because it has a center of inversion: no vibrational modes can be both IR and Raman active. So $\boxed{\text{structure 1 is inconsistent with observation.}}$ A more detailed explanation using group theoretical methods not discussed in this text, but which can be consulted in the author's Physical Chemistry, 8th edition, follows: The vibrational modes of structure **2** span $3A_1+A_2+2B_2$. (The full basis of 12 cartesian coordinates spans $4A_1+2A_2+2B_1+4B_2$; remove translations and rotations.) The C_{2v} character table says that five of these modes are IR active $(3A_1+2B_2)$ and all are Raman active. All of the modes of structure **3** are both IR and Raman active. (A look at the character table shows that both symmetry species are IR and Raman active, so determining the symmetry species of the normal modes does not help here.) Both structures **2** and **3** have more active modes than were observed. This is consistent with the observations. After all, group theory can only tell us whether the transition moment *must* be zero by symmetry; it does not tell us whether the transition moment is sufficiently strong to be observed under experimental conditions.

Solutions to projects

P19.34 For a spherical rotor, $P_J \propto (2J+1)^2 e^{-hcBJ(J+1)/kT}$ $[g_J=(2J+1)^2]$ and the greatest population occurs when

$$\frac{dP_J}{dJ} \propto \left(8J+4-\frac{hB(2J+1)^3}{kT}\right)e^{-hBJ(J+1)/kT}=0$$

which occurs when

$$4(2J+1)=\frac{hB(2J+1)^3}{kT} \quad \text{or at } J_{max}=\boxed{\left(\frac{kT}{hB}\right)^{1/2}-\frac{1}{2}}$$

P19.35 (a) Resonance Raman spectroscopy is preferable to vibrational spectroscopy for studying the O–O stretching mode because such a mode would be $\boxed{\text{infrared inactive}}$, or at best only weakly active. (The mode is sure to be inactive in free O_2, because it would not change the molecule's dipole moment. In a complex in which O_2 is bound, the O–O stretch may change the dipole moment, but it is not certain to do so at all, let alone strongly enough to provide a good signal.)

(b) The vibrational wavenumber is proportional to the frequency, and it depends on the effective mass as follows,

$$\tilde{\nu} \propto \left(\frac{k}{m_{eff}}\right)^{1/2}, \quad \text{so} \quad \frac{\tilde{\nu}(^{18}O_2)}{\tilde{\nu}(^{16}O_2)}=\left(\frac{m_{eff}(^{16}O_2)}{m_{eff}(^{18}\text{hmr}O_2)}\right)^{1/2}=\left(\frac{16.0\,u}{18.0\,u}\right)^{1/2}=0.943,$$

and $\tilde{\nu}(^{18}O_2)=(0.943)(844\,cm^{-1})=\boxed{796\,cm^{-1}}$

[hmrO_2 is oxygenated haemerythrin.] Note the assumption that the effective masses are proportional to the isotopic masses. This assumption is valid in the free molecule, where the effective mass of O_2 is equal to half the mass of the O atom; it is also valid if the O_2 is strongly bound at one end, such that one atom is free and the other is essentially fixed to a very massive unit.

(c) The vibrational wavenumber is proportional to the square root of the force constant. The force constant is itself a measure of the strength of the bond (technically of its stiffness, which correlates with strength), which in turn is characterized by bond order. Simple molecular orbital analysis of O_2, O_2^-, and O_2^{2-} results in bond orders of $\boxed{\text{2, 1.5, and 1 respectively}}$. Given decreasing bond order, one would expect decreasing vibrational wavenumbers (and vice versa).

(d) The wavenumber of the O–O stretch is very similar to that of the peroxide anion, suggesting $\boxed{Fe_2^{3+}O_2^{2-}}$.

(e) The detection of two bands due to $^{16}O^{18}O$ implies that the two O atoms occupy non-equivalent positions in the complex. Structures **6** and **7** are consistent with this observation, but structures **4** and **5** are not.

P19.36 These calculations, which use either the Møller-Plesset (MP2) model with two different basis sets or density functional theory (DFT) technique, were performed with Spartan '06™ running in Windows on a PC. The star indicates that the basis set adds d-type polarization functions for each atom other than hydrogen. Other basis sets give comparable results. 1 au = 27.2114 eV. In addition to reporting the vibrational wavenumbers of the normal modes, calculation of bond lengths, angles, ground-state energies, and dipole moments have been included.

(a) Water. The DFT calculation method is DFT(B3LYP).

H_2O Ground State					
	MP2/6-31G*	MP2/6-311G*	DFT/6-31G*	DFT/6-311G*	Exp.
Basis fns	19	24	19	24	
R / pm	96.9	95.7	96.8	96.3	95.8
E_0 / eV	−2073.4	−2074.5	−2074.6	−2079.9	
Angle / °	104.00	106.58	103.72	105.91	104.45
$\tilde{\nu}_1$ / cm^{-1}	3774.25	3858.00	3731.72	3764.70	3652
$\tilde{\nu}_2$ / cm^{-1}	1735.35	1739.88	1709.79	1705.47	1595
$\tilde{\nu}_3$ / cm^{-1}	3915.76	3994.30	3853.53	3877.60	3756
μ / D	n.s.	n.s.	2.0950	2.2621	1.854

Normal Modes and IR Spectrum:

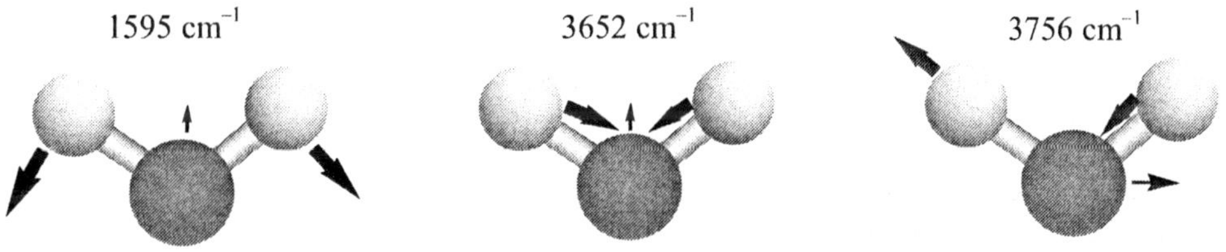

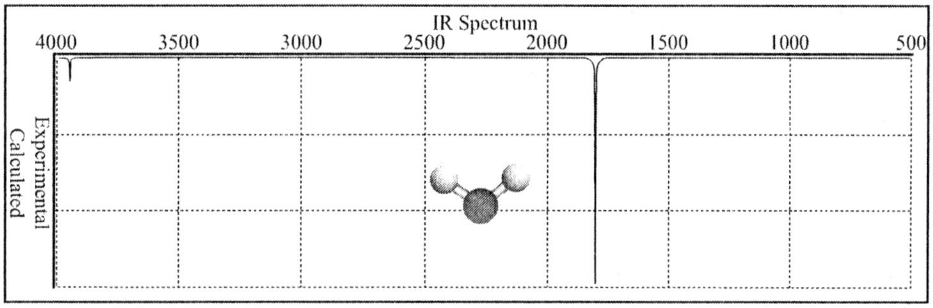

Carbon Dioxide. The DFT calculation method is DFT(B3LYP).

CO_2 Ground State					
	MP2/6-31G*	MP2/6-311G*	DFT/6-31G*	DFT/6-311G*	Exp.
Basis fns	45	54	45	54	
R / pm	118.0	116.9	116.9	116.0	116.3
E_0 / eV	−5118.7	−5121.2	−5131.6	−5133.2	
Angle /°	180.00	180.00	180.00	180.00	180
$\tilde{\nu}_1$/cm^{-1}	1332.82	1341.46	1373.05	1376.55	1388
$\tilde{\nu}_2$/cm^{-1}	636.22	657.60	641.47	666.39	667
$\tilde{\nu}_3$/cm^{-1}	2446.78	2456.16	2438.17	2437.85	2349
μ / D	n.s.	n.s.	0.0000	0.0000	0

Normal Modes and IR Spectrum:

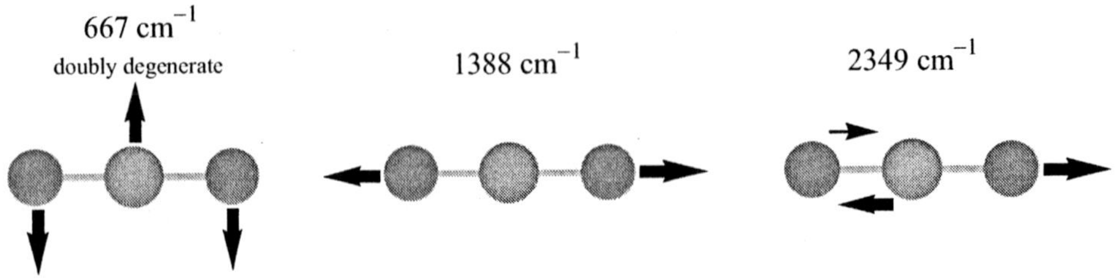

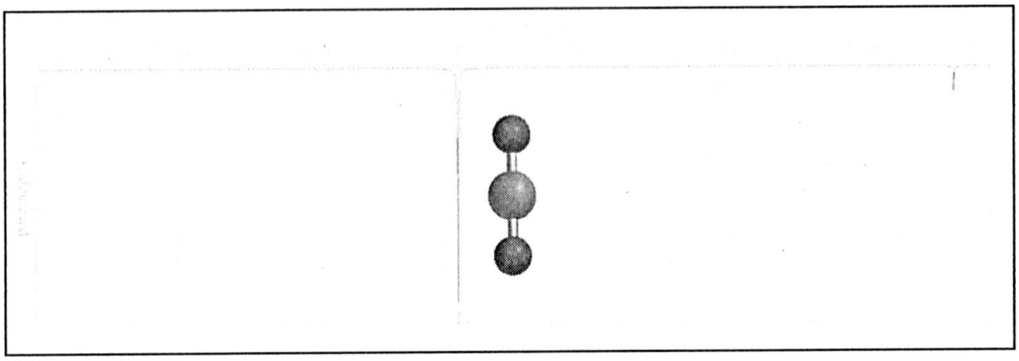

Methane, CH_4. This tetrahedral molecule has nine normal modes that you can view with your software. The following IR spectrum is calculated with the PM3 semi-empirical method.

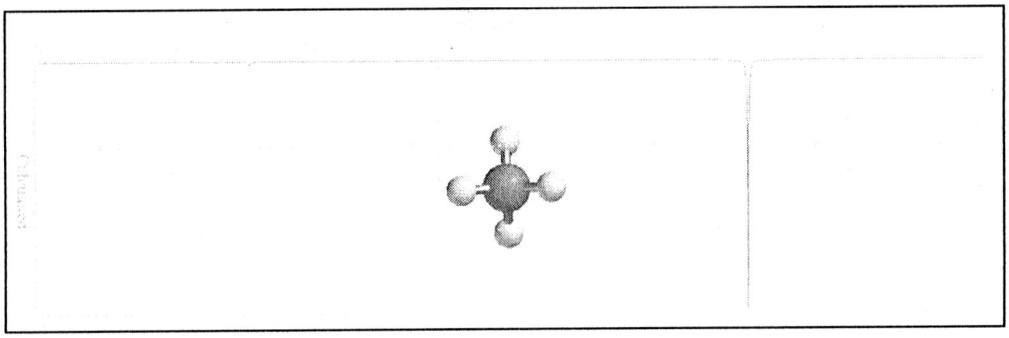

Except for the dipole moment, all calculations are typically within a reasonable 1–3% of the experimental value. The dipole moment is very sensitive to the distribution of charge density. The significant difference between the dipole moment calculations and the experimental dipole moment may indicate that the computation methods do not adequately account for charge distribution in the very polar water molecule.

(b) Water: all modes of water are infrared active. However, as seen in the above spectrum, not all have a strong intensity.

Carbon dioxide: the symmetric stretch is infrared inactive; the bend and asymmetric stretch are infrared active.

Methane: the vibrational modes that cause the molecular dipole to change as the molecule vibrates are infrared active; these are the six T_2 modes.

20 Electronic transitions and photochemistry

Answers to discussion questions

D20.1 The form of the Beer-Lambert law is justified in detail in *Further Information* 20.1 and will not be repeated here. The derivation shown assumes that the concentration of the absorbing species is uniform. If not, the law will not be obeyed. This situation might apply to a substance that associates or dissociates in solution. Since the molar absorption coefficient, ε, depends on wavelength, the absorbance A is wavelength dependent, and so if the radiation is not monochromatic, the Beer-Lambert law may not be obeyed.

D20.2 The Franck–Condon principle states that because electrons are so much lighter than nuclei an electronic transition occurs so rapidly compared to vibrational motions that the internuclear distance is relatively unchanged as a result of the transition. This implies that the most probable transitions $v_f \leftarrow v_i$ are vertical. This vertical line will, however, intersect any number of vibrational levels v_f in the upper electronic state. Hence transitions to many vibrational states of the excited state will occur with transition probabilities proportional to the Frank–Condon factors which are in turn proportional to the overlap integral of the wavefunctions of the initial and final vibrational states. A vibrational progression is observed, the shape of which is determined by the relative horizontal positions of the two electronic potential energy curves. The most probable transitions are those to excited vibrational states with wavefunctions having a large amplitude at the internuclear position R_e.

Question. You might check the validity of the assumption that electronic transitions are so much faster than vibrational transitions by calculating the time scale of the two kinds of transitions. How much faster is the electronic transition, and is the assumption behind the Franck–Condon principle justified?

D20.3 Color can arise by emission, absorption, or scattering of electromagnetic radiation by an object. Many molecules have electronic transitions that have wavelengths in the visible portion of the electromagnetic spectrum. When a substance emits radiation the perceived color of the object will be that of the emitted radiation and it may be an additive color resulting from the emission of more than one wavelength of radiation. When a substance absorbs radiation its color is determined by the subtraction of those wavelengths from white light. For example, absorption of red light results in the object being perceived as green. Scattering, including the diffraction that occurs when light falls on a material with a grid of variation in texture or refractive index having dimensions comparable to the wavelength of light, for example, a bird's plumage, may also form color.

D20.4 The overall process associated with fluorescence involves the following steps. The molecule is first promoted from the vibrational ground state of a lower electronic level to a higher vibrational-electronic energy level by absorption of energy from a radiation field. Because of the requirements of the Franck-Condon principle, the transition is to excited vibrational levels of the upper electronic state. See Figs. 20.14 & 20.15 of the text. Therefore, the absorption spectrum shows a vibrational structure characteristic of the upper state. The excited state molecule can now lose energy to the surroundings through radiationless transitions and decay to the lowest vibrational level of the upper state. A spontaneous radiative transition now occurs to the lower electronic level and this fluorescence spectrum has a vibrational structure characteristic of the lower state. The fluorescence spectrum is not the mirror image of the absorption spectrum because the vibrational frequencies of the upper and lower states are different due to the difference in their potential energy curves.

The first steps of phosphorescence are the same as in fluorescence: a singlet state absorbs energy in a transition to an excited singlet state and non-radiative vibrational relaxation occurs within the excited singlet state. The presence of a heavy atom, which can provide angular momentum via strong spin-orbit coupling, and a triplet state of energy similar to the excited singlet, provide the critical step needed for phosphorescence. It is the intersystem crossing step in which an electron undergoes a spin-flip from the excited singlet to the triplet state. The electron may remain in the triplet state for unusually long periods because the emission transition from a triplet state to a singlet state is spin-forbidden.

D20.5 See Sections 20.7 and 20.8 for a detailed description of both the theory and experiment involved in laser action. Here we restrict our discussion to only the most fundamental concepts. The basic requirement for a laser is that it has at least three energy levels. Of these levels, the highest lying state must be capable of being efficiently populated above its thermal equilibrium value by a pulse of radiation. A second state, lower in energy, must be a metastable state with a long enough lifetime for it to accumulate a population greater than its thermal equilibrium value by spontaneous transitions from the higher overpopulated state.

The metastable state must then be capable of undergoing stimulated transitions to a third lower lying state. This last requirements implies not only that the metastable state have more than its thermal equilibrium population, but also that it have a higher population than the third lower lying state, namely that it achieve population inversion. The amplification process occurs when low intensity radiation of frequency equal to the transition frequency between the metastable state and the lower lying state stimulates the transition to the lower lying state and many more photons (higher intensity of the radiation) of that frequency are created. Examples of practical lasers are featured in Figures 20.20–23 of the text.

Two important applications of lasers in chemistry have been to Raman spectroscopy and to the development of time resolved spectroscopy. Prior to the invention of lasers the source of intense monochromatic radiation required for Raman spectroscopy was a large spiral discharge tube with liquid mercury electrodes. The intense heat generated by the large current required to produce the radiation had to be dissipated by clumsy water-cooled jackets and exposures of several weeks were sometimes necessary to observe the weaker Raman lines. These problems have been eliminated with the introduction of lasers as the source of the required monochromatic radiation. As a consequence, Raman spectroscopy has been revitalized and is now almost as routine as infrared spectroscopy. See Chapter 19. Time resolved laser spectroscopy can be used to study the dynamics of chemical reactions. Laser pulses are used to obtain the absorption, emission, and Raman spectrum of reactants, intermediates, products, and even transition states of reactions. When we want to study the rates at which energy is transferred from one mode to another in a molecule, we need femotosecond and picosecond pulses. These time scales are available from mode-locked lasers and their development has opened up the possibility of examining the details of chemical reactions at a level that would have been unimaginable before.

D20.6 The primary quantum yield is associated with the primary photochemical event in the overall photochemical process which may involve secondary events as well. An example that illustrates both kinds of events is the photolysis of HI described in Section 20.9. The primary quantum yield is defined as the ratio of the number of primary events to the number of photons absorbed (eqn 20.8) and its value can never exceed one. However, in reactions described by complex mechanisms, the overall quantum yield, which is the number of reactant molecules consumed in both primary and secondary processes per photon absorbed, can easily exceed one. Experimental procedures for the determination of the overall quantum yield involve measurements of the intensity of the radiation used, defined here as the number of photons generated and directed at the reacting sample, and of the amount of product formed. This ratio is the overall quantum yield. See Example 20.2. In addition to chemical reactions, the concept of the quantum yield enters into the description of other kinds of photochemical processes, such as fluorescence and phosphorescence, and in each case there are techniques specific to the process for the determination of the quantum yield.

D20.7 Fluorescence is the most widely used form of spectroscopy in the study of biological substances. Fluorescence is a particularly powerful technique because there are many biological reactions, solvent rearrangements, and molecular motion processes that take place on the time scale of the fluorescent lifetime. Applications of fluorescence include the measurement of ligand binding to macromolecules, the use of fluorescent labels to macromolecules as environmental probes, the monitoring of conformational changes, and as a means of following a reaction. Cellular processes can be studied by detecting with a microscope the fluorescence emission from molecules used to tag biological macromolecules. In fluorescence microscopy, images of biological cells at work are obtained by attaching a large number of fluorescent molecules to proteins, nucleic acids, and membranes and the measuring of the distribution of fluorescence intensity within the illuminated area. Special techniques permit the observation of fluorescence from single molecules in cells. A common fluorescent label is the green fluorescent protein found in certain jellyfish. With proper filtering to remove light due to Rayleigh scattering of the incident beam, it is possible to collect light from a sample that contains only fluorescence from the label.

D20.8 The Marcus theory provides a framework for the discussion of electron transfer processes, $D + A \rightleftharpoons D^+ + A^-$, in both homogeneous solution (redox reactions) and at electrodes. The barrier to reaction, and the corresponding activation Gibbs energy, $\Delta^{\ddagger}G$, is the reorganization of the donor acceptor complex, DA, and the solvent molecules surrounding it. As a result of this reorganization, the energy levels of the electron in the DA and D^+A^- complexes match, and electron transfer is permitted by a tunneling process; the D^+A^- complex breaks apart and the D^+ and A^- ions diffuse into the solution. The rate constant, k_{et}, for the electron transfer from DA to D^+A^- is governed by $\Delta^{\ddagger}G$ which in turn is governed by the reorganization energy, λ. The observed rate constant, k_{obs}, for the overall process, $D + A \rightleftharpoons D^+ + A^-$, is related to k_{et}. Based on these considerations, R. A. Marcus was able to derive the following relationship for the observed rate constant for the electron transfer between species D and A in solution:

$$k_{obs} = (k_{DD}k_{AA}K)^{1/2}$$

In this Marcus cross-relation, k_{DD} and k_{AA} are the rate constants for the two self exchange processes and K is the equilibrium constant for the overall reaction. For a derivation of all the relationships referred to see Atkins, P. W., de Paula, J. C. (2006). *Physical Chemistry* (8th ed.). Oxford: Oxford University Press.

Solutions to exercises

E20.1

$$[J] = \frac{n}{V} = \frac{0.0172\ \text{g}/(502\ \text{g mol}^{-1})}{0.500\ \text{dm}^3} = 6.85\times10^{-5}\ \text{mol dm}^{-3}$$

(a) $A = \varepsilon[J]l \quad [20.4]$

Solve for ε,

$$\varepsilon = \frac{A}{[J]l} = \frac{1.011}{6.85\times10^{-5}\ \text{mol dm}^{-3}\times1.00\ \text{cm}}$$

$$= \boxed{1.48\times10^{4}\ \text{dm}^3\ \text{mol}^{-1}\ \text{cm}^{-1}}$$

(b) $\log T = -\varepsilon[J]l$

$$= -1.48\times10^{4}\ \text{dm}^3\ \text{mol}^{-1}\ \text{cm}^{-1}\times1.37\times10^{-4}\ \text{mol dm}^{-3}\times1.00\ \text{cm}$$

$$= -2.028$$

$$T = 0.00938 = \boxed{0.938\%}$$

E20.2 We proceed as in Example 20.1. With $A = -\log T$, eqn 20.4 becomes

$$\varepsilon = -\frac{\log T}{[\mathrm{J}]L} = -\frac{\log 0.22}{(0.080\ \mathrm{mol\ dm^{-3}})\times(1.5\ \mathrm{mm})} = \boxed{5.48\ \mathrm{dm^3\ mol^{-1}mm^{-1}}}$$

The absorbance is

$$A = -\log T = -\log 0.22 = \boxed{0.658}$$

For the longer cell we calculate

$$A = \varepsilon[\mathrm{J}]L = 5.48\ \mathrm{dm^3\ mol^{-1}mm^{-1}}\times 0.080\ \mathrm{mol\ dm^{-3}}\times 3.0\ \mathrm{mm} = \boxed{1.3}$$

The transmittance is then

$$T = 10^{-A} = 10^{-1.3} = \boxed{0.048}$$

E20.3

$$T_\mathrm{k} = T_\mathrm{u} \qquad \mathrm{k} = \text{Solution of known concentration}$$

$$\log T_\mathrm{k} = \log T_\mathrm{u} \qquad \mathrm{u} = \text{Solution of unknown concentration}$$

$$-\varepsilon[\mathrm{k}]l_\mathrm{k} = -\varepsilon[\mathrm{u}]l_\mathrm{u}$$

Solve for [u],

$$[\mathrm{u}] = [\mathrm{k}]\left(\frac{l_\mathrm{k}}{l_\mathrm{u}}\right) = 25\ \mu\mathrm{g\ dm^{-3}}\times\left(\frac{1.55\ \mathrm{cm}}{1.18\ \mathrm{cm}}\right) = \boxed{33\ \mu\mathrm{g\ dm^{-3}}}$$

E20.4 We may write $A = c_\mathrm{A}\varepsilon_\mathrm{A}l + c_\mathrm{B}\varepsilon_\mathrm{B}l$ at both wavelengths.

$$1.6 = c_\mathrm{A}\times(10.0\ \mathrm{dm^3\ mol^{-1}\ cm^{-1}})\times 0.200\ \mathrm{cm}$$
$$+\, c_\mathrm{B}\times(15.0\ \mathrm{dm^3\ mol^{-1}\ cm^{-1}})\times 0.200\ \mathrm{cm}$$

and

$$2.4 = c_\mathrm{A}\times(18.0\ \mathrm{dm^3\ mol^{-1}\ cm^{-1}})\times 0.200\ \mathrm{cm}$$
$$+\, c_\mathrm{B}\times(12.0\ \mathrm{dm^3\ mol^{-1}\ cm^{-1}})\times 0.200\ \mathrm{cm}$$

Solve these two simultaneous equations for c_A and c_B.

$$\boxed{c_\mathrm{A} = 0.56\ \mathrm{mol\ dm^{-3}}} \qquad \boxed{c_\mathrm{B} = 0.16\ \mathrm{mol\ dm^{-3}}}$$

E20.5 There are three isosbestic wavelengths (or wavenumbers). The presence of two or more isosbestic points is good evidence that $\boxed{\text{only two solutes in equilibrium with each other are present}}$. The solutes here being $\mathrm{Her(CNS)_8}$ and $\mathrm{Her(OH)_8}$.

E20.6 The transition dipole moment, μ_fi, between states i and f determines the intensity of a transition. If it has a high absolute value, the transition probability and intensity are high (see Section 19.3). We need to determine how it depends on the length of the chain. We assume that wavefunctions of the conjugated electrons in the linear polyene can be approximated by the wavefunctions of a particle in a one-dimensional box. Then, for a transition between the states n' and n

$$\mu_x = -e\int_0^L \psi_{n'}(x)\,x\,\psi_n(x)\mathrm{d}x, \qquad \psi_n = \left(\frac{2}{L}\right)^{1/2}\sin\left(\frac{n\pi x}{L}\right)$$

$$= -\frac{2e}{L}\int_0^L x\sin\left(\frac{n'\pi x}{L}\right)\sin\left(\frac{n\pi x}{L}\right)\mathrm{d}x$$

$$= \begin{cases} 0 & \text{if } n' = n+2 \\ \left(\dfrac{8eL}{\pi^2}\right)\dfrac{n(n+1)}{(2n+1)^2} & \text{if } n' = n+1 \end{cases}$$

Thus, the selection rule for radiation absorption is $\Delta n = +1$ and the transition integral is proportional to L.

Longer lengths of the dye's conjugated electronic π system are expected to yield greater absorption intensity.

To examine the effect that changing the length has on the apparent colour, consider the energy of the absorption.

$$h\nu = E_{n+1} - E_n = \frac{(n+1)^2 h^2}{8m_e L^2} - \frac{n^2 h^2}{8m_e L^2} = (2n+1)\frac{h^2}{8m_e L^2}$$

Therefore, since L appears in the denominator, increasing the length L of the polyene shifts the absorption to lower energy. This is a blue shift.

E20.7 A typical n-to-π^* transition at about 4 ev (32000 cm^{-1}) characterizes an absorption in which a carbonyl lone pair electron on oxygen is excited to a π^* orbital while an unconjugated alkene exhibits a π-to-π^* electron transition at about 7 eV (56000 cm^{-1}). Thus, the 30000 cm^{-1} absorption of $CH_3CH{=}CHCHO$ is a n-to-π^* transition and the 46950 cm^{-1} absorption is a π-to-π^* transition, which lays lower than the norm because of conjugation between the alkene and carbonyl π bonds.

E20.8 Tryptophan (Trp) and tyrosine (Tyr) show the characteristic absorption of a phenyl group at about 280 nm. Cysteine (Cys) and glycine (Gly) lack the phenyl group as is evident from their spectra.

E20.9 The fluorescence spectrum gives the vibrational splitting of the lower state. The wavelengths stated correspond to the wavenumbers 22 730, 24 390, 25 640, 27 030 cm^{-1}, indicating spacings of 1660, 1250, and 1390 cm^{-1}. The absorption spectrum spacing gives the separation of the vibrational levels of the upper state. The wavenumbers of the absorption peaks are 27 800, 29 000, 30 300, and 32 800 cm^{-1}. The vibrational spacings are therefore 1200, 1300, and 2500 cm^{-1}.

E20.10 We note that after some vibrational decay the benzophenone (which does absorb near 360 nm) can transfer its energy to naphthalene. The latter then emits the energy radiatively. This accounts for the difference in the A and B spectra.

E20.11 $E = eV = 1.602\times10^{-19}\,\text{C}\times10.0\,\text{kV} = \boxed{16.0\times10^{-19}\,\text{kJ}}$ (or 10.0 keV)

E20.12 Ionization energy = 10.0 eV = 1.60×10^{-18} J

$$\text{Energy of photon} = h\nu = 6.63\times10^{-34}\,\text{J s}\times\frac{3.00\times10^{8}\,\text{m s}^{-1}}{1.10\times10^{-7}\,\text{m}}$$

$$= 1.81\times10^{-18}\ \text{J}$$

The difference $E - I$ is the kinetic energy, KE, of the ejected electron.

$$E - I = \boxed{2.1\times10^{-19}\,\text{J}}$$

The kinetic energy is $\text{KE} = \frac{1}{2}m_e v^2$. Solving for v we have

$$v = \sqrt{\frac{2\text{KE}}{m_e}} = \sqrt{\frac{2\times2.1\times10^{-19}\ \text{J}}{9.11\times10^{-31}\ \text{kg}}} = \boxed{6.8\times10^{5}\ \text{m s}^{-1}}$$

E20.13 $I = h\nu - KE$ [20.7]

$= 21.21\ \text{eV} - KE$

$I = \boxed{10.20\ \text{eV},\ 12.98\ \text{eV, and } 15.99\ \text{eV}}$

See Figure 20.1 for a sketch of the ionizations.

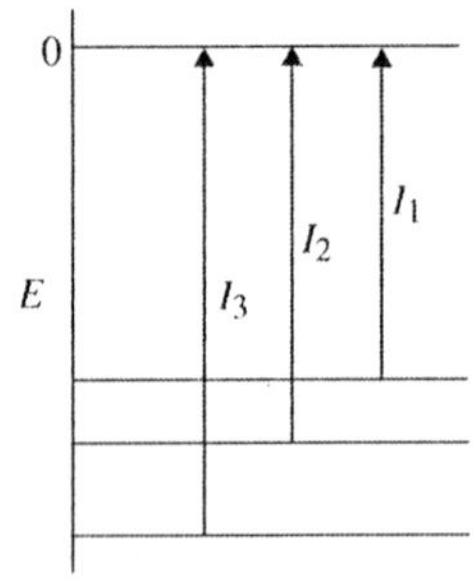

Figure 20.1

E20.14 The primary quantum yield, $\phi = \dfrac{v}{I_{\text{abs}}}$ [20.9], where v is the rate of radiation induced primary events and I_{abs} is the rate of photon absorption.

$$\phi = \frac{78\times10^{-6}\ \text{mol s}^{-1}}{0.14\ \text{mol h}^{-1}\times\dfrac{1\ \text{s}^{-1}}{3600\ \text{h}^{-1}}} = \boxed{2.0}$$

E20.15 For a source of power P and wavelength λ, the amount of photons (n_λ) generated in a time t is

$$n_\lambda = \frac{Pt}{h\nu N_A} = \frac{P\lambda t}{hcN_A} = \frac{(50\ \text{W})\times(35)\times(60\ \text{s})\times(390\times10^{-9}\ \text{m})}{(6.626\times10^{-34}\ \text{J s})\times(2.998\times10^{8}\ \text{m s}^{-1})\times(6.022\times10^{23}\ \text{mol}^{-1})}$$
$$= 0.342\ \text{mol}$$

The amount of photons absorbed is 40 percent of this incident flux, or 0.137 mol. Therefore,

$$\phi = \frac{0.384\ \text{mol}}{0.137\ \text{mol}} = \boxed{2.80}$$

Alternatively, expressing the amount of photons in einsteins [1 mol photons = 1 einstein],

$$\phi = 2.80\ \text{mol einstein}^{-1}$$

E20.16 See Example 20.2 in the text.

$$\text{Rate of photon production} = I_{\text{abs}} = \frac{P}{hc/\lambda} = \frac{P\lambda}{hc} = \frac{(40\ \text{J s}^{-1})\times(280\times10^{-9}\ \text{m})}{(6.62608\times10^{-34}\ \text{J s})\times(2.99792\times10^{8}\ \text{m s}^{-1})}$$
$$= 5.64\times10^{19}\ \text{s}^{-1}$$

The number of molecules destroyed per second is then 0.26 times this quantity, or $\boxed{1.47\times10^{19}\ \text{s}^{-1}}$. The chemical amount destroyed is then this latter quantity divided by N_A or $\boxed{2.4\times10^{-5}\ \text{mol s}^{-1}}$.

E20.17 The lifetime of the unimolecular photochemical reaction is $\tau = \dfrac{1}{k} = \dfrac{1}{1.7\times10^{4}\ \text{s}^{-1}} = 5.9\times10^{-5}\ \text{s}$. This is a much longer lifetime than that of fluorescence (1.0×10^{-9} s) so we conclude that the excited singlet state decays too rapidly to be the precursor of this photochemical reaction. The lifetime of phosphorescence (1.0×10^{-3} s) is longer than the reaction lifetime making the $\boxed{\text{triplet state}}$ the likely reaction precursor.

E20.18

$$\mathrm{A} \rightarrow 2\mathrm{R}\cdot \quad \mathcal{I}$$

$$\mathrm{A}+\mathrm{R}\cdot \rightarrow \mathrm{R}\cdot + \mathrm{B} \quad k_2$$

$$\mathrm{R}\cdot + \mathrm{R}\cdot \rightarrow \mathrm{R}_2 \quad k_3$$

$$\frac{\mathrm{d[A]}}{\mathrm{d}t} = \boxed{-\mathcal{I} - k_2[\mathrm{A}][\mathrm{R}\cdot]}, \qquad \frac{\mathrm{d[R\cdot]}}{\mathrm{d}t} = 2\mathcal{I} - 2k_3[\mathrm{R}\cdot]^2 = 0$$

The latter implies that $[\mathrm{R}\cdot] = \left(\frac{\mathcal{I}}{k_3}\right)^{1/2}$, and so

$$\frac{\mathrm{d[A]}}{\mathrm{d}t} = \boxed{-I - k_2\left(\frac{\mathcal{I}}{k_3}\right)^{1/2}[\mathrm{A}]}$$

$$\frac{\mathrm{d[B]}}{\mathrm{d}t} = k_2[\mathrm{A}][\mathrm{R}\cdot] = k_2\left(\frac{\mathcal{I}}{k_3}\right)^{1/2}[\mathrm{A}]$$

Therefore, only the combination $\frac{k_2}{k_3^{1/2}}$ may be determined if the reaction attains a steady state.

COMMENT. If the reaction can be monitored at short enough times so that termination is negligible compared to initiation, then $[\mathrm{R}] \approx 2\mathcal{I}t$ and $\frac{\mathrm{d[B]}}{\mathrm{d}t} \approx 2k_2\mathcal{I}t\,[\mathrm{A}]$. So monitoring B sheds light on just k_2.

E20.19 Number of photons absorbed $= \phi^{-1} \times$ number of molecules that react [Section 20.9]. Therefore,

$$\text{Number absorbed} = \frac{(1.14\times10^{-3}\ \mathrm{mol})\times(6.022\times10^{23}\ \mathrm{einstein}^{-1})}{2.1\times10^{2}\ \mathrm{mol\,einstein}^{-1}} = \boxed{3.3\times10^{18}}$$

E20.20

$$\mathrm{M} + h\nu_\mathrm{i} \rightarrow \mathrm{M}^*, \qquad I_\mathrm{abs}\ [\mathrm{M} = \text{benzophenone}]$$

$$\mathrm{M}^* + \mathrm{Q} \rightarrow \mathrm{M} + \mathrm{Q}, \qquad k_\mathrm{Q}$$

$$\mathrm{M}^* \rightarrow \mathrm{M} + h\nu_\mathrm{f}, \qquad k_\mathrm{f}$$

$$\frac{\mathrm{d[M^*]}}{\mathrm{d}t} = I_\mathrm{abs} - k_\mathrm{f}[\mathrm{M}^*] - k_\mathrm{Q}[\mathrm{Q}][\mathrm{M}^*] \approx 0 \text{ [steady state]}$$

and hence $[\mathrm{M}^*] = \dfrac{I_\mathrm{abs}}{k_\mathrm{f} + k_\mathrm{Q}[\mathrm{Q}]}$

Then $I_\mathrm{f} = k_\mathrm{f}[\mathrm{M}^*] = \dfrac{k_\mathrm{f} I_\mathrm{abs}}{k_\mathrm{f} + k_\mathrm{Q}[\mathrm{Q}]}$

and so $\boxed{\dfrac{1}{I_\mathrm{f}} = \dfrac{1}{I_\mathrm{abs}} + \dfrac{k_\mathrm{Q}[\mathrm{Q}]}{k_\mathrm{f} I_\mathrm{abs}}}$

If the exciting light is extinguished, $[\mathrm{M}^*]$, and hence I_f, decays as $\mathrm{e}^{-k_\mathrm{f}t}$ in the absence of a quencher. Therefore we can measure $k_\mathrm{Q}/k_\mathrm{f}I_\mathrm{abs}$ from the slope of $1/I_\mathrm{f}$ plotted against $[\mathrm{Q}]$, and then use k_f to determine k_Q.

We draw up the following table

$10^3[\mathrm{Q}]/\mathrm{M}$	1	5	10
$\frac{1}{I_\mathrm{f}}$	2.4	4.0	6.3

The points are plotted in Fig. 20.2.

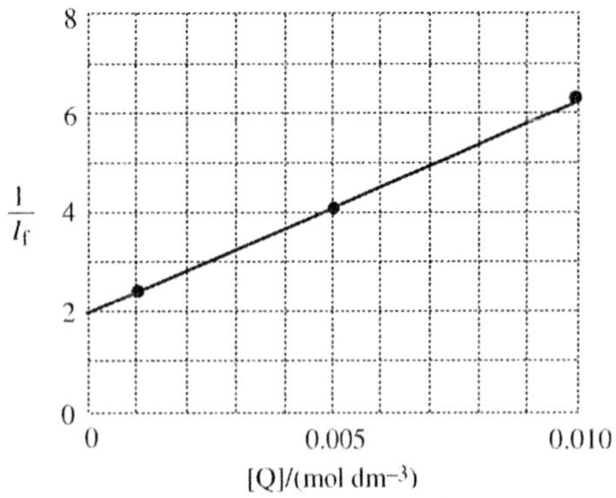

Figure 20.2

The intercept lies at 2.0, and so $I_{abs} = \frac{1}{2.0} = 0.50$. The slope is 430, and so

$$\frac{k_q}{k_f I_{abs}} = 430 \text{ dm}^3 \text{ mol}^{-1}$$

Then, since $I_{abs} = 0.50$ and $k_f = \frac{\ln 2}{t_{1/2}}$,

$$k_Q = (0.50)\times(430\,\text{dm}^3\text{ mol}^{-1})\times\left(\frac{\ln 2}{29\times10^{-6}\text{ s}}\right) = \boxed{5.1\times10^6 \text{ dm}^3 \text{ mol}^{-1}\text{s}^{-1}}$$

E20.21 $$\frac{\phi_f}{\phi} = \frac{\tau_{obs}}{\tau_{obs}(Q)} = 1 + \tau_{obs}k_Q[Q] \qquad \text{Stern-Volmer equation [20.15a]}$$

τ_{obs} is the lifetime in the absence of quenching; τ_{obs} (Q) is the lifetime in the presence of quenching. A plot of $\tau_{obs}/\tau_{obs}(Q)$ against $[O_2]$ with $\tau_{obs} = 2.6$ ns should be linear with slope equal to $\tau_{obs}k_Q$. We draw up a table for the plot, perform a linear regression fit with an intercept fixed at 1 (see Figure 20.3), and calculate k_Q.

$[O_2]$ / 10^{-2} mol dm^{-3}	2.3	5.5	8.0	10.8
$\tau_{obs}(Q)$ / ns	1.5	0.92	0.71	0.57
τ_{obs} / $\tau_{obs}(Q)$	1.73	2.83	3.66	4.56

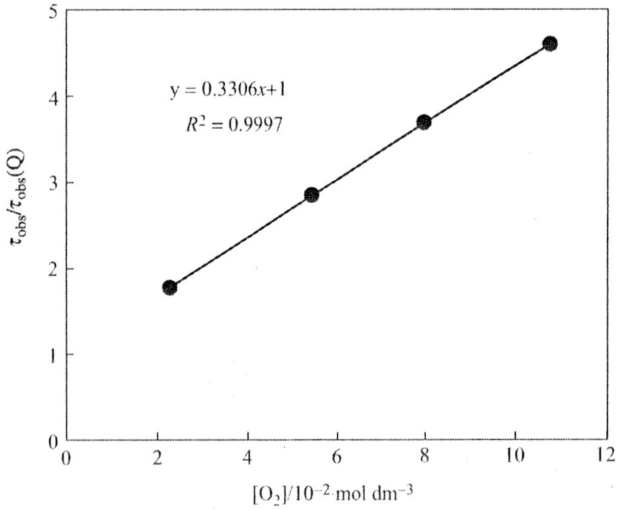

Figure 20.3

$$k_Q = \frac{\text{slope}}{\tau_{obs}} = \left(\frac{0.3306}{10^{-2}\ \text{mol dm}^{-3}}\right)\left(\frac{1}{2.6\times10^{-9}\ \text{s}}\right) = \boxed{1.2\bar{7}\times10^{10}\ \text{mol}^{-1}\ \text{dm}^3\ \text{s}^{-1}}$$

Note: There is an alternative to doing a linear regression fit in this particular case. The Stern-Volmer equation rearranges to $\dfrac{\dfrac{\tau_{obs}}{\tau_{obs}(Q)}-1}{[Q]} = \tau_{obs}k_Q$. Thus, the ratio $\dfrac{\dfrac{\tau_{obs}}{\tau_{obs}(Q)}-1}{[Q]}$ may be computed for the data set and, since it equals a constant, averaged. The quenching rate constant equals this average divided by the fluorescence lifetime in the absence of quenching agent.

E20.22 The Stern-Volmer equation may be written in the form

$$\frac{I_{abs}}{I_f} = 1 + \left(\frac{k_Q}{k_f}\right)[Q] \quad [20.15a]$$

This expression is essentially the same as eqn 20.15a if $k_f >> (k_{ISC} + k_{IC})$. Then eqn 20.13b reduces to $\tau_{obs} = \dfrac{1}{k_f}$ and the form of eqn 20.15a above follows.

This expression shows that, if we plot I_{abs}/I_f against the quencher concentration, which is called a Stern-Volmer plot, then we should get a straight line with the intercept 1 and slope k_Q/k_f. Furthermore, since

$$\frac{1}{\tau_{obs}(Q)} = \frac{1}{\tau_{obs}} + k_Q[Q] \quad [20.15b]$$

a plot of $1/\tau_{obs}(Q)$ against the quencher concentration should give a straight line with a slope equal to k_Q. [*Note:* $\tau_{obs}(Q)$ here and below is τ in the presence of the quencher.]

First, we plot I_{abs}/I_f against [Q] to determine the ratio k_Q/k_f, then we plot $1/\tau_{obs}(Q)$ against [Q] to determine k_Q from the slope. The half-life is determined from $t_{1/2} = \ln 2/k_f$. Draw up the following table of data.

$[Q]/(\text{mmol dm}^{-3})$	1.0	2.0	3.0	4.0	5.0
I_{abs}/I_f	3.2	5.5	7.7	10	12.3
$\left(\dfrac{1}{\tau_{obs}(Q)}\right)/\text{ns}^{-1}$	0.013	0.022	0.031	0.040	0.050

The two sets of data are plotted in Figure 20.4. The linear equations fitted to the data are also displayed. The slope of the plot of I_{abs}/I_f gives $k_Q/k_f = 2.27\ \text{dm}^3\ \text{mmol}^{-1}$.

The slope of the plot of $1/\tau_{obs}(Q)$ gives $\boxed{k_Q = 0.0092\ \text{dm}^3\ \text{mmol}^{-1}\ \text{ns}^{-1}}$. Thus,

$$k_Q = \frac{0.0092\ \text{dm}^3}{\text{mmol ns}} \times \frac{10^3\ \text{mmol}}{\text{mol}} \times \frac{10^9\ \text{ns}}{\text{s}}$$
$$= \boxed{9.2\times10^9\ \text{dm}^3\ \text{mol}^{-1}\ \text{s}^{-1}}$$

$$k_f = \frac{k_Q}{2.27\ \text{dm}^3\ \text{mmol}^{-1}} = \frac{0.0092\ \text{dm}^3\ \text{mmol}^{-1}\ \text{ns}^{-1}}{2.27\ \text{dm}^3\ \text{mmol}^{-1}} = \boxed{4.1\times10^6\ \text{s}^{-1}}$$

$$t_{1/2} = \frac{\ln 2}{4.1\times10^6\ \text{s}^{-1}} = \boxed{1.7\times10^{-7}\text{s}}$$

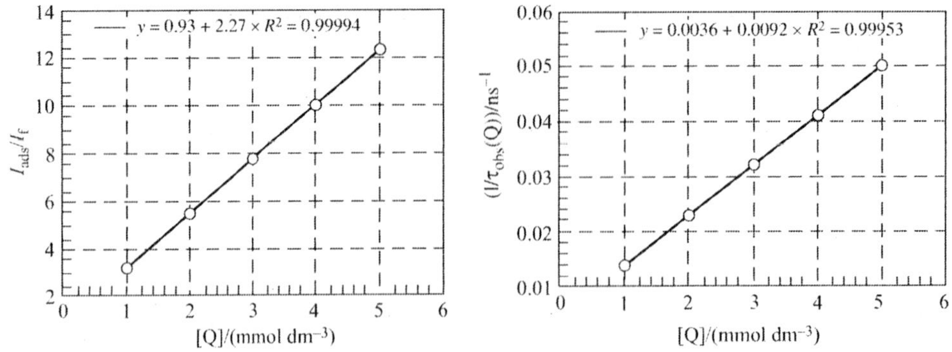

Figure 20.4

E20.23 We use eqn 20.16.

$$\eta_T = \frac{\tau_{obs} - \tau_{obs}(Q)}{\tau_{obs}} = \frac{1.4 - 0.8}{1.4} = \boxed{0.43}$$

E20.24
$$\eta_T = \frac{R_0^6}{R_0^6 + R^6} \quad \text{or} \quad \frac{1}{\eta_T} = 1 + (R/R_0)^6 \quad [20.17]$$

Since a plot of η_T^{-1} values against R^6 (Figure 20.5) appears to be linear with an intercept equal to 1, we conclude that eqn 20.17 adequately describes the data. Solving eqn 20.17 for R_0 gives $R_0 = R(\eta_T^{-1} - 1)^{-1/6}$. R_0 may be evaluated by taking the mean of experimental data in this expression. The two data points at lowest R must be excluded from the mean as they are highly uncertain. $\boxed{R_0 = 3.5\overline{2}\,\text{nm}}$ with a standard deviation of $0.17\overline{3}$ nm.

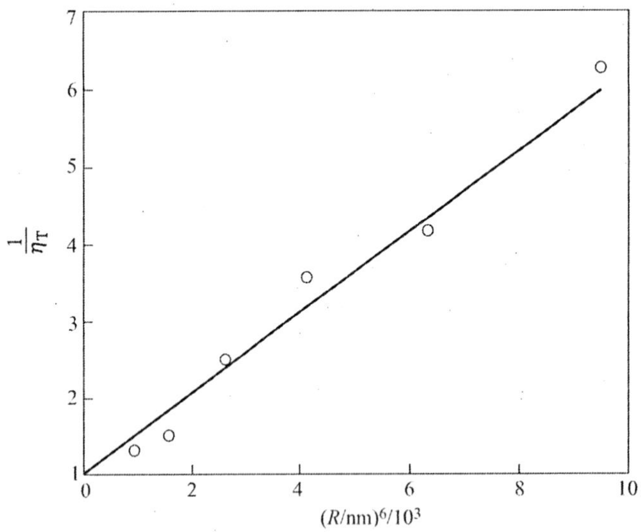

Figure 20.5

E20.25 The laser is delivering photons of energy

$$E = h\nu = \frac{hc}{\lambda} = \frac{(6.626\times10^{-34}\,\text{Js})\times(2.998\times10^{8}\,\text{ms}^{-1})}{488\times10^{-9}\,\text{m}} = 4.07\times10^{-19}\,\text{J}$$

Since the laser is putting out 1.0 mJ of these photons every second, the rate of photon emission is:

$$r = \frac{1.0\times10^{-3}\,\text{J s}^{-1}}{4.07\times10^{-19}\,\text{J}} = 2.5\times10^{15}\,\text{s}^{-1}$$

The time it takes the laser to deliver 10^6 photons (and therefore the time the dye remains fluorescent) is

$$t = \frac{10^6}{2.5\times10^{15}\,\text{s}^{-1}} = \boxed{4\times10^{-10}\,\text{s or } 0.4\,\text{ns}}$$

Solutions to projects

P20.26 (a) Percentage transmitted to the retina is

$$70\% - 0.25\times70\% - 0.09\times0.75\times70\% - 0.43\times(1-0.25-0.09\times0.75)\times70\%$$

$$= 70\% - (0.25+0.0675+0.293)\times70\%$$

$$= 27.2\%.$$

Number of photons focused on the retina in 0.1 s is

$$0.272\times40\ \text{mm}^2\times0.1s\times4\times10^3\ \text{mm}^{-2}\text{s}^{-1} = \boxed{4\times10^3}.$$

More than what one might have guessed.

(b) Trans-retinal has 10 conjugated carbon atoms. There are 9 carbon–carbon bonds, 4 single, and 5 double. There are 10 π-electrons associated with the double bonds. Two electrons occupy each level (by the Pauli principle), so levels E_1 to E_5 are filled. The minimum excitation energy is then $E_6 - E_5$.

$$E_n = \frac{n^2h^2}{8mL^2}.$$

$$\Delta E = E_6 - E_5 = (36-25)\frac{h^2}{8mL^2} = 11\frac{h^2}{8mL^2}.$$

In calculating L, we can add an extra half-bond length at each end of the box, so that the length of the 'box' is 10×140 pm $= 1.40 \times 10^{-9}$ m.

$$\Delta E = \frac{11\times(6.63\times10^{-34}\,\text{Js})^2}{8\times9.11\times10^{-31}\ \text{kg}\times(1.40\times10^{-9}\ \text{m})^2} = 3.38\times10^{-19}\,\text{J}.$$

$$\Delta E = \frac{hc}{\lambda} \quad \text{or} \quad \lambda = \frac{hc}{\Delta E}.$$

$$\lambda = \frac{6.63\times10^{-34}\times\text{Js}\times3.00\times10^8\,\text{m s}^{-1}}{3.38\times10^{-19}\,\text{J}} = 5.87\times10^{-7}\,\text{m}$$

$$= \boxed{587\text{nm}}.$$

This wavelength is in the visible region of the spectrum.

P20.27 (a) $$\eta_T = 1 - \frac{\phi}{\phi_f} = 1 - \frac{\tau_{obs}(Q)}{\tau_{obs}} \quad [20.16]$$

Note: τ here is the observed τ in the presence of the quencher.

$$\eta_T = \frac{R_0^6}{R_0^6 + R^6} \quad [20.17]$$

Equating these two expressions for η_T and solving for R gives:

$$\frac{R_0^6}{R_0^6+R^6}=1-\frac{\tau_{obs}(Q)}{\tau_{obs}}$$

$$\frac{R_0^6+R^6}{R_0^6}=\frac{1}{1-\dfrac{\tau_{obs}(Q)}{\tau_{obs}}}$$

$$\left(\frac{R}{R_0}\right)^6=\frac{1}{1-\dfrac{\tau_{obs}(Q)}{\tau_{obs}}}-1=\frac{\dfrac{\tau_{obs}(Q)}{\tau_{obs}}}{1-\dfrac{\tau_{obs}(Q)}{\tau_{obs}}} \quad \text{or} \quad R=R_0\left(\frac{\dfrac{\tau_{obs}(Q)}{\tau_{obs}}}{1-\dfrac{\tau_{obs}(Q)}{\tau_{obs}}}\right)^{1/6}$$

$T_{obs}(Q)/\tau_{obs} = 10\text{ ps} / 10^3\text{ ps} = 0.010$ and $R=5.6\text{ nm}\left(\dfrac{0.010}{1-0.010}\right)^{1/6}=\boxed{2.6\text{ nm}}$

(b) $$\underset{\text{Chlorophyll}}{C} + \underset{\text{Quinone}}{Q} \xrightarrow{h\nu} C^* + Q \xrightarrow[\text{transfer}]{\text{electron}} C^+ + Q^-$$

Direct electron transfer from the ground state of C is not spontaneous. It is spontaneous from the excited state. The difference between the ΔG's of the two processes is given by the expression:

$$\Delta(\Delta G)=\Delta G_{C^*}-\Delta G_C \approx U_C - U_{C^*} \approx -(U_{LUMO}-U_{HOMO})$$

where U_{LUMO} and U_{HOMO} are energies of the LUMO and HOMO of chlorophyll. Since $\Delta\Delta G < 0$, we see that electron transfer is exergonic and spontaneous when the electron is transferred from the excited state of chlorophyll.

21 Spectroscopy: magnetic resonance

Answers to discussion questions

D21.1 Discussions of the origins of the local, neighbouring group, and solvent contributions to the shielding constant can be found in Section 21.3. The local contribution is essentially the contribution of the electrons in the atom that contains the nucleus being observed. It can be expressed as a sum of a diamagnetic and paramagnetic parts, that is $\sigma(\text{local}) = \sigma_d + \sigma_p$. The diamagnetic part arises because the applied field generates a circulation of charge in the ground state of the atom. In turn, the circulating charge generates a magnetic field. The direction of this field can be found through Lenz's law which states that the induced magnetic field must be opposite in direction to the field producing it. Thus, it shields the nucleus. The diamagnetic contribution is roughly proportional to the electron density on the atom and it is the only contribution for closed shell free atoms and for distributions of charge that have spherical or cylindrical symmetry. The local paramagnetic contribution is somewhat harder to visualize since there is no simple and basic principle analogous to Lenz's law that can be used to explain the effect. The applied field adds a term to the hamiltonian of the atom which mixes in excited electronic states into the ground state and any theoretical calculation of the effect requires detailed knowledge of the excited state wave functions. It is to be noted that the paramagnetic contribution does not require that the atom or molecule be paramagnetic. It is paramagnetic only in the sense in that it results in an induced field in the same direction as the applied field.

The neighbouring group contributions arise in a manner similar to the local contributions. Both diamagnetic and paramagnetic currents are induced in the neighbouring atoms and these currents results in shielding contributions to the nucleus of the atom being observed. However, there are some differences: The magnitude of the effect is much smaller because the induced currents in neighbouring atoms are much farther away. It also depends on the anisotropy of the magnetic susceptibility (Section 17.9) of the neighbouring group. Only anisotropic susceptibilities result in a contribution.

Solvents can influence the local field in many different ways. Detailed theoretical calculations of the effect are difficult due to the complex nature of the solute-solvent interaction. Polar solvent–polar solute interactions are an electric field effect that usually causes deshielding of the solute protons. Solvent magnetic antisotropy can cause shielding or deshielding, for example, for solutes in benzene solution. In addition, there are a variety of specific chemical interactions between solvent and solute that can affect the chemical shift.

D21.2 Both spin–lattice and spin–spin relaxation are caused by fluctuating magnetic and electric fields at the nucleus in question and these fields result from the random thermal motions present in the solution or other form of matter. These random motions can be a result of a number of processes and it is hard to summarize all that could be important. In theory, every known nuclear interaction coupled with every type of motion can contribute to relaxation and detailed treatments can be exceedingly complex. However, they all depend on the magnetogyric ratio of the atom in question and the magnetogyric ratio of the proton is much larger than that of ^{13}C. Hence, the interaction of the proton with fluctuating local magnetic fields caused by the presence of neighboring magnetic nuclei will be greater, and the relaxation will be quicker, corresponding to a shorter relaxation time for protons. Another consideration is the structure of compounds containing carbon and hydrogen. Typically, the C atoms are in the interior of the molecule bonded to other C atoms, 99% of which are nonmagnetic, so the primary relaxation effects are due to bonded protons. Protons are on the outside of the molecule and are subject to many more interactions and hence faster relaxation.

D21.3 At, say, room temperature, the tumbling rate of benzene, the small molecule, in a mobile solvent, may be close to the Larmor frequency, and hence its spin-lattice relaxation time will be short. As the temperature increases, the tumbling rate may increase well beyond the Larmor frequency, resulting in an increased spin-lattice relaxation time.

For the larger oligopeptide at room temperature, the tumbling rate may be well below the Larmor frequency, but with increasing temperature, it will approach the Larmor frequency due to the increased thermal motion of the molecule combined with the decreased viscosity of the solvent. Therefore, the spin-lattice relaxation time may decrease.

D21.4 Spin–spin couplings in NMR are due to a polarization mechanism, which is transmitted through bonds. The following description applies to the coupling between the protons in H_X—C—H_Y group as is typically found in organic compounds. See Fig. 21.18 of the text. On H_X, the Fermi contact interaction causes the spins of its proton and electron to be aligned antiparallel. The spin of the electron from C in the H_X—C bond is then aligned antiparallel to the electron from H_X due to the Pauli exclusion principle. The spin of the C electron in the bond H_Y is then aligned parallel to the electron from H_X because of Hund's rule. Finally, the alignment is transmitted through the second bond in the same manner as the first. This progression of alignments (antiparallel × antiparallel × parallel × antiparallel × antiparallel) yields an overall energetically favourable parallel alignment of the two proton nuclear spins. Therefore, in this case the coupling constant, ${}^2J_{HH}$ is negative in sign.

D21.5 In the nuclear Overhauser effect (NOE) in NMR, spin relaxation processes are used to transfer the population difference typical of one species of nucleus X to another nucleus A, thereby enhancing the intensity of the signal produced by A. Eqn 21.23 shows that the signal enhancement is given by

$$\frac{I}{I_0} = 1 + \eta$$

where η is the enhancement parameter. In the case of maximal enhancement it is possible to show that

$$\eta = \frac{\gamma_X}{2\gamma_A}.$$

NOE can be used to determine interproton distances in biopolymers. This application makes use of the fact that when the dipole-dipole mechanism is not the only relaxation mechanism, the NOE is given by

$$\frac{I}{I_0} = 1 + \eta = 1 + \frac{\gamma_X}{2\gamma_A} \times \frac{T_1}{T_{1,\text{dip-dip}}}$$

where T_1 is the total relaxation time and $T_{1,\text{dip-dip}}$ is the relaxation time due to the dipole-dipole mechanism. Here A and X are both protons. The enhancement depends strongly on the separation, r, of the two spins, for the strength of the dipole-dipole interaction is proportional to $1/r^3$, and its effect depends on the square of that strength and therefore on $1/r^6$. This sharp dependence on separation is used to build up a picture of the conformation of the biopolymer by using NOE to identify which protons can be regarded as neighbors.

D21.6 The molecular orbital occupied by the unpaired electron in an organic radical can be identified through the observation of hyperfine splitting in the EPR spectrum of the radical. The magnitude of this splitting is proportional to the spin density of the unpaired electron at those positions in the radical having atoms with nuclear moments. In addition, the spin density on carbon atoms adjacent to the magnetic nuclei can be determined indirectly through the McConnell relation. Thus, for example, in the benzene negative ion, unpaired spin densities on both the carbon atoms and hydrogen atoms can be determined from the EPR hyperfine splittings. The next step then is to construct a molecular orbital which will theoretically reproduce these experimentally determined spin densities. A good match indicates that we have found a good molecular orbital for the radical.

D21.7 The mechanism for the hyperfine interaction of a proton in a methyl group attached to an aromatic ring is not the same as the mechanism responsible for the McConnell equation [21.28] that explains the hyperfine interaction of a proton attached to a ring carbon atom. However, we still expect a proportionality between the hyperfine splitting constant, a, and the spin density, ρ, on the ring carbon atom. The hyperfine interaction in this case is due to the overlap of the proton wavefunctions in the methyl group with the p_z orbital of the ring carbon atom which has unpaired electron density (spin density). The strength of the interaction will however be angularly dependent. The wavefunction for a p_z orbital indicates a distribution of spin density in the plane of the ring of the form $\psi_\theta^2 = \psi_0^2 \cos^2\theta$ where ψ_θ^2 and ψ_0^2 are the spin densities corresponding to the angles θ and 0°. On the assumption that the hyperfine splitting follows a similar relationship, one expects that the maximum splitting when the C—H bond is in a plane parallel to the p_z orbital and zero interaction in a plane perpendicular to it. So we can write for each proton in the methyl group the relationship $a = Q_{\mathrm{CCH}}\rho_{\mathrm{C}}\cos^2\theta$ and will be different for each proton of the group and will change as the methyl group rotates. In rapid rotation the three splittings will be averaged to one value.

Solutions to exercises

E21.1 $E_{m_s} = g_e\mu_B\mathcal{B}m_s$ [21.4] where $m_s = +\frac{1}{2}$ or $-\frac{1}{2}$ (α and β spin states, respectively)

$$\Delta E = E_\alpha - E_\beta = E_{m_s=1/2} - E_{m_s=-1/2} = g_e\mu_B\mathcal{B}\left(\frac{1}{2}-\left(-\frac{1}{2}\right)\right) = g_e\mu_B\mathcal{B} \quad [21.8]$$

$$= (2.0023)\times(9.274\times10^{-24}\ \mathrm{J\,T^{-1}})\times(0.250\ \mathrm{T}) = \boxed{4.64\times10^{-24}\ \mathrm{J}}$$

E21.2 $E_{m_I} = -\gamma_N\hbar\mathcal{B}m_I = -g_I\mu_N\mathcal{B}m_I$ [21.5, 21.7] $m_I = 3/2, 1/2, -1/2, -3/2$

$$E_{m_I} = -0.4289\times(5.051\times10^{-27}\ \mathrm{J\,T^{-1}})\times(6.000\ \mathrm{T})\times m_I$$

$$= \boxed{-1.300\times10^{-26}\,\mathrm{J}\times m_I}$$

E21.3 (a) Unit of γ_N = unit of $\left(\frac{g_I\mu_N}{\hbar}\right) = \frac{\mathrm{J\,T^{-1}}}{\mathrm{J\,s}} = \mathrm{T^{-1}\ s^{-1}} = \boxed{\mathrm{T^{-1}Hz}}$

(b) $1\ \mathrm{T} = 1\ \mathrm{kg\ s^{-2}\ A^{-1}}$

$\gamma_N = \mathrm{kg^{-1}\ s^2\ A \times s^{-1}} = \boxed{\mathrm{A\ s\ kg^{-1}}}$

E21.4 $\gamma_N\hbar = g_I\mu_N$ [21.5, 21.7]

Therefore,

$$g_I = \frac{\gamma_N\hbar}{\mu_N} = \frac{1.0840\times10^8\ \mathrm{T^{-1}\ s^{-1}}\times1.05457\times10^{-34}\,\mathrm{J\,s}}{5.051\times10^{-27}\ \mathrm{J\,T^{-1}}}$$

$$= \boxed{2.263}$$

E21.5 We assume a temperature of 300 K.

$$\frac{N_\beta - N_\alpha}{N} = \frac{N_\beta - N_\alpha}{N_\beta + N_\alpha} \simeq \frac{g_e\mu_B\mathcal{B}}{2kT} \quad \text{(See Derivation 21.1)}$$

(a) $$\frac{N_\beta - N_\alpha}{N} = \frac{(2.0023)\times(9.274\times10^{-24}\ \mathrm{J\,T^{-1}})\times(0.400\ \mathrm{T})}{2(1.381\times10^{-23}\ \mathrm{J\,K^{-1}})\times(300\ \mathrm{K})} = \boxed{8.96\times10^{-4}}$$

(b) $$\frac{N_\beta - N_\alpha}{N} = \frac{(2.0023)\times(9.274\times10^{-24}\ \mathrm{J\,T^{-1}})\times(1.2\ \mathrm{T})}{2(1.381\times10^{-23}\ \mathrm{J\,K^{-1}})\times(300\ \mathrm{K})} = \boxed{2.69\times10^{-3}}$$

E21.6 $\nu = \dfrac{g_e \mu_B \mathcal{B}}{h}$ [21.10] where $g_e = 2.0023$ and μ_B is the Bohr magneton

$$\nu = \frac{2.0023 \times 9.274 \times 10^{-24}\ \mathrm{J\,T^{-1}} \times 0.330\ \mathrm{T}}{6.626 \times 10^{-34}\ \mathrm{J\,s}}$$

$$= 9.248 \times 10^{9}\ \mathrm{s^{-1}} = \boxed{9.248\ \mathrm{GHz}}$$

$$\lambda = \frac{c}{\nu} = \frac{2.998 \times 10^{8}\ \mathrm{m\,s^{-1}}}{9.248 \times 10^{9}\ \mathrm{s^{-1}}} = \boxed{0.0324\ \mathrm{m}}$$

EPR employs microwave radiation, rather than the radio frequency radiation of NMR.

E21.7 We assume a temperature of 300 K.

$$\frac{N_\alpha - N_\beta}{N} = \frac{N_\alpha - N_\beta}{N_\alpha + N_\beta} \simeq \frac{\gamma_N \hbar \mathcal{B}}{2kT}\ [21.12] \text{ for spin-}\frac{1}{2} \text{ nuclei like } {}^1\mathrm{H} \text{ and } {}^{13}\mathrm{C}$$

Nuclear magnetogyric ratios are found in Table 21.2.

(a) ^{1}H: $$\frac{N_\alpha - N_\beta}{N} = \frac{(26.752 \times 10^{7}\ \mathrm{T^{-1}\,s^{-1}}) \times (1.055 \times 10^{-34}\ \mathrm{J\,s}) \times (8.5\ \mathrm{T})}{2(1.381 \times 10^{-23}\ \mathrm{J\,K^{-1}}) \times (300\ \mathrm{K})} = \boxed{2.9 \times 10^{-5}}$$

(b) ^{13}C: $$\frac{N_\alpha - N_\beta}{N} = \frac{(6.7272 \times 10^{7}\ \mathrm{T^{-1}\,s^{-1}}) \times (1.055 \times 10^{-34}\ \mathrm{J\,s}) \times (8.5\ \mathrm{T})}{2(1.381 \times 10^{-23}\ \mathrm{J\,K^{-1}}) \times (300\ \mathrm{K})} = \boxed{7.3 \times 10^{-6}}$$

E21.8 $$\nu = \frac{\gamma_N \mathcal{B}}{2\pi}\ [21.13] = \frac{2.5177 \times 10^{8}\ \mathrm{T^{-1}\,s^{-1}} \times 7.500\ \mathrm{T}}{2\pi}$$

$$= 3.005 \times 10^{8}\ \mathrm{s^{-1}} = \boxed{300.5\ \mathrm{MHz}}$$

E21.9 $$\nu = \frac{\gamma_N \mathcal{B}}{2\pi}\ [21.13] = \frac{g_I \mu_N}{h}\mathcal{B}$$

$$\nu = \frac{0.4036 \times 5.051 \times 10^{-27}\ \mathrm{J\,T^{-1}} \times 14.20\ \mathrm{T}}{6.626 \times 10^{-34}\ \mathrm{J\,s}}$$

$$= 4.369 \times 10^{7}\ \mathrm{s^{-1}} = \boxed{43.69\ \mathrm{MHz}}$$

E21.10 $\mathcal{B} = \dfrac{h\nu}{\gamma_N \hbar}$ [21.11] $= \dfrac{2\pi\nu}{\gamma_N}$ where $\gamma_N = 26.752 \times 10^{7}\ \mathrm{T^{-1}\,s^{-1}}$ for ^{1}H (Table 21.2)

$$= \frac{2\pi(800.0 \times 10^{6}\ \mathrm{Hz})}{26.752 \times 10^{7}\ \mathrm{T^{-1}\,s^{-1}}} = \boxed{18.79\ \mathrm{T}}$$

E21.11 $$\delta = \frac{\nu - \nu^{\circ}}{\nu^{\circ}} \times 10^{6}\ [21.17]$$

$$\text{shift} = \nu - \nu^{\circ} = \frac{\delta \times \nu^{\circ}}{10^{6}} = \frac{6.33 \times 500 \times 10^{6}\ \mathrm{Hz}}{10^{6}}$$

$$= 3.17 \times 10^{3}\ \mathrm{Hz} = \boxed{3.17\ \mathrm{kHz}}$$

E21.12 $\mathcal{B}_{loc} = (1-\sigma)\mathcal{B}$ [21.15]

$|\Delta\mathcal{B}_{loc}| = |(\Delta\sigma)|\mathcal{B} \approx |[\delta(\mathrm{CH_3}) - \delta(\mathrm{CHO})]|\mathcal{B}$ $\left[|\Delta\sigma| = |\Delta\delta \times 10^{-6}| \text{ which follows from eqns 21.16 \& 21.17}\right]$

$$= |(2.20 - 9.80)| \times 10^{-6}\mathcal{B} = 7.60 \times 10^{-6}\mathcal{B}$$

(a) $\mathcal{B} = 1.2\ \mathrm{T}$, $|\Delta\mathcal{B}_{loc}| = 7.60 \times 10^{-6} \times 1.2\ \mathrm{T} = \boxed{9.1\ \mu\mathrm{T}}$

(b) $\mathcal{B} = 5.0\ \mathrm{T}$, $|\Delta\mathcal{B}_{loc}| = 7.60 \times 10^{-6} \times 5.0\ \mathrm{T} = \boxed{38\ \mu\mathrm{T}}$

E21.13 $$\nu - \nu^{\circ} = \frac{\nu^{\circ}\delta}{10^6} \quad [21.17]$$

$$\Delta(\nu - \nu^{\circ}) = \frac{\nu^{\circ}\Delta\delta}{10^6}$$

(a) $\Delta(\nu - \nu^{\circ}) = 300\ \text{MHz} \times 10^{-6} \times (9.5 - 1.5) \approx \boxed{2.4\ \text{kHz}}$

(b) $\Delta(\nu - \nu^{\circ}) = 750\ \text{MHz} \times 10^{-6} \times (9.5 - 1.5) \approx \boxed{6.0\ \text{kHz}}$

E21.14 For identical nuclei with spin 1/2, there will be $N + 1$ lines from the splitting. In this case 8 lines. The lines will have relative intensities of $\boxed{1:7:21:35:35:21:7:1}$.

These relative intensities can be determined by extending Pascals' triangle shown in (**1**) of the text three more rows to $N + 1 = 8$. Alternatively the intensities can also be determined from the coefficients in the expansion of $(1 + x)^N$.

E21.15 Because each resonance is split into three lines by a single N nucleus, the result will be:

(a) $\boxed{\text{quintet } 1:2:3:2:1}$ and

(b) $\boxed{\text{septet } 1:3:6:7:6:3:1}$

Also see the solution to exercise 21.20 which discusses the procedure for determining the splitting pattern.

E21.16 $$E = -\gamma_N\hbar(1-\sigma_A)\,\mathcal{B}\,m_A - \gamma_N\hbar(1-\sigma_X)\,\mathcal{B}\,m_{X_1} - \gamma_N\hbar(1-\sigma_X)\,\mathcal{B}\,m_{X_2}$$

As m_{X_1} and m_{X_2} can each be $\pm\frac{1}{2}$, there are a total of 6 energy levels, two of which are two-fold degenerate, for a total of eight levels. These are shown on the left of Figure 21.1. The allowed transitions are indicated by arrows. There are 7 transitions, but only 2 transitions frequencies. This follows from the selection rule for magnetic resonance transitions, which is $\Delta(m_1 + m_2) = \pm 1$. The shorter arrows represent the X transitions, the larger arrows the A transitions. Spin-spin splitting perturbs these levels as follows:

$$E_{\text{spin-spin}} = hJm_A m_{X_1} + hJm_A m_{X_2}$$

	$\alpha\alpha_1\alpha_2$	$\alpha\alpha_1\beta_2$	$\alpha\beta_1\alpha_2$	$\alpha\beta_1\beta_2$
$E_{\text{spin-spin}}$	$\frac{1}{2}hJ$	0	0	$-\frac{1}{2}hJ$
	$\beta\beta_1\beta_2$	$\beta\alpha_1\beta_2$	$\beta\beta_1\alpha_2$	$\beta\alpha_1\alpha_2$

There are again a total of 6 energy levels (two of which are two-fold degenerate), but they are perturbed by the amounts in the above chart. The perturbed levels are shown on the right in the figure below. The frequencies of the X transitions are changed by $\pm\frac{1}{2}J$, the frequencies of the A transitions by $-J$, 0, $+J$. A stick diagram representing the spectrum is shown in Figure 21.2.

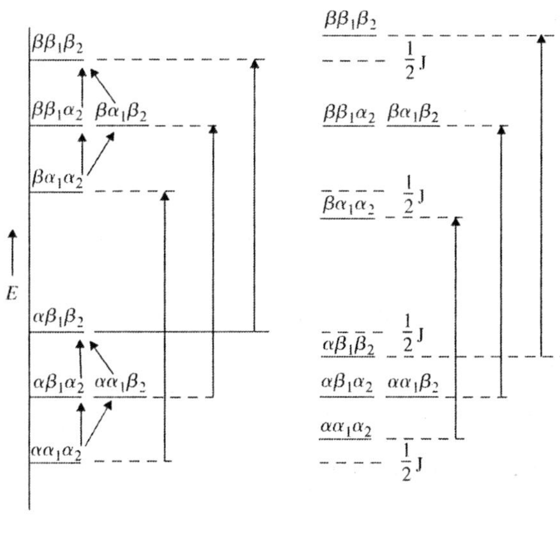

Figure 21.1

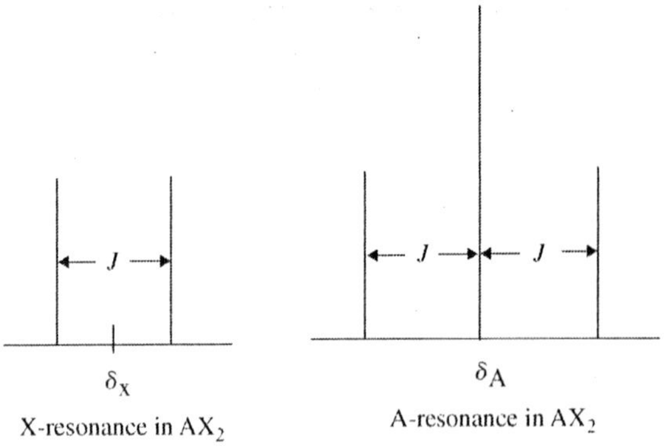

Figure 21.2

E21.17 $\quad \nu - \nu^{\circ} = \nu^{\circ}\, \delta \times 10^{-6}$

$$|\Delta\nu| \equiv (\nu - \nu^{\circ})(\mathrm{CHO}) - (\nu - \nu^{\circ})(\mathrm{CH_3})$$

$$= \nu(\mathrm{CHO}) - \nu(\mathrm{CH_3})$$

$$= \nu^{\circ}\, [\delta(\mathrm{CHO}) - \delta(\mathrm{CH_3})] \times 10^{-6}$$

$$= (9.80 - 2.20) \times 10^{-6}\, \nu^{\circ} = 7.60 \times 10^{-6}\, \nu^{\circ} \text{ [chemical shift values from exercise 21.12]}$$

(a) $\quad \nu^{\circ} = 300\ \mathrm{MHz} |\Delta\nu| = 7.60 \times 10^{-6} \times 300\ \mathrm{MHz} = \boxed{2.28\ \mathrm{kHz}}$

(b) $\quad \nu^{\circ} = 550\ \mathrm{MHz} |\Delta\nu| = 7.60 \times 10^{-6} \times 550\ \mathrm{MHz} = \boxed{4.18\ \mathrm{kHz}}$

(a) The spectrum is shown in Figure 21.3.

(b) When the frequency is changed to 550 MHz, the separation of the CH_3 and CHO resonance increases (4.18 kHz), the fine structure remains unchanged, and the intensity increases.

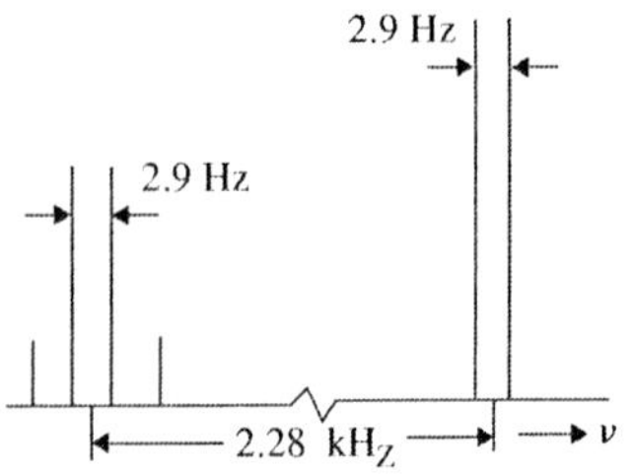

Figure 21.3

E21.18 The four equivalent ^{19}F nuclei ($I = \frac{1}{2}$) give a single line. However, the ^{10}B nucleus ($I = 3$, 19.6 percent abundant) splits this line into $2 \times 3 + 1 = 7$ lines and the ^{11}B nucleus ($I = \frac{3}{2}$, 80.4 percent abundant) into $2 \times \frac{3}{2} + 1 = 4$ lines. The splitting arising from the ^{11}B nucleus will be larger than that arising from the ^{10}B (because its magnetic moment is larger, by a factor of 1.5). Moreover, the total intensity of the four lines due to the ^{11}B nuclei will be greater (by a factor of $80.4/19.6 \approx 4$) than the total intensity of the seven lines due to the ^{10}B nuclei. The individual line intensities will be in the ratio $\frac{7}{4} \times 4 = 7$ ($\frac{4}{7}$ the number of lines and about four times as abundant). The spectrum is sketched in Figure 21.4.

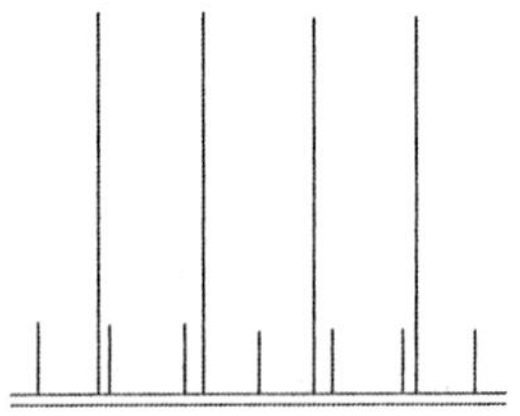

Figure 21.4

E21.19 The A, M, and X resonances lie in distinctly different groups. The A resonance is split into a 1:2:1 triplet by the M nuclei, and each line of that triplet is split into a 1:4:6:4:1 quintet by the X nuclei, (with $J_{AM} > J_{AX}$). The M resonance is split into a 1:3:3:1 quartet by the A nuclei and each line is split into a quintet by the X nuclei (with $J_{AM} > J_{MX}$). The X resonance is split into a quartet by the A nuclei and then each line is split into a triplet by the M nuclei (with $J_{AX} > J_{MX}$). The spectrum is sketched in the Figure 21.5.

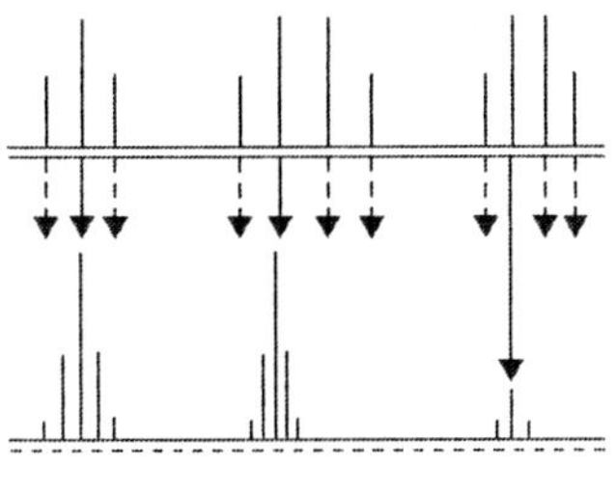

Figure 21.5

E21.20 See Example 21.3 and Fig. 21.34 for the approach to the solution to this exercise. That example and that figure are applied specifically to EPR spectra, but the process of determining the intensity pattern in the fine structure of the A resonance in an AX_N NMR spectrum is the same. See the table below for the version of Pascal's triangle for up to 5 spin-1 nuclei. Each number in the table is the sum of the three ($I = 1$, $2I + 1 = 3$) numbers in the three adjacent cells above it (one to the right, one in the middle, and one to the left).

					1					
				1	1	1				
			1	2	3	2	1			
		1	3	6	7	6	3	1		
	1	4	10	16	19	16	10	4	1	
1	5	15	30	45	51	45	30	15	5	1

E21.21 See Example 21.3 and Fig. 21.34 for the approach to the solution to this exercise. That example and that figure are applied specifically to EPR spectra, but the process of determining the intensity pattern in the fine structure of an NMR spectrum is the same. See the table below for the version of Pascal's triangle for up to 5 spin-3/2 nuclei. Each number in the table is the sum of the four ($I = 3/2$, $2I + 1 = 4$) numbers above it (2 to the right and 2 to the left).

															1															
												1		1		1		1												
									1		2		3		4		3		2		1									
						1		3		6		10		12		12		10		6		3		1						
			1		4		10		20		31		40		44		40		31		20		10		4		1			
1		5		15		35		65		101		135		155		155		135		101		65		35		15		5		1

E21.22 The frequency difference between the two signals is

$$\Delta\nu = \left[\nu^{\circ} + \nu^{\circ}\times10^{-6}\delta'\right] - \left[\nu^{\circ} + \nu^{\circ}\times10^{-6}\delta\right] = \nu^{\circ}\times10^{-6}(\delta' - \delta) \quad [21.18]$$

$$\tau = \frac{2^{1/2}}{\pi\Delta\nu}\ [21.22] = \frac{2^{1/2}}{\pi\nu^{\circ}\times10^{-6}(\delta' - \delta)}$$

$$= \frac{2^{1/2}}{\pi(550\times10^{6}\ \mathrm{s}^{-1})\times10^{-6}(4.8-2.7)} = 3.9\times10^{-4}\ \mathrm{s}$$

Therefore, the signals merge when the conversion rate is greater than about $1/\tau = \boxed{2.6\times10^{3}\ \mathrm{s}^{-1}}$

E21.23 The desired result is the linear equation:

$$[\mathrm{I}]_0 = \frac{[E]_0 \Delta\nu}{\delta\nu} - K_\mathrm{I},$$

so the first task is to express quantities in terms of $[\mathrm{I}]_0$, $[\mathrm{E}]_0$, $\Delta\nu$, $\delta\nu$, and K_I, eliminating terms such as [I], [EI], [E], ν_I, ν_EI, and ν. (*Note:* symbolic mathematical software is helpful here.) Begin with ν:

$$\nu = \frac{[\mathrm{I}]}{[\mathrm{I}]+[\mathrm{EI}]}\nu_\mathrm{I} + \frac{[\mathrm{EI}]}{[\mathrm{I}]+[\mathrm{EI}]}\nu_\mathrm{EI} = \frac{[\mathrm{I}]_0 - [\mathrm{EI}]}{[\mathrm{I}]_0}\nu_\mathrm{I} + \frac{[\mathrm{EI}]}{[\mathrm{I}]_0}\nu_\mathrm{EI},$$

where we have used the fact that total I (*i.e.*, free I plus bound I)is the same as initial I. Solve this subject to the condition that it must also be much greater than [EI]:

$$[\mathrm{EI}] = \frac{[\mathrm{I}]_0(\nu - \nu_\mathrm{I})}{\nu_\mathrm{EI} - \nu_\mathrm{I}} = \frac{[\mathrm{I}]_0 \delta\nu}{\Delta\nu},$$

where in the second equality we notice that the frequency differences that appear are the ones defined in the problem. Now take the equilibrium constant:

$$K_\mathrm{I} = \frac{[\mathrm{E}][\mathrm{I}]}{[\mathrm{EI}]} = \frac{([\mathrm{E}]_0 - [\mathrm{EI}])([\mathrm{I}]_0 - [\mathrm{EI}])}{[\mathrm{EI}]} \approx \frac{([\mathrm{E}]_0 - [\mathrm{EI}])[\mathrm{I}]_0}{[\mathrm{EI}]}.$$

We have used the fact that total I is much greater than total E (from the condition that $[\mathrm{I}]_0 \gg [\mathrm{E}]_0$), so it must also be much greater than $[\mathrm{EI}]$, even if all E binds I. Now solve this for $[\mathrm{E}]_0$:

$$[\mathrm{E}]_0 = \frac{K_\mathrm{I} + [\mathrm{I}]_0}{[\mathrm{I}]_0}[\mathrm{EI}] = \left(\frac{K_\mathrm{I} + [\mathrm{I}]_0}{[\mathrm{I}]_0}\right)\left(\frac{[\mathrm{I}]_0 \delta\nu}{\Delta\nu}\right) = \frac{(K_\mathrm{I} + [\mathrm{I}]_0)\delta\nu}{\Delta\nu}.$$

The expression contains the desired terms and only those terms. Solving for $[\mathrm{I}]_0$ yields:

$$\boxed{[\mathrm{I}]_0 = \frac{[\mathrm{E}]_0 \Delta\nu}{\delta\nu} - K_\mathrm{I}}$$

which would result in a straight line with slope $[\mathrm{E}]_0 \Delta\nu$ and *y*-intercept K_I if one plots $[\mathrm{I}]_0$ against $1/\delta\nu$.

E21.24 $$g = \frac{h\nu}{\mu_\mathrm{B}\mathcal{B}} \quad [21.25]$$

We shall often need the value

$$\frac{h}{\mu_\mathrm{B}} = \frac{6.62608\times10^{-34}\ \mathrm{J\,Hz^{-1}}}{9.27402\times10^{-24}\ \mathrm{J\,T^{-1}}} = 7.14478\times10^{-11}\ \mathrm{T\,Hz^{-1}}$$

Then, in this case

$$g = \frac{(7.14478\times10^{-11}\,\mathrm{T\,Hz^{-1}})\times(9.2231\times10^{9}\,\mathrm{Hz})}{329.12\times10^{-3}\,\mathrm{T}} = \boxed{2.0022}$$

E21.25 $$a = \mathcal{B}(\text{line }3) - \mathcal{B}(\text{line }2) = \mathcal{B}(\text{line }2) - \mathcal{B}(\text{line }1)$$

$$\left.\begin{array}{l}\mathcal{B}_3 - \mathcal{B}_2 = (334.8 - 332.5)\,\mathrm{mT} = 2.3\,\mathrm{mT} \\ \mathcal{B}_2 - \mathcal{B}_1 = (332.5 - 330.2)\,\mathrm{mT} = 2.3\,\mathrm{mT}\end{array}\right\} a = \boxed{2.3\ \mathrm{mT}}$$

Use the centre line to calculate g

$$g = \frac{h\nu}{\mu_\mathrm{B}\mathcal{B}}\ [21.24] = (7.14478\times10^{-11}\ \mathrm{T\,Hz^{-1}})\times\frac{9.319\times10^{9}\ \mathrm{Hz}}{332.5\times10^{-3}\ \mathrm{T}} = \boxed{2.002\bar{5}}$$

E21.26 We construct Figure 21.6(a) for CH_3 and Figure 21.6(b) for CD_3. The predicted intensity distribution is determined by counting the number of overlapping lines of equal intensity from which the hyperfine line is constructed.

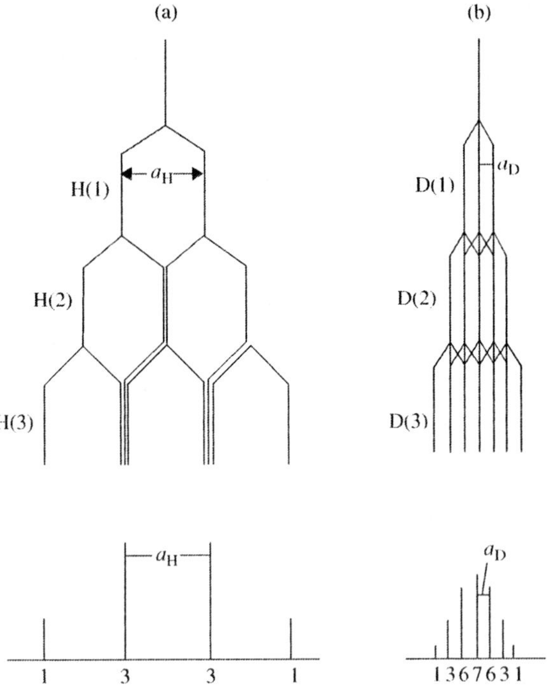

Figure 21.6

E21.27
$$\mathcal{B}=\frac{h\nu}{g\mu_B}\ [21.25]=\frac{7.14478\times10^{-11}}{2.0025}\ \mathrm{T\,Hz^{-1}}\times\nu\ [\text{Exercise}\,21.24]=35.68\,\mathrm{mT}\times(\nu/\mathrm{GHz})$$

(a) $\nu=9.302\,\mathrm{GHz},\quad \mathcal{B}_0=\boxed{331.9\ \mathrm{mT}}$

(b) $\nu=33.67\,\mathrm{GHz},\quad \mathcal{B}_0=1201\,\mathrm{mT}=\boxed{1.201\,\mathrm{T}}$

E21.28 If a radical contains N equivalent nuclei with spin quantum number I, then there are $2NI+1$ hyperfine lines in the EPR spectrum. For $N=2$ and five hyperfine lines, $\boxed{I=1}$. The intensity ratio for the lines is 1:2:3:2:1 as demonstrated in Figure 21.7 (also see Figures 21.33 and 21.34 of text).

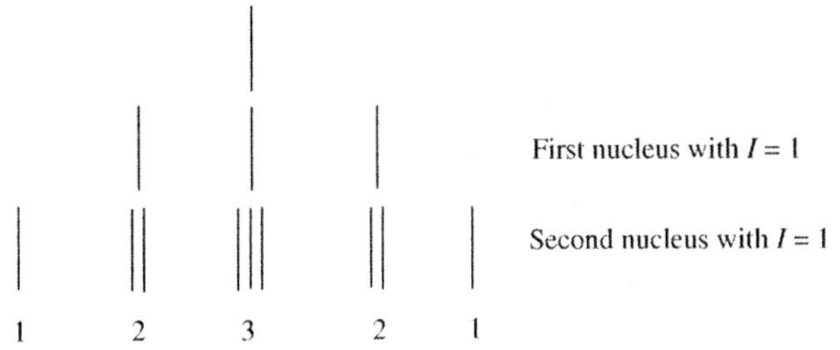

Figure 21.7

E21.29 See the table in the solution to Exercise 21.21 above. The Pascal "triangle" for N spin-3/2 nuclei is the same for the hyperfine structure from a collection of N spin-3/2 nuclei in an EPR spectrum or from the A resonance in an AX_N NMR spectrum.

E21.30 For $C_6H_6^-$, $a=Q\rho$ with $Q=2.25$ mT $[21.28]$. If we assume that the value of Q does not change from this value (a good assumption in view of the similarity of the anions), we may write

$$\rho=\frac{a}{Q}=\frac{a}{2.25\,\mathrm{mT}}$$

Hence, we can construct the spin density maps shown in Figure 21.8.

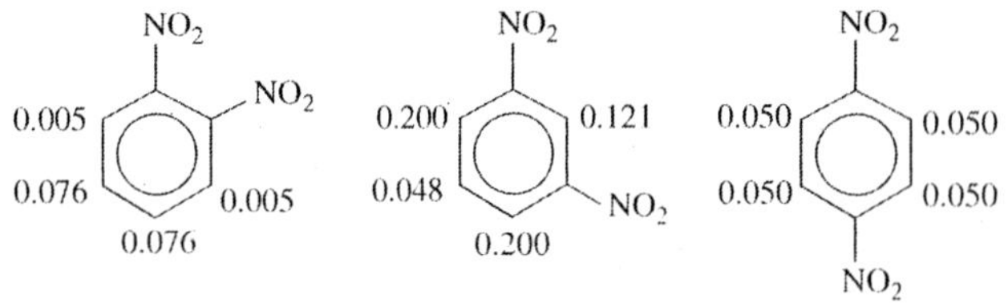

Figure 21.8

Solutions to projects

P21.31 $$^3J_{\mathrm{HH}} = A + B\cos\phi + C\cos 2\phi \quad [21.21]$$

$$\frac{\mathrm{d}}{\mathrm{d}\phi}\left({}^3J_{\mathrm{HH}}\right) = -B\sin\phi - 2C\sin 2\phi = 0$$

This equation has a number of solutions:

$$\phi = 0, \qquad \phi = n\pi, \qquad \phi = \pi - \arccos\left(\frac{B}{4C}\right) = \arccos\left(\frac{B}{4C}\right)$$

The first two are trivial solutions.

$$\text{If } \phi = \arccos\left(\frac{B}{4C}\right) \text{ then, } \sin\phi = \sqrt{1-\frac{B^2}{16C^2}}$$

$$\sin 2\phi = 2\sin\phi\cos\phi = 2\sqrt{1-\frac{B^2}{16C^2}}\left(\frac{B}{4C}\right)$$

$$B\sin\phi + 2C\sin 2\phi = B\sqrt{1-\frac{B^2}{16C^2}} + 4C\sqrt{1-\frac{B^2}{16C^2}}\left(\frac{B}{4C}\right) = 0$$

So $\dfrac{B}{4C} = \cos\phi$ clearly satisfies the condition for an extremum.

The second derivative is

$$\frac{\mathrm{d}^2}{\mathrm{d}\phi^2}\left({}^3J_{\mathrm{HH}}\right) = -B\cos\phi - 4C\cos 2\phi = -B\cos\phi - 4C(2\cos^2\phi - 1)$$

$$= -B\left(\frac{B}{4C}\right) - 4C\left(2\frac{B^2}{16C^2} - 1\right) = -\frac{B^2}{4C} - \frac{2B^2}{4C} + 4C$$

This quantity is positive if

$$16C^2 > 3B^2$$

This is certainly true for typical values of B and C, namely $B = -1$ Hz and C = 5 Hz. Therefore the condition for a minimum is as stated, namely, $\boxed{\cos\phi = B/4C}$.

P21.32 (a) We use $\nu = \dfrac{\gamma_{\mathrm{N}}\mathcal{B}_{\mathrm{loc}}}{2\pi} = \dfrac{\gamma_{\mathrm{N}}}{2\pi}(1-\sigma)\mathcal{B} \quad [21.16]$

where $\mathcal{B}$ is the applied field.

Because shielding constants are quite small (a few parts per million) compared to 1, we may write for the purposes of this calculation

$$\nu = \frac{\gamma_N \mathcal{B}}{2\pi}$$

$$\nu_L - \nu_R = 100\,\text{Hz} = \frac{\gamma_N}{2\pi}(\mathcal{B}_L - \mathcal{B}_R)$$

$$\mathcal{B}_L - \mathcal{B}_R = \frac{2\pi \times 100\ \text{s}^{-1}}{\gamma_N}$$

$$= \frac{2\pi \times 100\ \text{s}^{-1}}{26.752 \times 10^{7}\,\text{T}^{-1}\text{s}^{-1}} = 2.35 \times 10^{-6}\,\text{T}$$

$$= 2.35\ \mu\text{T}$$

The field gradient required is then

$$\frac{2.35\ \mu\text{T}}{0.08\ \text{m}} = \boxed{29\ \mu\text{T m}^{-1}}$$

Note that knowledge of the spectrometer frequency, applied field, and the numerical value of the chemical shift (because constant) is not required.

(b) Assume that the radius of the disk is 1 unit. The volume of each slice is proportional to length of slice multiplied by δx. See Figure 21.9(a).

Length of slice at $x = 2 \sin \theta$

$x = \cos \theta$

$\theta = \arccos x$

x ranges from -1 to $+1$

Length of slice at $x = 2 \sin (\arccos x)$

Plot $f(x) = 2 \sin (\arccos x)$ against x between the limits -1 and $+1$. The plot is shown Figure 21.9(b).

The volume at each value of x is proportional of $f(x)$ and the intensity of the MRI signal is proportional to the volume, so the figure represents the absorption intensity for the MRI image of the disk.

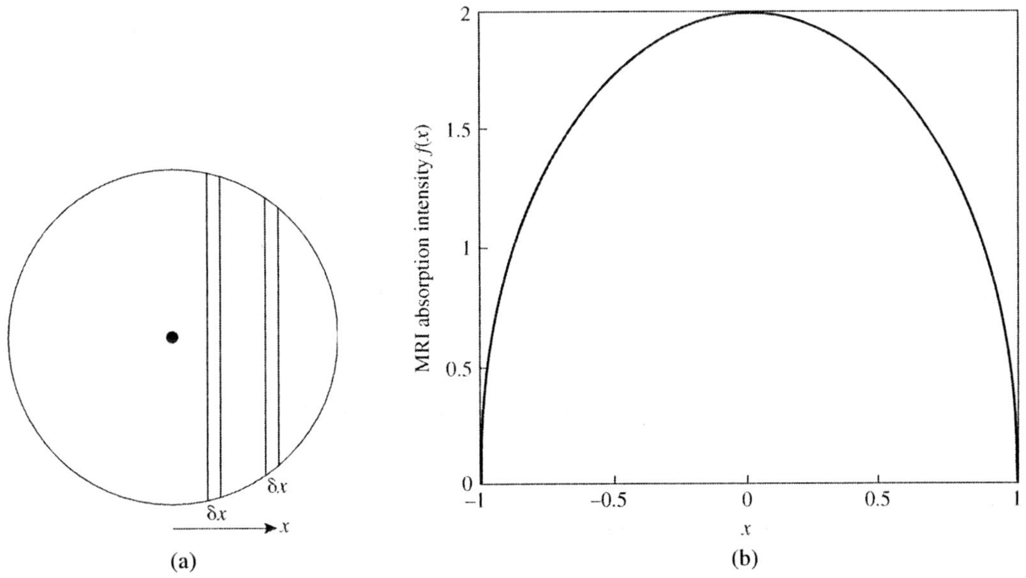

Figure 21.9

22 Statistical thermodynamics

Answers to discussion questions

D22.1 The complete derivation of the Boltzmann distribution formula is beyond the scope of this text. See P. Atkins and J. de Paula, *Physical Chemistry*, 8th edition, Oxford University Press, 2006, for the details. We can however summarize the basic steps of that derivation. We start with the concept of a distribution which is the arrangement of N molecules in the allowed states of the system, with N_1 molecules in state 1, N_2 molecules in state 2, etc. Then we look for the most probable of the allowed distributions, because the most probable distribution is for all practical purposes the only one that determines the actual properties of the system. The most probable distribution is the distribution with the maximum value of $W = \frac{N!}{N_1!N_2!...}$. W is the number of ways of arranging the N molecules within the allowed states. We must however look for the maximum value of W subject to the constraints that the total energy of the system and the number of molecules in the system be constant. This maximization technique subject to constraints is the so-called method of undetermined multipliers. It is explained in detail in the text referred to above. When these two constraints are taken into account we arrive at the formula for the Boltzmann distribution, eqn 22.1.

D22.2 Because this chapter focuses on the application of statistics to the distribution of physical states in systems that contain a large number of atoms or molecules, we begin with the statistical answer: the thermodynamic temperature is the one quantity that determines the most probable populations of those states in systems at thermal equilibrium (Sections 22.1 and 22.2). As a consequence, the temperature provides a necessary condition for thermal equilibrium; a system is at thermal equilibrium only if all of its sub-systems have the same temperature. Note that this is not a circular definition of temperature, for thermal equilibrium is not defined by uniformity of temperature: systems whose sub-systems can exchange energy tend toward thermal equilibrium. In this context, sub-systems can be different materials placed in contact (such as a block of copper in a beaker of water) or can be more abstract (such as rotational and vibrational modes of motion).

Finally, the equipartition theorem allows us to connect the temperature of statistical thermodynamics to the empirical concept of temperature developed in previous chapters. Temperature is a measure of the intensity of thermal energy, directly proportional to the mean energy for each quadratic contribution to the energy (provided that the temperature is sufficiently high; section 22.5).

D22.3 Consider the value of the partition function at the extremes of temperature. The limit of q as T approaches zero, is simply g_0, the degeneracy of the ground state. As T approaches infinity, each term in the sum is simply the degeneracy of the energy level. If the number of levels is infinite, the partition function is infinite as well. In some special cases where we can effectively limit the number of states, the upper limit of the partition function is just the number of states. In general, we see that the molecular partition function gives an indication of the average number of states thermally accessible to a molecule at the temperature of the system.

D22.4 The words "identical" and "indistinguishable" have related, but slightly different meanings. Particles of the same composition can be regarded as distinguishable when they are localized as in a crystal lattice where we can assign a set of coordinates to each particle. Strictly speaking, it is the lattice site that carries the set of coordinates, but as long as the particle is fixed to the site, it too can be considered distinguishable. In the gas phase, however, particles of the same composition would be indistinguishable. As explained in section 19.2, nuclear statistics play a role in determining whether particles of the same chemical composition are identical. *Ortho-* and *para*-hydrogen are of the same chemical composition, but are not identical. Stereoisomers have the same chemical composition, but are not identical. Other such examples can be found.

D22.5 See Figure 22.3 of text, Example 22.1 of text, and Self-test 22.1 for the discussion of the partition function of a two state system. For convenience, the lower state is assigned an energy equal to 0 with the upper state having an energy that is $hc\tilde{\nu}$ higher. Each state is assumed to have a degeneracy equal to 1. Furthermore, to explore the variation of both the internal energy and entropy with temperature, it is convenient to use the definition for a unitless 'temperature', $x = kT/hcv$. Being proportional, x and T vary in the same way. For this simple system of non-interacting states the average energy E of the occupied states and the internal energy are identical. Thus,

$$U = E = \frac{NkT^2}{q}\frac{\mathrm{d}q}{\mathrm{d}T} \quad \text{[22.10 and Derivation 22.2]}$$

$$= \frac{NKT^2}{1+\mathrm{e}^{-hc\tilde{\nu}/kT}}\frac{\mathrm{d}}{\mathrm{d}T}\left(1+\mathrm{e}^{-hc\tilde{\nu}/kT}\right) \quad \text{[Example 22.1]},$$

$$U = \frac{NkT^2}{1+\mathrm{e}^{-1/x}}\frac{\mathrm{d}x}{\mathrm{d}T}\frac{\mathrm{d}}{\mathrm{d}x}\left(1+\mathrm{e}^{-1/x}\right) = \left(\frac{NkT^2}{1+\mathrm{e}^{-1/x}}\right)\times\left(\frac{k}{hc\tilde{\nu}}\right)\times\left(\frac{1}{x^2}\mathrm{e}^{-1/x}\right)$$

$$= Nhc\tilde{\nu}\left(\frac{\mathrm{e}^{-1/x}}{1+\mathrm{e}^{-1/x}}\right),$$

$$\boxed{U = \frac{Nhc\tilde{\nu}}{1+\mathrm{e}^{-1/x}}}.$$

The equation for $U(x)$ indicates that, at the absolute zero of temperature (i.e., $x = 0$), $U = 0$. At absolute zeto the upper state is unpopulated. All particles are in the ground state. At very high temperature the term $\mathrm{e}^{1/x}$ approaches $\mathrm{e}^0 = 1$ and the internal energy approaches $Nhc\tilde{\nu}/2$. The two levels are equally populated as T approaches infinity and the internal energy equals the average of the two state energies.

The entropy variation of distinguishable particles is explored with eqn 22.14a.

$$S = \frac{U-U(0)}{T} + Nk\,\mathrm{In}\,q\ [22.14\mathrm{a}] = \frac{U}{T} + Nk\,\mathrm{In}\,q$$

$$= \frac{Nhc\tilde{\nu}/T}{1+\mathrm{e}^{1/x}} + Nk\,\mathrm{In}\left(1+\mathrm{e}^{1/x}\right),$$

$$\boxed{S = Nk\left\{\frac{1}{x\left(1+\mathrm{e}^{1/x}\right)} + \ln\left(1+\mathrm{e}^{-1/x}\right)\right\}}.$$

The equation for $S(x)$ indicates that, at the absolute zero of temperature (i.e., $x = 0$) $S = 0$ and the system is perfectly ordered with all particles in the lowest level. At very high temperature the term $\mathrm{e}^{1/x}$ approaches $\mathrm{e}^0 = 1$ and the internal energy approaches $Nk\ln 2$. Comparing this result with the Boltzmann formula for entropy ($S = k\ln W$ [22.13]) reveals that $W = 2^N$. This happens because each particle has an equal probability of being in either the lower or the upper level at very high temperature.

D22.6 The thermodynamic entropy is defined in terms of the quantity $\mathrm{d}S = \frac{\mathrm{d}q_{\mathrm{rev}}}{T}$ where $\mathrm{d}q_{\mathrm{rev}}$ is the infinitesimal quantity of energy supplied as heat to the system reversibly at a temperature T. The statistical entropy is defined in terms of the Boltzmann formula for the entropy: $S = k\ln W$ where k is the Boltzmann constant and W is the number of microstates, the total number of ways in which the molecules of the system can be arranged to achieve the same total energy of the system. These two definitions turn out to be equivalent provided the thermodynamic entropy is taken to be zero at $T = 0$.

The concept of the number of microstates makes quantitative the ill-defined qualitative concepts of 'disorder' and 'dispersal of matter and energy' that are used widely to introduce the concept of entropy: a more 'disorderly' distribution of energy and matter corresponds to a greater number of microstates associated with the same total energy. The more molecules that can participate in the distribution of energy, the more microstates there are for a given total energy and the greater the entropy than when the energy is confined to a smaller number of molecules.

The molecular interpretation of entropy given by the Boltzmann formula also suggests the thermodynamic definition. At high temperatures where the molecules of a system can occupy a large number of available energy levels, a small additional transfer of energy as heat will cause only a small change in the number of accessible energy levels, whereas at low temperatures the transfer of the same quantity of heat will increase the number of accessible energy levels and microstates significantly. Hence, the change in entropy upon heating will be greater when the energy is transferred to a cold body than when it is transferred to a hot body. This argument suggests that the change in entropy should be inversely proportional to the temperature at which the transfer takes place as in indicated in the thermodynamic definition.

D22.7 Residual entropy is due to the presence of some disorder in the system even at $T = 0$. It is observed in systems where there is very little energy difference, or none, between alternative arrangements of the molecules at very low temperatures. Consequently, the molecules cannot lock into a preferred orderly arrangement and some disorder persists.

D22.8 See Further Information 22.2 for a derivation of the general expression (eqn 22.19) for the equilibrium constant in terms of the partition functions and difference in molar energy, ΔE, of the products and reactants in a chemical reaction. The partition functions are functions of temperature and the ratio of partition functions in eqn 22.19 will therefore vary with temperature. However, the most direct effect of temperature on the equilibrium constant is through the exponential term $e^{-\Delta E/RT}$. The manner in which both factors affect the magnitudes of the equilibrium constant and its variation with temperature is described in detail for a simple $R \rightleftharpoons P$ gas phase equilibrium in Example 22.5.

The molecular partition function gives an indication of the number of states that are thermally accessible to a molecule at the temperature of the system. The fact that a ratio of partition functions for products and reactants appears in eqn 22.19 reveals the important method of comparing molecular distributions over available energy levels when evaluating an equilibrium constant. This ratio accounts for the role of entropy in the establishment of equilibrium. The exponential factor accounts for the role of energy. See Section 22.8 for a detailed discussion of these features.

Solutions to exercises

E22.1 We apply the method of the *Brief Illustration* following eqn 22.3.

$$\frac{N_{\text{stretched}}}{N_{\text{coil}}} = e^{-\Delta E/RT} = e^{-\frac{2.4\times10^3\ \text{J mol}^{-1}}{(8.3145\ \text{J K}^{-1}\ \text{mol}^{-1})\times(293\ \text{K})}}$$

$$= \boxed{0.37}$$

E22.2 The energy difference is $\Delta E = \gamma_N \hbar \mathcal{B}$ [21.11]

$$\gamma_N = \gamma_{\text{hydrogen-1}} = 26.752\times10^7\ \text{T}^{-1}\ \text{s}^{-1} \quad \text{(Table 21.2)}$$

$$\frac{\Delta E}{kT} = \frac{\gamma_N \hbar \mathcal{B}}{kT} = \frac{26.752\times10^7\ \text{T}^{-1}\text{s}^{-1}\times1.05457\times10^{-34}\ \text{J s}\ \mathcal{B}}{1.38066\times10^{-23}\ \text{J K}^{-1}\times293.15\ \text{K}} = \left(6.97\times10^{-6}\ \text{T}^{-1}\right)\times\mathcal{B}$$

(a) $$\frac{N_\beta}{N_\alpha} = e^{-\Delta E/kT} = e^{-6.97\times10^{-6}\ \text{T}^{-1}\times1.5\ \text{T}}$$

$$= \boxed{0.9999895}$$

(b) $\dfrac{N_\beta}{N_\alpha} = e^{-6.97\times10^{-6}\,T^{-1}\times15\,T}$

$= \boxed{0.9998955}$

E22.3 $\Delta E = g_e\mu_B\mathcal{B}$ [21.8] where $g_e = 2.0023$ and μ_B is the Bohr magneton

$$\frac{N_\alpha}{N_\beta} = e^{-\Delta E/kT} = e^{-g_e\mu_B\mathcal{B}/kT} = e^{-\frac{2.0023\times9.274\times10^{-24}\,J\,T^{-1}\times0.33\,T}{1.38066\times10^{-23}\,J\,K^{-1}\times293\,K}}$$

$= \boxed{0.99849}$

E22.4 $E_J = hBJ(J+1)$ [19.3]

$$\Delta E = E_4 - E_2 = hB[4(4+1) - 2(2+1)]$$
$$= 14\,hB$$

$$\frac{N_4}{N_2} = \frac{g_4}{g_2}e^{-\Delta E/kT} = \frac{g_4}{g_2}e^{-14hB/kT} \qquad g_J = 2J+1$$

$$= \frac{9}{5}\,e^{-\frac{14\times6.626\times10^{-34}\,J\,s\times11.70\times10^{9}\,s^{-1}}{1.38066\times10^{-23}\,J\,K^{-1}\times298\,K}}$$

$= \boxed{1.753}$

E22.5 $\dfrac{N_4}{N_2} = \dfrac{g_4}{g_2}e^{-28\,hB/kT} \qquad g_J = (2J+1)^2$

$$= \frac{81}{25}\,e^{-\frac{14\times6.62\times10^{-34}\,J\,s\times157\times10^{9}\,s^{-1}}{1.38066\times10^{-23}\,J\,K^{-1}\times298\,K}}$$

$$= \left(\frac{81}{25}\right)0.702 = \boxed{2.27}$$

E22.6 (a) $q = \sum_i g_i e^{-E_i/kT}$ [22.2] with degeneracy factor inserted

$$q = g_0e^{-E_0/kT} + g_1e^{-E_1/kT} + g_2e^{-E_2/kT}$$
$$= \boxed{1 + 6e^{-2\varepsilon/kT} + 3e^{-5\varepsilon/kT}}$$

(b) Since $\lim_{T\to0}\left(e^{-\varepsilon/kT}\right) = 0$, $\boxed{q = 1}$ in the limit of the absolute zero of temperature.

(c) Since $\lim_{T\to\infty}\left(e^{-\varepsilon/kT}\right) = 1$, $q = 1 + 6 + 3 = \boxed{10}$

E22.7 $q = \sum_i g_i e^{-E_i/kT} = \sum_i g_i e^{-hc\tilde{\nu}_i/kT}$

$$\frac{hc}{k} = \frac{6.626\times10^{-34}\,J\,s\times2.998\times10^{8}\,m\,s^{-1}}{1.381\times10^{-23}\,J\,K^{-1}}$$
$$= 0.014387\,m\,K = 1.4387\,cm\,K$$

$$q = 1 + 3\,e^{-1.4387\,cm\,K\times16.4\,cm^{-1}/T} + 5e^{-1.4387\,cm\,K\times43.5\,cm^{-1}/T} = 1 + 3e^{-23.59\,K/T} + 5e^{-62.58\,K/T}$$

(a) $q = 1 + 3e^{-23.59\,K/10\,K} + 5e^{-62.58\,K/10\,K}$

$= \boxed{1.29}$

(b) $q = 1 + 3e^{-23.59\,K/298\,K} + 5e^{-62.58\,K/298\,K}$

$= \boxed{7.82}$

E22.8

$$q = \sum_i g_i e^{-E_i/kT} = \sum_i g_i e^{-hc\tilde{\nu}_i/kT}$$

$$= \sum_i g_i e^{-1.4387\text{ cm K}(\tilde{\nu}_i/T)} \text{ [solution to Exercise 22.7]}$$

(a) $q = (2\times 2 + 1) = \boxed{5}$

(b) $q = 5 + 3e^{-1.4387\text{ cm K }(158.5\text{ cm}^{-1}/298\text{ K})} + e^{-1.4387\text{ cm K }(226.5\text{ cm}^{-1}/298\text{ K})} = \boxed{6.731}$

E22.9 After converting frequency to wavenumber, Eqn 22.5 can be rewritten as

$$q = \frac{1}{1 - e^{-hc\tilde{\nu}/kT}}$$

$$= \frac{1}{1 - e^{-1.4387\text{ cm K}(\tilde{\nu}/T)}} = \frac{1}{1 - e^{-1.4387\text{ cm K}\times 2649\text{ cm}^{-1}/T}} = \frac{1}{1 - e^{-3811\text{ K}/T}} = \frac{1}{1 - e^{-3811/298}}$$

$$= \boxed{1.000}$$

The high temperature approximation is $q = \dfrac{kT}{h\nu} = \dfrac{kT}{hc\tilde{\nu}} = \dfrac{T}{1.4387\text{ cm K }\tilde{\nu}} = \dfrac{T}{3811\text{ K}}$.

In order to determine the temperature at which the high temperature approximation for q is only 10% less than the exact value of q we set

$$0.9\left(\frac{1}{1 - e^{-3811\text{ K}/T}}\right) = \frac{T}{3811\text{ K}}$$

and solve for T. This equation is most easily solved with mathematical software. Here we used MathCad® which gave the result $\boxed{T = 17762\text{ K}}$. Above this temperature the high temperature approximation is in error by less than 10%. That is a very high temperature! The high temperature approximation is not very useful for vibrational partition functions.

E22.10 The partition function for a mode of molecular vibration is $q_{\text{mode}} = \dfrac{1}{1 - e^{-hc\tilde{\nu}/kT}}$ [22.5 after converting frequency to wavenumber], and the overall vibrational partition function is the product of the partition functions of the individual modes.

(a) At 500 K we draw up the following table:

mode	1	2	3	4
$\tilde{\nu}$ / cm^{-1}	667	667	1388	2349
$hc\tilde{\nu}/kT$	1.919	1.919	3.994	6.759
q_{mode}	1.172	1.172	1.019	1.001

The overall vibrational partition function at 500 K is

$$q_{\text{total}} = 1.172 \times 1.172 \times 1.019 \times 1.001 = \boxed{1.401}.$$

(b) We repeat the process at 1000 K and obtain the result $q_{\text{total}} = \boxed{3.147}$.

E22.11 We use eqn 22.7 with $T = 300$ K and $V = 10 \times 10^{-6}$ m^3.

For N_2, $m = 28.02\text{ u} \times 1.66054 \times 10^{-27}\text{ kg/u} = 4.653 \times 10^{-26}$ kg, and similarly for CS_2, $m = 76.13\text{ u} \times 1.66054 \times 10^{-27}\text{ kg/u} = 1.264 \times 10^{-25}$ kg.

For N_2 we calculate

$$q = \frac{(2\pi mkT)^{3/2}V}{h^3}$$
$$= \frac{(2\pi \times 4.653\times10^{-26}\ \text{kg}\times1.3806\times10^{-23}\ \text{J K}^{-1}\times300\ \text{K})^{3/2}\times10.0\times10^{-6}\ \text{m}^3}{(6.626\times10^{-34}\ \text{J s})^3}$$
$$= \boxed{1.448\times10^{27}}$$

Similarly for CS_2, after substituting for its mass, we obtain $q = \boxed{6.485\times10^{27}}$. This value is considerably larger than the partition function for N_2. The reason is that the mass of CS_2 is larger than the mass of N_2; hence the energy levels of CS_2 are closer together than those of N_2 and therefore at a given temperature there are more accessible states in CS_2 than in N_2. The partition function is a measure of the number of accessible states.

E22.12 $V = \tfrac{4}{3}\pi r^3 = \tfrac{4}{3}\pi\ (0.5\ \text{nm})^3 = 0.52\ \text{nm}^3 = 5.2\times10^{-28}\ \text{m}^3$

(a) $$q = \frac{(2\pi mkT)^{3/2}\ V}{h^3}\ [22.7]$$

$$q = \frac{(2\pi\times16.04\times1.66054\times10^{-27}\ \text{kg}\times1.381\times10^{-23}\ \text{J K}^{-1}\times298\ \text{K})^{3/2}\times5.2\times10^{-28}\ \text{m}^3}{(6.626\times10^{-34}\ \text{J s})^3}$$
$$= \boxed{3.2\times10^4}$$

(b) $V = 100\ \text{cm}^3 = 1.00\times10^{-4}\ \text{m}^3$

$$q = \boxed{6.2\times10^{27}}$$

E22.13 The partition function for a nonsymmetrical (AB) linear rotor such as HBr is [see *Further Information* 22.1]

$$q^{\text{R}} = \sum_J (2J+1)\text{e}^{-hBJ(J+1)/kT}$$

Converting the rotational constant to wavenumbers this may be rewritten

$$q^{\text{R}} = \sum_J (2J+1)\text{e}^{-hc\tilde{B}J(J+1)/kT}$$

At 298.15 K, $hc/kT = 4.8256\times10^{-3}$ cm. Then we may write

$$q^{\text{R}} = \sum_J (2J+1)\text{e}^{-4.8256\times10^{-3}\ \text{cm}\tilde{B}J(J+1)}\ \text{with}\ \tilde{B} = 8.465\ \text{cm}^{-1}$$

(a) The direct summation of this quantity converges to the value $\boxed{24.816}$ at $J = 15$.

(b) The high temperature approximation is $q^{\text{R}} = \dfrac{kT}{hc\tilde{B}} = \dfrac{207.226\ \text{cm}^{-1}}{8.465\ \text{cm}^{-1}} = \boxed{24.480}$. The difference from the exact value for HBr at 298 K is not large.

E22.14 No closed form for the exact value of the rotational partition function exists, so this problem is most easily solved using a spreadsheet such as Excel® or Mathematical software such as MathCad®.

In order to determine the temperature at which the high temperature approximation for q^{R} is 10% less than the exact value of q^{R} we set

$$0.9\sum_J (2J+1)\text{e}^{-hc\tilde{B}J(J+1)/kT} = \frac{kT}{hc\tilde{B}}$$

and solve for T. After substituting values for the constants h, c, k, and $\tilde{B}$ this expression becomes

$$0.9\left(\sum_{J=0}^{J=15}(2J+1)\text{e}^{-(12.1786\ \text{K})J(J+1)/T}\right) = \frac{T}{12.1786\ \text{K}}.$$

This equation is most easily solved with mathematical software. Here we used MathCad® which gave the result $\boxed{T = 38.96\ \text{K}}$. Above this temperature the high temperature approximation is in error by less than 10%.

E22.15 (a) $$q = \frac{kT}{\sigma hB}\ [22.8] = \frac{1.381\times10^{-23}\ \text{J K}^{-1}\times 298\ \text{K}}{1\times 6.626\times10^{-34}\ \text{J s}\times 318\times10^{9}\ \text{s}^{-1}} = \boxed{19.5}$$

(b) $$q = \frac{1.381\times10^{-23}\ \text{J K}^{-1}\times 298\ \text{K}}{2\times 6.626\times10^{-34}\ \text{J s}\times 11.70\times10^{9}\ \text{s}^{-1}} = \boxed{265}$$

E22.16 N_2O is linear, but not symmetrical. The structure is NNO. Thus, the symmetry number is one for NNO, but 2 for CO_2.

E22.17 $$q = \sum_i g_i e^{-E_i/kT} = 1e^{-0\times\varepsilon/kT} + 5e^{-1\times\varepsilon/kT} + 3e^{-3\times\varepsilon/kT} = \boxed{1 + 5e^{-\varepsilon/kT} + 3e^{-3\varepsilon/kT}}$$

E22.18 (a) From the solution to Exercise 22.7 we have

$$q = 1 + 3e^{-23.59\text{K}/T} + 5e^{-62.58\text{K}/T}$$

$$E = \frac{NkT^2}{q}\frac{dq}{dT}\ \text{[Derivation 22.2]}$$

$$\frac{dq}{dT} = 3\left(\frac{23.59\text{K}}{T^2}\right)e^{-23.59\text{K}/T} + 5\left(\frac{62.58}{T^2}\right)e^{-62.58\text{K}/T}$$

$$E = \frac{R(70.77\ \text{K})e^{-23.59\text{K}/T} + R(312.9\ \text{K})e^{-62.58\text{K}/T}}{1 + 3e^{-23.59\text{K}/T} + 5e^{-62.58\text{K}/T}}$$

The plot of E against T is shown in Figure 22.1.

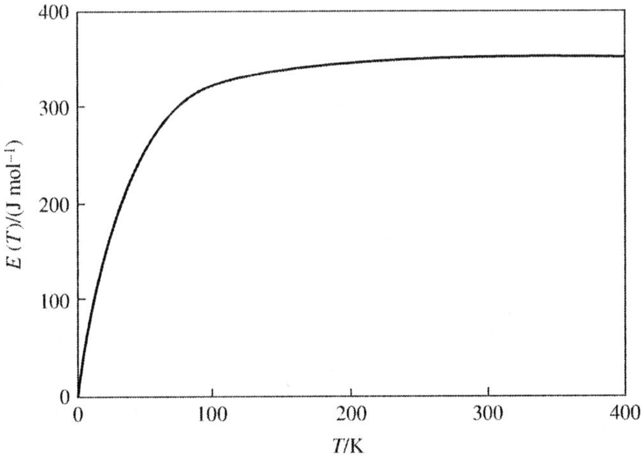

Figure 22.1

(b) Substitute $T = 298$ K into the expression for E and obtain $E = \boxed{339\ \text{J mol}^{-1}}$

E22.19 (a) From Exercise 22.8 we have

$$q = 5 + 3e^{-228.0\text{K}/T} + e^{-325.9\text{K}/T}$$

$$E = \frac{NkT^2}{q}\frac{dq}{dT}$$

$$\frac{dq}{dT} = 3\left(\frac{228.0\ \text{K}}{T^2}\right)e^{-228.0\text{K}/T} + \left(\frac{325.9\ \text{K}}{T^2}\right)e^{-325.9\text{K}/T}$$

$$E = \frac{R\left(684.0\ \text{K}\ e^{-228.0\text{K}/T} + 325.9\ \text{K}\ e^{-325.9\text{K}/T}\right)}{5 + 3e^{-228.0\text{K}/T} + e^{-325.9\text{K}/T}}$$

$$C_{V,\mathrm{m}} = \frac{\mathrm{d}U}{\mathrm{d}T} = \frac{\mathrm{d}E}{\mathrm{d}T}$$

The expression for $C_{V,\mathrm{m}}$ is derived below and plots are presented of $E_\mathrm{m}(T)$ in Figure 22.2 and $C_{V,\mathrm{m}}(T)$ in Figure 22.3.

$$a = 228\ \mathrm{K} \qquad b = 325.9\ \mathrm{K} \qquad R = 8.3145\ \mathrm{J\ K^{-1}\ mol^{-1}}$$

$$E_\mathrm{m}(T) = R\frac{\left(3\cdot a\cdot e^{\frac{-3\cdot a}{T}} + b\cdot e^{\frac{-b}{T}}\right)}{\left(5+3\cdot e^{\frac{-a}{T}} + e^{\frac{-b}{T}}\right)}$$

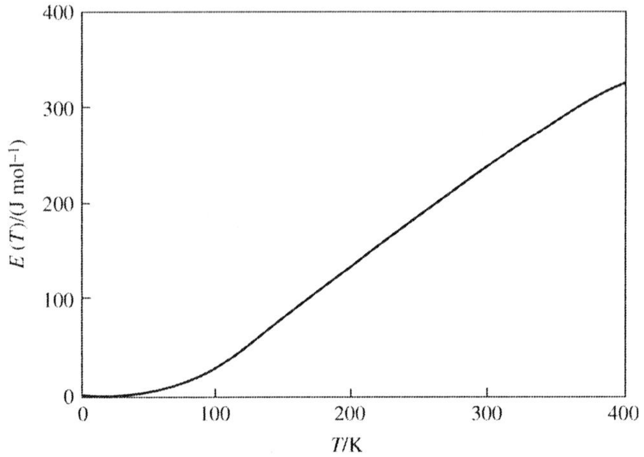

Figure 22.2

$$C_{V,\mathrm{m}}(T) = R\frac{\left(9\cdot\frac{a^2}{T^2}\cdot\exp\left(-3\cdot\frac{a}{T}\right)+\frac{b^2}{T^2}\cdot\exp\left(\frac{-b}{T}\right)\right)}{\left(5+3\cdot\exp\left(\frac{-a}{T}\right)+\exp\left(\frac{-b}{T}\right)\right)}$$

$$-R\frac{\left(3\cdot a\cdot\exp\left(-3\cdot\frac{a}{T}\right)+b\cdot\exp\left(\frac{-b}{T}\right)\right)}{\left(5+3\cdot\exp\left(\frac{-a}{T}\right)+\exp\left(\frac{-b}{T}\right)\right)^2}\cdot\left(3\cdot\frac{a}{T^2}\cdot\exp\left(\frac{-a}{T}\right)+\frac{b}{T^2}\cdot\exp\left(\frac{-b}{T}\right)\right)$$

(b) $C_{V,\mathrm{m}}(298\ \mathrm{K}) = \boxed{0.993\ \mathrm{J\ K^{-1}\ mol^{-1}}}$

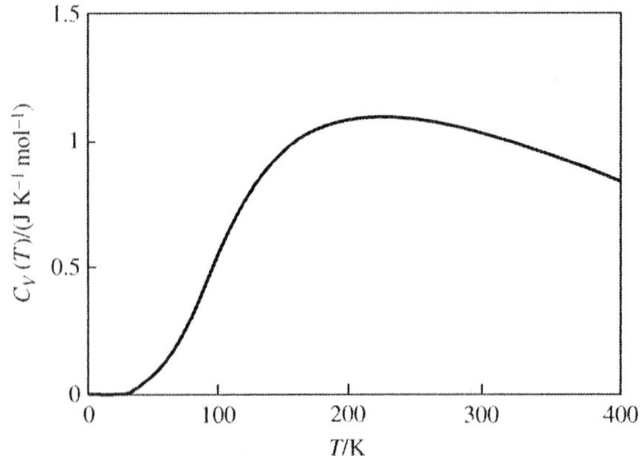

Figure 22.3

E22.20 $$S_m = k\ln 4^{N_A} = N_A k\ln 4 = R\ln 4$$
$$= \boxed{11.5\ \mathrm{J\,K^{-1}\,mol^{-1}}}$$

E22.21 $$S = k\ln W\ [22.13] = k\ln 4^N = Nk\ln 4 = (5\times10^8)\times(1.38\times10^{-23}\ \mathrm{J\,K^{-1}})\times\ln 4 = \boxed{9.57\times10^{-15}\ \mathrm{J\,K^{-1}}}$$

Question. Is this a large residual entropy? The answer depends on what comparison is made. Multiply the answer by Avogadro's number to obtain the molar residual entropy, 5.76×10^9 J K^{-1} mol^{-1}, surely a large number—but then DNA is a macromolecule. The residual entropy per mole of base pairs may be a more reasonable quantity to compare to molar residual entropies of small molecules. To obtain that answer, divide the molecule's entropy by the number of base pairs before multiplying by N_A. The result is 11.5 J K^{-1} mol^{-1}, a quantity more in line with examples discussed in Section 22.6.

E22.22 We use eqn 22.14b for indistinguishable molecules.

$$S_m = \frac{U-U(0)}{T} + N k \ln q_m - N k(\ln N - 1) \text{ with } N = N_A.$$

For N_2 (g) at 298 K we assume that the high-temperature limit (equipartion value) for the energy applies. Then $U_m - U_m(0) = 3/2\,RT + RT = 5/2\,RT$ (see Self-test 22.2).

$$S_m = 5/2\,R + R\ln q_m - R\ln N_A + R = 5/2\,R + R\left(\ln\frac{q_m}{N_A}+1\right)$$

$$q_m = q_m^T q^R$$
$$q_m^T = (2\pi mkT)^{3/2}V_m/h^3$$
$$q^R = \frac{kT}{\sigma hB}$$

We replace V_m by $RT/p^{\ominus}$ and substitute in the values of the various constants and obtain

$$\frac{q_m^T}{N_A} = 2.561\times10^{-2}\left(\frac{T}{\mathrm{K}}\right)^{5/2}(M/g\ \mathrm{mol}^{-1})^{3/2}$$

$$q^R = \frac{0.6950}{\sigma}\times\frac{T/K}{(\tilde{B}/\mathrm{cm}^{-1})} \qquad \tilde{B} = 1.987\ \mathrm{cm}^{-1} \text{ for } N_2$$

Then

$$\frac{q_m^T}{N_A} = (2.561\times10^{-2})\times(298)^{5/2}\times(28.02)^{3/2} = 5.823\times10^6$$

$$q^R = \frac{1}{2}\times0.6950\times\frac{298}{1.987} = 51.81$$

and

$$\frac{q_m^{\ominus}}{N_A} = (5.823\ \times10^6)\times51.81 = 3.02\times10^8$$

$$S_m = \tfrac{5}{2}\,R + R(\ln 3.02\times10^8 + 1) = 23.03\ \mathrm{R}$$
$$= \boxed{191.4\ \mathrm{J\,K^{-1}\,mol^{-1}}}$$

E22.23 The general qualitative rule is that the greater the complexity of the system, the greater the standard entropy.

(a) Monatomic gases with greater numbers of particles and greater molar mass have greater standard translational entropies. More energy levels are accessible and the partition function is greater. Hence, $\boxed{S^{\ominus}(\mathrm{Xe}) > S^{\ominus}(\mathrm{Ne})}$.

(b) All contributions, translational, rotational, and vibrational, to the molecular partition functions depend on the mass of the molecule. The greater the mass, the greater the partition functions and the greater the entropy. Hence, $\boxed{S^{\ominus}(D_2O) > S^{\ominus}(H_2O)}$.

(c) Diamond has a very rigid and orderly lattice. Graphite has a layer-like structure in which the layers can slide past each other, resulting in much more disorder in the graphite structure; hence

$$\boxed{S^{\ominus}(\text{Graphite}) > S^{\ominus}(\text{Diamond})}$$

E22.24 The change in entropy per micelle is

$$\Delta S = 100\, k \ \text{In} \ \frac{V_{\text{solution}}}{V_{\text{micelle}}}$$

In the absence of specific data on the volumes involved we can arrive at a rough value of ΔS by making some reasonable estimates. Let us assume that the micelle is spherical in shape. Let us also assume that the radius of this sphere is roughly the same as the length of the hydrocarbon chain of the amphiphile. Assume that the chain consists of 10 zig-zag carbon atoms with an average C-C-C length of 250 pm, or about 125 pm per carbon atom. Then the radius of the micelle is about 1.25 nm and the volume is

$$V_{\text{micelle}} = \frac{4}{3}\pi r^3 \approx 4r^3 = 4 \times \left(1.25\times10^{-9}\ \text{m}\right)^3$$

$$\approx 1\times10^{-26}\ \text{m}^3$$

Assume that the volume of the solution is that of a typical 100 cm^3 beaker. 100 cm^3 = 10^{-4} m^3.

$$\Delta S = 100\ \text{k}\ \ln\left(\frac{10^{-4}\,\text{m}^3}{10^{-26}\ \text{m}^3}\right)$$

$$= 100\times1.38\times10^{-23}\,\text{J K}^{-1}\times\ln\ 10^{22}$$

$$= 7\times10^{-20}\,\text{J K}^{-1}\ \text{(per micelle)}$$

For one mole of micelles

$$\Delta S \approx N_{\text{A}}\times7\times10^{-20}\,\text{J K}^{-1}$$

$$\approx 4\times10^{4}\,\text{J K}^{-1}\ \text{mol}^{-1}$$

$$\approx \boxed{40\ \text{kJ K}^{-1}\ \text{mol}^{-1}}$$

This value seems high and is probably a result of underestimating V_{micelle} for a micelle consisting of 100 amphiphiles. End effects contributing to the length of the amphiphile were neglected and the number of carbon atoms for amphiphiles that could form a micelle of 100 amphiphiles may be larger than the 10 assumed. Experimental evidence for the volume of micelles with 100 amphiphiles indicates typical values of about 1×10^{-25} m^3. For that volume

$$\Delta S = 4\ \text{kJ K}^{-1}\ \text{mol}^{-1}$$

E22.25

$$G^{\ominus}_{\text{m}} - G^{\ominus}_{\text{m}}(0) = -RT\ \ln\frac{q^{\ominus}_{\text{m}}}{N_{\text{A}}}\ [22.18]$$

If we ignore the electronic and spin contributions to the total standard molar partition function we can write eqn 22.9 as $q^{\ominus}_{\text{m}} = q^{\text{T}}_{\text{m}} q^{\text{R}} q^{\text{V}}$.

As in exercise 22.22 we have

$$q^{\text{T}}_{\text{m}} = (2\pi mkT)^{3/2}\ V_{\text{m}}\,/\,h^3$$

$$q^{\text{R}}_{\text{m}} = \frac{kT}{\sigma hB}$$

We replace V_m by $RT / p^{\ominus}$ and substitute in the values of the various constants and obtain

$$\frac{q_m^T}{N_A} = 2.561\times10^{-2}\left(\frac{T}{K}\right)^{5/2} (M / g\ mol^{-1})^{3/2}$$

$$q_m^R = \frac{0.6950}{\sigma}\times\frac{T / K}{(\tilde{B}/\text{cm}^{-1})} \qquad \tilde{B} = \frac{B}{c} = \frac{11.70\times10^9\ \text{s}^{-1}}{2.9979\times10^{10}\ \text{cm s}^{-1}} = 0.3903\ \text{cm}^{-1}$$

Then

$$\frac{q_m^T}{N_A} = (2.561\times10^{-2})\times(298)^{5/2}\times(44.01)^{3/2} = 11.462\times10^6$$

$$q^R = \frac{1}{2}\times0.6950\times\frac{298}{0.3903} = 763.5$$

To obtain q^V we proceed as in Exercise 22.10.

The partition function for a mode of molecular vibration is $q_{\text{mode}} = \frac{1}{1-e^{-hc\tilde{\nu}/kT}}$ [22.5 after converting frequency to wavenumber], and the overall vibrational partition function is the product of the partition functions of the individual modes.

At 298 K we draw up the following table:

mode	1	2	3	4
$\tilde{\nu}$ / cm^{-1}	667	667	1388	2349
$hc\tilde{\nu} / kT$	3.210	3.210	6.701	11.341
q_{mode}	1.042	1.042	1.001	1.000

The overall vibrational partition function at 500 K is then

$$q^V = 1.042 \times 1.042 \times 1.001 \times 1.000 = \boxed{1.087}.$$

Finally, $q_m^{\ominus} = q_m^T q^R q^V = 11.462\times10^6\times763.5\times1.087 = 9.511\times10^9$.

Then,

$$G_m^{\ominus} - G_m^{\ominus}(0) = -8.3145\ \text{J K}^{-1}\ \text{mol}^{-1}\times298\ \text{K} \times \ln(9.511\times10^9)$$
$$= \boxed{-56.9\ \text{kJ mol}^{-1}}$$

E22.26 We may write the expression for the equilibrium constant for the reaction

$$N_2(g) + 3H_2(g) \rightleftarrows 2NH_3(g)$$

in a manner analogous to eqn 22.19 in the text. The principal difference is the stoichiometry of the reaction and the fact that all three species have translational, rotational, and vibrational motions, and therefore translational, rotational, and vibrational partition functions. Because all three species are closed shell systems with non degenerate ground states all electronic partition functions $q^E = 1$. The total partition function for all three species can then be written in the form

$$q_m^{\ominus} = q_m^T q_m^R q_m^V.$$

The expression for the equilibrium constant becomes

$$\boxed{K = \frac{(q_{NH_3,m}^{\ominus} / N_A)^2}{(q_{N_2,m}^{\ominus} / N_A)(q_{H_2,m}^{\ominus} / N_A)^3} e^{-\Delta E/RT}}$$

E22.27 Follow the procedure of Example 22.5 in the text.

$$K = \frac{q_m^{\ominus}(\mathrm{Na}^+)q_m^{\ominus}(\mathrm{e}^-)}{q_m^{\ominus}(\mathrm{Na})N_A}\,\mathrm{e}^{-\Delta E/RT}$$

$$q_m^{\ominus}(\mathrm{e}^-) = \frac{2(2\pi m_e kT)^{3/2}\,RT}{p^{\ominus}h^3}$$

$$q_m^{\ominus}(\mathrm{Na}^+) = \frac{(2\pi m_{\mathrm{Na}} kT)^{3/2}\,RT}{p^{\ominus}h^3}$$

$$q_m^{\ominus}(\mathrm{Na}) = 2\,q_m^{\ominus}(\mathrm{Na}^+)$$

$$K = \frac{(2\pi m_e kT)^{3/2}\,kT}{p^{\ominus}h^3}\,\mathrm{e}^{-I/RT} \qquad I = \Delta E = 495.8 \text{ kJ mol}^{-1}$$

$$= \frac{(2\pi m_e)^{3/2}(kT)^{5/2}}{p^{\ominus}h^3}\,\mathrm{e}^{-I/RT}$$

$$= \frac{(2\pi\times 9.11\times10^{-31}\text{ kg})^{3/2}(1.381\times10^{-23}\text{ J K}^{-1}\times1000\text{ K})^{5/2}}{10^5\text{ Pa}\times(6.626\times10^{-34})^3}\times \mathrm{e}^{-(4.958\times10^5/8.3145\times1000)}$$

$$= \boxed{1.37\times10^{-25}}$$

E22.28 Assume $\Delta E \approx \Delta H(\text{I–I}) = 151 \text{ kJ mol}^{-1}$ [Table 3.5]

$$K = \left[\frac{(q_{\mathrm{I,m}}^{\ominus})^2}{(q_{\mathrm{I_2,m}}^{\ominus})N_A}\right]\mathrm{e}^{-\Delta E/RT}$$

$q_{\mathrm{I,m}}^{\ominus} = q_m^{\mathrm{Trans}}(\mathrm{I})q^{\mathrm{Elec}}(\mathrm{I})$ $\qquad q^{\mathrm{Elec}}(\mathrm{I}) = 4$ [The degeneracy of the ground state electronic state of I]

$(q_{\mathrm{I_2,m}}^{\ominus}) = q_m^{\mathrm{Trans}}(\mathrm{I_2})q^{\mathrm{Rot}}(\mathrm{I_2})q^{\mathrm{Vib}}(\mathrm{I_2})q^{\mathrm{Elec}}(\mathrm{I_2})$ $\qquad q^{\mathrm{Elec}}(\mathrm{I_2}) = 1$

See the solution to Exercises 22.10, 22.22, and 22.25 for the partition function formulas.

$$\frac{q_m^{\mathrm{Trans}}}{N_A} = 2.561\times10^{-2}(T/K)^{5/2}\times(M/\text{g mol}^{-1})^{3/2}$$

$$\frac{q_m^{\mathrm{Trans}}(\mathrm{I_2})}{N_A} = 2.561\times10^{-2}\times500^{5/2}\times 253.8^{3/2} = 5.79\times10^8$$

$$\frac{q_m^{\mathrm{Trans}}(\mathrm{I})}{N_A} = 2.561\times10^{-2}\times500^{5/2}\times126.9^{3/2} = 2.05\times10^8$$

$$q^{\mathrm{Rot}}(\mathrm{I_2}) = \frac{0.6950}{\sigma}\times\frac{T/K}{\tilde{B}/\text{cm}^{-1}} = \frac{1}{2}\times0.6950\times\frac{500}{0.0373} \quad \text{[CRC Handbook for value of } \tilde{B}\text{]}$$

$$= 4.66\times10^3$$

$$q^{\mathrm{Vib}}(\mathrm{I}_2) = \frac{1}{1-\mathrm{e}^{-a}} \qquad a = 1.4388\ \frac{\tilde{\nu}/\mathrm{cm}^{-1}}{T/K}$$

$$= \frac{1}{1-\mathrm{e}^{-1.4388\times 214.36/500}} = 2.17 \ \text{[CRC Handbook for value of } \tilde{\nu}\text{]}$$

$$K = \frac{\left(2.05\times10^{8}\right)^{2}\times 4^{2}\ \times\ \mathrm{e}^{-151\times10^{3}/8.3145\times500}}{5.79\times10^{8}\times4.66\times10^{3}\times2.17}$$

$$= \boxed{1.93\times10^{-11}}$$

E22.29 (a) $q = \dfrac{1}{1-\mathrm{e}^{-h\nu/kT}}$ [22.5; $E_0 \equiv 0$] $\qquad E_{\upsilon=0,1,2,\cdots} = \upsilon h\nu$ [12.21; $E_0 \equiv 0$]

$$E = \frac{N}{q}\sum_{\upsilon=0} E_{\upsilon}\mathrm{e}^{-E_{\upsilon}/kT}\ [22.10] = \frac{N}{q}\sum_{\upsilon=0}\upsilon h\nu\,\mathrm{e}^{-\upsilon h\nu/kT} = \frac{Nh\nu}{q}\left(0\times\mathrm{e}^{-0\times h\nu/kT}+1\times\mathrm{e}^{-1\times h\nu/kT}+2\times\mathrm{e}^{-2\times h\nu/kT}+\cdots\right)$$

$$= \frac{Nh\nu\,\mathrm{e}^{-h\nu/kT}}{q}\left(1+2\mathrm{e}^{-h\nu/kT}+3\mathrm{e}^{-2h\nu/kT}+4\mathrm{e}^{-3h\nu/kT}+\cdots\right) = \frac{Nh\nu\,\mathrm{e}^{-h\nu/kT}}{q}\left\{\frac{1}{\left(1-\mathrm{e}^{-h\nu/kT}\right)^{2}}\right\}$$

$$= \frac{Nh\nu\,\mathrm{e}^{-h\nu/kT}}{q}\left(\frac{q}{1-\mathrm{e}^{-h\nu/kT}}\right) = \boxed{\frac{Nh\nu}{\mathrm{e}^{h\nu/kT}-1}}$$

Since $\mathrm{e}^{x} = 1+x+\frac{1}{2!}x^{2}+\frac{1}{3!}x^{3}\cdots \simeq 1+x$ when $|x| \ll 1$, which means $T \gg \frac{h\nu}{k}$

$$\lim_{T\to\infty}\left(\mathrm{e}^{\frac{h\nu}{kT}}\right) = 1+\frac{h\nu}{kT} \qquad \text{and} \qquad \lim_{T\to\infty} E = \lim_{T\to\infty}\left(\frac{Nh\nu}{\mathrm{e}^{\frac{h\nu}{kT}}-1}\right) = \frac{Nh\nu}{1+\frac{h\nu}{kT}-1} = \boxed{NkT}$$

As demonstrated in the above derivation the high temperature approximation is reliable when $\boxed{T \gg \frac{h\nu}{k}}$.

(b) It is shown in part (a) above that the mean energy of N harmonic oscillators is

$$E = \frac{Nh\nu}{\mathrm{e}^{h\nu/kT}-1}$$

For a system on non-interacting harmonic oscillators the internal energy U equals the mean energy. Consequently,

$$C_V = \frac{\mathrm{d}U}{\mathrm{d}T}\text{at constant } V = \frac{\mathrm{d}}{\mathrm{d}T}\left(\frac{Nh\nu}{\mathrm{e}^{h\nu/kT}-1}\right) = Nh\nu\left\{\frac{-\left(-\frac{h\nu}{kT^{2}}\right)\mathrm{e}^{h\nu/kT}}{\left(\mathrm{e}^{h\nu/kT}-1\right)^{2}}\right\} = Nk\left(\frac{h\nu}{kT}\right)^{2}\left(\frac{\mathrm{e}^{h\nu/2kT}}{\mathrm{e}^{h\nu/kT}-1}\right)^{2}$$

$$= \boxed{Nk\left(\frac{h\nu}{kT}\right)^{2}\times\left(\frac{\mathrm{e}^{-h\nu/2kT}}{1-\mathrm{e}^{-h\nu/kT}}\right)^{2}}$$

A plot of C_V against T is presented in Figure 22.4.

Since $\mathrm{e}^{x} = 1+x+\frac{1}{2!}x^{2}+\frac{1}{3!}x^{3}\cdots \simeq 1+x$ when $|x| \ll 1$

$$\lim_{T\to\infty}\left(\mathrm{e}^{-\frac{h\nu}{kT}}\right) = 1-\frac{h\nu}{kT} \qquad \text{when } \boxed{T \gg h\nu/k}$$

$$\lim_{T\to\infty} C_V = \lim_{T\to\infty}\left(Nk\left(\frac{h\nu}{kT}\right)^2\left(\frac{e^{-h\nu/2kT}}{1-e^{-h\nu/kT}}\right)^2\right) = Nk\left(\frac{h\nu}{k}\right)^2 \lim_{T\to\infty}\left(\frac{1}{T^2}\left(\frac{1-\frac{h\nu}{kT}}{1-1+\frac{h\nu}{kT}}\right)^2\right)$$

$$= Nk\left(\frac{h\nu}{k}\right)^2 \lim_{T\to\infty}\left(\frac{1}{T^2}\left(\frac{1}{\frac{h\nu}{kT}}\right)^2\right) = \boxed{Nk}$$

Thus, the high temperature limit of C_V is Nk and the condition for "high" temperature is $T \gg h\nu/k$. Furthermore, if $N = N_A$,

$$\lim_{T\to\infty} C_V = N_A k = \boxed{R}$$

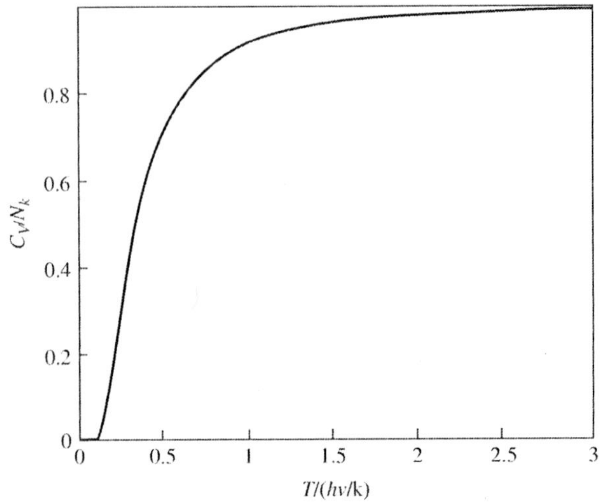

Figure 22.4

E22.30 No solution.